PURE AND APPLIED MATHEMATICS

A Program of Monographs, Textbooks, and Lecture Notes

Zariski Surfaces and Differential Equations in Characteristic $p > 0$

MONOGRAPHS AND TEXTBOOKS IN PURE AND APPLIED MATHEMATICS

1. *K. Yano,* Integral Formulas in Riemannian Geometry (1970)*(out of print)*
2. *S. Kobayashi,* Hyperbolic Manifolds and Holomorphic Mappings (1970) *(out of print)*
3. *V. S. Vladimirov,* Equations of Mathematical Physics (A. Jeffrey, editor; A. Littlewood, translator) (1970) *(out of print)*
4. *B. N. Pshenichnyi,* Necessary Conditions for an Extremum (L. Neustadt, translation editor; K. Makowski, translator) (1971)
5. *L. Narici, E. Beckenstein, and G. Bachman,* Functional Analysis and Valuation Theory (1971)
6. *S. S. Passman,* Infinite Group Rings (1971)
7. *L. Dornhoff,* Group Representation Theory (in two parts). Part A: Ordinary Representation Theory. Part B: Modular Representation Theory (1971, 1972)
8. *W. Boothby and G. L. Weiss (eds.),* Symmetric Spaces: Short Courses Presented at Washington University (1972)
9. *Y. Matsushima,* Differentiable Manifolds (E. T. Kobayashi, translator) (1972)
10. *L. E. Ward, Jr.,* Topology: An Outline for a First Course (1972) *(out of print)*
11. *A. Babakhanian,* Cohomological Methods in Group Theory (1972)
12. *R. Gilmer,* Multiplicative Ideal Theory (1972)
13. *J. Yeh,* Stochastic Processes and the Wiener Integral (1973) *(out of print)*
14. *J. Barros-Neto,* Introduction to the Theory of Distributions (1973) *(out of print)*
15. *R. Larsen,* Functional Analysis: An Introduction (1973) *(out of print)*
16. *K. Yano and S. Ishihara,* Tangent and Cotangent Bundles: Differential Geometry (1973) *(out of print)*
17. *C. Procesi,* Rings with Polynomial Identities (1973)
18. *R. Hermann,* Geometry, Physics, and Systems (1973)
19. *N. R. Wallach,* Harmonic Analysis on Homogeneous Spaces (1973) *(out of print)*
20. *J. Dieudonné,* Introduction to the Theory of Formal Groups (1973)
21. *I. Vaisman,* Cohomology and Differential Forms (1973)
22. *B.-Y. Chen,* Geometry of Submanifolds (1973)
23. *M. Marcus,* Finite Dimensional Multilinear Algebra (in two parts) (1973, 1975)
24. *R. Larsen,* Banach Algebras: An Introduction (1973)
25. *R. O. Kujala and A. L. Vitter (eds.),* Value Distribution Theory: Part A; Part B: Deficit and Bezout Estimates by Wilhelm Stoll (1973)
26. *K. B. Stolarsky,* Algebraic Numbers and Diophantine Approximation (1974)
27. *A. R. Magid,* The Separable Galois Theory of Commutative Rings (1974)
28. *B. R. McDonald,* Finite Rings with Identity (1974)
29. *J. Satake,* Linear Algebra (S. Koh, T. A. Akiba, and S. Ihara, translators) (1975)

30. *J. S. Golan,* Localization of Noncommutative Rings (1975)
31. *G. Klambauer,* Mathematical Analysis (1975)
32. *M. K. Agoston,* Algebraic Topology: A First Course (1976)
33. *K. R. Goodearl,* Ring Theory: Nonsingular Rings and Modules (1976)
34. *L. E. Mansfield,* Linear Algebra with Geometric Applications: Selected Topics (1976)
35. *N. J. Pullman,* Matrix Theory and Its Applications (1976)
36. *B. R. McDonald,* Geometric Algebra Over Local Rings (1976)
37. *C. W. Groetsch,* Generalized Inverses of Linear Operators: Representation and Approximation (1977)
38. *J. E. Kuczkowski and J. L. Gersting,* Abstract Algebra: A First Look (1977)
39. *C. O. Christenson and W. L. Voxman,* Aspects of Topology (1977)
40. *M. Nagata,* Field Theory (1977)
41. *R. L. Long,* Algebraic Number Theory (1977)
42. *W. F. Pfeffer,* Integrals and Measures (1977)
43. *R. L. Wheeden and A. Zygmund,* Measure and Integral: An Introduction to Real Analysis (1977)
44. *J. H. Curtiss,* Introduction to Functions of a Complex Variable (1978)
45. *K. Hrbacek and T. Jech,* Introduction to Set Theory (1978)
46. *W. S. Massey,* Homology and Cohomology Theory (1978)
47. *M. Marcus,* Introduction to Modern Algebra (1978)
48. *E. C. Young,* Vector and Tensor Analysis (1978)
49. *S. B. Nadler, Jr.,* Hyperspaces of Sets (1978)
50. *S. K. Segal,* Topics in Group Rings (1978)
51. *A. C. M. van Rooij,* Non-Archimedean Functional Analysis (1978)
52. *L. Corwin and R. Szczarba,* Calculus in Vector Spaces (1979)
53. *C. Sadosky,* Interpolation of Operators and Singular Integrals: An Introduction to Harmonic Analysis (1979)
54. *J. Cronin,* Differential Equations: Introduction and Quantitative Theory (1980)
55. *C. W. Groetsch,* Elements of Applicable Functional Analysis (1980)
56. *I. Vaisman,* Foundations of Three-Dimensional Euclidean Geometry (1980)
57. *H. I. Freedman,* Deterministic Mathematical Models in Population Ecology (1980)
58. *S. B. Chae,* Lebesgue Integration (1980)
59. *C. S. Rees, S. M. Shah, and C. V. Stanojević,* Theory and Applications of Fourier Analysis (1981)
60. *L. Nachbin,* Introduction to Functional Analysis: Banach Spaces and Differential Calculus (R. M. Aron, translator) (1981)
61. *G. Orzech and M. Orzech,* Plane Algebraic Curves: An Introduction Via Valuations (1981)
62. *R. Johnsonbaugh and W. E. Pfaffenberger,* Foundations of Mathematical Analysis (1981)
63. *W. L. Voxman and R. H. Goetschel,* Advanced Calculus: An Introduction to Modern Analysis (1981)
64. *L. J. Corwin and R. H. Szcarba,* Multivariable Calculus (1982)
65. *V. I. Istrătescu,* Introduction to Linear Operator Theory (1981)
66. *R. D. Järvinen,* Finite and Infinite Dimensional Linear Spaces: A Comparative Study in Algebraic and Analytic Settings (1981)

67. *J. K. Beem and P. E. Ehrlich,* Global Lorentzian Geometry (1981)
68. *D. L. Armacost,* The Structure of Locally Compact Abelian Groups (1981)
69. *J. W. Brewer and M. K. Smith, eds.,* Emmy Noether: A Tribute to Her Life and Work (1981)
70. *K. H. Kim,* Boolean Matrix Theory and Applications (1982)
71. *T. W. Wieting,* The Mathematical Theory of Chromatic Plane Ornaments (1982)
72. *D. B. Gauld,* Differential Topology: An Introduction (1982)
73. *R. L. Faber,* Foundations of Euclidean and Non-Euclidean Geometry (1983)
74. *M. Carmeli,* Statistical Theory and Random Matrices (1983)
75. *J. H. Carruth, J. A. Hildebrant, and R. J. Koch,* The Theory of Topological Semigroups (1983)
76. *R. L. Faber,* Differential Geometry and Relativity Theory: An Introduction (1983)
77. *S. Barnett,* Polynomials and Linear Control Systems (1983)
78. *G. Karpilovsky,* Commutative Group Algebras (1983)
79. *F. Van Oystaeyen and A. Verschoren,* Relative Invariants of Rings: The Commutative Theory (1983)
80. *I. Vaisman,* A First Course in Differential Geometry (1984)
81. *G. W. Swan,* Applications of Optimal Control Theory in Biomedicine (1984)
82. *T. Petrie and J. D. Randall,* Transformation Groups on Manifolds (1984)
83. *K. Goebel and S. Reich,* Uniform Convexity, Hyperbolic Geometry, and Nonexpansive Mappings (1984)
84. *T. Albu and C. Năstăsescu,* Relative Finiteness in Module Theory (1984)
85. *K. Hrbacek and T. Jech,* Introduction to Set Theory, Second Edition, Revised and Expanded (1984)
86. *F. Van Oystaeyen and A. Verschoren,* Relative Invariants of Rings: The Noncommutative Theory (1984)
87. *B. R. McDonald,* Linear Algebra Over Commutative Rings (1984)
88. *M. Namba,* Geometry of Projective Algebraic Curves (1984)
89. *G. F. Webb,* Theory of Nonlinear Age-Dependent Population Dynamics (1985)
90. *M. R. Bremner, R. V. Moody, and J. Patera,* Tables of Dominant Weight Multiplicities for Representations of Simple Lie Algebras (1985)
91. *A. E. Fekete,* Real Linear Algebra (1985)
92. *S. B. Chae,* Holomorphy and Calculus in Normed Spaces (1985)
93. *A. J. Jerri,* Introduction to Integral Equations with Applications (1985)
94. *G. Karpilovsky,* Projective Representations of Finite Groups (1985)
95. *L. Narici and E. Beckenstein,* Topological Vector Spaces (1985)
96. *J. Weeks,* The Shape of Space: How to Visualize Surfaces and Three-Dimensional Manifolds (1985)
97. *P. R. Gribik and K. O. Kortanek,* Extremal Methods of Operations Research (1985)
98. *J.-A. Chao and W. A. Woyczynski, eds.,* Probability Theory and Harmonic Analysis (1986)
99. *G. D. Crown, M. H. Fenrick, and R. J. Valenza,* Abstract Algebra (1986)
100. *J. H. Carruth, J. A. Hildebrant, and R. J. Koch,* The Theory of Topological Semigroups, Volume 2 (1986)

101. *R. S. Doran and V. A. Belfi,* Characterizations of C*-Algebras: The Gelfand-Naimark Theorems (1986)
102. *M. W. Jeter,* Mathematical Programming: An Introduction to Optimization (1986)
103. *M. Altman*, A Unified Theory of Nonlinear Operator and Evolution Equations with Applications: A New Approach to Nonlinear Partial Differential Equations (1986)
104. *A. Verschoren*, Relative Invariants of Sheaves (1987)
105. *R. A. Usmani,* Applied Linear Algebra (1987)
106. *P. Blass and J. Lang,* Zariski Surfaces and Differential Equations in Characteristic $p > 0$ (1987)

Other Volumes in Preparation

Zariski Surfaces and Differential Equations in Characteristic $p > 0$

PIOTR BLASS
University of Northern Florida
Jacksonville, Florida

JEFFREY LANG
University of San Francisco
San Francisco, California

With sections by
DAVID JOYCE • WILLIAM E. LANG
and with the collaboration of
RAYMOND HOOBLER • JOSEPH LIPMAN
MARC LEVINE • THORSTON EKEDAHL

MARCEL DEKKER, INC. **New York and Basel**

Library of Congress Cataloging-in-Publication Data

Blass, Piotr.
Zariski surfaces and differential equations in characteristic p > 0.

(Pure and applied mathematics ; 106)
"Includes all of J. Lang's 1981 Purdue thesis"--Pref.
Includes bibliographies, references and index.
1. Zariski surfaces. 2. Differential equations.
I. Lang, Jeffrey, [date]. II. Title. III. Series:
Pure and applied mathematics (Marcel Dekker) ; v. 106.
QA573.B6 1987 516.3'52 86-19898
ISBN 0-8247-7637-2

MARCEL DEKKER, INC.
270 Madison Avenue, New York, New York 10016

Current printing (last digit):
10 9 8 7 6 5 4 3 2 1

PRINTED IN THE UNITED STATES OF AMERICA

To the memory of my mother, Judith Blass, to my father, Bronislaw Blass, and to my two teenage sons, James and Oscar.

Peter Blass

To my wife, Ragia, and my daughter, Jameelah.

Jeffrey Lang

Preface

This book represents the current (1985) state of knowledge about Zariski surfaces and related topics in differential equations in characteristic $p > 0$.

The book is aimed at research mathematicians and graduate and advanced undergraduate students of mathematics (and computer science). It can be read alongside Hartshorne's popular textbook *Algebraic Geometry*. The chapters on computations in characteristic $p > 0$ should be of interest to computer scientists and coding theorists.

A special feature of this book which will make it, we hope, of particular interest to all of the above is the fact that we present a wide range of techniques of algebraic geometry using relatively simple and understandable examples and explicit computations. Zariski surfaces afford a unique opportunity for this kind of presentation since they are both relatively simple and explicitly described by equations, but at the same time, their geometry can be quite subtle, and it sometimes requires the most sophisticated tools and techniques of classical and modern algebraic geometry.

Some of the techniques used and explained in this book are very classical, for example, Italian surface theory, adjoints, arithmetic and geometric genus, irregularity, plurigenera, and Enriques classification of surfaces. We often use the birational point of view of Zariski. We explicitly resolve singularities and use differential forms in the style of Zariski, thus giving the reader an introduction to the ideas of Zariski, Hironaka, and Abhyankar. This has already proven useful to several readers who wanted to learn how to work with resolution. We use the theory of schemes and cohomology of coherent sheaves throughout some parts of the book. We also use results from étale cohomology, crystalline cohomology, and the de Rham-Witt complex, thus giving the reader a gentle initiation into these very modern topics and, we hope, the motivation to study these topics more thoroughly.

The book includes all of J. Lang's 1981 Purdue thesis in which algorithms are developed for computing Picard groups of Zariski surfaces. The purely inseparable descent theory developed by Samuel in his 1964 Tata[†] notes is approached algorithmically via a differential equation. This equation is then analyzed using p-linear algebra and matrix algebra. D. Joyce has adopted Lang's program in this book. We hope that in the future the availability of the computer program will enable us to run some interesting statistical tests in the study of Picard groups of Zariski surfaces.

We hope that this overview has already convinced the reader of the wide scope of techniques used in the study of Zariski surfaces. For a complete list of topics covered, we refer to the Introduction. The whole theory has been developed (starting in 1970) over the past fifteen years, and we are confident that it will develop even more rapidly in the future. The main aim of this book is to stimulate further research.

Piotr Blass
Jeffrey Lang

[†]Samuel, P., Lectures on Unique Factorization Domains, Tata Lecture Notes, 1964. See also Fossum, R., *The Divisor Class Group of a Krull Domain*, Springer-Verlag, New York, 1973, and Jacobson, N., *Lectures in Abstract Algebra*, Vol. III, *Theory of Fields and Galois Theory*, Van Nostrand, New York, 1964.

Contents

0

Introduction[†]

This book is a result of some 13 years of work on a class of algebraic surfaces which we call Zariski surfaces since Oscar Zariski and his famous student, Heisuke Hironaka, introduced us to this project at Harvard in 1970-1971. These surfaces were discovered in the 1950s by Zariski, who discusses them briefly at the end of his paper on Castelnuovo's criterion (Zariski, 1958), and were also considered by Abhyankar, who studied them locally as important examples for the resolution problem in characteristic $p > 0$. Although the theory today is not and perhaps never will be definitive, we feel that the time has come to set down some of the most important currents and problems that emerge in the study of Zariski surfaces.

Historically, from the way that the subject has developed, there are first the problems of birational classification. A Zariski surface (ZS) is defined as a smooth surface birationally equivalent to a hypersurface $z^p = f(x,y)$ over an algebraically closed field of characteristic $p > 0$.

A sufficient condition for rationality is given by the vanishing of a single number p_2, the bigenus of the surface. It is also known that the vanishing of p_g, the geometric genus, is not enough, at least for $p = 2$ and $p = 3$. It seems to us that a very basic problem is to characterize those polynomials $f(x,y)$ that give rise to rational Zariski surfaces birationally equivalent to $z^p = f(x,y)$. We are still very far from such a characterization.

Also, it would be very nice to find examples of irrational Zariski surfaces with $p_g = 0$ in characteristic $p \geq 5$ (Zariski's

---

[†]By Piotr Blass.

problem). This challenge has puzzled us for years. We hope that one of the effects of this book will be for someone to find such examples.

What makes us think that there never will be a complete theory of Zariski surfaces is the fact that this family of surfaces is so rich. Although an Enriques-type classification has been initiated, more work remains to be done on the question as to precisely what types of surfaces can be Zariski surfaces for a fixed characteristic $p > 0$. This is probably a rather crude problem but is necessary as a preamble to more serious work on moduli.

In studying ZS we were forced, at least until now, into the following predicament. We are forced to study mainly *generic Zariski surfaces*. Also, the meaning of "generic" has not remained the same over the years as we have progressed in our work. The general idea is to choose the polynomial $f(x,y)$ to be in some sense "nice" and yet "general." Our aim is to have the surface $z^p = f(x,y)$ with only the simplest singularities that are absolutely unavoidable at finite distance and also to have either no singularity at infinity on a suitable compactification or again to have such singularities be as mild as possible.

One clear reason for insisting on such a generic $f(x,y)$ is that otherwise the singularities are difficult to resolve, and we were afraid to use the full power of Abhyankar's resolution algorithms, partly because of our ignorance of them.

Be that as it may, we have so far worked with two definitions of generic, one when $p \mid \deg f(x,y)$, and a slightly different one for $\deg f = p + 1$ (see Chapters 1 and 2). The idea of a generiz ZS, i.e., a surface defined by such a generic f, is explored in Chapter 1, Section 4, but a far more profound use of this notion and the related notion of families of generic ZSs plus monodromy-type ideas is used in Chapter 4 [which is really a joint effort with Deligne (1973)] in which we study and determine the Picard group of generic Zariski surfaces. We do this in the case $p \mid \deg f$. We feel that a suitable and analogous notion of generic ZS can and should be studied when $p \mid \deg f$, and perhaps the results of Chapter 4 can be generalized.

Early in our study of Zariski surfaces we became interested in understanding their Picard groups. Let us discuss this problem at least in the generic case when $p|\deg f$, which should become the model for all other degrees of f. First we obtain some obvious cycles by resolving singularities, etc. This is $\mathrm{Pic}^{ob} \subset \mathrm{Pic}$. The quotient group $\mathrm{Pic}/\mathrm{Pic}^{ob}$ has an arithmetic meaning. It is the divisor class group of the affine surface $z^p = f(x,y)$. Using the purely inseparable descent theory of Cartier, Grothendieck, and Samuel, one can understand this group. Jeffrey Lang has created some strong computational tools in his thesis (see Chapter 3), such as the Lang-Ganong† formula to compute this group. In essence he found an algorithm that reduces the computation to a system of linear and p-linear equations. This was then implemented on a computer using Pascal by David Joyce (for surfaces defined over a finite field). On the other hand, using monodromy techniques plus Lang's ideas, we have proven that there is an open and dense set of polynomials $f(x,y)$ for which $\mathrm{Pic} = \mathrm{Pic}^{ob}$.

At a crucial step in the proof we had to use a theorem of W. E. Lang's which we had conjectured, namely, that Pic has no p-torsion ($\deg f|p$). In order to progress in the whole theory, one has to generalize this result of W. E. Lang's as well as the techniques of Chapter 4 to $(\deg f, p) = 1$.

We have always wanted to construct moduli spaces, preferably a *fine* module space for Zariski surfaces. Clearly, we think, we must begin with the generic ones. Let us describe a conjectural theory in at least one case, again in the hope that it may generalize to other degrees of $f(x,y)$. Fix $p \geq 5$. Consider all generic polynomials $f(x,y)$ of degree exactly p. Generic here means that f_x, f_y, and $f_{xx}f_{yy} - f_{xy}^2$ have no points in common in $\mathbb{A}_k^2$ and that f_x and f_y intersect in the maximum number of possible points in $\mathbb{A}_k^2$, namely, $p^2 - 3p + 3$. k represents the algebraically closed field of characteristic p over which we are working (see the introduction to

†This name is due to P. Blass. The formula was conjectured by Ganong and proven by Lang.

Appendix 1 of Chapter 1). What we are really after is to construct a module space for smooth projective models of surfaces defined by $z^p = f(x,y)$ for all generic f's. However, it seems easier first to deal with *singular* Zariski surfaces or just the surfaces in $\mathbb{P}^3$ defined by the homogeneous equation $z^p = f(x,y,x_0)$. Let us denote such a surface as S^h_f. A basic question, we think, is the following: When is S_f isomorphic to S_g (f,g generic)? The answer is (see Chapter 2, theorem 9) that they are isomorphic if and only if the forms $f(x,y,x_0)$ and $g(x,y,x_0)$ can be obtained one from the other by the following operations:

1. A linear change of the variables x, y, and x_0.
2. The addition of linear combinations of monomials x_0^p, x^p, and y^p.

It is easy to see that this defines an equivalence relation on the set of all generic forms of degree p. Let us call the equivalence β-equivalence and write fβg to mean that f and g are in the same equivalence class.

We have checked that the orbit space M_1 = generic forms of degree p/(β-equivalence) exists as an open scheme. M_1 is clearly a good candidate for a coarse module space for singular Zariski surfaces of degree p. We have not yet checked whether a universal family of singular Zariski surfaces F_1 exists over M_1 and whether we in fact obtain a *fine* module space. This is probably just a mental block.

A more serious problem is whether or not the family $F_1 \to M_1$ admits a simultaneous resolution. We conjecture that this is true only after a base change, not necessarily separable.

$$\begin{array}{ccc} F_2 & \longrightarrow & F_1 \\ \downarrow & & \downarrow \\ M_2 & \longrightarrow & M_1 \end{array}$$

Now M_2 should serve conjecturally as a module space for smooth Zariski surfaces of degree p but with some additional structure (perhaps an ordering of the exceptional curves). Until these problems are resolved, it seems premature even to speculate about compacting these module spaces.

This book is really a collection of several articles, and thus it naturally falls into several parts. We give a detailed introduction to each of the parts later. Let us now only give an overview of the structure of the book.

Chapter 1 is P. Blass's thesis. It defines Zariski surfaces and to some extent develops their birational theory using the theory of adjoints. A long counterexample is given to Zariski's question in characteristic 2. The theory of generic Zariski surfaces of degree $p + 1$ is also described. Some questions about Picard groups are raised for such surfaces.

Chapter 2 serves as a link between Chapter 1 and J. Lang's thesis (Chapter 3). It includes a description of the differential equation (JBHS) that he uses, and a basic exact sequence which links solving the differential equation with computing Picard groups of ZSs. We work here with a new notion of generic where $p \mid \deg f$. It is in this chapter that a description of modules is attempted in a rudimentary way.

Chapter 3 is J. Lang's thesis. It provides the basic computational tools and algorithms for understanding the Picard groups of Zariski surfaces. Lang works in the affine case, but due to the exact sequence in Chapter 2, everything that he does can be immediately applied to projective ZSs.

Chapter 4 is really the joint work of Blass and Deligne. We prove, using monodromy and SGA VII-type techniques, that the (generic) Zariski surface of degree d where $p \mid d$ has the obvious Picard group denoted $\mathrm{Pic}^{\mathrm{ob}}$. We can then pass from *generic* to *general* using Chapter 3. This yields one of the main theorems of this book, the Blass-Deligne-Lang theorem (BDL).

Chapter 5 gives D. Joyce's PASCAL program, which is based on Chapter 3 and gives a powerful method for computing Picard groups of Zariski surfaces defined over finite fields.

Chapter 6, a joint work with M. Levine, deals with families of Zariski surfaces.

Chapter 7 is about the unirationality of Enriques surfaces. It is intimately related to the theory of ZSs.

Chapter 8 is by Ekedahl. In it he applies the theory of the DeRham-Witt complex and of dominoes to study the crystalline cohomology of and one forms on generic Zariski surfaces.

Chapter 9 is a joint work of Blass and Hoobler. In it are studied the Brauer groups and Picard groups of Zariski schemes, which are higher-dimensional analogs of Zariski surfaces. A higher-dimensional analog of the Blass-Deligne-Lang theorem is proven and some results are obtained about Brauer groups. An important tool there is the Cartier-Yuan exact sequence.

Chapter 10 is a section reprinted from W. E. Lang's Ph.D. thesis (Harvard, 1978). It contains a counterexample to a question posed by Zariski and a beautiful example of a Zariski surface with H^1 nontrivial.

REFERENCES

Abhyankar, S. S., Local uniformation on algebraic surfaces over ground fields of characteristics $p \neq 0$, *Ann. Math. 63* (1956), 491-526.

Barsotti, I. Repartitions on Abelian varieties, *Ill. J. Math., 2* (1958), 43-69 (especially pp. 58-59).

Blass, P., Some geometric applications of a differential equation in characteristics $p > 0$ to the theory of algebraic surfaces, *Contemporary Mathematics*, AMS 13 (1982) (with the cooperation of James Sturnfield and Jeffrey Lang).

Cartier, P. Questions de Rationalité des diviseurs en geometrie algébrique, *Bull. Soc. Math. France, 86* (1958), 177-251.

Deligne, P., and N. Katz, *SGA VII, Part II*, Lecture Notes in Mathematics, No. 340, Springer-Verlag, New York, 1973.

Grothendieck, A. Technique de descente et theoremes d'existence en geometrie algébrique I, Géneralitiés, Descentes por morphismes fidelement plats, *Seminair Bourbaki 190*, W. A. Benjamin Inc., New York and Amsterdam, 1966.

Grothendieck, A., *SGA I*, Lecture Notes in Mathematics, No. 224, Springer-Verlag, New York, 1971.

Hartshorne, R., Equivalence relations on algebraic cycles and subvarieties of small codimension, *Pure Math.*, AMS 29 (Arcata), American Mathematical Society, Providence, R.I., 1975, corollary 3.5.

Hartshorne, R., *Algebraic Geometry*, Graduate Texts in Mathematics, Springer-Verlag, New York, 1977.

Hironaka, H., Resolution of singularities of an algebraic variety over a field of characteristic zero, *Ann. Math. 79* (1964), 109-326.

Hochschild, G. Simple algebras with purely inseparable splitting fields of exponent one, *Trans. AMS*, *79* (1955), 477.

Jacobson, N. Abstract derivations and Lie algebras, *Trans. AMS*, *47* (1937), 206.

Katz, N. Nilpotent connections and the monodromy theorem: Applications of a result of Turrittin, *Publication Mathematique IHES*, *39* (1970), 355.

Katz, N. Algebraic solutions of differential equations (-curvature and the Hodge filtration), *Inv. Math.*, *18* (1972), 1-119.

Safarevic, I., and A. N. Rudakov, Supersingular K3 surfaces over fields of characteristic two, *Math. USSR, Izvestija*, *13* (1979), No. 1.

Lang, J., Ph.D. thesis, Purdue University, 1981.

Lang, W. E., Remarks on p-torsion of algebraic surfaces, *Compositio Math.* (to appear).

Rudakov, A. N., and I. R. Šafarevič, Supersingular K3 surfaces over fields of characteristic two, *Math. USSR Izv. 13*, No. 1 (1979), 147-165.

Samuel, P., Lectures on Unique Factorization Domains, Tata Lecture Notes, 1964.

Steenbrink, J., On the Picard group of certain smooth surfaces in weighted projective spaces, preprint.

Zariski, O., On Castelnuovo's criterion of rationality $p_a = p_g = 0$ of an algebraic surface, *Ill. J. Math. 2*, No. 3 (1958), 303.

Zariski, O., Introduction to the problem of minimal models in the theory of algebraic surfaces, *Am. J. Math., Publ. Math. Soc. Japan*, No. 4 (1958).

Zariski, O., The problem of minimal models in the theory of algebraic surfaces, *Am. J. Math. 80* (1958), 153.

Zariski, O., *An Introduction to the Theory of Algebraic Surfaces*, Lecture Notes in Mathematics, No. 83, Springer-Verlag, New York, 1969.

1

Basic Theory of Zariski Surfaces†

INTRODUCTION

In the classical case all unirational surfaces are rational. The situation is different in characteristic $p > 0$. This was first realized by Oscar Zariski (see Zariski, 1958c, p. 314). Prompted by Hironaka's suggestion, I began an investigation of the types of surfaces introduced by Zariski in that paper. The research was done in 1970-1971 at Harvard with the advice of Hironaka and Zariski, and continued during 1974-1977 under the direction of J. S. Milne and M. Hochster at the University of Michigan.

A smooth algebraic surface X defined over an algebraically closed field k of characteristic $p > 0$ is called a Zariski surface (ZS) iff there exist two elements x, y in the function field of X, denoted k(X), that are algebraically independent over k and such that

$$k(x,y) \subseteq k(X) \subseteq k(x^{1/p},y^{1/p})$$

The main results of the thesis are as follows. First, Section 3 answers a question posed by Zariski in 1970-1971. He asked whether a ZS with vanishing geometric genus p_g is necessarily rational. A long counterexample is given in Section 3. Second, it is shown in

---

†This is a revised version of Piotr Blass's Ph.D. thesis, University of Michigan, 1977. The thesis adviser was J. S. Milne. This chapter is reprinted from *Dissertationes Mathematicae*, Vol. 200, PWN, Warsaw, Poland, 1983.

Section 5 that the value of p_g is unbounded from above over any algebraically closed field of characteristic larger than or equal to 5. This, together with some other results of Section 5, illustrates the richness of the class of ZS.

In Section 4 a more detailed study is made of a particularly simple subclass of ZSs which we call "generic" ZSs. [A generic ZS is a smooth minimal model of the function field of a hypersurface given by $z^p = f(x,y)$, where f has degree p + 1 and the hypersurface has only the simplest singularities.] We determine p_a, p_g, P_2, K^2 for a generic Zariski surface as well as the rank of the Néron-Severi group ρ and the étale Betti numbers, b_i. Using generic Zariski surfaces, we give examples of ZSs which are of general type and K3. Trivially, there also exist rational ZSs.

All Zariski surfaces are unirational and consequently supersingular, i.e., $\rho = b_2$. Thus the richness of the class of ZSs in characteristic $p > 0$ is in sharp contrast to the situation in characteristic 0, where every unirational surface is well known to be rational.

The principal technical tool used in the thesis is the theory of adjoints and multiadjoints. This theory deals with the influence of singularities on differential forms and is classical. However, since no reference could be found for the results that we needed, a self-contained exposition of the facts from the theory of adjoints that we used is given in Section 2. We develop the theory of adjoints for a normal, two-dimensional hypersurface in affine or projective 3-space. The results about adjoints are proved over an arbitrary algebraically closed ground field, including fields of characteristic 0.

It seems to me that ZSs are an interesting subclass of unirational surfaces. Among the many open problems concerning ZSs, let me select two very concrete ones. The first is to answer Zariski's question in characteristic $p > 2$. The second, a rather puzzling problem, is whether $H^1(X,\mathcal{O}_X)$ can be nontrivial for a Zariski surface X.†

NOTATION

k algebraically closed field of characteristic $p > 0$, unless stated to the contrary

$\mathbf{A}^n$ affine n space over k

$\mathbf{P}^n$ projective n space over k

Surface irreducible, reduced, two-dimensional, quasi-projective variety over k

If X denotes a surface, k(X) will denote its function field.

For a smooth surface X we will write as usual:

p_g geometric genus of $X = \dim H^2(X,\mathcal{O}_X)$

p_a arithmetic genus $= p_g - \dim H^1(X,\mathcal{O}_X)$

P_i $\dim H^0(X,K^{\otimes i})$, where K is the canonical line bundle on X

$P_1 = P_g$, by Serre duality

q dim of the Picard variety of X = dim of the Albanese variety of X

The notation $F\colon f(x,y,z) = 0$ means

$$F = \operatorname{Spec} \frac{k[x,y,z]}{(f(x,y,z))} \qquad F \subseteq A^3$$

If $F(x,y,z,z_0)$ is a homogeneous form, $\bar{F}\colon F(x,y,z,z_0) = 0$ means

$$\bar{F} = \operatorname{Proj} \frac{k[x,y,z,z_0]}{(F(x,y,z,z_0))} \qquad \bar{F} \subseteq P^3$$

If X is a surface, we denote by Sing(X) the subscheme consisting of singular points of X. X' denotes X - Sing(X). A *desingularization* of X is a proper, birational surjective morphism π, $\tilde{X} \xrightarrow{\pi} X$ satisfying the following definitions:

1. $\tilde{X}$ is smooth.
2. $\pi^{-1}(X') \xrightarrow{\pi|\pi^{-1}(X')} X'$ is an isomorphism.

If X is a surface and $p \in X$ a closed point, we refer to blowing up the point p as "quadratic transformation with center p."

---

†Added in revision: In his 1978 Harvard thesis, W. E. Lang (see Lang, 1978) gave an example of a ZS with H^1 nontrivial. He also settled Zariski's question in characteristic 3.

Let I be an ideal in a ring R. $r \in R$ is said to be *integrally dependent* on I if there exists an equation

$$r^n + a_1 r^{n-1} + a_2 r^{n-2} + \cdots + a_n = 0$$

with $a_i \in I^i$.

I is *integrally closed* if it contains all elements of R integrally dependent on I (see Nagata, 1962, p. 34).

For the definition and basic properties of rational singularities, we refer the reader to Artin (1966, p. 129) and Lipman (1969).

1. ZARISKI SURFACES: DEFINITION AND GENERAL PROPERTIES

In this preliminary section we define Zariski surfaces and give some of their simplest properties. Throughout this section X denotes a smooth projective surface over k, k(X) is the function field of X, $k = \bar{k}$, and $\mathrm{char}(k) = p > 0$.

DEFINITION 1.1.0 A smooth projective surface X is called a *Zariski surface* if there exist two elements x, y in k(X) that are algebraically independent over k such that $k(x,y) \subseteq \kappa(X) \subseteq k(x^{1/p}, y^{1/p})$.

REMARK 1.1.1 Unless one of the inclusions in definition 1.1.0 is an equality, the extensions k(X) over k(x,y) and $k(x^{1/p}, y^{1/p})$ over k(X) are both purely inseparable of degree p.

REMARK 1.1.2 We propose to call a surface a *singular* Zariski surface if it satisfies definition 1.1.0 but is not necessarily smooth.

REMARK 1.1.3 Our definition is birational, i.e., it depends only on the function field; thus any smooth projective surface birationally equivalent[†] to a Zariski surface is itself a Zariski surface.

REMARK 1.1.4 A Zariski surface is unirational by definition.

[†]Two surfaces are birationally equivalent if they have isomorphic function fields over k.

NOTATION 1.1.5 We shall sometimes abbreviate "Zariski surface" as "ZS."

PROPOSITION 1.2.0 X is a Zariski surface if and only if X is birationally equivalent to a hypersurface in $\mathbf{A}^3$ defined by an irreducible equation of the form $z^p - f(x,y) = 0$, where $f(x,y) \in k[x,y]$, the polynomial ring in two variables over k.

Proof: Suppose first that X is birationally equivalent to

$$\operatorname{Spec} \frac{k[x,y,z]}{(z^p - f(x,y))}$$

with $z^p - f(x,y)$ irreducible. Then $k(X) \cong k(\bar{x},\bar{y},\bar{z})$, where $\bar{x}$, $\bar{y}$, and $\bar{z}$ are the residue classes in $k[x,y,z]/(z^p - f(x,y))$ of x, y, and z, respectively. We have, by standard field theory, $k(x,y) = k(\bar{x},\bar{y}) \subsetneq k(\bar{x},\bar{y},\bar{z})$. Since $\bar{z} = f_1(\bar{x}^{1/p},\bar{y}^{1/p})$, where f_1 is obtained from f by taking pth roots of all the coefficients, we have $k(\bar{x},\bar{y},\bar{z}) \subseteq k(\bar{x}^{1/p},\bar{y}^{1/p})$. Counting degrees of extensions, we easily see that $k(\bar{x},\bar{y}) \subsetneq k(\bar{x},\bar{y},\bar{z}) \subsetneq k(\bar{x}^{1/p},\bar{y}^{1/p})$ so that X is a ZS.

Suppose next that X is a Zariski surface. If $k(X) = k(x,y)$ or if $k(X) = k(x^{1/p},y^{1/p})$, then X is birationally equivalent, for example, to the irreducible hypersurface

$$F = \operatorname{Spec} \frac{k[x,y,z]}{(z^p - y)}$$

since $k(F) \approx k(x,y^{1/p}) \approx k(x,y) \approx k(x^{1/p},y^{1/p})$. Thus we may assume that $k(x,y) \subsetneq k(X) \subsetneq k(x^{1/p},y^{1/p})$.[†] Let $z \in k(x) - k(x,y)$. We have that $z^p = g(x,y) \in k(x,y)$ or $z^p = w(x,y)/v(x,y)$ with w,v polynomials in x,y and w/v not a pth power in k(x,y). $z^p = w/v$ implies that $z^p v^p = wv^{p-1}$. Set $z_1 = z_v$; $z_1 \in k(X) - k(x,y)$. We have $z_1^p = wv^{p-1}$ and wv^{p-1} is not a pth power in k[x,y]. Since $k(x,y) \subsetneq k(x,y,z_1) \subseteq k(x,y,z)$ and k(x,y,z) is an extension of k(x,y) of degree p, we have $k(x,y,z_1) = k(x,y,z) = k(X)$. It follows easily that X is birationally equivalent to

[†]The part of the proof that follows is due to O. Zariski.

$$\operatorname{Spec} \frac{k[\bar{x},\bar{y},\bar{z}_1]}{(\bar{z}_1^p - w(\bar{x},\bar{y})v(\bar{x},\bar{y})^{p-1}}$$

where $\bar{x}$, $\bar{y}$, $\bar{z}_1$ are indeterminates and the equation is irreducible.

Note: The notation F: $z^p = f(x,y)$ will mean that F = Spec $k[x,y,z]/(z^p - f(x,y))$, etc.

PROPOSITION 1.3.0 Given any irreducible, reduced hypersurface in $\mathbb{A}^3$, F: $z^p = f(x,y)$, there exists a Zariski surface $\tilde{F}$ birationally equivalent to F.

Proof: Let $\bar{F}$ be the closure of F in projective 3-space. Let $\tilde{F}$ be a desingularization of $\bar{F}$ [exists by Abhyankar (1956)]. Then $\tilde{F}$ is a projective smooth surface which is birationally equivalent to F, so it is a ZS by proposition 1.2.0.

REMARK 1.3.1 F, $\bar{F}$, $\tilde{F}$ will often be given the foregoing meaning.

REMARK 1.3.2 $\tilde{F}$ is not unique. However, if we require $\tilde{F}$ to be a relatively minimal model of k(F), then $\tilde{F}$ will be unique unless $\tilde{F}$ is ruled or rational (Zariski, 1958a,b). We shall prove later that q = dim Alb($\tilde{F}$) = 0 (see lemma 1.6.0), where Alb($\tilde{F}$) denotes the Albanese variety of $\tilde{F}$. It follows that $\tilde{F}$ ruled $\Rightarrow$ $\tilde{F}$ rational.

PROPOSITION 1.4.0 Let t_1,t_2 be indeterminates. Let L be a field with $k(t_1,t_2) \subseteq L \subseteq k(t^{1/p},t^{1/p})$. Then there exists a ZS X with $k(X) \underset{\text{over } k}{\approx} L$.

Proof: It is well known that there exists an irreducible hypersurface $F \subseteq \mathbb{A}^3$ with k(F) = L. We construct a smooth ZS $\tilde{F}$ as in proposition 1.3.0. Then $k(\tilde{F}) = k(F) = L$.

PROPOSITION 1.5.0 Let k be as above, and t_1,t_2 two indeterminates over k. Let $\mathcal{L}$ be the set of all fields L such that $k(t_1,t_2) \subseteq L \subseteq k(t_1^{1/p},t_2^{1/p})$. Let $\mathcal{ZS}$ be the set of all (projective smooth) Zariski surfaces over k. We have a one-to-one correspondence

$\mathcal{L}$/(field isomorphism over k) $\approx$ $\mathcal{ZS}$/(birational equivalence)

Proof: Follows easily from 1.4.0 and the definitions.

REMARK 1.5.1 We shall use proposition 1.5.0 to show that the set L/(field isomorphism over k) is infinite in some cases (see 5.1.2).

LEMMA 1.6.0 The dimension of the Albanese variety of a Zariski surface is 0. In fact, the Albanese variety is trivial.
Proof: Since Zariski surfaces are unirational, the proof follows from Lang (1959, corollary, p. 25), where a unirational variety is called semipure.

LEMMA 1.7.0 The Picard variety of a Zariski surface is trivial.
Proof: Let $\tilde{F}$ be a Zariski surface. By lemma 1.6.0 the Albanese variety of $\tilde{F}$, Alb($\tilde{F}$), is trivial. The Picard variety of $\tilde{F}$ is the same as that of Alb($\tilde{F}$) (Lang, 1959, p. 148, theorem 1), but the Picard variety of the trivial variety is trivial.

PROPOSITION 1.8.0 If a nonsingular Zariski surface $\tilde{F}$ satisfies $P_2(\tilde{F}) = 0$, then $\tilde{F}$ is rational.
Proof[†]*:* We know by lemmas 1.6.0 and 1.7.0 that dim Alb($\tilde{F}$) = dim Picard variety of $\tilde{F} = 0$. Also, $P_2(\tilde{F}) = 0$ implies that $p_g = 0$. But Nakai has shown that if $p_g = 0$, then the dimension of the Picard variety is equal to the dimension of $H^1(\tilde{F},\mathcal{O}_{\tilde{F}})$ (see Nakai, 1957, theorem 5, or Mumford, 19 , lecture 27, 2°, pp. 195-198). Thus $p_g - p_a = \dim H^1(\tilde{F},\mathcal{O}_{\tilde{F}}) = 0$ and hence $p_g = p_a = 0$. Also, $P_2 = 0$ by assumption and the rationality of $\tilde{F}$ follows from Castelnuovo's criterion of rationality (see Zariski, 1958c).

PROPOSITION 1.9.0 Let X be a ZS. Let b_i = rank of $H^i_{et}(X,Q_\ell)$, where $\ell \neq p$. Then $b_0 = b_4 = 1$ and $b_1 = b_3 = 0$.
Proof: Since q = dim Alb(X) = 0, the proposition follows from Milne, 1980, Chaps. 5 and 6.

REMARK 1.9.1 1.6.0, 1.7.0, 1.8.0, and 1.9.0 remain valid if we replace the words "Zariski surface" by "smooth unirational surface."

[†]This proof is adapted from Zariski (1958c, p. 314) and is given here only for the convenience of the reader.

REMARK 1.10.0 Using the method of proof of Grothendieck (1971, p. 286), one can show that the étale fundamental group of a Zariski surface is trivial.

2. THE THEORY OF ADJOINTS

Introduction

The theory of adjoints is well known and classical for algebraic curves. For surfaces it was developed by Clebsch, M. Noether, the Italian geometers, and Zariski. However, we found that a completely rigorous treatment of the theory of adjoints for a two-dimensional normal hypersurface was missing in the literature. Without a doubt that theory has been well known to several writers in the theory of algebraic surfaces, but no reference could be found for the proofs of the results they used. It is the purpose of this chapter to fill this gap in the literature.† We also discuss briefly the notion of multiadjoints and connections with plurigenera. This was a favorite topic with Enriques, but again a modern reference is missing. The theory of adjoints and multiadjoints is our principal tool in all the later sections.

Throughout this section, $k = \bar{k}$ will be an algebraically closed field of any characteristic. All surfaces will be quasi-projective. We shall denote surfaces by capital letters X, Y, W, Z, S, etc. All surfaces are assumed irreducible and reduced. k(X) denotes the function field. $\Omega^1_{k(X)/k}$, $\Omega^2_{k(X)/k}$ are the Kähler differentials of k(X) over k (respectively, 2-forms). The letters p, q always denote *closed* points. We will need some facts about differentials.

PROPOSITION 2.1.0 $\Omega^2_{k(X)/k}$ is a one-dimensional vector space over k(X). If $a,b \in k(X)$ are a separating transcendence basis, $da\, db \neq 0$

†Added in revision: Our treatment here relies only on rudiments of sheaf theory. The reader who wishes to see a more homological treatment of adjoints is referred to our joint note with J. Lipman (Blass and Lipman, 1979).

in $\Omega^2_{k(X)/k}$ and da db is a basis over k(X) [d: $k(X) \to \Omega^1$ denotes the usual differential map].

Proof: Well known.

REMARK 2.1.1 $a \in k(X)$, $\alpha \in \Omega^2_{k(X)/k}$, $a \neq 0$, $\alpha \neq 0$, implies that $a\alpha \neq 0$ in $\Omega^2_{k(X)/k}$.

Let X be a surface. We write $X' = X - \mathrm{Sing}(X)$. If $p \in X'$ is a closed point, we define a subset of $\Omega^2_{k(X)/k}$ which we shall call $\mathrm{Reg}_X(p)$ as follows:

DEFINITION 2.1.2 $\alpha \in \mathrm{Reg}_X(p)$ iff $\alpha = \Sigma\, a_i\, db_i\, dc_i$, with a_i, b_i, $c_i \in (\mathcal{O}_X)_p$, where $(\mathcal{O}_X)_p$ is the local ring of p. If $\alpha \in \mathrm{Reg}_X(p)$, we say that α is *regular* at p.

PROPOSITION 2.2.0 If $p \in X'$, then $\mathrm{Reg}_X(p)$ is an $(\mathcal{O}_X)_p$ module. It is free of rank 1. If t_p, t'_p are a regular system of parameters at p, then $dt_p\, dt'_p$ is a generator of $\mathrm{Reg}_X(p)$ as an $(\mathcal{O}_X)_p$-module.

Proof: Well known.

Using the notions above, we now proceed to define a sheaf ω_X on X which if X is smooth is isomorphic to the sheaf of Kähler 2-forms. In the general case considered here $\omega_X \cong j_*$ (sheaf of Kähler 2-forms on X'), where j: $X' \hookrightarrow X$ is the open immersion.

First we define a presheaf of subsets of $\Omega^2_{k(X)/k}$. Let $U \subset X$ be open; we define

$$\omega_X(U) = \bigcap_{p \in U'} \mathrm{Reg}_X(p)$$

If $V \subset U$, $\omega_X(V) \supset \omega_X(U)$ and we define the restriction ρ^U_V: $\omega_X(U) \to \omega_X(V)$ to be the inclusion map.

PROPOSITION 2.3.0 ω_X is a sheaf of $\mathcal{O}_X$-modules. All restriction maps are monomorphisms and k-linear.

Proof: Omitted.

DEFINITION 2.4.0 Let $S \subset X'$ be any set of smooth closed points. We define

$$\omega_X(S) = \bigcap_{q \in S} \mathrm{Reg}_X(q)$$

PROPOSITION 2.5.0 If $a \in \bigcap_{q \in S} (\mathcal{O}_X)_q$, then $a\omega_X(S) \subseteq \omega_X(S)$.

Proof: Clear.

Behavior of ω_X under Restriction to Open Subschemes

PROPOSITION 2.6.0 Suppose that $U \subset X$ is open, and U, X surfaces, $U \neq \emptyset$. We identify $k(U) = k(X)$, $\Omega^2_{k(U)/k} = \Omega^2_{k(X)/k}$. Then we have $\omega_U(V) = \omega_X(V)$ for each $V \subset U$.

Proof: Under our identifications, if $p \in V$, then $(\mathcal{O}_U)_p = (\mathcal{O}_X)_p$. Thus $\mathrm{Reg}_X(p)$ and $\mathrm{Reg}_U(p)$ do not depend on whether p is considered as a point of U or X, so that $\mathrm{Reg}_X(p) = \mathrm{Reg}_U(p)$ and the proposition follows from the definitions of the sheaves ω_U and ω_X.

PROPOSITION 2.6.1 Let $\{U_i\}$, $i = 0, 1, \ldots, k$, be a covering of X; then $\omega_X(X) = \bigcap_{i=0}^k \omega_X(U_i) = \bigcap_{i=0}^k \omega_{U_i}(U_i)$.

Proof: The first equality is trivial; the second follows from proposition 2.6.0.

Behavior of ω_X under Certain Birational Maps

Let W, Z be surfaces. Let $W \xrightarrow{\pi} Z$ be a birational map satisfying the following condition

$$\text{(2.7.0)} \quad \pi^{-1}(Z') \xrightarrow{\pi|\pi^{-1}(Z')} Z' \text{ is an isomorphism}$$

Then we have the following isomorphism: $k(Z) \xrightarrow[\sim]{\pi_1} k(W)$ of function fields over k. (We sometimes refer to π_1 as a "pullback" and occasionally we will use the notation π^{-1} for π_1.) π_1 induces $\Omega^2_{k(Z)/k} \xrightarrow[\sim]{\pi_1} \Omega^2_{k(W)/k}$, an isomorphism of k-vector spaces. We denote the inverse isomorphism by π_2, so that $\Omega^2_{k(W)/k} \xrightarrow[\sim]{\pi_1} \Omega^2_{k(Z)/k}$.

PROPOSITION 2.8.0 Let $W \xrightarrow{\pi} Z$ be as above with condition 2.7.0 satisfied. π_1, π_2 are as above. Then $\pi_2(\omega_W(\pi^{-1}(U))) \subseteq \omega_Z(U)$ for every open $U \subseteq Z$.

Proof: Omitted.

REMARK 2.8.1 π_2: $\omega_W(\pi^{-1}(U)) \to \omega_Z(U)$ is a monomorphism of K-vector spaces.

REMARK 2.8.2 In fact, π_2 defines a monomorphism of $\mathcal{O}_Z$-modules $\pi_*\omega_W \xhookrightarrow{\pi_2} \omega_Z$.

Affine Case, Adjoints

2.8.3 Let us assume that $F \subseteq A^3$ is a normal, irreducible, reduced hypersurface.

Throughout this section F will have this meaning. Let us describe ω_F.

PROPOSITION 2.9.0 There exists a differential $\sigma_F \in \omega_F(F)$ such that for every $U \subseteq F$ open, $\omega_F(U) = \mathcal{O}_F(U)\sigma_F \subseteq \Omega^2_{k(F)/k}$ is a free $\mathcal{O}_F(U)$-module on one generator, σ_F, and such that for every point $p \in F'$, $\sigma_F \in \mathrm{Reg}_F(p)$ and $\mathrm{Reg}_F(p)$ is a free $(\mathcal{O}_F)_p$-module on one generator, namely, σ_F.

The differential σ_F is unique up to multiplication by a unit of $\mathcal{O}_F(F)$. We shall call it a canonical generating differential (abbreviated cgd) on F. Moreover, if we write

$$F = \mathrm{Spec}\,\frac{k[x_1,x_2,x_3]}{(f(x_1,x_2,x_3))} \qquad \text{and} \qquad \frac{\partial f}{\partial x_i} \neq 0$$

we may take

$$\sigma_F = \frac{dx_j\,dx_k}{\partial f/\partial x_i} \qquad i,\ j,\ k \text{ all distinct}$$

Proof: Well known; see, e.g., Šafarevič (1974, chap. 3, sec. 5.4), where nonsingularity is assumed but is not essential in the proof.

REMARK 2.10.0 $\sigma_F \neq 0$.

LEMMA 2.11.0 $a \notin (\mathcal{O}_F)_p \Rightarrow a\sigma_F \notin \mathrm{Reg}_F(p)$.
Proof: Suppose that $a\sigma_F \in \mathrm{Reg}_F(p)$. Then $a\sigma_F = a'\sigma_F$, $a' \in \mathcal{O}_F(p)$. Therefore, $(a - a')\sigma_F = 0$. So $a - a' = 0$, $a = a'$, so that $a \in (\mathcal{O}_F)_p$.

DEFINITION 2.12.0 Let F be as in 2.8.3. Let $p \in F$ be an isolated singular point. Let $\tilde{F} \xrightarrow{\pi} F$ be any map such that

(a) π is birational and proper.
(b) $\pi|\pi^{-1}(F')$ is an isomorphism.
(c) $\pi^{-1}(p) \subseteq \tilde{F}'$.

We shall call such a map a *resolution* of p in F.

DEFINITION 2.13.0 Let F, p be as above. Let σ_F be a canonical generating differential on F. We call $a \in (\mathcal{O}_F)_p$ an *adjoint* (locally) at p iff there exists a resolution $\tilde{F} \xrightarrow{\pi} F$ of p in F such that $\pi_1(a\sigma_F)$ is regular on $\pi^{-1}(p)$ [or, equivalently, $\pi_1(a\sigma_F) \in \mathrm{Reg}_{\tilde{F}}(\pi^{-1}(p))$].

REMARK 2.13.1 Let σ'_F be another canonical generating differential on F. We have $\sigma'_F = u\sigma_F$, $u \in \mathcal{O}_F(F)$ a unit; then $\pi_1(u\sigma'_F) = \pi_1(u) \times \pi_1(a\sigma_F)$. Since $\pi_1(u)$ is regular on $\pi^{-1}(p)$, $\pi_1(a\sigma'_F)$ is regular on $\pi^{-1}(p)$. Thus definition 2.13.0 does not depend on the choice of σ_F.

PROPOSITION 2.14.0 (Independence of the resolution) Again let F, p be as above, and let $a \in \mathcal{O}_p$ be an adjoint. Let $G \xrightarrow{\rho} F$ be any resolution of p in F; then for any choice of a canonical generating differential σ_F, we have $\rho_1(a\sigma_F) \in \mathrm{Reg}(\rho^{-1}(p))$.
Proof: Consider an open neighborhood V of p such that $V = V' \cup \{p\}$. Let $\tilde{F} \xrightarrow{\pi} F$ be a resolution as in 2.13.0, which arises from a being an adjoint. Let $V_1 = \rho^{-1}(V)$, $V_2 = \pi^{-1}(V)$; V_1, V_2 are smooth. We have the following diagram where Z is a desingularization of $(V_1 \times V_2)_{red}$. It is easy to see that r and p are both birational and proper. Suppose that $\rho_1(a\sigma_F) \notin \mathrm{Reg}(\rho^{-1}(p))$. Then $\rho_1(a\sigma_F)$ has a

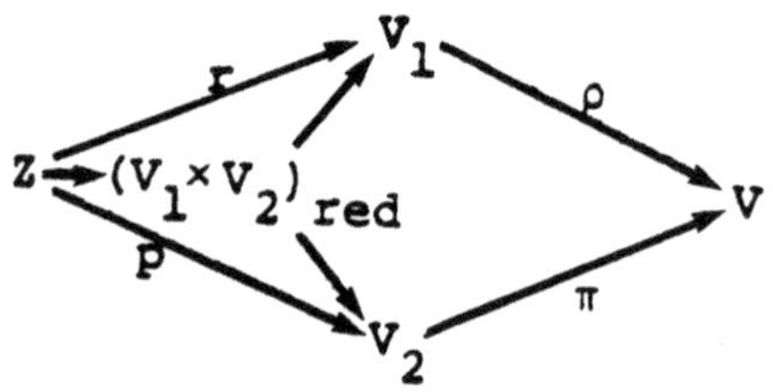

polar curve on V_1. By Šafarevič (1966, p. 55), r is a composition of finitely many quadratic transforms with point centers. Thus we easily conclude that $r_1\rho_1(a\sigma_F)$ cannot be regular on Z.

However, since $\pi_1(a\sigma_F)$ is regular on V_2 and p is everywhere defined, we easily see that $p_1\pi_1(a\sigma_F)$ is regular on Z. But $r_1\rho_1(a\sigma_F) = p_1\pi_1(a\sigma_F)$, so we have a contradiction.

REMARK 2.14.1 The argument above is adapted from the one used by Grauert and Riemenschneider in the analytic case (see Grauert and Riemenschneider, 1970, p. 271).

REMARK 2.14.2 If G is isomorphic to an affine normal surface $F \subseteq A^3$, we can define in the obvious way what is meant by an adjoint at an isolated singularity of G. One easily sees that this notion is independent of the choice of F and of the isomorphism.

THEOREM 2.15.0 Let G be isomorphic to $F \subseteq A^3$, F as in 2.8.3. Let $p \in G$ be an isolated singularity. Then the adjoints in $(\mathcal{O}_G)_p$ form an ideal of finite colength. Moreover, if G = F: $f(x_1,x_2,x_3) = 0$, then $\partial f/\partial x_i$ is adjoint for every i = 1, 2, 3.

Proof: We may assume that G = F and

$$F = \operatorname{Spec} \frac{k[x_1,x_2,x_3]}{(f(x_1,x_2,x_3))}$$

Let $\tilde{F} \xrightarrow{\pi} F$ be a resolution of p in F. The partial $\partial f/\partial x_i$ is adjoint if $\partial f/\partial x_i \equiv 0$. If $\partial f/\partial x_i \not\equiv 0$, then $\sigma_F^{(i)} = dx_j\, dx_k/(\partial f/\partial x_i)$ is a canonical generating differential on F, for $0 \le i,j,k \le 3$ all distinct. Then we have $(\partial f/\partial x_i)\sigma_F^{(i)} = dx_j\, dx_k$. Since $x_j, x_k \in (\mathcal{O}_F)_p$,

$$\pi_1(x_j),\pi_1(x_k) \in \bigcap_{q\in\pi^{-1}(p)} (\mathcal{O}_{\tilde{F}})_q$$

Thus

$$\pi_1\left(\frac{\partial f}{\partial x_i}\sigma_F^{(i)}\right) = d\pi_1(x_j)d\pi_1(x_k) \in \mathrm{Reg}_{\tilde{F}}(\pi^{-1}(p))$$

which shows that $\partial f/\partial x_i$ is adjoint.

Let us show that the adjoints in $(\mathcal{O}_F)_p$ form an ideal. The proof is standard. Suppose that $a \in (\mathcal{O}_F)_p$ and $b \in (\mathcal{O}_F)_p$, and that b is adjoint. For any canonical generating differential σ_F, $\pi_1(ab\sigma_F) = \pi_1(a)\pi_1(b\sigma_F)$, but $\pi_1(a)$ is regular on $\pi^{-1}(p)$ and $\pi_1(b\sigma_F) \in \mathrm{Reg}_{\tilde{F}} \times (\pi^{-1}(p))$ by independence of the resolution (2.14.0). Thus $\pi_1(ab\sigma_F) \in \mathrm{Reg}_{\tilde{F}}(\pi^{-1}(p))$ and ab is adjoint.

If a is adjoint and b is adjoint in $(\mathcal{O}_F)_p$, then a + b is trivially so. Thus the adjoints form an ideal in $(\mathcal{O}_F)_p$ which we will call Adj(p). Since $p \in F$ is an isolated singularity, it is well known that the ideal $J_p = (\partial f/\partial x_1, \partial f/\partial x_2, \partial f/\partial x_3)$ has finite colength but $J_p \subseteq \mathrm{Adj}(p)$, so Adj(p) has finite colength. (This argument is due to M. Hochster.)

We propose to call $\dim_k(\mathcal{O}_F)_p/\mathrm{Adj}(p) = e_p$ = the *Noether-Enriques number* of the singularity. It is often denoted p and sometimes called the "genus." For surfaces this invariant seems first to have been studied extensively by Max Noether and then taken up by the Italian school; Enriques discussed a number of special cases in his book *Le Superficie Algebriche* (Enriques, 1949).

PROPOSITION 2.16.0 Let $G \subset X$ be an open subscheme isomorphic to $F \subseteq A^3$, where F is as in 2.8.3. Let σ_G be a canonical generating differential. Let X be a desingularization of X, $\tilde{X} \xrightarrow{\pi} X$, $\pi^{-1}(G) = \tilde{G}$. Let $\mathrm{Sing}(G) = \{p_1,\ldots,p_2\}$. Then

$$\pi_2(\omega_X(G)) = (\mathcal{O}_{\tilde{X}}(\tilde{G}) \cap \bigcap_{i=1}^{s} \mathrm{Adj}(p_i))\sigma_G$$

Proof: Again we easily reduce to the case where $G = X$, $G \subseteq A^3$ is affine, and $\tilde{G} \xrightarrow{\pi} G$ is a desingularization. We have to show

$$\pi_2(\omega_{\tilde{G}}(\tilde{G})) = (\mathcal{O}_G(G) \cap \bigcap_{i=1}^{s} \mathrm{Adj}(p_i))\sigma_G$$

Suppose that $\alpha \in$ RHS. Then $\alpha = a\sigma_G$, $a \in \mathcal{O}_G(G)$, where a is adjoint at all the singular points of G. Then $\pi_1(\alpha) \in \omega_{\tilde{G}}(\tilde{G})$, so $\alpha = \pi_2(\pi_1(\alpha))$ implies that $\alpha \in \pi_2(\omega_{\tilde{G}}(\tilde{G}))$.

Suppose now that $\beta \in$ LHS. Then $\pi_1(\beta) \in \omega_{\tilde{G}}(\tilde{G})$. Therefore β is regular on G', so $\beta \in \omega_G(G') = \mathcal{O}_G(G')\sigma_G$. But since G is normal and G - G' is a finite set of points, we have $\mathcal{O}_G(G') = \mathcal{O}_G(G)$. Therefore, $\beta = a\sigma_G$ with $a \in \mathcal{O}_G(G)$. Moreover, since $\pi_1(\beta)$ is regular on $\pi^{-1}(p_j)$, j = 1, 2, ..., s, a is adjoint at all the singular points of G. We conclude that $a \in \mathcal{O}_G(G) \cap \bigcap_{j=1}^{s} \mathrm{Adj}(p_j)$, so that $\beta \in (\mathcal{O}_G(G) \cap \bigcap_{j=1}^{s} \mathrm{Adj}(p_j)\sigma_G)$, so $\beta \in$ RHS.

For future reference in Section 3, we introduce one more definition, that of an adjoint surface.

DEFINITION 2.16.1 Let

$$F = \mathrm{Spec}\ \frac{k[x_1,x_2,x_3]}{(f(x_1,x_2,x_3))}$$

be as in 2.8.3. Let $p \in F$ be an isolated singularity. Let

$$H = \mathrm{Spec}\ \frac{k[x_1,x_2,x_3]}{(A(x_1,x_2,x_3))}$$

be another closed subscheme of $\mathbf{A}^3$. We call H an *adjoint surface* to F at p iff $A(x_1,x_2,x_3)$ considered as an element of k[F] is an adjoint element of the local ring of F at P.

REMARK 2.16.2 An "adjoint surface" is not assumed to be irreducible or reduced. It need not be a surface in the sense of this thesis.

Projective Case

Let us assume

2.17.0 $$X = \mathrm{Proj}\ \frac{k[X_0,X_1,X_2,X_3]}{(F(X_0,X_1,X_2,X_3))}$$

where the degree of F is $n \geq 4$; X is assumed to be reduced, irreducible, and normal. Let us denote by $k[X] = k[x_0,x_1,x_2,x_3]$ the graded ring

$$\frac{k[X_0,X_1,X_2,X_3]}{(F(X_0,X_1,X_2,X_3))}$$

and by $k[X]_{(d)}$, d = 0, 1, 2, ..., the dth graded piece of k[X].

As is usual, we identify k(X) with the set of ratios a/b, where $a \in k[X]$, $b \in k[X]$, and a and b are of the same degree. X is covered by four affines.

$$X_i = \mathrm{Spec}(k[X]_{x_i})_{(0)} = \mathrm{Spec}\ k\left[\frac{x_0}{x_i},\frac{x_1}{x_i},\frac{x_2}{x_i},\frac{x_3}{x_i}\right]$$

It is important for us to note that each X_i is isomorphic to an affine surface. Explicitly, we have, for example,

$$X_0 \approx H_0 = \mathrm{Spec}\ \frac{k[Y_1,Y_2,Y_3]}{(F(1,Y_1,Y_2,Y_3))} = \mathrm{Spec}\ k[y_1,y_2,y_3]$$

the isomorphism of $k[H_0]$ with $k[x_1/x_0,x_2/x_0,x_3/x_0]$ being given by $y_1 \to x_1/x_0$; $y_2 \to x_2/x_0$; $y_3 \to x_3/x_0$; and similarly for i = 2, 3, 4.

Since X is reduced and normal, at least two of the partial derivatives $\partial F/\partial X_i$ are not $\equiv 0$. After renumbering the variables, we may assume the following condition.

2.17.1 $\partial F/\partial X_0 \not\equiv 0$ and $\partial F/\partial X_1 \not\equiv 0$ in the polynomial ring $k[X_0,X_1,X_2,X_3]$, where the X_i are indeterminates. Then it is easy to show that $\partial F/\partial X_0$ and $\partial F/\partial X_1 \in k[X]_{(n-1)}$ and are also $\not\equiv 0$ in k[X].

We fix the following canonical generating differentials on X_0, X_2, X_3, and X_1:

$$\sigma_0 = \frac{d(x_2/x_0)\,d(x_3/x_0)}{\frac{\partial F}{\partial X_1}(1,x_1/x_0,x_2/x_0,x_3/x_0)} = \frac{d(x_2/x_0)\,d(x_3/x_0)}{\frac{\partial F}{\partial X_1}/x_0^{n-1}}$$

$$\sigma_2 = \frac{d(x_3/x_2)\,d(x_0/x_2)}{\frac{\partial F}{\partial X_1}/x_2^{n-1}} = \frac{d(x_1/x_2)\,d(x_3/x_2)}{\frac{\partial F}{\partial X_0}/x_2^{n-1}}^{\dagger}$$

$$\sigma_3 = \frac{d(x_0/x_3)\,d(x_2/x_3)}{\frac{\partial F}{\partial X_1}/x_3^{n-1}}$$

and

$$\sigma_1 = \frac{d(x_3/x_1)\,d(x_2/x_1)}{\frac{\partial F}{\partial X_0}/x_1^{n-1}}$$

respectively.

The following observation is of primary importance to us.

KEY COMPUTATION LEMMA 2.18.0

$$\sigma_i = \left(\frac{x_i}{x_j}\right)^{n-4} \sigma_j \qquad 0 \le i,\ j \le 3, \text{ in } \Omega^2_{k(X)/k}$$

Proof[††]*:* First we compare σ_0 and σ_3.

$$\frac{\sigma_0}{(x_0/x_3)^{n-1}} = \frac{d(x_2/x_0)\,d(x_3/x_0)}{\frac{\partial F}{\partial X_1}(x_0,x_1,x_2,x_3)\cdot\frac{1}{x_3^{n-1}}}$$

$$= \frac{d\left(\frac{x_2/x_3}{x_0/x_3}\right) d\left(\frac{1}{x_0/x_3}\right)}{\frac{\partial F}{\partial X_1}(x_0,x_1,x_2,x_3)\cdot\frac{1}{x_3^{n-1}}}$$

$$= \frac{1}{(x_0/x_3)^4}\,\frac{(x_0/x_3)\,d(x_2/x_3) - (x_2/x_3)\,d(x_0/x_3)\,(-d(x_0/x_3))}{\frac{\partial F}{\partial X_1}(x_0,x_1,x_2,x_3)\cdot\frac{1}{x_3^{n-1}}}$$

[†] The second equality can be easily justified; see, e.g., Šafarevič (1974, chap. 7, sec. 5.4).

[††] For more details, see Blass (1977, pp. 25-28) or compare Šafarevič (1974, chap. 7, sec. 5.4).

$$= \frac{1}{(x_0/x_3)^3} \frac{d(x_2/x_3)\,d(x_0/x_3)}{\frac{\partial F}{\partial X_1}(x_0,x_1,x_2,x_3) \cdot \frac{1}{x_3^{n-1}}}$$

$$= \frac{1}{(x_0/x_3)^3} \sigma_3$$

so that $\sigma_0 = (x_0/x_3)^{n-4}\sigma_3$ in $\Omega^2_{k(X)/k}$. Similarly, one proves that $\sigma_0 = (x_0/x_2)^{n-4}\sigma_2$.

Finally, we wish to compare σ_0, σ_2, and σ_3 with σ_1. This is easily done by recalling that

$$\sigma_2 = \frac{d(x_1/x_2)\,d(x_3/x_2)}{\frac{\partial F}{\partial X_0}/x_2^{n-1}}$$

$$\sigma_1 = \frac{d(x_3/x_1)\,d(x_2/x_1)}{\frac{\partial F}{\partial X_0}/x_1^{n-1}}$$

and then we proceed as earlier in this proof to show that

$$\sigma_2 = \left(\frac{x_2}{x_1}\right)^{n-4} \sigma_1$$

We have shown that

$$\sigma_0 = \left(\frac{x_0}{x_3}\right)^{n-4} \sigma_3 = \left(\frac{x_0}{x_2}\right)^{n-4} \sigma_2 = \left(\frac{x_0}{x_2}\right)^{n-4} \cdot \left(\frac{x_2}{x_1}\right)^{n-4} \sigma_1 = \left(\frac{x_0}{x_1}\right)^{n-4} \sigma_1$$

This implies that $\sigma_i = (x_i/x_j)^{n-4}\sigma_j$, $0 \le i,j \le 3$

DEFINITION 2.19.0 Assume X to be as in 2.17.0. Let $p \in X$ be an isolated singularity, and let $A \in k[X]_{(d)}$. Then A is called an *adjoint form of degree* d *at* p iff for every i such that $p \in X_i$ we have that A/x_i^d is adjoint at p. [We note that $A/x_i^d \in (\mathcal{O}_{X_i})_p$.]

DEFINITION 2.20.0 $A \in k[X]_{(d)}$ is called an *adjoint form* iff A is adjoint at every singular point of X.

LEMMA 2.21.0 $\bigcap_{i=0}^{3}(\mathrm{Adj}(X_i))\sigma_i$ = adjoint forms of degree (n - 4) as k-vector spaces.

Proof: Suppose that A is an adjoint form of degree n - 4 in k[X]. We have $(A/x_i^{n-4})\sigma_i = (A/x_i^{n-4})(x_i/x_j)^{n-4}\sigma_j = (A/x_j^{n-4})\sigma_j$ for all $0 \le i,j \le 3$. Denote by $\sigma(A)$ the differential in $\Omega^2_{k(X)/k}$ that is equal to

$$\frac{A}{x_0^{n-4}}\sigma_0 = \frac{A}{x_1^{n-4}}\sigma_1 = \frac{A}{x_2^{n-4}}\sigma_2 = \frac{A}{x_3^{n-4}}\sigma_3$$

Clearly, $\sigma(A) \in \bigcap_{i=0}^{3}(\mathrm{Adj}(X_i))\sigma_i$ and σ is seen to define a k-linear injective map from adjoint forms of degree n - 4 to $\bigcap_{i=0}^{3} \mathrm{Adj}(X_i)\sigma_i$.

Let $\beta \in \bigcap_{i=0}^{3}(\mathrm{Adj}(X_i))\sigma_i$. Write $\beta = a_0\sigma_0 = a_1\sigma_1 = a_2\sigma_2 = a_3\sigma_3$, $a_i \in \mathrm{Adj}(X_i)$. We have $a_i\sigma_i = a_i(x_i/x_j)^{n-4}\sigma_j = a_j\sigma_j$, so that $a_i(x_i/x_j)^{n-4} = a_j$. By a well-known result of Serre (1955) this implies the existence of a *unique* form A of degree n - 4 in k[X] such that $A/x_k^{n-4} = a_i$. Then $\sigma(A) = a_i\sigma_i = \beta$, i = 0, 1, 2, 3. Thus σ is surjective, hence it is an isomorphism of k-vector space.

LEMMA 2.22.0 Let $\tilde{X} \xrightarrow{\pi} X$ be any desingularization of X. Then π_2: $\Omega^2_{k(\tilde{X})/k} \to \Omega^2_{k(X)/k}$ defines an isomorphism of k-vector spaces.

$$\pi_2\colon\ \omega_{\tilde{X}}(\tilde{X}) \tilde{\to} \bigcap_{i=0}^{3} (\mathrm{Adj}(X_i))\sigma_i$$

Proof: Let $\tilde{X}_i = \pi^{-1}(X_i)$. Since π_2 is a monomorphism, the only thing to compute is $\pi_2(\omega_{\tilde{X}}(\tilde{X}))$.

$$\pi_2(\omega_{\tilde{X}}(\tilde{X})) = \pi_2\left(\bigcap_{i=0}^{3} \omega_{\tilde{X}}(\tilde{X}_i)\right)$$

$$= \bigcap_{i=0}^{3} \pi_2(\omega_{\tilde{X}}(\tilde{X}_i)) = \bigcap_{i=0}^{3} (\mathrm{Adj}(X_i))\sigma_i$$

THEOREM 2.23.0 The following two k-vector spaces are isomorphic:

$$\omega_{\tilde{X}}(\tilde{X}) \simeq \text{adjoint forms of degree n - 4 in k[X]}$$

Proof: Put together the foregoing two lemmas.

PROPOSITION 2.24.0 Let $X \subset P^3$ be as above. Assume 2.17.1 after renumbering variables. Let $\tilde{X} \xrightarrow{\pi} X$ be any desingularization of X. Let $p \in X$ be an isolated singularity. Let $A \in k[X]_{(d)}$. Let $\tilde{X}_i = \pi^{-1}(X_i)$. The following conditions are equivalent:

(a) A is adjoint at p.
(b) For all i = 0, 1, 2, 3 such that $p \in X_i$, $\pi_1((A/x_i^d)\sigma_i) \in \mathrm{Reg}_{\tilde{X}}(\pi^{-1}(p))$.
(c) For some i = 0, 1, 2, or 3 such that $p \in X_i$, $\pi_1((A/x_i^d)\sigma_i) \in \mathrm{Reg}_{\tilde{X}}(\pi^{-1}(p))$.
(d) For some i = 0, 1, 2, or 3 such that $p \in X_i$, A/x_i^d is adjoint at p as an element of $(\mathcal{O}_{X_i})_p$.

COROLLARY 2.24.1 Retain the notation of 2.24.0. Assume that $p \in X_0$; then A is adjoint at p iff $A(1,y_1,y_2,y_3)$ is adjoint at the point of

$$H_0 = \mathrm{Spec}\,\frac{k[y_1,y_2,y_3]}{(F(1,y_1,y_2,y_3))}$$

that corresponds to p; similarly, if $p \in X_1$, X_2, or X_3, with H_0 replaced by H_1, H_2, or H_3, and A suitably dehomogenized.

REMARK 2.24.2 2.24.1 means that in order to check whether A is an adjoint form, we can dehomogenize the equation of X and A and reduce to an affine hypersurface case. We use this fact extensively in this chapter.

Proof of proposition 2.24.0: Let $\tilde{X}_i = \pi^{-1}(X_i)$. (a) ⇒ (b): Assume (a). For every i such that $p \in X_i$, $\pi_i((A/x_i^d)\sigma_i) \in \mathrm{Reg}_{\tilde{X}_i}(\pi^{-1}(p)) = \mathrm{Reg}_{\tilde{X}}(\pi^{-1}(p))$. (b) ⇒ (a) also follows since $\mathrm{Reg}_{\tilde{X}_i}(\pi^{-1}(p)) = \mathrm{Reg}_{\tilde{X}}(\pi^{-1}(p^1))$. (b) ⇒ (c) trivially from the definition of adjoints. (c) ⇒ (d) trivially as above. (c) ⇒ (b): Assume (c). Let $j \neq i$, $p \in X_j$.

$$\pi_1\left(\frac{A}{x_j^d}\sigma_j\right) = \pi_i\left(\left(\frac{x_j}{x_i}\right)^{n-4}\frac{A}{x_j^d}\sigma_i\right)$$

$$= \pi_1\left(\left(\frac{x_j}{x_i}\right)^{n-4-d}\left(\frac{A}{x_i^d}\sigma_i\right)\right)$$

$$= \pi_1\left(\left(\frac{x_j}{x_i}\right)^{n-4-d}\right)\pi_1\left(\frac{A}{x_i^d}\sigma_i\right)$$

Since we assume (c), it is enough by 2.5.0 to see that $\pi_1((x_j/x_i)^{n-4-d})$ is regular on $\pi^{-1}(p)$. But $(x_j/x_i)^{n-4-d} \in (\mathcal{O}_X)_p$ since $p \in X_i \cap X_j$, so the last statement follows.

REMARK 2.24.3 If g is any form of degree 1 such that $g(p) \neq 0$, A is adjoint at p iff A/g^d is adjoint to X_g. Hence the definition of adjoint forms is invariant under a linear change of coordinates.

Theory of ℓ-Adjoints for $\ell \geq 1$ (An Outline)

We shall use notations from the theory of adjoints. We consider $\Omega^{2}_{k(X)/k}{}^{\otimes\ell}$, $\otimes$ over k(X). We define $\ell\text{-Reg}_X(p) \overset{\text{def}}{=\!=}$ {elements of $\Omega^{2}_{k(X)/k}{}^{\otimes\ell}$ of the form $\Sigma\, c_i\mu_i$, where $\mu_i = (da_{i1}db_{i1}) \otimes (da_{i2}db_{12}) \otimes \cdots \otimes (da_{i\ell}db_{i\ell})$ and c_i, a_{ij}, $b_{ij} \in \mathcal{O}_p$}. We define for each $\ell \geq 1$ a sheaf $\ell\text{-}\omega_X(U) = \bigcap_{q\in U'} \ell\text{-Reg}_X(q)$.

PROPOSITION 2.25.0 $\ell\text{-}\omega_X(U)$ is an $\mathcal{O}_X$-module.

PROPOSITION 2.26.0 $\ell\text{-}\omega_X$ is canonically isomorphic to ω_X^{ℓ}. From now on we will use $\ell\text{-}\omega_X$ and $\omega_X^{\otimes\ell}$, and $\ell\text{-Reg}_X(p)$ and $\text{Reg}_X{}^{\otimes\ell}(p)$ interchangeably.

$\omega_X^{\otimes\ell}$ has an analogous theory to $\omega_X^{\otimes\ell}$. Propositions 2.3.0-2.8.9 have obvious analogs for $\omega_X^{\otimes\ell}$. In particular, we note that for $F \subsetneq A^3$ as in 2.8.3 and 2.9.0, we have, for $U \subsetneq F$:

PROPOSITION 2.27.0 $\omega_F^{\otimes\ell}(U) = \mathcal{O}_F(U)\sigma_F^{\otimes\ell}$.

DEFINITION 2.28.0 Let $F \subset A^3$ and $p \in F$ be as in 2.13.0, $\tilde{F} \to F$ a resolution of p. An element $a \in (\mathcal{O}_F)_p$ is said to be ℓ-adjoint iff $\pi_1(a\sigma_F^{\otimes\ell}) = \pi_1(a) \cdot \pi_1(\sigma_F^{\otimes\ell}) = \pi_1(a)(\pi_1(\sigma_F))^{\otimes\ell}$ is regular along $\pi^{-1}(p)$.

Again propositions 2.13.0-2.16.0 have obvious analogies for ℓ-adjoints.

A new fact is the following.

LEMMA 2.28.1 Let $p \in F \subsetneq A^3$ be as in 2.9.0. If a_1 is ℓ_1-adjoint and a_2 is ℓ_2-adjoint at p, then a_1a_2 is $\ell_1 + \ell_2$ adjoint.
Proof: Obvious.

Passing to the projective case, let $X \subsetneq P^3$ be as in 2.17.0.

DEFINITION 2.29.0 $A \in k[X]_{(d)}$ is *ℓ-adjoint at* $p \in X$ iff A/x_i^d is ℓ-adjoint on the scheme X_i at p.

DEFINITION 2.30.0 A is *ℓ-adjoint to* X iff A is ℓ-adjoint at all the singular points of X.

Again we have propositions analogous to 2.24.0-2.24.3. The principal difference is that

$$(2.31.0) \quad \sigma_1^{\otimes\ell} = \left(\frac{x_i}{x_j}\right)^{\ell(n-4)} \sigma_j^{\otimes\ell}$$

which follows from lemma 2.18.0.

Exactly as in the theory of adjoints, we prove:

THEOREM 2.32.0 $H^0(\tilde{X},\omega_X^{\otimes\ell}) \approx [\ell\text{-adjoint forms of degree } \ell(n-4)]$ as k-vector spaces.

Using Lemma 2.28.1, we prove the following:

PROPOSITION 2.33.0 Let A_1 be an ℓ_1-adjoint form of degree d_1 (at p), A_2 and ℓ_2-adjoint form of degree d_2 (at p). Then A_1A_2 is an $\ell_1 + \ell_2$ adjoint form (at p) of degree $d_1 + d_2$.

Valuation Theory for Differentials

We need only deal with the case of a normal hypersurface $F \subseteq A^3$,

$$F = \text{Spec}\ \frac{k[x,y,z]}{(f(x,y,z))} = \text{Spec R}$$

Choose a canonical generating differential σ_F. Let v be a discrete valuation of k(F) that corresponds to a height 1 prime ideal of R. We extend v to $\Omega^2_{k(F)/k}$ by setting

(2.34.0) $v(\alpha) = v(a)$ where $\alpha = a\sigma_F$

This is clearly independent of the choice of σ_F, as v is equal to 0 on units of R.

PROPOSITION 2.34.1 $\alpha \in \Omega^2_{k(F)/k}$ is regular on F iff for every valuation v corresponding to a height 1 prime of R, $v(\alpha) \geq 0$.

Proof: Set $\alpha = a\sigma_F$. α is regular on F iff $a \in R$ iff $v(a) \geq 0$ for all v, as in the proposition; but $v(\alpha) = v(a)$ for all such valuations, by definition of $v(\alpha)$.

LEMMA 2.35.0 Let X be a smooth surface. Then any point $p \in X$ has an affine neighborhood U(p) with the following properties:

(a) U(p) = Spec R, $R \subseteq k(X)$, R has fraction field k(X); R is regular.

(b) $\omega_X|U(p)$ is free on one generator $\sigma_R \in \Omega^2_{k(X)/k}$; i.e., if $V \subseteq U$, then $\omega_X(V) = \mathcal{O}_X(V)\sigma_R$ is a free $\mathcal{O}_X(V)$-module on one generator. In particular, $\omega_X(U(p)) = R\sigma_R$.

Proof: In this case ω_X is the same as the sheaf of Kähler differentials and the lemma is well known (see Mumford, 19 , p. 335).

PROPOSITION 2.36.0 Let F be isomorphic to a hypersurface in A^3; F is normal. Let $p \in F$ be an isolated singularity. Then the ideal of adjoints $\text{Adj}(p) \subseteq (\mathcal{O}_F)_p$ is integrally closed (see Notation).

Proof: Let $\tilde{F} \xrightarrow{\pi} F$ be a desingularization. Suppose that Adj(p) is not integrally closed. Then we have for some r, $r \in (\mathcal{O}_F)_p - \text{Adj}(p)$ and

$$r^n + a_1 r^{n-1} + a_2 r^{n-2} + \cdots + a_1 r^{n-i} + \cdots + a_n = 0 \tag{1}$$

where $a_i \in [\mathrm{Adj}(p)]^i$. Let σ_F be a canonical generating differential on F. $\pi_1(r\sigma_F)$ fails to be regular at some point $q \in \pi^{-1}(p)$. Choose a neighborhood $U(q) = \mathrm{Spec}\, R$ of q in $\tilde{F}$ with $R \subseteq k(\tilde{F})$ as in lemma 2.35.0; then

$$\pi_1(\sigma_F) = \gamma\sigma_R \quad \text{where } \gamma \in k(\tilde{F})$$

$$\pi_1(r\sigma_F) = \pi_1(r)\gamma\sigma_R$$

so

$$\pi_1(r) \cdot \gamma \notin R$$

On the other hand, if $a \in \mathrm{Adj}(p)$, then

$$\pi_1(a) \cdot \gamma \in R$$

Since R is normal, there exists a discrete valuation v such that

$$v(\pi_1(r) \cdot \gamma) = v(\pi_1(r)) + v(\gamma) < 0 \tag{2}$$

whereas $v(\pi_1(a)) + v(\gamma) \geq 0$ if $a \in \mathrm{Adj}(p)$.

If $a_i \in [\mathrm{Adj}(p)]^i$,

$$v(\pi_1(a_i)) + iv(\gamma) \geq 0 \tag{3}$$

We prove that (1), (2), and (3) are incompatible. Let us apply π_1 to (1).

$$(\pi_1(r))^n = -\pi_1(a_1)(\pi_1(r))^{n-1} - \pi_1(a_2)(\pi_1(r))^{n-2} \cdots + \pi_1(a_n)$$

Applying v to both sides gives

$$\begin{aligned} nv(\pi_1(r)) &\geq v(\pi_1(a_{i_0})(\pi_1(r))^{n-i_0} \\ &= v(\pi_1(a_{i_0})) + (n - i_0)v(\pi_1(r)) \end{aligned}$$

for some $i \leq i_0 \leq n$; so

$$i_0 v(\pi_1(r)) \geq v(\pi_1(a_{i_0}))$$

Using (3) gives us

$$v(\pi_1(a_{i_0})) \geq -i_0 v(\gamma)$$

so

$$i_0 v(\pi_i(r)) \geq -i_0 v(\gamma)$$

or

$$v(\pi_1(r)) \geq -v(\gamma)$$

so $v(\pi_1(r)) + v(\gamma) \geq 0$, which contradicts (2).

REMARK 2.37.0 Similarly, we can extend valuation theory to $\Omega^2_{k(F)/k}{}^{\otimes 2}$ by setting $v(a\sigma_F^{\otimes 2}) = v(a)$. For $\alpha \in \Omega^2_{k(F)/k}{}^{\otimes 2}$ we again have that α is regular on F iff $v(\alpha) \geq 0$ for all valuations, as in proposition 2.34.1.

3. AN EXAMPLE RELATING TO A QUESTION OF O. ZARISKI

Introduction

We give an example of a surface defined over a field of characteristic 2 with the following properties: F is an affine surface with equation of the form

$$F: \quad z^2 = f(x,y)$$

in affine 3-space. The closure of $\tilde{F}$ in projective space, denoted $\bar{F}$, has a nonsingular model F with

$$p_g(\tilde{F}) = 0 \quad \text{but} \quad P_2(\tilde{F}) \geq 1$$

so that F is nonrational. The example answers in characteristic 2 a question suggested to the author by Zariski in 1971. (See the last paragraph of this Introduction.)

Our construction in characteristic 2 follows very closely the classical construction of Enriques surfaces as double planes. However, everything had to be done anew in characteristic 2 to verify that the classical computations will still go through.

Part I is new. On the other hand, our resolution of the singularity at infinity very closely parallels that given in Šafarevič's seminar on surfaces (Šafarevič, 1965a) [for the English translation see Šafarevič (1965b), pp. 148-165].

The author acknowledges with pleasure the help of his advisers J. S. Milne and M. Hochster in working out the details of this example. Conversations with and advice from O. Zariski, D. Mumford, A. Beauville, H. Hironaka, S. Bloch, and S. Abhyankar are also gratefully acknowledged.

3.0.0 [Zariski's Question (1971)] Suppose that F is defined by $z^p = f(x,y)$ over an algebraically closed field k of characteristic $p > 0$. Let $\tilde{F}$ be a nonsingular projective model of k(F). Is it true that $p_g(\tilde{F}) = 0$ implies F rational?

AUTHOR'S REMARK (1977) The question remains open for $p > 2$. Also, it would be interesting to find an example of a hypersurface F: $z^p = f(x,y)$ such that the corresponding Zariski surface $\tilde{F}$ has $P_2(\tilde{F}) > 0$ and $p_g(\tilde{F}) = 0$, where the degree of the defining equation of F is $p + 1$ and there are no singularities at infinity.

Part I: Equation of the Surface Singularities in Characteristic 2

In this part of the section we state the equation of the affine surface F and later of its closure in projective space $\bar{F}$.

All the nonrational singularities of $\bar{F}$ will be determined. This will be done for a "generic choice" of the coefficients entering in the equation of F. It will turn out that $\bar{F}$ has only four nonrational singularities.

The equation is F: $z^2 = RQ$ over an algebraically closed field k of characteristic 2. Here

$$R(x,y) = xy(x^2 + y^2 + pxy + 1)$$

and

$$Q(x,y) = Ax^4 + A'y^4 + Bxy^3 + B'yx^3 + C(xy)^2$$
$$+ xy^2 + yx^2 + Ax^2 + A'y^2 + Dxy$$

This is a "picture" of the curve $R(x,y) = 0$:

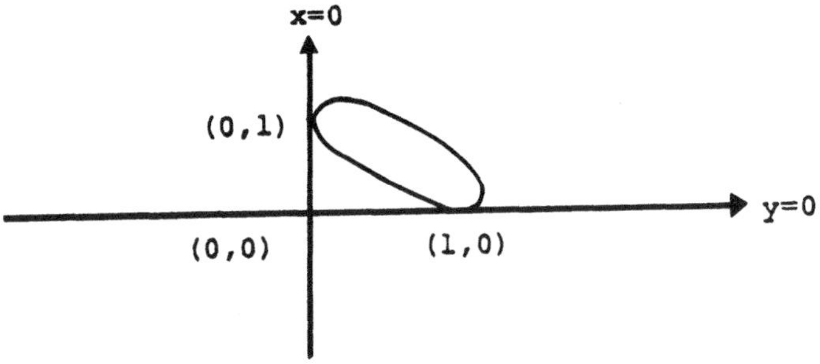

The curve $Q = 0$ (not shown) is tangent to the conic at (1,0) and (0,1), and it has a node at (0,0).

We now prove the following:

PROPOSITION 3.0.1 For a generic choice of coefficients p, A, A', B, B', C, D, the affine surface F has only three nonrational singularities: (0,0,0); (0,1,0); (1,0,0).

Proof: We shall prove later (3.3.1, 3.5.2, 3.5.3) that the three points are nonrational singularities.

Here we shall only prove that F has no other nonrational singularities. It is easy to show that if (x,y,z) is a nonrational singular point of F, we must have (see 3A.3.0†)

(1) $(RQ)_x = \frac{\partial(RQ)}{\partial x} = 0$

(2) $(RQ)_y = \frac{\partial(RQ)}{\partial y} = 0$

(3) $(RQ)_{xy} = \frac{\partial(RQ)}{\partial xy} = 0$

These are equivalent to

---

†See the appendix (Part 3) to this section, hereafter referred to as 3A.

(1)' $R_xQ + Q_xR = 0$

(2)' $R_yQ + Q_yR = 0$

(3)' $R_{xy}Q + Q_{xy}R + R_xQ_y + Q_xR_y = 0$

Let us consider

$$\Delta = \det\begin{vmatrix} R_x & Q_x \\ R_y & Q_y \end{vmatrix} = R_xQ_y + Q_xR_y$$

We will show that $\Delta = 0$ at a point satisfying (1)', (2)', (3)'. Suppose that $\Delta \neq 0$; then by the theory of linear equations applied (1)' and (2)' we have $R = Q = 0$; then by (3)', $R_xQ_y + R_yQ_x = 0$, so that $\Delta = 0$, a contradiction. Therefore,

(4) $\Delta = 0$ is a consequence of (1)', (2)', (3)'. But Δ turns out to be easy to compute:

$$\Delta = \det\begin{vmatrix} R_x & Q_x \\ R_y & Q_y \end{vmatrix} = R_xQ_y + R_yQ_x$$

To evaluate Δ, we observe that

(5) $R_x = (x^3y + xy^3 + p(xy)^2 + xy)_x$

$= x^2y + y^3 + y = y(x^2 + y^2 + 1) = y(x + y + 1)^2$

(6) $R_y = x(x^2 + y^2 + 1) = x(x + y + 1)^2$

(7) $Q_x = By^3 + B'yx^2 + y^2 + Dy = y(By^2 + B'x^2 + y + D)$

(8) $Q_y = B'x^3 + Bxy^2 + x^2 + Dx = x(B'x^2 + By^2 + x + D)$

so that

(9) $Q_{xy} = By^2 + B'x^2 + D$

and therefore,

$$\Delta = R_xQ_y + R_yQ_x = (x \pm y + 1)^2[yQ_y + xQ_x]$$

$$= (x + y + 1)^2(xy)[By^2 \pm B'x^2 + y + D + B'x^2 + By^2 + x + D]$$

and finally,

$$\Delta = (x + y + 1)^2(xy)(x + y)$$

Thus for a singularity satisfying (1)', (2)', (3)' there are four possible cases:

(a) $x = 0$

(b) $y = 0$

(c) $x + y = 0$

(d) $x + y + 1 = 0$

Case (a): Since $x = 0$, $R = 0$ and $Q_y = 0$, so that (1)' and (2)' reduce to $R_xQ = 0$ and $R_yQ = 0$, respectively. If $Q = 0$, then using $x = 0$, we get

$$A'y^4 + A'y^2 = A'(y^2 + y)^2 = 0$$

so if we assume that $A' \neq 0$, then $y = 0$ or $y = 1$, and we get only (0,0,0) or (0,1,0). If $Q \neq 0$, $R = R_x = R_y = 0$ again gives only (0,0,0) or (0,1,0).

Case (b): Similarly, since $y = 0$, if we assume that $A \neq 0$, we only get (0,0,0) or (1,0,0) because the equation of F is symmetric in x and y.

Case (c): Here $x = y$. Let us use (1)' and (3)', namely

$$R_xQ + Q_xR = 0$$

$$R_{xy}Q + Q_{xy}R = 0$$

Since $x = y$,

$$R_{xy} = 1 \qquad Q_{xy} = (B + B')x^2 + D$$

$$R_x = x \qquad R_y = x$$

$$Q(x,x) = (A + A' + B + B' + C)x^4 + (A + A' + D)x^2$$

$$R(x,x) = x^2(px^2 + 1)$$

$$Q_x(x,x) = x[(B + B')x^2 + x + D]$$

$$Q_y(x,x) = x[(B + B')x^2 + x + D]$$

$$R_xQ + Q_xR = x[(A + A' + B + B' + C)x^4 + (A + A' + D)x^2]$$
$$+ x[(B + B')x^2 + x + D]x^2(px^2 + 1) = 0$$

and

$$R_{xy}Q + Q_{xy}R = (A + A' + B + B' + C)x^4 + (A + A' + D)x^2$$
$$+ [(B + B')x^2 + D][x^2(px^2 + 1)] = 0$$

If $x = 0$, then also $y = 0$, so there is no new solution. If $x \neq 0$, divide by x^3 to get $(A + A' + B + B' + C)x^2 + (A + A' + D) + [(B + B')x^2 + x + D](px^2 + 1) = 0$, and divide the second equation by x^2 to get $(A + A' + B + B' + C)x^2 + (A + A' + D) + [(B + B')x^2 + D][px^2 + 1] = 0$. Subtracting the two equations, we get $x(px^2 + 1) = 0$; hence since $x \neq 0$, $px^2 + 1 = 0$, and hence $p \neq 0$; but this implies that $(A + A' + B + B' + C)x^2 + A + A' + D = 0$ and $px^2 = 1$, so that $A + A' + B + B' + C = p(A + A' + D)$. This last equality will not be true for a general choice of coefficients. Thus, in general, the only solution of (1)', (2)', (3)' with $x = y$ is $x = y = 0 = z$.

Case (d): Here $x + y + 1 = 0$, so $y = x + 1$ (characteristic 2). $R_x = R_{xy} = 0$ and $\Delta = 0$, so that (1)' and (3)' become $Q_xR = 0$ and $Q_{xy}R = 0$, respectively. There are two cases. In one case, $R = 0$, so that $x(x + 1)(px^2 + px) = 0$, and hence $px^2(x + 1)^2 = 0$. If we assume that $p \neq 0$, then $x = 0$ or $x = 1$. $x = 0$ gives $(0,1,0)$ and $x = 1$ gives $(1,0,0)$. In the other case, $R \neq 0$. Then $Q_x = 0$ and $Q_{xy} = 0$. If $y = x + 1$, then

$$Q_x = (x + 1)(Bx^2 + B'x^2 + x + D + B + 1)$$

and

$$Q_{xy} = Bx^2 + B'x^2 + B + D$$

We cannot have $x + 1 \neq 0$, because then

$$Bx^2 + B'x^2 + x + D + B + 1 = 0$$

and

$$Bx^2 + B'x^2 + B + D = 0$$

Adding these, we get $x + 1 = 0$. This is a contradiction. Therefore, $x + 1 = 0$ and $y = x + 1$. Hence $y = 0$, and we only get the point $(1,0,0)$.

This completes the proof of proposition 3.0.1.

Next we determine the singularities at infinity. Let us write the equation of F in homogeneous form†:

$$\bar{F}: \quad z_0^6 z^2 = xy(x^2 + y^2 + pxy + z_0^2)Q(x,y,z_0)$$

where

$$Q(x,y,z_0) = Ax^4 + A'y^4 + Bxy^3 + B'yx^3 + C(xy)^2 + z_0(xy^2 + yx^2) + z_0^2(Ax^2 + A'y^2 + Dxy)$$

The right-hand side of the equation of $\bar{F}$ is a homogeneous form of degree 8 in x, y, z_0. We shall denote it $F_8(x,y,z_0)$. Also,

$$z_0^6 z^2 = f_8(x,y) + z_0 F_7(x,y,z_0)$$

where

$$f_8(x,y) = xy(x^2 + y^2 + pxy)[Ax^4 + A'y^4 + Bxy^3 + B'yx^3 + C(xy)^2]$$

and F_7 is a homogeneous form of degree 7.

We look for singularities at infinity, i.e., at the hyperplane $z_0 = 0$. We note that $\partial/\partial z \equiv 0$. At such a singularity

$$(1) \quad \frac{\partial}{\partial x} = \frac{\partial f_8}{\partial x}(x,y) = 0$$

$$(2) \quad \frac{\partial}{\partial y} = \frac{\partial f_8}{\partial y}(x,y) = 0$$

---

†This means that

$$\bar{F} = \text{Proj}\, \frac{k[x,y,z,z_0]}{(z_0^6 z^2 - xy(x^2 + y^2 + pxy + z_0^2)Q(x,y,z,z_0))}$$

$\bar{F}$ is the closure of F in $\mathbb{P}^3$.

(3) $f_8(x,y) = 0$

For a general choice of p, A, A', B, B', and C, the form f_8 will split into distinct linear factors.

Thus (1), (2), (3) give only $x = y = 0$ as a solution; so we get only one singular point at infinity with $z_0 = x = y = 0$ and $z = 1$. The local equation is

$$F_0: \quad z_0^6 = f_8(x,y) + z_0F_7(x,y,z_0) = F_8(x,y,z_0)$$

with F_8 homogeneous of degree 8.

REMARK 3.0.2 For a generic choice of coefficients, Sing($\bar{F}$) consists of a finite number of isolated singularities.

Proof: There is only one singular point on $\bar{F} - F$. By proposition 3.0.1, F has only three singularities satisfying (1), (2), and (3). All the other singularities of F are isolated by 3A.3.0. Thus all the singularities of F are isolated. Then also Sing($\bar{F}$) is a finite set because it is closed in the Zariski topology of $\bar{F}$.

REMARK 3.0.3 For a generic choice of the coefficients, $\bar{F}$ is a normal surface.

Proof: Follows from 3.0.2 since a projective hypersurface of dimension 2 with isolated singularities is well known to be normal.

Part 2

The singularities of F are resolved, and adjoints and biadjoints are determined locally at each of the three nonrational singular points (corollaries 3.3.1, 3.3.2, and 3.5.1-3.5.6).

LEMMA 3.1.0 Suppose that

$$w^2 = f_4(u,v) + f_5(u,v) + \cdots + f_n(u,v) \tag{1}$$

with $n \geq 5$, is an equation over an algebraically closed ground field k of characteristic 2. Suppose that the singularity at the point $u = v = w = 0$ is isolated. Suppose also that the following condition, referred to later as I* or 3.1.1, is satisfied.

3.1.1 Neither the system

$$\frac{\partial f_4}{\partial x}(u,1) = 0$$

$$f_5(u,1) = 0$$

nor the system

$$\frac{\partial f_4}{\partial v}(1,v) = 0$$

$$f_5(1,v) = 0$$

has a solution. Then

(a) Equation (1) defines an irreducible hypersurface

$$S = \text{Spec}\,\frac{k[u,v,w]}{(w^2 - f_4 - f_5 - \cdots - f_n)} = \text{Spec}\,k[S]$$

(b) The singularity at (0,0,0) is resolved by a quadratic transformation followed by blowing up a double line which is the exceptional locus.

Proof: (a) To show that S is irreducible, it is enough to see that the right-hand side of (1) is not a square since $f_5 \neq 0$.

(b) First we perform a quadratic transformation. We blow up the ideal (u,v,w). Since w is integrally dependent on (u,v), by lemma 3A.2.0 the following two affine schemes cover QS, the blown-up variety:

$$\text{Spec}\,R' = \text{Spec}\,k\left[u,\frac{v}{u},\frac{w}{u}\right] = \text{Spec}\,k[u',v',w']$$

and

$$\text{Spec}\,k\left[\frac{u}{v},v,\frac{w}{v}\right] = \text{Spec}\,k[u'',v'',w''] = \text{Spec}\,R''$$

Because of lemma 3A.2.0 these schemes can be identified with hypersurfaces having equations

$$w'^2 = u'^2 f_4(1,v') + \cdots + u'^{n-2} f_n(1,v') \qquad (2)$$

$$w''^2 = v''^2 f_4(u'',1) + \cdots + v''^{n-2} f_n(u'',1) \tag{3}$$

respectively. The gluing together is by the isomorphism of open sets $(\mathrm{Spec}\ R')_{(v')} = (\mathrm{Spec}\ R'')_{(u'')}$ that is given by $k[u,v/u,w/u]_{(v/u)} = k[u/v,v,w/v]_{(u/v)}$.

Next we blow up the "double line," i.e., the coherent sheaf of ideals given by $(w',u') = (w/u,u)$ on Spec R' and $(w'',v'') = (w/v,v)$ on Spec R''. Since in the ring of the intersection $k[u,v/u,w/u]_{(v/u)}$, $(w/u,u) = (w/v,v)$, the sheaf of ideals is well defined on QS. Blowing it up gives a scheme covered by two affines since w' is integrally dependent on the ideal (u') and w'' on the ideal (v'').

The new scheme $\tilde{S}$ is thus covered by the spectra of the following two rings:

$$R_1 = k\left[u',v',\frac{w'}{u'}\right] = k\left[u,\frac{v}{u},\frac{w}{u^2}\right] = k[u_1,v_1,w_1]$$

and

$$R_2 = k\left[u'',v'',\frac{w''}{v''}\right] = k\left[\frac{u}{v},v,\frac{w}{v^2}\right] = k[u_2,v_2,w_2]$$

Again, by lemma 3A.2.0, R_1 and R_2 are coordinate rings of the following hypersurfaces:

(a) $\tilde{S}_1$: $w_1^2 = f_4(1,v_1) + u_1 f_5(1,v_1) + \cdots + u_1^{n-4} f_n(1,v_1)$

(b) $\tilde{S}_2$: $w_2^2 = f_4(u_2,1) + v_2 f_5(u_2,1) + \cdots + v_2^{n-4} f_n(u_2,1)$

The gluing together is given by $(\mathrm{Spec}\ R_1) \cap (\mathrm{Spec}\ R_2) = \mathrm{Spec}(R_1)_{(v_1)} = \mathrm{Spec}(R_2)_{(u_2)} = \mathrm{Spec}(\text{compositum } R_1R_2 \text{ in } k(S))$. We note that $k(S) = k(u,v,w)$. $\tilde{S}$ has a natural projection down to S. Call it

$$\pi: \tilde{S} \to S$$

That projection restricted to Spec $R_1 = \tilde{S}_1$ and Spec $R_2 = \tilde{S}_2$ corresponds to the natural inclusions

$$k[u,v,w] \hookrightarrow R_1 = k\left[u, \frac{v}{u}, \frac{w}{u^2}\right] = k[u_1, v_1, w_1]$$

and

$$k[u,v,w] \hookrightarrow R_2 = k\left[v, \frac{u}{v}, \frac{w}{v^2}\right] = k[v_2, u_2, w_2]$$

To find the fiber over the point (0,0,0) on S, we observe that in R_1 the ideal (u,v,w) is equal to (u) = (u_1); and in R_2 (u,v,w) is equal to (v) = (v_2). Thus the exceptional fiber is the curve u_1 = 0 on Spec R_1 glued together with the curve $v_2 = 0$ on Spec R_2. Equations (1) and (2) together with condition I* (3.1.1) suffice to show, using the jacobian criterion, that all the points of the exceptional fiber on $\tilde{S}$ are simple for $\tilde{S}$. Hence the singularity is resolved by

$$\pi: \tilde{S} \to S$$

LEMMA 3.2.0 Let S: $w^2 = f_4(u,v) + f_5(u,v) + \cdots + f_n(u,v) = f(u,v)$ be as in lemma 3.1.0, satisfying condition 3.1.1. In addition, we assume condition 3.2.1 (or II*).

3.2.1 $f_4(1,u)$ and $f_4(v,1)$ are not perfect squares (characteristic 2).

Then a monomial $u^a v^b w^c$ is adjoint at the isolated singular point $u = v = w = 0$ iff $a + b + 2c \geq 1$ and it is biadjoint iff $a + b + 2c \geq 2$.

Proof: We adopt the notation of lemma 3.1.0. $\tilde{S} \xrightarrow{\pi} S$ is the resolution. $\tilde{S}$ is covered by $\tilde{S}_1$ and $\tilde{S}_2$, two affine hypersurfaces with equations

$$\tilde{S}_1: \quad w_1^2 = f_4(1,v_1) + u_1 f_5(1,v_1) + \cdots + u_1^{n-4} f_n(1,v_1)$$
$$= g(u_1,v_1)$$

and

$$\tilde{S}_2: \quad w_2^2 = f_4(u_2,1) + v_2 f_5(u_2,1) + \cdots + v_2^{n-4} f_n(u_2,1)$$
$$= g_2(u_2,v_2)$$

We denote the restriction of π to $\tilde{S}_i$ by π_i. π_1, π_2 are dual to the ring inclusions π_1^*, π_2^* given by

$$k[u,v,w] \overset{\pi_1^*}{\hookrightarrow} k\left[u,\frac{v}{u},\frac{w}{u^2}\right] = k[u_1,v_1,w_1]$$

$$k[u,v,w] \overset{\pi_2^*}{\hookrightarrow} k\left[\frac{u}{v},v,\frac{w}{v^2}\right] = k[u_2,v_2,w_2]$$

To determine adjoints we compute the pullback of the canonical generating differential σ_S to $\tilde{S}_1$ and $\tilde{S}_2$.

$$\sigma_S = \frac{du\ dw}{\partial f/\partial v} = \frac{dv\ dw}{\partial f/\partial u}$$

We have

$$\pi_1^*(u) = u_1 \qquad \pi_1^*(v) = u_1 v_1 \qquad \pi_1^*(w) = u_1^2 w_1$$

and

$$\pi_2^*(v) = v_2 \qquad \pi_2^*(u) = u_2 v_2 \qquad \pi_2^*(w) = v_2^2 w_2$$

so

$$\pi^{-1}(\sigma_S) = \frac{u_1^2\ du_1\ dw_1}{\frac{\partial f}{\partial v}(u_1,u_1 v_1)}$$

and

$$\pi^{-1}(\sigma_S) = \frac{v_2^2\ dv_2\ dw_2}{\frac{\partial f}{\partial u}(v_2 w_2,v_2)}$$

But we have

$$g_1(u_1,v_1) = \frac{1}{u_1^4} f(u_1,u_1 v_1) \qquad g_2(u_2,v_2) = \frac{1}{v_2^4} f(u_2 v_2,v_2)$$

Using the chain rule gives us

$$\frac{\partial g_1}{\partial v_1} = \frac{1}{u_1^4}\left(\frac{\partial f}{\partial v}\right)(u_1,u_1 v_1)u_1 = \frac{1}{u_1^3}\left(\frac{\partial f}{\partial v}\right)(u_1,u_1 v_1)$$

Similarly,

$$\frac{\partial g_2}{\partial u_2}(u_2,v_2) = \frac{1}{v_2^3}\left(\frac{\partial f}{\partial u}\right)(u_2,u_2v_2)$$

Therefore,

$$\pi_1^{-1}(\sigma_S) = \frac{1}{u_1}\frac{du_1\ dw_1}{\frac{\partial g_1}{\partial v_1}(u_1,v_1)} = \frac{1}{u_1}\sigma_{\tilde{S}_1}$$

and

$$\pi_2^{-1}(\sigma_S) = \frac{1}{v_2}\frac{dv_2\ dw_2}{\frac{\partial g_2}{\partial u_2}(u_2,v_2)} = \frac{1}{v_2}\sigma_{\tilde{S}_2}$$

where $\sigma_{\tilde{S}_1}$, $\sigma_{\tilde{S}_2}$ are canonical generating differentials on $\tilde{S}_1$, $\tilde{S}_2$ regular along the exceptional fiber.

The exceptional fiber of the map π: $\tilde{S} \to S$ is glued together from the curve $u_1 = 0$ on $\tilde{S}_1$ and $v_2 = 0$ on $\tilde{S}_2$. By lemma 3A.1.0, using condition II* (3.2.1), we know that these curves on $\tilde{S}_1$ and $\tilde{S}_2$ define discrete valuations ν_1 of $k(\tilde{S}_1)$ and ν_2 of $k(\tilde{S}_2)$ such that

$$\nu_1(u_1^{\alpha}v_1^{\beta}w_1^{\gamma}) = \alpha$$

$$\nu_2(u_2^{\alpha}v_2^{\beta}w_2^{\gamma}) = \beta$$

It is easy to check that under the canonical identification of $k(\tilde{S}_1) = k(\tilde{S}_2) = k(S) = k(u,v,w)$, $\nu_1 = \nu_2$. (The valuation measures the order of pole or zero along the exceptional fiber.) We write $\nu = \nu_1 = \nu_2$.

We can apply this valuation to

$$\pi_1^{-1}(\sigma_S) \qquad \text{and} \qquad \pi_2^{-1}(\sigma_S)$$

We obtain the value -1 since $\nu(\sigma_{\tilde{S}_1}) = \nu(\sigma_{\tilde{S}_2}) = 0$ (see 2.34.0). Thus to be adjoint at (0,0,0) a monomial $u^a v^b w^c$ has to pull back on $\tilde{S}_1$ and $\tilde{S}_2$ to a function with value ≥ 1. We have

$$\pi_1^{-1}(u^a v^b w^c) = u_1^{a+b+2c} v_1^b w_1^c$$

$$\pi_2^{-1}(u^a v^b w^c) = v_2^{a+b+2c} u_2^a w_2^c$$

Using lemma 3A.1.0, we see easily that $u^a v^b w^c$ is adjoint iff $a + b + 2c \geq 1$. Similarly, by considering the pullback of $\omega_S^{\otimes 2}$ we conclude that $u^a v^b w^c$ is biadjoint iff $a + b + 2c \geq 2$ (see 3A.1.1).

COROLLARY 3.2.2 $P(u,v,w) \in k[u,v,w] = k[S]$ is adjoint at $(0,0,0)$ iff $p(0,0,0) = 0$.

Proof: Obvious.

COROLLARY 3.2.3 $p(u,v,w)$ is biadjoint iff p is a linear combination of monomials $u^a v^b w^c$ satisfying $a + b + 2c \geq 2$.

Proof: We need only prove the "only if" part. The only monomials not satisfying $a + b + 2c \geq 2$ are u, v constant. We can therefore write

$$p(u,v,w) = p_1(u,v,w) + c_1 u_1 + c_2 v + c_3$$

where p_1 is biadjoint and c_1, c_2, c_3 belong to k.

It is enough to show that p being biadjoint implies that $c_1 = c_2 = c_3 = 0$. If p is biadjoint, then p is adjoint, so $c_3 = 0$ by corollary 3.2.2. Thus $c_1 u + c_2 v$ is biadjoint. Computing the pullback,

$$\pi_1^{-1}(c_1 u + c_2 v) = c_1 u_1 + c_2 u_1 v_1 = u_1(c_1 + c_2 v_1)$$

We must have

$$\nu(u_1(c_1 + c_2 v_1)) \geq 2$$

so

$$\nu(c_1 + c_2 v_1) \geq 1$$

but if $(c_1, c_2) \neq (0,0)$, this is not true. To prove the last statement we use the equation of $\tilde{S}_1$, which is

$$\tilde{S}_1:\ w_1^2 = f_4(1,v_1) + u_1 f_5(1,v_1) + \cdots + u_1^{n-4} f_n(1,v_1)$$

We have that the degree of $c_1 + c_2 v_1 <$ degree of $w_1^2 + f_4(1,v_1)$. By lemma 3A.1.0 this implies that $(c_1 + c_2 v_1) = 0$ unless $c_1 = c_2 = 0$. Therefore, we conclude that $(c_1, c_2) = (0,0)$.

PROPOSITION 3.3.0 Take the equation of F,

$$z^2 = xy(x^2 + y^2 + pxy + 1)Q$$

where

$$Q = Ax^4 + A'y^4 + Bxy^3 + B'yx^3 + C(xy)^2 + xy^2 + yx^2$$
$$+ Ax^2 + A'y^2 + Dxy$$

For a generic choice of coefficients the origin (0,0,0) 3.1.0 is an isolated singularity satisfying conditions I* and II* of lemmas 3.1.0 and 3.2.0.

Proof: By 3.0.2 we can assume that the origin is an isolated singularity. We have $z^2 = xy(Ax^2 + A'y^2 + Dxy) + xy(xy^2 + yx^2) +$ higher-order terms. We use the notation of lemma 3.1.0 with $u = x$, $v = y$, $w = z$. Then we have

$$f_4(x,1) = x(Ax^2 + A' + Dx)$$
$$f_5(x,1) = x(x + x^2) = x^2(x + 1)$$
$$\frac{\partial}{\partial x} f_4(x,1) = Ax^2 + A'$$

Now if A, A' are generically chosen, the last polynomial has no common roots with $f_5(x,1) = 0$. By symmetry the same conclusion holds for $f_5(1,y)$ and $\partial f_4(1,y)/\partial y$. If $AA' \neq 0$, the origin also satisfies condition II*, i.e., $f_4(x,1)$ and $f_4(1,y)$ are not perfect squares.

We restate corollary 3.2.2 in terms of adjoint surfaces (see 2.16.1 for the definition of an adjoint surface).

COROLLARY 3.3.1 Adjoint surfaces to our surface F pass through the singular point (0,0,0). Conversely, any surface that passes through (0,0,0) is adjoint.

Proof: Corollary 3.2.2.

COROLLARY 3.3.2 xy is biadjoint at (0,0,0).
Proof: x is adjoint and y is adjoint, and use 2.28.1.

PROPOSITION 3.4.0 The singularity of F at (0,1,0) is resolved as follows. After a single quadratic transform one obtains in the exceptional locus only one isolated singular point. It is a quadruple point satisfying conditions I* and II* of lemmas 3.1.0 and 3.2.0.

This is true for a general choice of parameters in the equation of F. The same fact is true for the singularity of F at (1,0,0).
Proof: The last statement follows by symmetry of the equation of F in terms of x and y and by replacing x's by y's everywhere in the proof given below and also interchanging A with A', B with B', etc.

Let us concentrate on (0,1,0). For a generic choice of coefficients the singularity is isolated. To get the local equation of F, replace y by $\bar{y} + 1$ to get $z^2 = x(\bar{y} + 1)(x^2 + \bar{y}^2 + px\bar{y} + px)\bar{Q}$, where

$$\begin{aligned}\bar{Q} &= Ax^4 + A'\bar{y}^4 + Bx\bar{y}^3 + B'\bar{y}x^3 + C(x\bar{y})^2 + x\bar{y}^2 \\ &\quad + \bar{y}x^2 + Ax^2 + A'\bar{y}^2 + Dx\bar{y} + Bx(\bar{y}^2 + \bar{y} + 1) \\ &\quad + B'x^3 + Cx^2 + x + x^2 + Dx = Q(x,\bar{y}) \\ &\quad + x(B + D + 1) + x^2(C + 1) + B'x^3 + Bx(\bar{y}^2 + \bar{y})\end{aligned}$$

We blow up the ideal $(x,\bar{y},z)$ to obtain a new variety QF. Since z is integrally dependent on $(x,\bar{y})$, QF is (by lemma 3A.2.0) covered by two affines, E_1 and E_2. The first affine chart, E_1, is given by (see lemma 3A.2.0)

$$E_1 = \text{Spec } R_1$$

where

$$R_1 = k\left[x, \frac{\bar{y}}{x}, \frac{z}{x}\right] = k[x', \bar{y}', z']$$

E_1 is then a hypersurface with equation

$$z'^2 = x'(\bar{y}'x' + 1)(x' + \bar{y}'^2x' + px'\bar{y}' + p)(B + D + 1)$$
$$+ x\tilde{Q}(x',\bar{y}')$$

where $\tilde{Q}(x',\bar{y}')$ is some polynomial.

Along the exceptional fiber given by $x' = 0$, $\partial/\partial x'$ is easily computed and equals $p(B + D + 1)$. This is $\neq 0$ if we assume that $p \neq 0$ and $B + D + 1 \neq 0$. Under these assumptions there are no singularities along the exceptional locus in this chart.

The second affine chart, E_2, is given, again by lemma 3A.2.0, by

$$E_2 = \operatorname{Spec} R_2$$

where

$$R_2 = k\left[\frac{x}{\bar{y}},\bar{y},\frac{z}{\bar{y}}\right] = k[x'',\bar{y}'',z'']$$

Again E_2 is a hypersurface. It is given by

$$E_2:\quad z''^2 = x''\bar{y}''(\bar{y}'' + 1)(x''^2\bar{y}'' + \bar{y}'' + px''\bar{y}'' + px'')$$
$$\times\ [(B + D + 1)x'' + A'\bar{y}'' + (B + D)x''\bar{y}'' + \bar{y}''Q''(x'',\bar{y}'')]$$

where $Q''(x'',\bar{y}'')$ contains only monomials of degree 2 or higher.

Along the exceptional locus, $\bar{y}'' = 0$, and therefore $\partial/\partial\bar{y}''$ is given by $x''px''(B + D + 1)x''$; therefore, again if we assume that $p(B + D + 1) \neq 0$, $\bar{y}'' = x'' = z'' = 0$ is the only singular point. For a generic choice of coefficients, the singular point will be isolated since the original singularity is isolated. It is quadruple, so let us only check conditions I* and II* from lemmas 3.1.0 and 3.2.0. In the proof we drop the primes and bars.

We have

$$z^2 = f_4(x,y) + f_5(x,y) + \text{higher-order terms}$$

where

$$f_4(x,y) = xy(px + y)(\beta x + A'y)$$

(we set $\beta = B + D + 1$). Also,

$$f_5(x,y) = xy^2[(px + y)(\beta x + A'y) + px(\beta x + A'y) + (px + y)(B + D)x]$$

Consequently,

$$f_4(x,1) = x(px + 1)(\beta x + A')$$

and

$$f_4(1,y) = y(p + y)(\beta + A'y)$$

The partial derivatives give

$$\frac{\partial f_4}{\partial y}(1,y) = A'y^2 + p\beta \tag{1}$$

and

$$\frac{\partial f_4}{\partial x}(x,1) = p\beta x^2 + A' \tag{2}$$

Also,

$$\begin{aligned} f_5(1,y) &= y^2[(p + y)(\beta + A'y) + p(\beta + A'y) + (B + D)(p + y)] \\ &= y^2[A'y^2 + y(pA' + \beta + pA' + B + D) + p\beta + p\beta + p(B + D)] \\ &= y^2[A'y^2 + y(\beta + \beta + 1) + p(B + D)] \\ &= y^2[A'y^2 + y + p(\beta + 1)] \qquad (\text{using } B + D = \beta + 1) \qquad (1)' \end{aligned}$$

Similarly,

$$\begin{aligned} f_5(x,1) &= x[(px + 1)(\beta x + A') + px(\beta x + A') + (B + D) \cdot x(px + 1)] \\ &= x[x^2(p\beta + p\beta + pB + pD) + x(A'p + \beta + A'p + B + D) + A'] \\ &= x[x^2p(\beta + 1) + x + A'] \qquad (2)' \end{aligned}$$

We have to show that for a general choice of parameters, (1) and (1)' have no common root, and the same for (2) and (2)'.

First we deal with (1) and (1)'.

$$A'y^2 + p\beta = 0$$

$$y^2[A'y^2 + y + p(\beta + 1)] = 0$$

For a general choice of parameters we can assume that $y \neq 0$; therefore,

$$A'y^2 + p\beta = 0$$

$$A'y^2 + y + p(\beta + 1) = 0$$

Adding the two equations, we get

$$y + p = 0 \quad \text{or} \quad y = p$$

so that

$$A'p^2 = p\beta = p(B + D + 1)$$

but this is not true for a general choice of coefficients.

Similarly, for (2) and (2)',

$$p\beta x^2 + A' = 0$$

$$x[p(\beta + 1)x^2 + x + A'] = 0$$

Again $x \neq 0$, so we divide the last equation by x and obtain

$$p\beta x^2 + A' = 0$$

$$p(\beta + 1)x^2 + x + A' = 0$$

Adding, we get

$$px^2 + x = 0 \quad \text{or} \quad x(px + 1) = 0$$

Since $x \neq 0$, $x = 1/p$; therefore,

$$x^2 = \frac{1}{p^2}$$

but from (2),

$$x^2 = \frac{A'}{p\beta}$$

Therefore,

$$\frac{A'}{p\beta} = \frac{1}{p^2}$$

i.e.,

$$A' = \frac{\beta}{p} = \frac{B + D + 1}{p}$$

but the latter is not true for a general choice of parameters.

Thus I* is proved. To check condition II* we observe that in the notations above, $f_4(1,y) = y(y + p)(\beta + A'y)$ is not a perfect square if $A' \neq 0$ in characteristic 2. Also, $f_4(x,1) = x(px + 1) \times (\beta x + A')$ is not a square if $p \neq 0$ and $B + D + 1 \neq 0$.

PROPOSITION 3.5.0 Let $\bar{y} = y + 1$ and let F be our surface. A monomial $x^a \bar{y}^b z^c$ is adjoint at (0,1,0) iff $b + 2a + 3c \geq 1$ and biadjoint iff $b + 2a + 3c \geq 2$.

Proof: From resolution (proposition 3.4.0) we know that if we write F in terms of x and $\bar{y}$, then F: $z^2 = f_3(x,\bar{y}) + f_4(x,\bar{y}) + \cdots + f_8(x,\bar{y}) = f(x,\bar{y})$. Also, if we blow up $(x,\bar{y},z)$, we obtain QF covered by two affine hypersurfaces: $E_1 = \operatorname{Spec} k[x,\bar{y}/x,z/x] = \operatorname{Spec} k[x', \bar{y}',z']$, where

$$E_1\colon\quad z'^2 = \frac{1}{x'^2} f(x',\bar{y}'x') = g_1(x',\bar{y}')$$

and

$$E_2 = \operatorname{Spec} k\left[\frac{x}{\bar{y}},\bar{y},\frac{z}{\bar{y}}\right] = \operatorname{Spec} k[x'',\bar{y}'',z'']$$

with equation

$$E_2\colon\quad z''^2 = \frac{1}{\bar{y}''^2} f(x''\bar{y}'',\bar{y}'') = g_2(x'',\bar{y}'')$$

E_1 is nonsingular along the exceptional fiber and E_2 has only one singular point $x'' = \bar{y}'' = z'' = 0$ on the exceptional fiber. That singular point is an isolated singularity of E_2 satisfying conditions I* and II* of lemmas 3.1.0 and 3.2.0.

By lemma 3.2.0 we know that a function $x''^{\alpha}\bar{y}''^{\beta}z''^{\gamma}$ is adjoint iff $\alpha + \beta + 2\gamma \geq 1$ and biadjoint iff $\alpha + \beta + 2\gamma \geq 2$.

We first compute the pullback of the canonical generating differential on F to E_1. The projection of E_1 to F is given by

$$k[x,\bar{y},z] \hookrightarrow k\left[x,\frac{\bar{y}}{x},\frac{z}{x}\right] = k[x',\bar{y}',z']$$

so that $x \to x'$, $y \to \bar{y}'x'$, $z \to z'x'$.

We write

$$\sigma_F = \frac{dx\ dz}{\partial f/\partial y}$$

The pullback to E_1 is

$$\frac{x'\ dx'\ dz'}{\frac{\partial f}{\partial \bar{y}}(x',\bar{y}'x')}$$

but

$$\frac{\partial q'}{\partial \bar{y}'}(x',\bar{y}') = \frac{1}{x'^2}\left(\frac{\partial f}{\partial \bar{y}}\right)(x',y'x')x'$$

$$= \frac{1}{x'}\left(\frac{\partial f}{\partial \bar{y}}\right)(x',\bar{y}'x')$$

Therefore, the pullback of σ_F to E_1 is

$$\frac{dx'\ dz'}{\frac{\partial g_1}{\partial \bar{y}'}(x',\bar{y}')}$$

which is a canonical generating differential on E_1, σ_{E_1}, regular along the exceptional fiber, which is nonsingular.

Similarly, σ_F pulls back to

$$\frac{d\bar{y}''\ dz''}{\frac{\partial g_2}{\partial \bar{y}''}(x'\bar{y}'',\bar{y}'')}$$

the generating differential σ_{E_2} on E_2.

Because of the computations above and since E_1 is already nonsingular, it is clear that a function $p(x,\bar{y},z) \in k[x,\bar{y},z]$ is adjoint at $(0,0,0)$ iff its pullback to E_2 is locally adjoint at $x'' = \bar{y}'' = z'' = 0$.

Taking a monomial $x^{a-b}\bar{y}^{b}z^{c}$, its pullback is $x''^{a}(\bar{y}'')^{a+b+c}z''^{c}$. This is adjoint by lemma 3.2.0 iff $a + (a + b + c) + 2c \geq 1$, so

$b + 2a + 3c \geq 1$. Similarly, $x^a y^b z^c$ is biadjoint iff

$$b + 2a + 3c \geq 2$$

COROLLARY 3.5.1 Let $p \in k[F] = k[x,y,z]$, let $\bar{y} = y + 1$, and write $p = p(x,\bar{y},z)$; then $p(x,\bar{y},z)$ is adjoint at $x = \bar{y} = z = 0$ iff $p(0,0,0) = 0$.

Proof: The "if" part is obvious by proposition 3.5.0. To prove "only if," note that a constant $\neq 0$ is never adjoint, also by proposition 3.5.0.

We restate corollary 3.5.1 using the notion of adjoint surfaces (see 2.16.1).

COROLLARY 3.5.2 Adjoint surfaces to our surface F pass through (1,0,0). Conversely, any surface passing through (1,0,0) is adjoint to F at that point.

COROLLARY 3.5.3 Adjoint surfaces pass through (0,1,0) and any surface passing through (0,1,0) is adjoint to F at that point.

Proof: Symmetry in x, y.

COROLLARY 3.5.4 Using notations of corollary 3.5.1, $p(x,\bar{y},z) \in k[x,\bar{y},z]$ is biadjoint at (0,1,0) iff p is a linear combination of monomials $x^a \bar{y}^b z^c$ satisfying $2a + b + 3c \geq 2$.

Proof: The only monomials not satisfying the above are the constants and $\bar{y}$. It is enough to show that $c_1 + c_2\bar{y}$ is biadjoint iff $c_1 = c_2 = 0$. A biadjoint is adjoint, so by corollary 3.5.1, $c_1 = 0$. Therefore, $c_2\bar{y}$ is biadjoint. Pulling back to E_2 (see the proof of proposition 3.5.0), $c_2\bar{y}''$ is biadjoint at $x'' = \bar{y}'' = z'' = 0$, but this contradicts lemma 3.2.0.

COROLLARY 3.5.5 $xy = x\bar{y} + x$ is biadjoint at (0,1,0).

COROLLARY 3.5.6 xy is biadjoint at (1,0,0).

Proof: Symmetry.

REMARK 3.5.7 Corollaries 3.3.1, 3.3.2, and 3.5.1-3.5.6 are true for a generic choice of the coefficients of the equation of F.

Part 3: The Singularity of $\bar{F}$ at Infinity

The Italian geometers knew that the singular point at infinity on a "double plane" given by $z^2 = f_{2m}(x,y)$ forces the adjoint surfaces of degree 2m - 4 to consist of the plane "at infinity" counted m - 1 times together with a cylinder over the (x,y) plane over a curve of degree m - 3†.

In this section we prove rigorously a special case of this for our surface F in characteristic 2. The coefficients are chosen "generically" to satisfy various conditions. Then the singularity is resolved, and it is shown that the adjoint forms of degree 8 - 4 = 4 have to be of the form

$$z_0^3(ax + by + cz_0)$$

For the nonsingular model $\tilde{F}$ using Part 2, we can then show that $p_g = 0$. However, the form xyz_0^6 is shown to be biadjoint everywhere, so that $P_2 \geq 1$. The computations are very similar to those given in the Šafarevič seminar (Šafarevič, 1965b, pp. 148-165). The theory of adjoints and biadjoints has been developed in Section 2 (and is, of course, classical).

Introductory Remarks It was proved above that generically there is only one singular point of F at infinity (i.e., at the hyperplane $z_0 = 0$).

The equation of the surface in the neighborhood of the singular point at infinity $x = y = z_0 = 0$, $z = 1$, is

$$F_0:\quad z_0^6 = xy(x^2 + y^2 + pxy + z_0^2)Q(x,y,z_0)$$

where

$$Q(x,y,z_0) = Ax^4 + A'y^4 + Bxy^3 + B'yx^3 + C(xy)^2 + z_0(xy^2 + yx^2) + z_0^2(Ax^2 + A'y^2 + Dxy)$$

---

†See, e.g., Enriques (1949, pp. 77-79).

We introduce various notations for the right-hand side, namely,

$$F_0\colon\ z_0^6 = F_8(x,y,z_0) = f_8(x,y) + f_7(x,y)z_0 + z_0^2F_6(x,y,z_0)$$

where $f_8(x,y)$ is the homogeneous form of degree 8 given by

$$f_8(x,y) = xy(x^2 + y^2 + pxy)[Ax^4 + A'y^4 + Bxy^3 + B'yx^3 + C(xy)^2]$$

and

$$f_7(x,y) = xy(x^2 + y^2 + pxy)(xy^2 + yx^2)$$

We will show that, provided A, A', B, B', C are chosen generically, we can assume that:

(a) $f_8(1,y)$ and $f_8(x,1)$ are of degree ≥ 7.

(b) $f_8(1,y)$ and $f_8(x,1)$ are not squares in $k[y]$ and $k[x]$, respectively.

(c) The two systems of equations

$$\frac{\partial}{\partial y} f_8(1,y) = 0 \qquad \text{and} \qquad \frac{\partial}{\partial x} f_8(x,1) = 0$$
$$f_7(1,y) = 0 \qquad\qquad\qquad f_7(x,1) = 0$$

have no solutions in y and x, respectively.

(d) $(\partial f_8/\partial y)(1,y) = 0$ has no common roots with $f_8(1,y) = 0$, and $(\partial f_8/\partial x)(x,1) = 0$ has no common roots with $f_8(x,1) = 0$.

Under the assumptions (a)-(d) we shall resolve the singularity and find conditions satisfied by adjoint and biadjoint polynomials.

Now we verify conditions (a)-(d) for a generic choice of coefficients.

(a) is trivially true if $A \neq 0 \neq A'$.

(b) $f_8(1,y) = y(1 + y^2 + py)(A + A'y^4 + By^3 + B'y + Cy^2)$ will not be a square provided that p, A, A', B, B', C are chosen so that all the roots of $f_8(1,y)$ are distinct; similarly for $f_8(x,1)$.

(c) $f_7(1,y) = y(y^2 + py + 1)(y^2 + y)$ will have no roots in common with $(\partial f_8/\partial y)(1,y)$ for a generic choice of coefficients.

Proof of (c):

$$f_8(1,y) = y(y^2 + py + 1)(A'y^4 + By^3 + Cy^2 + B'y + A)$$

$$\frac{\partial f_8}{\partial x} = (y^2 + py + 1)(A'y^4 + By^3 + Cy^2 + B'y + A)$$
$$+ py(A'y^4 + By^3 + Cy^2 + B'y + A)$$
$$+ y(y^2 + py + 1)(By^2 + B')$$

Suppose that y_0 is a root of f_7. Then there are three cases:

(i) $y_0 = 0$

(ii) $y_0 = 1$

(iii) $y_0^2 + py_0 + 1 = 0$

In case (i), $(\partial f_8/\partial y)(y_0) = A$, so choose $A \neq 0$.

In case (ii), $\partial f_8/\partial y = p(A' + B + C + B' + A) + p(A' + B + C + B' + A) + p(B + B') = p(B + B')$, so choose p, B, B' to satisfy $p(B + B') \neq 0$.

In case (iii), $\partial f_8/\partial y = py_0(A'y_0^4 + By_0^3 + Cy_1^2 + B'y_0 + A)$ and $y_0 \neq 0$. Therefore,

$$A'y_0^4 + By_0^3 + Cy_0^2 + B'y_0 + A = 0$$

but we can avoid that by a suitable choice of A', B, C, B', and A.

(d) will be satisfied if $f_8(1,y)$ and $f_8(x,1)$ has no multiple roots. This will clearly be true for a generic choice of the coefficients.

From now on we shall assume that assumptions (a)-(d) are satisfied for the choice of coefficients under consideration.

Resolution of the Singularity at Infinity:

$$F_0\colon\ z_0^6 = F_8(x,y,z_0)$$

$$k[F_0] = k[x,y,z_0] = R_0$$

In the process of resolving the singularity we shall consider several rings $R_i \subseteq k(x,y,z_0)$. We denote $F_i = \mathrm{Spec}\ R_i$.

First we blow up the maximal ideal (x,y,z_0) of R_0. Using "lemma blowup" 3A.2.0, we know that since z_0 is integrally dependent (see Notation) on the ideal (x,y), the scheme obtained B_1F_0 is covered by two affines, F_1 and F_2, with rings

$$R_1 = k\left[x,\frac{y}{x},\frac{z_0}{x}\right] = k[x_1,y_1,z_1]$$

and

$$R_2 = k\left[\frac{x}{y},y,\frac{z_0}{y}\right] = k[x_2,y_2,z_2]$$

respectively. Also by lemma blowup 3A.2.0, F_1 and F_2 are hypersurfaces with equations

$$F_1:\quad z_1^6 = x_1^2F_8(1,y_1,z_1)$$

and

$$F_2:\quad z_2^6 = y_2^2F_8(x_2,1,z_2)$$

F_1 and F_2 are glued together via the isomorphism

$$(R_1)_{y/x} = (R_2)_{x/y} = k\left[x,\frac{y}{x},\frac{z_0}{x},\frac{x}{y}\right] = \text{compositum of } R_1 \text{ and } R_2 \text{ in } k(x,y,z_0)$$

In general, we shall use in the proof of resolution the following fact (Mumford, 1966, p. 72): If U, W are open subsets of a separated variety V with coordinate rings R(U) and R(W), respectively, the intersection $U \cap W$ is an affine with coordinate ring $R(U \cap W)$ = compositum of R(U) and R(W) in K(V).

To continue with the resolution, we construct a coherent sheaf of ideals on B_1F_0 by taking $(x_1,z_1) = (x,z_0/x)$ in R_1 and $(y_2,z_2) = (y,z_0/y)$ in R_2. These ideals induce the same ideal on $F_1 \cap F_2$ = Spec $k[x,y/x,z_0/x,x/y]$, so the sheaf is well defined. We blow up that sheaf of ideals to obtain a new scheme B_2F_0 covered by four affines F_3, F_4, F_5, F_6 with coordinate rings and equations given by

$$R_3 = k\left[\frac{x^2}{z_0}, \frac{y}{x}, \frac{z_0}{x}\right] = k[x_3, y_3, z_3]^{\dagger}$$

$$R_4 = k\left[x, \frac{y}{x}, \frac{z_0}{x^2}\right] = k[x_4, y_4, z_4]$$

$$R_5 = k\left[\frac{y^2}{z_0}, \frac{x}{y}, \frac{z_0}{y}\right] = k[y_5, x_5, z_5]$$

$$R_6 = k\left[y, \frac{x}{y}, \frac{z_0}{y^2}\right] = k[y_6, x_6, z_6]$$

The gluing together is given by

$$F_i \cap F_j = \text{Spec(compositum of } R_i \text{ and } R_j \text{ in } k(x,y,z_0))$$

We have

$$F_3\colon\ z_3^4 = x_3^2 F_8(1, y_3, z_3)$$

$$F_4\colon\ z_4^6 x_4^4 = F_8(1, y_4, z_4 x_4)$$

$$F_5\colon\ z_5^4 = y_5^2 F_8(x_5, 1, z_5)$$

$$F_6\colon\ z_6^6 y_6^4 = F_8(x_6, 1, z_6 y_6)$$

As we shall prove, at the end of resoltuion, F_4, F_6 are already nonsingular along the exceptional loci.

On B_2F we construct a new sheaf of ideals given by the unit ideals on F_4 and F_6, the ideal $(z_3, x_3) = (z_0/x, x^2/z_0)$ on F_3, and $(z_5, y_5) = (z_0/y, y^2/z_0)$ on F_5. We check agreement on $F_i \cap F_j$, $3 \le i$, $j \le 6$. On

$$F_3 \cap F_5 = \text{Spec } k\left[\frac{x^2}{z_0}, \frac{y}{x}, \frac{z_0}{x}, \frac{x}{y}\right]$$

we have $(z_0/x, x^2/z_0) = (z_0/y, y^2/z_0)$, so we have agreement. On

---

†The notation means $x_3 = x^2/z_0$, $y_3 = y/x$, $z_3 = z_0/x$, etc.

$$F_3 \cap F_4 = \text{Spec } k\left[\frac{x^2}{z_0}, \frac{y}{x}, \frac{z_0}{x}, \frac{z_0}{x^2}\right] \qquad \left(\frac{z_0}{x}, \frac{x^2}{z_0}\right) = \text{unit ideal}$$

so the ideals match. Similarly, we have agreement on $F_5 \cap F_6$ by symmetry in x, y. On

$$F_3 \cap F_6 = \text{Spec } k\left[\frac{x^2}{z_0}, \frac{y}{x}, \frac{z_0}{x}, y, \frac{x}{y}, y, \frac{z_0}{y^2}\right] \qquad \left(\frac{z_0}{x}, \frac{x^2}{z_0}\right) = \left(\frac{z_0}{y}, \frac{y^2}{z_0}\right) = \text{unit ideal}$$

and so again the ideals match. Again we get agreement on $F_4 \cap F_5$ by symmetry in x and y. Therefore, the sheaf is well defined.

We blow up that sheaf to obtain a new variety B_3F_0. It is covered by six affine schemes, namely, F_4, F_6, which are not affected by the blowing-up process, and in addition F_7, F_8, F_9, F_{10} with rings:

$$R_4 = k\left[x, \frac{y}{x}, \frac{z_0}{x^2}\right] = k[x_4, y_4, z_4]$$

$$R_6 = k\left[y, \frac{x}{y}, \frac{z_0}{y^2}\right] = k[y_6, x_6, z_6]$$

$$R_7 = k\left[\frac{x^3}{z_0^2}, \frac{y}{x}, \frac{z_0}{x}\right] = k[x_7, y_7, z_7]$$

$$R_8 = k\left[\frac{x^2}{z_0}, \frac{y}{x}, \frac{z_0^2}{x^3}\right] = k[x_8, y_8, z_8]$$

$$R_9 = k\left[\frac{y^3}{z_0^2}, \frac{x}{y}, \frac{z_0}{y}\right] = k[y_9, x_9, z_9]$$

$$R_{10} = k\left[\frac{y^2}{z_0}, \frac{x}{y}, \frac{z_0^2}{y^3}\right] = k[y_{10}, x_{10}, z_{10}]$$

By lemma blowup 3A.2.0, the F_i's, where $F_i = \text{Spec } R_i$, are affine hypersurfaces with equations as follows: F_4, F_6 as above, and further,

$$F_7\colon \quad z_7^2 = x_7^2 F_8(1, y_7, z_7)$$

$$F_8\colon \quad z_8^4 x_8^2 = F_8(1, y_8, z_8 x_8)$$

$$F_9: \quad z_9^2 = y_9^2 F_8(x_9, 1, z_9)$$

$$F_{10}: \quad z_{10}^4 y_{10}^2 = y_{10}^2 F_8(x_{10}, 1, z_{10} y_{10})$$

Again we shall check later that F_8, F_{10} are nonsingular along the exceptional loci. The gluing together is given by $F_i \cap F_j =$ Spec $(R_i \cdot R_j)$ = Spec of compositum $R_i R_j$ in $k(x,y,z_0)$.

We finally blow up the sheaf of ideals on $B_3 F_0$ given by the unit ideal on F_4, F_6, F_8, F_{10} and by $(x^3/z_0^2, z_0/x)$ on F_7 and $(y^3/z_0^2, z_0/y)$ on F_9. We first check that the sheaf of ideals is well defined. The unit ideals certainly glue together, so it is enough to check agreement on intersections involving F_7 and F_9. By symmetry it is enough to check $F_7 \cap F_9$, $F_7 \cap F_4$, $F_7 \cap F_6$, $F_7 \cap F_8$, and $F_7 \cap F_{10}$. On

$$F_7 \cap F_9 = \text{Spec } k\left[\frac{x^3}{z_0^2}, \frac{y}{x}, \frac{z_0}{x}, \frac{x}{y}\right]$$

$(x^3/z_0^2, z_0/x) = (y^3/z_0^2, z_0/y)$, so the ideals agree. As to the remaining intersections, since $F_7 \cap F_j$ = Spec $R_7 \cdot R_j$, it is enough to check that $(x^3/z_0^2, z_0/x)$ becomes the unit ideal in each of the rings $R_7 \cap R_j = R_7$ compositum R_j, for j = 4, 6, 8, 10. First in

$$R_7 \cdot R_4 = k\left[\frac{x^3}{z_0^2}, \frac{y}{x}, \frac{z_0}{x}, x, \frac{z_0}{x^2}\right]$$

we have $(x^3/z_0^2, z_0/z) \supset (x^2/z_0, z_0/x)$ = unit ideal, as asserted.

Similarly, in

$$R_7 \cdot R_6 = k\left[\frac{x^3}{z_0^2}, \frac{y}{x}, \frac{z_0}{x}, \frac{x}{y}, y, \frac{z_0}{y^2}\right]$$

we have

$$\left(\frac{x^3}{z_0^2}, \frac{z_0}{x}\right) = \left(\frac{y^3}{z_0^2}, \frac{z_0}{y}\right) \supset \left(\frac{y^2}{z_0}, \frac{z_0}{y}\right) = \text{unit ideal}$$

In

$$R_7 \cdot R_8 = k\left[\frac{x^3}{z_0^2},\frac{y}{x},\frac{z_0}{x},\frac{z_0^2}{x^3}\right]$$

we have again $(x^3/z_0^2, z_0/x)$ = unit ideal. Finally, in

$$R_7 \cdot R_{10} = k\left[\frac{x^3}{z_0^2},\frac{y}{x},\frac{z_0}{x},\frac{x}{y},\frac{y^2}{z_0},\frac{z_0^2}{y^3}\right]$$

$(x^3/z_0^2, z_0/x) = (y^3/z_0^2, z_0/y)$ = unit ideal. Thus we have checked that the sheaf of ideals is well defined. Once again we blow up that sheaf of ideals and we obtain B_4F_0, the resolution. B_4F_0 is covered by the affine schemes F_4, F_6, F_8, F_{10}, which are not affected by the last blowing up and in addition by only two other affine schemes, which we will call F_{11}, F_{12}. That is true because in R_7, z_7 is integrally dependent on (x_7), and in R_9, z_9 is integrally dependent on (y_9). We have

$$R_{11} = k\left[\frac{x^3}{z_0^2},\frac{y}{x},\frac{z_0^3}{x^4}\right] = k[x_{11},y_{11},z_{11}]$$

$$R_{12} = k\left[\frac{y^3}{z_0^2},\frac{x}{y},\frac{z_0^3}{y^4}\right] = k[y_{12},x_{12},z_{12}]$$

$$F_{11}\colon\ z_{11}^2 = F_8(1,y_{11},z_{11}x_{11})$$

$$F_{12}\colon\ z_{12}^2 = F_8(x_{12},1,z_{12}y_{12})$$

Let $\pi_4\colon\ B_4F_0 \to F_0$ be the natural projection. The restriction of π to the F_i's is given by the natural inclusion of $k[F_0] \to R_i$. We call $\pi_4^{-1}(p)$ the exceptional locus. We shall now check that F_4, F_6, F_8, F_{10}, F_{11}, F_{12} are nonsingular along the exceptional loci. To show that F_4, F_6, F_8, F_{10}, F_{11}, F_{12} are nonsingular along the exceptional loci, we write the original equation of F_0 as follows:

$$z_0^6 = f_8(x,y) + z_0f_7(x,y) + z_0^2F_6(x,y,z_0)$$

We need only to discuss F_4, F_8, F_{11} because F_6, F_{10}, F_{12} will then follow as a result of symmetry in x's and y's. We have

$$F_4:\quad z_4^6x_4^4 = F\,(1,y_4,z_4x_4) = f_8(1,y_4) + (z_4x_4)f_7(1,y_4) + (z_4x_5)^2F_6(1,y_4,z_4x_4)$$

$$F_8:\quad z_8^4x_8^2 = F_8(1,y_8,z_8x_8) = f_8(1,y_8) + z_8x_8f_7(1,y_8) + z_8^2x_8^2F_6(1,y_8,z_8x_8)$$

$$F_{11}:\quad z_{11}^2 = F_8(1,y_{11},z_{11}x_{11}) = f_8(1,y_{11}) + z_{11}x_{11}f_7(1,y_{11}) + z_{11}^2x_{11}^2F_6(1,y_{11},z_{11}x_{11})$$

It is easy to see that the exceptional locus on F_4 is given by $x_4 = 0$; thus $f_8(1,y_4) = 0$, but $\partial/\partial y_4 = 0$ means $(\partial f_8/\partial y_4)(1,y_4) = 0$. By assumption (d), these have no roots in common; thus F_4 is nonsingular along the exceptional locus. Analogously, F_8 is nonsingular along the exceptional locus, which is given by $x_8z_8 = 0$.

As to F_{11}, the exceptional locus is $x_{11}z_{11} = 0$; at a singular point with $x_{11}z_{11} = 0$, we must have $(\partial f_8/\partial y)(1,y_{11}) = 0$ and taking $\partial/\partial x_{11}$, also $z_{11}f_7(1,y_{11}) = 0$. If $z_{11} = 0$, also $f_8(1,y_{11}) = 0$, which contradicts assumption (d). If $z_{11} \neq 0$, $f_7(1,y_{11}) = 0$, but that contradicts assumption (c).

Thus we conclude that there are no singular points on the exceptional locus, $x_{11} = 0$, in F_{11}. As mentioned above, identical arguments with x's replaced by y's hold for F_6, F_{10}, and F_{12}.

So B_4F_0, which is covered by F_4, F_6, F_8, F_{10}, F_{11}, F_{12}, has no singularities on the exceptional locus $\pi^{-1}((0,0,0))$, where π: $B_4F_0 \to F_0$ is the natural projection. Therefore, $B_4F_0 \xrightarrow{\pi} F_0$ is a resolution of the singularity of F_0 at the origin.

Behavior of Differentials at Infinity under the Resolution We are still working under assumptions (a)-(d), so we can use the resolution constructed in the preceding subsection.

PROPOSITION 3.6.0 Suppose that $A \in k[F_0] = k[x,y,z_0]$ has a representation

$$A = \Sigma\, c_{abc} x^a y^b z_0^c$$

where

$$c_{abc} \in k \quad \text{and} \quad a + b + c \leq 4$$

Then if A is adjoint at the singular point (0,0,0) in F_0, A has a representation in $k[F_0]$:

$$A = z_0^3(c_1 x + c_2 y + c_3 z_0) \quad \text{with } c_1,\ c_2,\ c_3 \in k$$

PROPOSITION 3.7.0 xyz_0^6 is biadjoint at (0,0,0) on F_0.

Proof of proposition 3.6.0: Let $A = \Sigma\, c_{abc} x^a y^b z_0^c$, $a + b + c \leq 4$. Let $\sigma_{F_0} = dx\, dz_0/(\partial F_8/\partial y)$ be the canonical generating differential of F_0 with isolated singularity at (0,0,0). If A is adjoint, then $A\sigma_{F_0}$ pulls back to a regular differential along the exceptional fiber on the resolution $B_4F_0 \xrightarrow{\pi} F_0$ constructed above.

When we were passing from B_3F_0 to B_4F_0 we considered the chart $F_7 = \operatorname{Spec} R_1$, where

$$R_7 = k\left[\frac{x^3}{z_0^2}, \frac{y}{x}, \frac{z_0}{x}\right] = k[x_7, y_7, z_7]$$

and

$$F_7\colon \quad z_7^2 = x_7^2 F_8(1, y_7, z_7)$$

We blew up the ideal $(x_7, z_7) = (x^3/z_0^2, z_0/x)$. The blowing up gives rise to two affine charts in the final resolution, B_4F_0, namely F_{11} discussed in the proof of the resolution and another chart, which we shall call $F_{13} = \operatorname{Spec} R_{13}$. F_{13} was not mentioned in the proof of the resolution because $F_{13} \subseteq F_{11}$ as a result of the fact that z_7 is integrally dependent on the ideal (x_7). However, F_{13} will be particularly convenient for the proof of 3.6.0.

We have

$$F_{13}\colon \quad 1 = x_{13}^2 F_8(1, y_{13}, z_{13})$$

and

$$R_{13} = k\left[\frac{x^4}{z_0^3}, \frac{y}{x}, \frac{z_0}{x}\right] = k[x_{13}, y_{13}, z_{13}]$$

Also, if we write the original equation of F_0,

$$F_0: \quad z_0^6 = F_8(x,y,z_0) = f_8(x,y) + z_0F_7(x,y,z_0)$$

with F_8 and f_8 homogeneous of degree 8 and F_7 of degree 7, we obtain

$$F_{13}: \quad 0 = 1 + x_{13}^2 f_8(1,y_{13}) + z_{13}x_{13}^2 F_7(1,y_{13},z_{13})$$
$$= g_{13}(x_{13},y_{13},z_{13})$$

The projection π: $B_4F_0 \to F_0$ restricted to F_{13} is given by the canonical inclusion

$$k[F_0] = k[x,y,z_0] \hookrightarrow k\left[\frac{x^4}{z_0^3}, \frac{y}{x}, \frac{z_0}{x}\right] = k[x_{13},y_{13},z_{13}]$$

so that

$$x = x_{13}z_{13}^3 \qquad y = y_{13}x_{13}z_{13}^3 \qquad z_0 = x_{13}z_{13}^4$$

Let us compute the pullback of $A\sigma_{F_0}$ to F_{13}. We obtain the following differential; since $dx\,dz_0$ pulls back to $x_{13}z_{13}^6\,dx_{13}\,dz_{13}$ by an easy computation using characteristic 2, and since

$$F_8(1,y_{13},z_{13}) = \frac{F_8(x_{13}z_{13}^3, y_{13}x_{13}z_{13}^3, x_{13}z_{13}^4)}{(x_{13}z_{13}^3)^8}$$

we have, using the chain rule,

$$\frac{F_8(1,y_{13},z_{13})}{y_{13}} = \frac{1}{(x_{13}z_{13}^3)^8}\left(\frac{\partial F_8}{\partial y}\right)(x_{13}z_{13}^3, y_{13}x_{13}z_{13}^3, x_{13}z_{13}^4)\,x_{13}z_{13}^2$$

$$= \frac{1}{(x_{13}z_{13}^3)^7}\left(\frac{\partial F_8}{\partial y}\right)(x_{13}z_{13}^3, y_{13}x_{13}z_{13}^3, x_{13}z_{13}^4)$$

Therefore, the pullback of $dx\, dz_0/\partial F_8/\partial y$ is

$$\frac{x_{13}z_{13}^6\, dx_{13}\, dz_{13}}{x_{13}^7 z_{13}^{21}[\partial F_8(1,y_{13},z_{13})/\partial y_{13}]} = \frac{1}{x_{13}^6 z_{13}^{15}}\sigma_{F_{13}}$$

where $\sigma_{F_{13}}$ is a canonical generating differential on F_{13} regular along the exceptional fiber $z_{13} = 0$, which is nonsingular in F_{13}. We note here that x_{13} is a unit in $k[F_{13}]$.

The pullback of A is

$$A_{13} = \Sigma\, c_{abc} x_{13}^{a+b+c} y_{13}^{b} z_{13}^{3a+3b+4c}$$

To find conditions of regularity of the pullback of $A\sigma_{F_0}$, we shall use lemma valuation 3A.1.0. The pullback of $A\sigma_{F_0}$ is $(1/x_{13}^6 z_{13}^{15})A_{13} \times \sigma_{F_{13}}$. If A is adjoint, this differential is regular along the exceptional fiber $z_{13} = 0$. We note that from our assumptions it follows that $g(x_{13},y_{13},0) = x_{13}^2 f_8(1,y_{13}) + 1$ is irreducible since $f_8(1,y_{13})$ is not a square [assumption (b)]; also, the degree of $g(x_{13},y_{13},0)$ is 9.

By lemma valuation 3A.1.0 there is a discrete valuation ν of $k(F_{13})$ such that $\nu(x_{13}^{\alpha} y_{13}^{\beta} z_{13}^{\gamma}) = \gamma$ and such that for any nonzero polynomial $p(x_{13},y_{13})$ of degree ≤ 8, $\nu(p) = 0$. Let us write

$$A_{13} = \Sigma\, z_{13}^{3a+3b+4c} P_{3a+3b+4c}(x_{13},y_{13})$$

where the sum is taken over all possible values of $3a + 3b + 4c$ in the expansion of A.

First, let us note that all monomials in $P_{3a+3b+4c}$ are of degree ≤ 8 and that $P_{3a+3b+4c} \ne 0$. The monomials appearing in $P_{3a+3b+4c}$ come from monomials $x^a y^b z_0^c$ in the expansion of A via $x^a y^b z_0^c \to x_{13}^{a+b+c} y_{13}^{b} z_{13}^{3a+3b+4c}$, which gives rise to $x_{13}^{a+b+c} y_{13}^{b}$ in $P_{3a+3b+4c}$.

We have $a + b + c \le 4$, so $a + b + c + b \le 8$. To see that $P_{3a+3b+4c} \ne 0$ in $k[F_{13}]$, it is enough to show that distinct monomials in A cannot give rise to the same monomial in $P_{3a+3b+4c}$.

If we have $x^a y^b z^c$, $x^{a'} y^{b'} z^{c'}$ give rise to the same monomial. We must have

$$3a + 3b + 4c = 3a' + 3b' + 4c'$$
$$a + b + c = a' + b' + c'$$
$$b = b'$$

but this implies that $a = a'$, $b = b'$, $c = c'$. We also note that because of the fact that the degree of $g(x_{13}, y_{13}, 0)$ is 9, *distinct* monomials in x_{13}, y_{13} of degree ≤ 8 are linearly independent over k in $k[F_{13}]$. Thus we have $P_{3a+3b+3c} \neq 0$ and

$$\nu(P_{3a+3b+3c}) = 0$$

by lemma valuation 3A.1.0(iii).

To conclude the proof, let m be the least value of $3a + 3b + 4c$ over all monomials $x^a y^b z_0^c$ appearing in A.

$$\nu(\text{pullback of } A) = \nu(A_{13})$$
$$= \nu(\Sigma\ z_{13}^{3a+3b+4c} P_{3a+3b+4c}(x_{13}, y_{13})) = m$$

Also, $\nu(x_{13}) = \nu(x_{13}^1 y_{13}^0 z_{13}^0) = 0$ and $\nu(\sigma_{F_{13}}) = 0$ because $\sigma_{F_{13}}$ is the canonical generating differential along $z_{13} = 0$ (see 2.34.0). Hence

$$\nu(\text{pullback of } A\sigma_{F_0}) = \nu\left(\frac{1}{x_{13}^6 z_{13}^{15}} A_{13}\sigma_{F_{13}}\right) = m - 15$$

Therefore, by lemma 3A.1.0(ii), the pullback of $A\sigma_{F_0}$ being regular implies that $m \geq 15$. Hence we must have for all monomials $x^a y^b z^c$ appearing in A, $3a + 3b + 4c \geq 15$. However, this together with $a + b + c \leq 4$ means that $a + b + c = 4$ and $c \geq 3$ for all such monomials. Therefore,

$$A = z_0^3(c_1 x + c_2 y + c_3 z_0)$$

Proof of proposition 3.7.0: To prove that xyz_0^6 is biadjoint at (0,0,0) on F_0, we shall prove that xz_0^3 and yz_0^3 are both adjoint at (0,0,0) and then apply 2.28.1. We also note that xz_0^3 and yz_0^3 are $\neq 0$ in $k[F_0]$ since the degree of the defining equation of F_0 is 8.

Consequently, $xyz_0^6 \neq 0$ since $k[F_0]$ is a domain. To prove that xz_0^3 and yz_0^3 are adjoint, it is sufficient to show that if $\sigma_{F_0} = dx\, dz_0/\partial F_8/\partial y = \omega$ is the canonical generating differential on F_0 and π: $B_4F_0 \to F_0$ is the resolution constructed before, then $\pi^{-1}(xz_0^3\omega)$ and $\pi^{-1}(yz_0^3\omega)$ are regular along the exceptional locus $\pi^{-1}((0,0,0))$. B_4F_0 is covered by six affine schemes, F_4, F_8, F_{11}, F_6, F_{10}, and F_{12}. By symmetry of equations in x's and y's it is enough to consider F_4, F_8, F_{11}.

First F_4:

$$F_4:\quad z_4^6x_4^4 = F_8(1,y_4,z_4x_4)$$

$$k\left[x,\frac{y}{x},\frac{z_0}{x^2}\right] = k[x_4,y_4,z_4]$$

$\pi|F_4$ is given by $k[x,y,z_0] \hookrightarrow k[x,y/x,z_0/x^2]$, the natural inclusion; thus

$$x = x_4 \qquad y = y_4x_4 \qquad z_0 = z_4x_4^2$$

To find the pullback of $\omega = dx\, dz_0/\partial F_8/\partial y$,

$$\begin{aligned}\frac{\partial F_8(1,y_4,z_4x_4)}{\partial y_4} &= \frac{1}{x_4^8}\left(\frac{\partial F_8}{\partial y}\right)(x_4,y_4x_4,z_4x_4^2)\,x_4\\ &= \frac{1}{x_4^7}\left(\frac{\partial F_8}{\partial y}\right)(x_4,y_4,z_4x_4^2)\end{aligned}$$

Also, $dx\, dz_0$ pulls back to $x_4^2\, dx_4\, dz_4$ (characteristic 2), so the pullback of

$$\omega = \frac{1}{x_4^5}\,\frac{dx_4\, dz_4}{\partial F_8(1,y_4,z_4x_4)/\partial y_4} = \frac{1}{x_4^5}\,\omega_4$$

where ω_4 is the generating differential on F_4 regular along the exceptional locus $x_4 = 0$. But the pullback of xz_0^3 is $x_4^7z_4^3$ and the pullback of yz_0^3 is $y_4x_4^7z_4^3$. Hence the pullbacks of $xz_0^3\omega$ and $yz_0^3\omega$ are regular on F_4 along $x_4 = 0$.

We handle F_8 and F_{11} quite similarly. On

$$F_8:\quad z_8^4x_8^2 = F_8(1,y_8,z_8x_8)$$

$$k[F_8] = k\left[\frac{x^2}{z_0},\frac{y}{x},\frac{z_0^2}{x^3}\right] = k[x_8,y_8,z_8]$$

$$x = x_8^2z_8 \qquad y = y_8x_8^2z_8 \qquad z_0 = x_8^3z_8^2$$

Consequently,

$$dx\ dz_0 = x_8^4z_8^2\ dx_8\ dz_8$$

$$\frac{\partial F_8(1,y_8,z_8x_8)}{\partial y_8} = \left(\frac{\partial}{\partial y_8}\right)\frac{F_8(x_8^2z_8,y_8x_8^2z_8,x_8^3z_8^2)}{(x_8^2z_8)^8}$$

$$= \frac{1}{x_8^{16}z_8^8}\left(\frac{\partial F_8}{\partial y}\right)(x_8^2z_8,y_8x_8^2z_8,x_8^3z_8^2)\,x_8^2z_8$$

$$= \frac{1}{x_8^{14}z_8^7}\left(\frac{\partial F_8}{\partial y}\right)(x_8^2z_8,y_8x_8^2z_8,x_8^3z_8^2)$$

so the pullback of ω is

$$\frac{1}{x_8^{10}z_8^5}\cdot\frac{dx_8\ dz_8}{\partial F_8(1,y_8,z_8x_8)/\partial y_8} = \frac{1}{x_8^{10}z_8^5}\,\omega_8$$

with ω_8 regular along the exceptional locus. The pullback of xz_0^3 is $x_8^{11}z_8^7$ and the pullback of yz_0^3 is $y_8x_8^{11}z_8^7$. Hence the pullbacks of $xz_0^3\omega$ and $yz_0^3\omega$ are regular along the exceptional locus $x_8z_8 = 0$.

Finally, on F_{11}, $F_{11} = \operatorname{Spec} R_{11}$,

$$F_{11}:\quad z_{11}^2 = F_8(1,y_{11},z_{11}x_{11})$$

$$R_{11} = k\left[\frac{x^3}{z_0^2},\frac{y}{x},\frac{z_0^3}{x^4}\right]$$

Hence

$$x = x_{11}^3 z_{11}^2 \qquad y = y_{11} x_{11}^3 z_{11}^2 \qquad z_0 = x_{11}^4 z_{11}^3$$

Therefore,

$$dx\, dz_0 = x_{11}^6 z_{11}^4\, dx_{11}\, dz_{11} \quad \text{(characteristic 2)}$$

Also,

$$\frac{\partial F_8(1, y_{11}, z_{11}x_{11})}{\partial y_{11}}$$

$$= \frac{1}{(x_{11}^3 z_{11}^2)^8} \left(\frac{\partial F}{\partial y}\right) (x_{11}^3 z_{11}^2, y_{11} x_{11}^3 z_{11}^2, x_{11}^4 z_{11}^3) x_{11}^3 z_{11}^2$$

so the pullback of ω is

$$\frac{dx_{11}\, dz_{11}}{x_{11}^{15} z_{11}^{10} (\partial F_8/\partial y_{11})(1, y_{11}, z_{11}x_{11})} = \frac{1}{x_{11}^{15} z_{11}^{10}} \omega_{11}$$

where ω_{11} is regular along the exceptional locus. But the pullback of xz_0^3 is $x_{11}^{15} z_{11}^{11}$, so the pullback of yz_0^3 is $y_{11} x_{11}^{15} z_{11}^{11}$. Therefore, the pullbacks of $xz_0^3\omega$ and $yz_0^3\omega$ are regular along the exceptional locus $x_{11} z_{11} = 0$. F_6, F_{10}, F_{12} are handled by exactly the same computations, with x's replaced by y's, and vice versa.

Thus we conclude that the pullbacks of $xz_0^3\omega$ and $yz_0^3\omega$ to the resolution of the singularity of F_0 at (0,0,0), B_4F_0, are both regular differentials along the exceptional locus. Therefore, xz_0^3, yz_0^3 are adjoint at (0,0,0). Hence $xyz_0^6 = (xz_0^3) \cdot (yz_0^3)$ is biadjoint (by 2.28.1).

Part 4: Conclusion

THEOREM 3.7.1 Let $\tilde{F}$ be a desingularization of $\bar{F}$. We have $p_g(\tilde{F}) = 0$ and $P_2(\tilde{F}) \geq 1$.

Proof: By 2.23.0 we have $p_g(\tilde{F})$ = dimension of the space of adjoint forms to $\bar{F}$ of degree $8 - 4 = 4$. Let $A(x,y,z,z_0)$ be an adjoint form. Then by our definition of adjoint forms (2.19.0), $A(x,y,z,1)$ is adjoint at every point of F and $A(x,y,1,z_0)$ is adjoint at the point $x = y = z_0 = 0$ on F_0.

Therefore, by proposition 3.7.0,

$$A(x,y,z,z_0) = A(x,y,1,z_0) = z_0^3(ax + by + cz_0)$$

So $A(x,y,z,1) = ax + by + c$. Now $A(x,y,z,1)$ is adjoint to F and hence by 3.3.1, 3.5.2, and 3.5.3, we must have

$$A(0,0,0,1) = A(0,1,0,1) = A(1,0,0,1) = 0$$

The equalities above give that $a = b = c = 0$, hence $A(x,y,z,z_0) = 0$. Therefore, the only adjoint form of degree 4 is the zero form. Hence p_g of $\tilde{F}$ is equal to 0.

To see that $P_2 \geq 1$, let us show that xyz_0^6 is a biadjoint form to $\bar{F}$ of degree $2(8 - 4) = 8$. It is enough to show that xy is biadjoint to F and that xyz_0^6 is biadjoint at the singularity on F_0 which corresponds to the nonrational singularity at infinity. But we have proved both of the foregoing facts (see 3.3.2, 3.5.5, 3.5.6, 3.7.0). The rational singularities of F cause no difficulty since all forms are ℓ-adjoint at them for all ℓ (see proposition 3A.3.1).

Let us also note that $xyz_0^6 \neq 0$ in the graded ring of $\bar{F}$ since that ring is a domain and obviously x, y, z_0 are $\neq 0$ in it.

REMARK 3.7.2 F is not a rational surface since P_2 of $\tilde{F}$ is ≥ 1 and P_2 is a birational invariant (see Šafarevič, 1966, p. 141).

REMARK 3.7.3 The desingularization $\tilde{F}$ in 3.7.5 exists by Abhyankar (1956); another proof that $\tilde{F}$ exists and in fact can be explicitly constructed follows from the fact that we resolved every one of the nonrational singularities of $\bar{F}$ by blowing up sheaves of ideals which induce the unit deal at every point of $\bar{F}$ that was not being resolved. It follows from this that all the resolutions glue together to give a surface $F_1 \xrightarrow{\pi_1} \bar{F}$. F_1 still may have singularities corresponding to the rational singularities of $\bar{F}$; they are rational and easily resolved, each by a single blowing up with point center (see lemma 3A.3.0). The resulting surface, $\tilde{F}$, is nonsingular. We have

$$\tilde{F} \xrightarrow{\pi_2} F_1 \xrightarrow{\pi_1} \bar{F}$$

and $\tilde{F} \xrightarrow{\pi_1 \cdot \pi_2} \bar{F}$ is a desingularization of $\bar{F}$.

REMARK 3.7.4 We hope to study the minimal model of $\tilde{F}$ in a future paper. We should like to know if that minimal model is a "classical Enriques surface" in the sense of Mumford-Bombieri.

Part 5: Appendix

LEMMA VALUATION 3A.1.0 Let S: $f(x,y,z) = 0$ be a normal, irreducible hypersurface in $\mathbb{A}^3$ (affine 3-space over an algebraically closed field k). Assume that $g(x,y) = f(x,y,0)$ is irreducible in $k[x,y]$, the polynomial ring in two variables; assume also that $g(x,y) \neq x$, $g(x,y) \neq y$. Then there exists a discrete valuation ν of $k(S)$ such that $\nu|k[S]$ satisfies the following conditions:

(i) $\nu(x^\alpha y^\beta z^\gamma) = \gamma$ for all nonnegative integers α, β, γ.

(ii) $\nu|k[S]$ is positive only on the ideal (z).

(iii) $\nu(p(x,y) = 0$ if the degree of $p(x,y)$ is less than the degree of $g(x,y)$ and $p(x,y) \neq 0$.

Proof: We have a natural inclusion

$$k[x,y] \overset{\alpha}{\hookrightarrow} k[S]$$

It is easy to check that α induces an isomorphism

$$\alpha': \frac{k[x,y]}{(g(x,y))} \xrightarrow{\sim} \frac{k[S]}{(z)}$$

This proves that (z) is prime in $k[S]$. Also, $k[S]_{(z)} \subset k(S)$ is a one-dimensional normal, local domain, so it is a d.v.r. Let ν be the valuation such that $\nu(z) = 1$. We have

$$(z)k[S]_{(z)} \cap k[S] = (z)k[S]$$

so ν is positive only on the ideal (z) in $k[S]$. This proves (ii). Since α' is an isomorphism,

$$(k[x,y]) \cap (z) = (g(x,y))$$

from our assumption

$$x, y \notin (g(x,y))$$

so $\nu(x) = 0$, $\nu(y) = 0$, and (i) follows. If $p(x,y)$ has degree less than degree $g(x,y)$ and $p(x,y) \neq 0$, then $p(x,y) \notin (g(x,y))$, so $\nu(p(x,y)) = 0$. This proves (iii).

REMARK 3A.1.1 We need not assume that S is normal because the existence of the valuation follows from the fact that the ideal

$$(z)k[S]_{(z)}$$

is principal.

LEMMA 3A.2.0 (valuation, part 2) In the notation of lemma 3A.2.0, let $\alpha \in \Omega^2(S)$ be a differential such that $(\alpha) < 0$. Then α is not regular at any of the points of the curve $C \subsetneq S$ given by $z = 0$.

Proof: Let $p \in C$ be a closed point and $k[S]_p$ the local ring at p. Then

$$k[S]_p \subsetneq k[S]_{(z)}$$

Let us write $\alpha = a\sigma_S$, where σ_S is a canonical generating differential on S and $a \in k(S)$. We have $\nu(a) < 0$, so $a \notin k[S]_{(z)}$; therefore, $a \notin k[S]_p$, which means that α is not regular at p.

REMARK 3A.2.1 Similarly, if $\alpha \in \Omega^2(S)^{\otimes 2}$ and $\nu(\alpha) < 0$, then α is not regular at any of the points of the curve $C \subset S$ given by $z = 0$.

LEMMA 3A.2.0 (blowup) Let X_1, X_2, X_3 be indeterminates over k. Let

$$S = \operatorname{Spec} \frac{k[X_1,X_2,X_3]}{(f(X_1,X_2,X_3))}$$

with $f(X_1,X_2,X_3)$ irreducible. We assume that $f \neq X_1, X_2, X_3$. As usual we write

$$k[S] = \frac{k[X_1,X_2,X_3]}{(f(X_1,X_2,X_3))} = k[x_1,x_2,x_3]$$

Let I be the ideal $(x_1,x_2,x_3) \subsetneq k[S]$. Let BIS be the scheme obtained by "blowing up the sheaf of ideals defined by I," i.e.,

$$BIS = \mathrm{Proj}(k[S] \oplus IT \oplus I^2T^2 \oplus \cdots)$$

Then BIS is covered by three open affines:

$$S_1 = \mathrm{Spec}\ k\left[x_1,\frac{x_2}{x_1},\frac{x_3}{x_1}\right] = \mathrm{Spec}\ R_1$$

$$S_2 = \mathrm{Spec}\ k\left[\frac{x_1}{x_2},x_2,\frac{x_3}{x_2}\right] = \mathrm{Spec}\ R_2$$

$$S_3 = \mathrm{Spec}\ k\left[\frac{x_1}{x_3},\frac{x_2}{x_3},x_3\right] = \mathrm{Spec}\ R_3$$

The gluing together of the S_i's is induced by the isomorphisms $(R_1)_{x_2/x_1} = (R_2)_{x_1/x_2}$, etc.

Each of the S_i's is isomorphic to an affine hypersurface. For example, $S_1 \approx H_1$: $\tilde{f}(U_1,U_2,U_3) = 0$, where the U_i's are indeterminates, and the polynomial $\tilde{f}$ is obtained as follows. Write $f(U_1,U_2U_1,U_3U_1) = U_1^c\tilde{f}(U_1,U_2,U_3)$, where U_1 does not divide $\tilde{f}$. The isomorphism of

$$k[H_1] = \frac{k[U_1,U_2,U_3]}{(f(U_1,U_2,U_3))} = k[u_1,u_2,u_3]$$

with $k[x_1,x_2/x_1,x_3/x_1]$ is given by $u_1 \to x_1$, $u_2 \to x_2/x_1$, and $u_3 \to x_3/x_1$. Analogous statements hold for S_2 and S_3.

Proof: The assertions concerning the covering of BIS by the S_i and the gluing of S_i are standard (EGA 2). We do not prove them here. As for the remaining assertions, first, it is not hard to show that $\tilde{f}(U_1,U_2,U_3) \in k[U_1,U_2,U_3]$ is irreducible. We omit the proof.

Let us now show that $R_1 = k[x_1,x_2/x_1,x_3/x_1]$ is isomorphic to

$$\frac{k[U_1,U_2,U_3]}{(f(U_1,U_2,U_3))} = k[u_1,u_2,u_3]$$

We have the following commutative diagram of maps:

$$\begin{array}{ccc} k[X_1, X_2, X_3] & \longrightarrow & k[x_1, x_2, x_3] \\ \downarrow & & \downarrow \\ k\left[X_1, \frac{X_2}{X_1}, \frac{X_3}{X_1}\right] & \longrightarrow & k\left[x_1, \frac{x_2}{x_1}, \frac{x_3}{x_1}\right] \end{array}$$

We identify $k[U_1,U_2,U_3]$ with $k[X_1,X_2/X_1,X_3/X_1]$ and obtain a surjective map m: $k[U_1,U_2,U_3] \to k[x_1,x_2/x_1,x_3/x_1]$ such that $m(U_1) = x_1$, $m(U_2) = x_2,x_1$, $m(U_3) = x_3/x_1$. Because $k[x_1,x_2/x_1,x_3/x_1]$ has dimension 2 and is a domain, the kernel of m is a principal prime ideal.

Let us show that $\tilde{f}(U_1,U_2,U_3) \in \ker m$. We observe that $m(f(U_1, U_2U_1,U_3U_1)) = f(m(U_1),m(U_2)m(U_1),m(U_3)m(U_1)) = f(x_1,x_2,x_3) = 0$. Hence $f(U_1,U_2U_1,U_3U_1) \in \ker f$. However, $f(U_1,U_2U_1,U_3U_1) = U_1^c\tilde{f}(U_1, U_2,U_3)$. Since $m(U_1) = x_1 \neq 0$ in R_1, then $U_1 \notin \ker m$ and therefore since $\ker m$ is prime, $\tilde{f}(U_1,U_2,U_3) \in \ker m$.

Therefore, $\tilde{f}$ generates $\ker m$ and finally m induces the desired isomorphism,

$$k[u_1,u_2,u_3] \approx \frac{k[U_1,U_2,U_3]}{(f(U_1,U_2,U_3))} \xrightarrow{\bar{m}} k\left[x_1,\frac{x_2}{x_1},\frac{x_3}{x_1}\right]$$

with $\bar{m}(u_1) = x_1$, $\bar{m}(u_2) = x_2/x_1$, $\bar{m}(u_3) = x_3/x_1$.

LEMMA 3A.2.0 (blowup, part 2) If in the assumptions of 3A.2.0 we assume that x_3 is integrally dependent on the ideal (x_1,x_2) in $k[S]$, then S_3 is contained as an open subscheme in the union of S_1 and S_2.

Proof: Let

$$x_3^n + a_1x_3^{n-1} + \cdots + a_n = 0 \qquad (1)$$

with $a_i \in (x_1,x_2)^i$ be the equation of integral dependence. We have $a_i/x_3^i \in$ the ideal $(x_1/x_3,x_2/x_3) \subseteq k[x_1/x_3,x_2/x_3,x_3] = R_3$. Therefore, since dividing (1) by x_3^n results in

$$1 + \frac{a_1}{x_3^1} + \frac{a_2}{x_3^2} + \cdots + \frac{a_n}{x_3^n} = 0$$

then $1 \in$ the ideal $(x_1/x_3, x_2/x_3) \subsetneq R_3$.

Therefore, $\operatorname{Spec}(R_3)_{x_1/x_3}$ and $\operatorname{Spec}(R_3)_{x_2/x_3}$ form an open covering of Spec R_3. But since $\operatorname{Spec}(R_3)_{x_1/x_3}$ is identified with $\operatorname{Spec}(R_1)_{x_3/x_1}$, and $\operatorname{Spec}(R_3)_{x_2/x_3}$ is identified with $\operatorname{Spec}(R_2)_{x_3/x_2}$, then Spec R_3 is identified with an open subscheme of $S_1 \cup S_2 =$ Spec $R_1 \cup$ Spec R_2.

LEMMA 3A.2.0 (blowup, part 3) Let S, f, $k[S] = k[x_1,x_2,x_3]$, $k[X_1, X_2,X_3]$, U_1, U_2, U_3 be as above. Let $I_0 = (x_1,x_2)$ and

$$\text{BIS} = \operatorname{Proj}(k[S] \oplus I_0T \oplus I_0^2T^2 \oplus \cdots)$$

Then BIS is covered by two open affines,

$$T_1 = \operatorname{Spec} k\left[x_1, \frac{x_2}{x_1}, x_3\right] \qquad \text{and} \qquad T_2 = \operatorname{Spec} k\left[\frac{x_1}{x_2}, x_2, x_3\right]$$

The T_i's are glued together via

$$k\left[x_1, \frac{x_2}{x_1}, x_3\right]_{x_2/x_1} = k\left[\frac{x_1}{x_2}, x_2, x_3\right]_{x_1/x_2}$$

Also, T_1 is isomorphic to a hypersurface F_1: $f_1(U_1,U_2,U_3) = 0$, where f_1 is not divisible by U_1 and $f(U_1,U_2U_1,U_3) = U_1^c f_1(U_1,U_2,U_3)$. The isomorphism of $k[F_1] = k[U_1,U_2,U_3]$ with $k[x_1,x_2/x_1,x_3]$ is given by $u_1 \to x_1$, $u_2 \to x_2/x_1$, $u_3 \to x_3$.

If x_2 is integrally dependent on (x_1) in $k[x_1,x_2,x_3]$, we again have $T_2 \subsetneq T_1$ as an open subscheme.

Proof: The proof is the same as for parts 1 and 2.

LEMMA 3A.3.0 (We assume characteristic 2.) Suppose that S: $z^2 = f(x,y)$ has a singularity and suppose that at the singular point $\partial f/\partial x\partial y \neq 0$. Then the singularity is isolated and is a rational singularity.

Proof: After a change of coordinates we can assume that singularity is at (0,0,0) and is given by

$$z^2 = ax^2 + bxy + cy^2 + \text{terms of order} \geq 3$$

We have $b \neq 0$ and so because of characteristic 2, we know that $ax^2 + bxy + cy^2$ is not a perfect square.

After a further change of coordinates, we can therefore assume that

$$S: \quad z^2 = xy + \text{terms of degree} \geq 3$$

We denote the coordinate ring of S by $k[x,y,z]$. Since

$$\frac{\partial f}{\partial x} = y + (\text{terms of order} \geq 2)^{\dagger}$$

and

$$\frac{\partial f}{\partial y} = x + (\text{terms of order} \geq 2)^{\dagger}$$

$\partial f/\partial x$, $\partial f/\partial y$ are a system of parameters of the local ring $k[x,y,z]_{(x,y,z)}$ and the singularity is isolated.

Now we resolve the singularity. Blow up the ideal (x,y,z) to obtain QS. z is integrally dependent on (x,y), so by lemma 3A.2.0, QS is covered by two affine charts S_1 and S_2 with equations

$$S_1 = \text{Spec } k\left[x, \frac{y}{x}, \frac{z}{x}\right] = \text{Spec } k[x_1, y_1, z_1]$$

and

$$S_2 = \text{Spec } k\left[\frac{x}{y}, y, \frac{z}{y}\right] = \text{Spec } k[x_2, y_2, z_2]$$

We have

$$S_1: \quad z_1^2 = y_1 + x_1 \cdot (\text{terms of degree} \geq 1)^{\dagger}$$

and

$$S_2: \quad z_2^2 = x_2 + y_2 \cdot (\text{terms of degree} \geq 1)^{\dagger}$$

The exceptional fiber if given by $x_1 = 0$ on S_1 and $y_2 = 0$ on S_2 and using the jacobian criterion, we see that it contains no singularities. Therefore, QS is a resolution of the singularity.

---

† Or the term identically zero.

The singularity is resolved by a quadratic transform alone, so it is rational by, e.g., Lipman (1969) and is easy to varify directly.

PROPOSITION 3A.3.1 (Characteristic k = 2) Let $S \subseteq A^3$ be normal and given by

$$S: \quad z^2 = xy + f_3(x,y) + \cdots + f_n(x,y)$$

or S: $z^2 = f(x,y)$ with f_i forms of degree i. Let p be the point (0,0,0) of F. Then p is an isolated singularity, and adjoints at p as well as ℓ-adjoints at p coincide with $(\mathcal{O}_S)_p$.

Proof: We know by 3A.3.0 that p is an isolated singularity and also we know that a resolution of p in S,

$$\tilde{S} \xrightarrow{\pi} S$$

$\tilde{S}$ is covered by

$$S_1 = \operatorname{Spec} k\left[x, \frac{y}{x}, \frac{z}{x}\right] = \operatorname{Spec} k[x_1, y_1, z_1]$$

and

$$S_2 = \operatorname{Spec} k\left[\frac{x}{y}, y, \frac{z}{y}\right] = \operatorname{Spec} k[x_2, y_2, z_2]$$

by lemma blowup.

S_1 and S_2 are isomorphic to hypersurfaces as follows:

$$S_1: \quad z_1^2 = y_1 + x_1 f_3(1,y_1) + x_1^2 f_4(1,y_2) + \cdots$$

$$S_2: \quad z_2^2 = x_2 + y_2 f_3(x_2,1) + y_2^2 f_4(x_2,1) + \cdots$$

Let us consider the canonical generating differential

$$\sigma_S = \frac{dx\ dz}{\partial f/\partial y} = \frac{dy\ dz}{\partial f/\partial x}$$

on S. We compute the pullback of σ_S on $\tilde{S}$; we get

$$\pi_1(\sigma_S) = \frac{x_1\ dx_1\ dz_1}{\frac{\partial f}{\partial y}(x_1, y_1 x_1)}$$

but

$$\frac{\partial}{\partial y_1}(f(x_1,y_1x_1)) = \left(\frac{\partial f}{\partial y}\right)(x_1,y_1x_1)x_1$$

so

$$\frac{x_1\, dx_1\, dz_1}{\frac{\partial f}{\partial y}(x_1,y_1x_1)} = \frac{x_1^2\, dx_1\, dz_1}{\frac{\partial f}{\partial y_1}(f(x_1,y_1x_1))} = \frac{dx_1\, dz_1}{\frac{\partial f}{\partial y_1}\ \frac{f(x_1,y_1x_1)}{x_1^2}}$$

This is a canonical generating differential on $\tilde{S}_1$ and similarly, we can check that $\pi_1(\sigma_S)$ is a canonical generating differential on $\tilde{S}_2$.

Thus $\pi_1(\sigma_S)$ is regular along the exceptional locus of π. Therefore, $\pi_1(a\sigma_S^{\otimes \ell})$ is also regular along $\pi^{-1}(p)$ for every $a \in (\mathcal{O}_S)_p$, and hence all elements of $(\mathcal{O}_S)_p$ are ℓ-adjoint for all ℓ.

REMARK 3A.3.2 We note that $\pi_1(\sigma_S)$ also has no zero curves that meet $\pi^{-1}(p)$.

4. GENERIC ZARISKI SURFACES

We describe here the simplest features of the geometry of a Zariski surface F that arises from a hypersurface with equation F: $z^p = f_{p+1}(x,y)$, where f_{p+1} is a generic polynomial of degree p + 1. We deal with the case $p \geq 3$. We shall call such a ZS a "generic ZS." As we know, $\tilde{F}$ is unirational. It is well known that a unirational surface is also supersingular; i.e., the second étale Betti number of $\tilde{F}$, b_2, is equal to the rank ρ of the Néron-Severi group of $\tilde{F}$.[†] We give an alternative proof of this for $\tilde{F}$ and determine the common value of $\rho = b_2$. We also find explicitly ρ independent cycles in the Néron-Severi group of $\tilde{F}$.

We work out the classical invariants p_a, p_g, $K_{\tilde{F}}^2$ for $\tilde{F}$. Finally, we compute the zeta function of $\tilde{F}$ when $\tilde{F}$ is defined over a finite field.

[†]See, e.g., T. Shioda (1974, p. 235, corollary 2).

Throughout this section let k be an algebraically closed field of characteristic $p \geq 3$. We start with some preliminary material about singularities.

THEOREM 4.1.0 Suppose that X is a hypersurface in A^3 given by

$$X: \quad z^{\ell} + xy + f(x,y) = 0$$

where $\ell > 1$ is an odd integer and all the terms of f have degree ≥ 3. Let $p = (0,0,0)$ in X. Then p is an isolated singularity and there exists a resolution of the singularity of X at p by means of a sequence of quadratic transformations

$$\tilde{X} \xrightarrow{\pi} X$$

such that $\pi^{-1}(p) = E_1 \cup \cdots \cup E_{\ell-1}$ where each E_i is isomorphic to P^1 and the intersection numbers are as follows:

$$E_i \cdot E_j = \begin{cases} 1 & \text{if } |i - j| = 1 \\ 0 & \text{if } 0 \neq |i - j| \neq 1 \end{cases}$$

This can be pictured as follows:

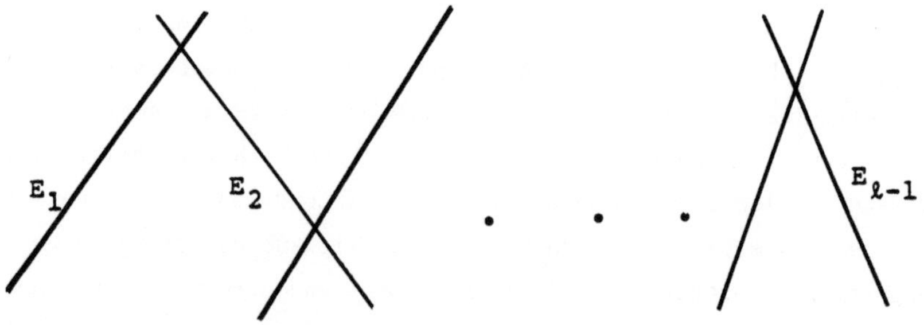

REMARK 4.1.1 $\tilde{X}$ is obtained by blowing up a sheaf of ideals which, except at p, induces a unit ideal at every other closed point of X.

LEMMA 4.2.0 Let S be a scheme defined over k which contains an open subscheme S_0 which is an affine hypersurface with equation of the form

$$S_0: \quad z^m + xy + zg(x,y,z) = 0$$

where m is an odd integer and $g = \Sigma\, a_{\alpha\beta\gamma} x^\alpha y^\beta z^\gamma$ with $a_{\alpha\beta\gamma} \in k$, $\alpha + \beta \geq 3$. Let

$$\ell_x^m: \quad z = x = 0 \qquad \text{and} \qquad \ell_y^m: \quad z = y = 0$$

$$\ell_x^m \subseteq S_0 \qquad \ell_y^m \subseteq S_0$$

and let $p_0 = (0,0,0)$. Let us assume that the singularity at p_0 is isolated. Then there exists a resolution $\tilde{S} \xrightarrow{\rho} S$ of the singularity p_0 in S by means of a sequence of quadratic transformations with $\rho^{-1}(p_0) = F_1 \cup F_2 \cup \cdots \cup F_{m-1}$, where $F_i \approx P^1$ and again

$$F_i \cdot F_j = \begin{cases} 1 & \text{iff } |i - j| = 1 \\ 0 & \text{iff } 0 \neq |i - j| \neq 1 \end{cases}$$

In addition, the proper transform of ℓ_x^m and ℓ_y^m, denoted by $pr(\ell_x^m)$ and $pr(\ell_y^m)$, respectively, is a rational curve meeting $\rho^{-1}(p_0)$ at one smooth point, belonging to F_1 and F_{m-1}, respectively, with intersection multiplicity 1.

Proof: We proceed by induction on m. If m = 1, the only thing to check is that ℓ_x^1 and ℓ_y^1 are smooth and $\simeq A^1$, but this is clear.

If m > 0, we consider the quadratic transform $S_1 \xrightarrow{\pi_1} S$ with center p_0 · S_1 is covered by three open sets.

$$\bar{S}_1: \quad \text{Spec } k\left[\frac{x}{z},\frac{y}{z},z\right] = \text{Spec } k[x_1,y_1,z_1]$$

where k[x,y,z] is the coordinate ring of $\bar{S}_0$,

$$\bar{S}_1' = \text{Spec } k\left[x,\frac{y}{z},\frac{z}{z}\right] = \text{Spec } k[x_1',y_1',z_1']$$

and

$$\bar{S}_1'' = \text{Spec } k\left[\frac{x}{y},y,\frac{z}{y}\right] = \text{Spec } k[x_1'',y_1'',z_1'']$$

with equations

$$\bar{S}_1: \quad (z_1)^{m-2} + x_1y_1 + z_1\bar{g}(x_1,y_1,z_1) = 0$$

$$\bar{S}_1': \quad (z_1')^m(x_1')^{m-2} + y_1' + z_1'x_1'\bar{g}'(x_1',y_1',z_1') = 0$$

$$\bar{S}_1'': \quad (z_1'')^m(y_1'')^{m-2} + x_1'' + z_1''y_1''\bar{g}''(x_1'',y_1'',z_1'') = 0$$

with $\bar{g}$ satisfying $\bar{g} = \Sigma\, a_{\alpha\beta\gamma}x_1^\alpha y_1^\beta z_1^\gamma$ with $\alpha + \beta \geq 3$. The exceptional fiber is

$$\ell_{x_1}^{m-2}: \quad x_1 = z_1 = 0 \qquad \ell_{y_1}^{m-2}: \quad y_1 = z_1 = 0 \text{ on } \bar{S}_1$$

together with

$$x_1' = y_1' = 0 \text{ on } \bar{S}_1'$$

and

$$y_1'' = x_1'' = 0 \text{ on } \bar{S}_1''$$

The only singular point of the exceptional fiber is $p_1 \in \bar{S}_1$, $p_1 = (0,0,0)$. It is an isolated singularity of S_1 because the original singularity was isolated. The line $\ell_{x_1}^{m-1}$ glues together with

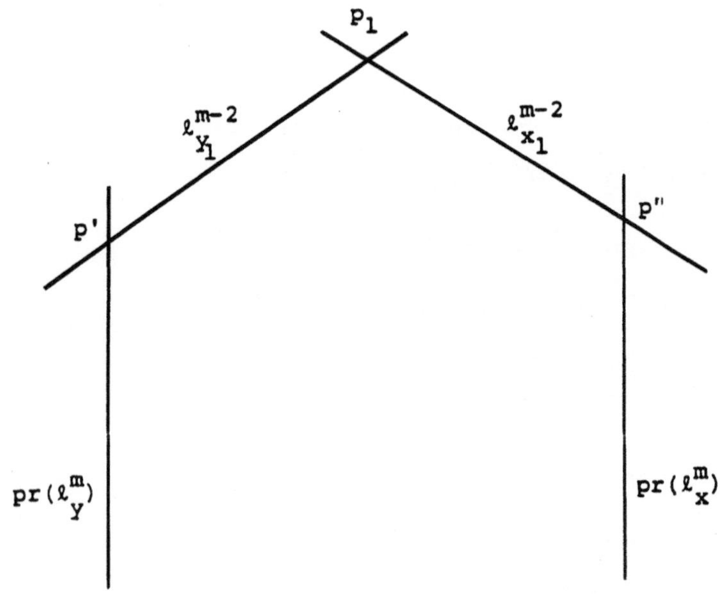

the line $y_1'' = x_1'' = 0$. The closure $\overline{\ell_{x_1}^{m-1}}$ contains the point p'': $y_1'' = x_1'' = z_1'' = 0$ and is smooth. Similarly, $\overline{\ell_{y_1}^{m=2}} = \ell_{x_1}^{m-2} \cup \{p'\}$, where p' is the point $x' = y' = z' = 0$ on $\bar{S}_1'$ • p' and p'' are smooth for $\overline{\ell_{x_1}^{m-2}}$, $\overline{\ell_{y_1}^{m-2}}$ and for S_1. Thus $\overline{\ell_{y_1}^{m-2}}$, $\overline{\ell_{x_1}^{m-2}}$ are rational smooth curves, so they are isomorphic to p^1. The proper transform of ℓ_x^m is $z_1'' = x_1'' = 0$ on $\bar{S}_1''$. This line meets the exceptional locus only at the point p'', and there they meet transversally; similarly for the proper transform of the line ℓ_y^m.

By induction we now resolve the singularity $p_1 \in \bar{S}_1$ and the lemma easily follows.

LEMMA 4.2.1 All assumptions and notations are as in lemma 4.2.0. We also assume that S_0 is normal. Let σ_{S_0} be a canonical generating differential on S_0. Then $\rho_1(\sigma_{S_0})$ is regular and $\neq 0$ at all points $q \in \rho^{-1}(p)$.

Proof: If $m = 1$, then $\rho_1(\sigma_{S_0}) = \sigma_{S_0}$ and the lemma is clear. If $m \geq 3$, assume that the lemma is true for $m - 2$. Let us write the equations of S_0, $\bar{S}_1$, $\bar{S}_1'$, $\bar{S}_1''$ as follows:

$$S_0:\quad f(x,y,z) = 0$$

$$\bar{S}_1:\quad \bar{h}(x_1,y_1,z_1) = 0$$

$$\bar{S}_1':\quad \bar{h}'(x_2,y_2,z_2) = 0$$

$$\bar{S}_1'':\quad \bar{h}''(x_3,y_3,z_3) = 0$$

[Here $\bar{h}(x_1,y_1,z_1) = f(x_1z_1,y_1z_1,z_1)/z_1^2$, etc. Also, $x_2 = x_1'$, etc., $x_3 = x_1''$, etc.] Let $S_1 \xrightarrow{\pi_1} S$ be as in 4.2.0. Let $\sigma_{S_0} = dx\,dz/(\partial f(x,y,z)/\partial y)$. $(\pi_1)_1(\sigma_{S_0})$ is given by

$$\frac{d(x_1z_1)\ dz_1}{\frac{\partial f}{\partial y}(x_1z_1,y_1z_1,z_1)} \quad \text{on } \bar{S}_1$$

but

$$\frac{\partial f}{\partial y}(x_1z_1, y_1z_1, z_1) = z_1 \frac{\partial \bar{h}}{\partial y_1}(x_1, y_1, z_1)$$

so

$$(\pi_1)_1(\sigma_{S_0}) = \frac{dx_1\ dz_1}{\frac{\partial \bar{h}}{\partial y_1}(x_1, y_1, z_1)}$$

This is a canonical generating differential on $\bar{S}_1$. An analogous computation shows that the pullback of σ_{S_0} is a canonical generating differential on $\bar{S}_1'$ and $\bar{S}_1''$. We now resolve the singularity of $\bar{S}_1$ and apply the induction hypotheses to $(\pi_1)_1(\sigma_{S_0})$ on $\bar{S}_1$. The lemma follows.

Proof of theorem 4.1.0: (The singularity is isolated by the same reasoning as in 3A.3.0.) Blow up $p \in X$ to obtain

$$X_1 \xrightarrow{\pi_1} X$$

X_1 is covered by three affines:

$$\text{Spec } R_1 = \text{Spec } k\left[\frac{x}{z}, \frac{y}{z}, z\right]$$

$$\text{Spec } A_1 = \text{Spec } k\left[x, \frac{y}{x}, \frac{z}{x}\right]$$

$$\text{Spec } T_1 = \text{Spec } k\left[\frac{x}{y}, y, \frac{z}{y}\right]$$

($k[x,y,z]$ denotes the coordinate ring of X.) Let

$$S = \text{Spec } R_1$$

$$S:\quad z_1^{\ell-2} + x_1y_1 + z_1g_1(x_1, y_1, z_1) = 0$$

The following two facts are easy to see.

(i) $\pi_1^{-1}(p)$ consists of two projective lines:

$$L_1 = \ell_x^{m-2}:\quad \{x_1 = z_1 = 0 \text{ in } S\} \cup \{\text{one point}\}$$

$$L_2 = \ell_y^{m-2}:\quad \{y_1 = z_1 = 0 \text{ in } S\} \cup \{\text{one point}\}$$

with graph

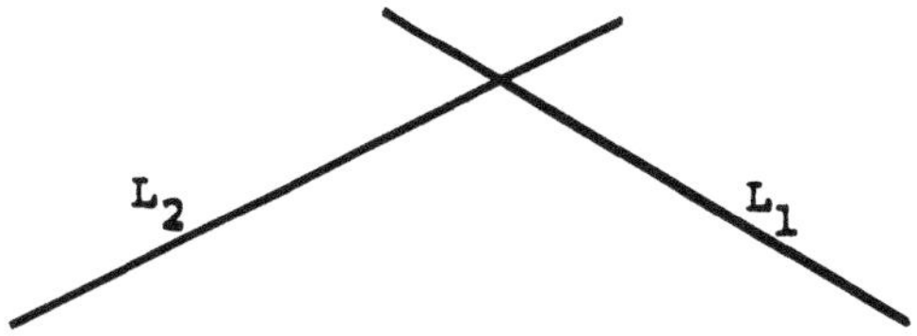

The common point is $p_1 = (0,0,0)$ in S.

(ii) All points of the exceptional locus except possibly for p_1 are smooth for X_1 and for the L_i's. Therefore, since the original singularity was isolated, so is p_1 and we see that S above satisfies lemma 4.2.0. By lemma 4.2.0, $p_1 = (0,0,0)$ in X_1 can be resolved. Thus we obtain

$$\tilde{X} \xrightarrow{\pi'} X_1 \xrightarrow{\pi_1} X$$

so that if $\pi = \pi_1 \cdot \pi^1$, we get

$$\tilde{X} \xrightarrow{\pi} X$$

It is easy to see that $\pi^{-1}(p)$ has the desired properties. [The only point to note is that the proper transform of L_1 (and L_2) by π^1 is a smooth rational curve; thus $\approx P^1$.]

COROLLARY 4.2.1.1 Suppose that k is of characteristic $p \neq 2$. Suppose that F is a hypersurface in $\mathbf{A}^3$ defined by $z^p = f(x,y)$. Suppose that $(x_0,y_0,z_0) \in F$ is a point which is singular but the hessian of f at (x_0,y_0) is $\neq 0$. Then (x_0,y_0,z_0) is a rational isolated singularity of F.

Proof: After a linear change of coordinates, we may assume that $(x_0,y_0,z_0) = (0,0,0)$ and also because the hessian is $\neq 0$,

$$f(x,y) = xy + (\text{higher-order terms in } x,\ y)$$

It follows from the proof of theorem 4.1.0 that the singularity can be resolved by quadratic transformations alone and is isolated. Thus it is a rational singularity by Lipman (1969).

PROPOSITION 4.2.2 We retain the notation of 4.1.0 and its proof. In addition, X is assumed to be normal. Let σ_X be a canonical generating differential on X, and let $\tilde{X} \xrightarrow{\pi} X$ be as in the theorem. Then $\pi_1(\sigma_X)$ is regular at all points of $\pi^{-1}(p)$ and $\neq 0$ there.

Proof: We construct $X_1 \xrightarrow{\pi_1} X$ as in the proof of theorem 4.1.0. X_1 is covered by Spec R_1, Spec A_1, Spec T_1. We easily compute that $\pi_1(\sigma_X)$ is a canonical generating differential on all three. Then we apply lemma 4.2.1 to Spec R_1 and proposition 4.2.2 follows.

COROLLARY 4.2.3 (Notation of 4.1.0) Every element of $(\mathcal{O}_X)_p$ is ℓ-adjoint for all ℓ.

Proof: For $a \in (\mathcal{O}_X)_p$ we have $\pi_1(a\sigma_X^{\otimes\ell}) = \pi_1(a)\pi_1(\sigma_X^{\otimes\ell})$, which is regular at all points of $\pi^{-1}(p)$.

As a special case, which, however, is typical for the geometry of Zariski surfaces, let us study

$$(4.3.0)\quad F:\quad z^p = f_{p+1}(x,y) = \bar{f}_{p+1}(x,y) + \bar{f}_p(x,y) + \cdots + \bar{f}_0(x,y)$$

where the $\bar{f}_i$ are homogeneous forms of degree i in x, y and the polynomial $f_{p+1}(x,y)$ is "generic" in the following sense:

(a) $\bar{f}_{p+1}$ splits into distinct linear factors.

(b) $\partial f_{p+1}/\partial x = \partial f_{p+1}/\partial y = 0$ implies that the hessian of f_{p+1} is $\neq 0$. The locus "at infinity" is $z_0 = 0$, so that $\bar{f}_{p+1} = 0$. Thus the locus at infinity consists of p + 1 distinct projective lines $L_1 \cdots L_{p+1}$ meeting transversally (easy proof) at the single point $x = y = z_0 = 0$, $z = 1$. Moreover, it is easy to show that the intersection index of L_i and L_j is 1 if $i \neq j$.

LEMMA 4.4.0 A "generic surface" in the foregoing sense gives rise to an $\bar{F}$ with no singularities at infinity and with exactly p^2 rational singularities at finite distance $s_1, \ldots, s_{p^2}$ which are of the type described in theorem 4.1.0.

Proof: We show only that there are precisely p^2 singularities at finite distance.

Consider the two plane curves C_x: $\partial f_{p+1}/\partial x = 0$ and C_y: $\partial f_{p+1}/\partial y = 0$ (we use the notation of 4.3.0). Clearly, it suffices to show that C_x and C_y have p^2 points in common. Assumption (a) gives the fact that C_x and C_y are both of degree p and also that they do not meet at infinity. Thus C_x and C_y have no common components and they meet at finitely many points at finite distance. Let q be one such point. Using (b) and changing coordinates, we may assume that $q = (0,0)$ and that $f_{p+1}(x,y) = xy$ + higher-order terms. Clearly, C_x and C_y are smooth at q and have distinct tangent lines. Thus the intersection number at any common point q is 1. By Bezout's theorem, C_x and C_y have exactly p^2 points in common.

COROLLARY 4.5.0 There exists a nonsingular model $\tilde{F}$ of F, $\tilde{F} \xrightarrow{\pi} F$ such that for the p^2 singularities $s_1, s_2, \ldots, s_{p^2}$ of F, we have

$$\pi^{-1}(s_i) = E_{i,1} \cup \cdots \cup E_{i,p-1}$$

where each $E_{i,j}$ is isomorphic to the projective line P^1. Moreover, the intersection graph of the $E_{i1}, E_{i2}, \ldots, E_{i,p-1}$ is as follows:

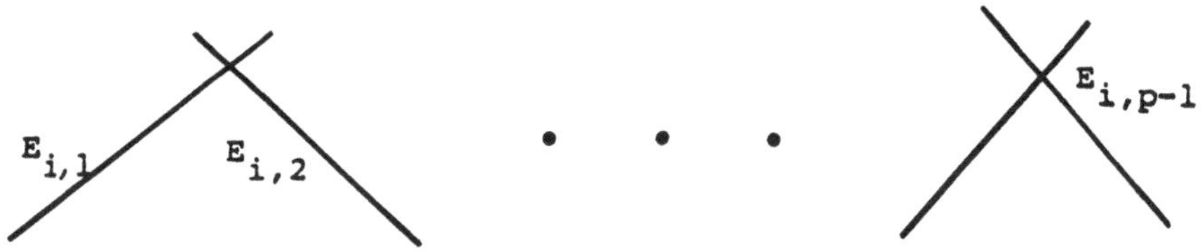

and the intersection matrix is

$$\begin{bmatrix} -2 & 1 & & & & & 0 \\ 1 & -2 & 1 & & & & \\ & 1 & \cdot & \cdot & & & \\ & & \cdot & \cdot & \cdot & & \\ & & & \cdot & \cdot & \cdot & \\ & 0 & & & \cdot & \cdot & 1 \\ & & & & & 1 & -2 \end{bmatrix} \quad (p - 1) \times (p - 1) \text{ matrix}$$

with determinant = p.

The fact that $E_{ij}^2 = -2$ will be shown below.

Proof: Let us write the equation of $\bar{F}$: $g(x_0,x_1,x_2,x_3)$, where we have relabeled the variables x, y, z, z_0 so that $\partial g/\partial x_0$, $\partial g/\partial x_1 \not\equiv 0$. Now we can apply results of Section 2, the projective case, directly (because 2.17.1 is satisfied). We still consider $\tilde{F} \to \bar{F}$ constructed by p^2 applications of theorem 4.1.0. Let σ_0, σ_1, σ_2, σ_3 be as in theorem 2.18.0.

Let A be any form in $k[\bar{F}]$ of degree d. Let us first observe that any form $A \in k[\bar{F}]$ is adjoint, indeed is ℓ-adjoint for all ℓ. To see this, let $s \in \bar{F}$ be a singular point; assume that $s \in F_i$. We know by proposition 4.2.2 that $\pi_1(\sigma_i)$ is regular along $\pi^{-1}(s)$ and $\neq 0$ there. Therefore, $\pi_1((A/x_i^d)\sigma_i)$ is also regular along $\pi^{-1}(s)$, so A is adjoint at s by 2.24.0. A similar argument using $\sigma_i^{\otimes \ell}$ shows that A is ℓ-adjoint for all ℓ. Let d = p - 3, and let us for any such A construct $\sigma(A) \in \omega_{\tilde{F}}(\bar{F})$ as in 2.21.0.

$$\sigma(A) = \frac{A}{x_0^d}\sigma_0 = \frac{A}{x_1^d}\sigma_1 = \frac{A}{x_2^d}\sigma_2 = \frac{A}{x_3^d}\sigma_3$$

We know from the proof of 2.21.0 and 2.22.0 that $\pi_1(\sigma(A))$ is a regular form on $\tilde{F}$.

Let us choose A and another form A', also of degree p - 3, so that (1) A and A' do not vanish at the singular points of $\bar{F}$, and (2) the surfaces $\bar{F}$, A = 0, and A' = 0 are in general position. Let us observe that the divisor of $\pi_1(\sigma(A))$, denoted $(\pi_1(\sigma(A))$ (see Zariski, 1964), and of $\pi_1(\sigma(A'))$ satisfy:

(i) $\pi_1(\sigma(A)) \cdot E_{ij} = 0.$

(ii) $\pi_1(\sigma(A)) \cdot \pi_1(\sigma(A')) = (p - 3)(p - 3)(p + 1).$

[We leave out the precise justification of (ii).]

Thus if $K_{\tilde{F}}$ denotes the canonical line bundle on $\tilde{F}$, we have

$$K_{\tilde{F}}^2 = (p - 3)^2(p + 1)$$

Also,

$$p_a(E_{ij}) = 0 = \frac{E_{ij} \cdot (K_{\tilde{F}} + E_{ij})}{2} + 1$$

(see Zariski, 1969, p. 50), which gives

$$1 + \frac{E_{ij} \cdot K_{\tilde{F}} + E_{ij}^2}{2} = 0$$

or

$$E_{ij}^2 = -2$$

LEMMA 4.6.0 Let H be a plane section of F which does not contain any of the singular points s_i, $i = 1, 2, \ldots, p^2$. Let $\tilde{H}$ be the pullback of H to $\tilde{F}$. We have $\tilde{H}^2 = p + 1$.

Proof: Let us choose H', H" linearly equivalent to H and in "general position." We see easily that $\tilde{H}' \cdot \tilde{H}'' = p + 1 =$ (number of intersections of F with a generic line). But $\tilde{H}^2 = \tilde{H}' \cdot \tilde{H}''$.

LEMMA 4.7.0 Let

$$\tilde{L}_i = \pi^{-1}(L_i)$$

Then

$$\tilde{L}_i^2 = 1 - p$$

Proof: $\tilde{L}_i \cdot \tilde{H} = 1$ (clear).

$$\tilde{H} \equiv \tilde{L}_1 + \tilde{L}_2 + \cdots + \tilde{L}_{p+1}$$

so

$$\begin{aligned}\tilde{L}_i \cdot \tilde{H} &= \tilde{L}_1 \cdot \tilde{L}_i + \tilde{L}_2 \cdot \tilde{L}_i + \cdots + \tilde{L}_i^2 + \cdots + \tilde{L}_{p+1} \cdot \tilde{L}_i \\ &= p + \tilde{L}_i^2 = 1\end{aligned}$$

so $\tilde{L}_i^2 = 1 - p$.

THEOREM 4.8.0 $\tilde{F}$ has the following invariants:

(a) $p_a = p_g = p(p - 1)(p - 2)/6$.

(b) $K_{\tilde{F}}^2 = (p - 3)^2(p + 1)$.

(c) ρ = rank of the Néron-Severi group of $\tilde{F} = \dim H_{et}^2(\tilde{F},Q_\ell) = b_2 = p^3 - p^2 + p + 1$.

(d) $|D|$ = |discriminant of the intersection form on the Néron-Severi group of $\tilde{F}$| is an even power of p less than or equal to $p^2 + p$.

Proof: (a) Since F has only rational singularities, it is known (see Artin, 1966; Lipman, 1969) that $p_a(\tilde{F}) = p_a(\bar{F})$, but

$$p_a(\bar{F}) = \frac{p(p - 1)(p - 2)}{6}$$

by Serre (1955), because $\bar{F}$ is a hypersurface of degree p + 1 in $\mathbf{P}^3$. Also,

$$p_a(\tilde{F}) = \dim H^2(\tilde{F},\mathcal{O}_{\tilde{F}}) - \dim H^1(\tilde{F},\mathcal{O}_{\tilde{F}})$$

$$p_g = \dim H^2(\tilde{F},\mathcal{O}_{\tilde{F}})$$

We know by the theory of adjoints that $p_g \leq$ dim of forms of degree $p - 3$ in $k[\bar{F}] \leq p(p - 1)(p - 2)/6$. Thus

$$\frac{p(p - 1)(p - 2)}{6} \geq p_g \geq p_a = \frac{p(p - 1)(p - 2)}{6}$$

(b) Proved above.

(c) Using étale Betti numbers, we have $b_1 = b_3 = 0$ since $q = 0$ (Milne, 1980, chap. 5). Also, $b_0 = b_4 = 1$.

The Noether formula says, therefore (see Milne, 1980, V.3.13), that $\chi_{et} + K_{\tilde{F}}^2 = 12(p_a + 1)$; hence $b_2 + 2 + K_{\tilde{F}}^2 = 12(p_a + 1)$ and $b_2 = p^3 - p^2 + p + 1$ follows from parts (a) and (b).

We have $\rho \leq b_2$ (Igusa's inequality; see Milne, 1980, V3.28). So it is enough to show that $\rho \geq p^3 - p^2 + p + 1$. But we have the following curves on $\tilde{F}$: E_{ij}, $1 \leq i \leq p^2$, $i \leq j \leq p - 1$, which gives $p^3 - p^2$ curves and also the curves $\tilde{L}_1, \ldots, \tilde{L}_2, \ldots, \tilde{L}_{p+1}$, so altogether, $p^3 - p^2 + p + 1$ curves.

Let us compute the intersection matrix of the E_{ij}'s and $\tilde{L}_i$'s. Since $\tilde{L}_i \cdot E_{ij} = 0$ and

$$E_{ij}E_{i'j'} = 0 \quad \text{if } i \neq i'$$

the matrix consists of $p^2 + 1$ blocks, namely, p^2 blocks as in 4.5.0 and the block of the L_i's

$$L = \begin{bmatrix} 1-p & & & & & \\ & 1-p & & & 1 & \\ & & \cdot & & & \\ & & & \cdot & & \\ & 1 & & & \cdot & \\ & & & & & 1-p \end{bmatrix} \qquad L \text{ is a } (p+1) \times (p+1) \text{ matrix}$$

A simple manipulation with determinants shows that the det $L = -p^p$, so the determinant of the whole matrix is

$$-p^{p^2} p^p = -p^{p^2+p}$$

Thus the curves are independent and

$$\rho = b_2 = p^3 - p^2 + p + 1$$

(d) Let $G \subsetneq NS(\tilde{F})$ be the subgroup generated by the E_{ij}'s and $\tilde{L}_i$'s. We know that G is of finite index t. If D is the discriminant of the intersection form on $NS(\tilde{F})$, we have by a standard property of bilinear forms

$$D \cdot t^2 = -p^{p(p+1)}$$

because we have shown that the intersection form restricted to G has discriminant $p^{p(p+1)}$. Therefore, $|D|$ and t are powers of p and

$$|D| = \frac{p^{p(p+1)}}{t^2} = p^{\alpha}$$

where α is even[†] and

$$\alpha < p(p+1)$$

[†]The fact that α has to be even follows from the fact proved by Milne that $|D|$ has to be a square or twice a square for a supersingular surface (Milne, 1976, p. 196).

REMARK 4.8.1 $\tilde{F}$ contains no exceptional curves of the first kind.

Proof: We use the notation of 4.5.0 and 4.8.0. First let us observe that the intersection form restricted to G is even (we assume that $p \geq 3$) since

$$E_{ij}^2 = -2$$

$$\tilde{L}_1^2 = 1 - p$$

Also, G is a subgroup of the Néron-Severi group of F of index which is a power of the odd prime p. Suppose that $E \in NS(\tilde{F})$ and $E^2 = -1$; then for some integer n,

$$p^n E \in G$$

so the self-intersection number $(p^n E \cdot p^n E) = p^{2n} E^2 = -p^{2n}$ has to be even. This is a contradiction. Therefore, $\tilde{F}$ contains no curves with self-intersection -1, so no exceptional curves of the first kind.

REMARK 4.8.2 $\tilde{F}$ is the minimal model of k(F).

Proof: Since $p \geq 3$, $p_g(\tilde{F}) > 0$. Therefore, $k(\tilde{F})$ is not the function field of a ruled surface. Therefore, $k(\tilde{F}) = k(F)$ has a unique minimal model (Zariski, 1958a, p. 79). Since $\tilde{F}$ contains no exceptional curves of the first kind, $\tilde{F}$ is a relatively minimal model and is isomorphic to the minimal model of k(F) (see Zariski, 1958b, proposition 3.1, and remarks after corollary 1.8, p. 153).

REMARK 4.9.0 (In the notation of 4.1.0) In the case when k is the algebraic closure of a finite field and X is defined over a finite field $k_0 \subsetneq k$, it is clear from the proof of theorem 4.1.0 that $\tilde{X}$ is also defined over k_0. Moreover, the curves E_i are also defined over k_0. Since the curves E_i are isomorphic to P^1 over k, they are also isomorphic to the projective line $P^1_{k_0}$.

It follows that if F is a surface as in 4.3.0 and if F together with all its singularities is defined over k_0, the desingularization

$\tilde{F}$ in 4.5.0 is also defined over k_0. Moreover, all the curves E_{ij} (4.5.0) are isomorphic to the projective line over k_0.

THEOREM 4.10.0 Let us assume here that k is the algebraic closure of a finite field. Let F be a surface as in 4.3.0 and $\bar{F}$ its closure in P^3. Let $\tilde{F}$ be the minimal desingularization of F as in 4.5.0. Let $k_1 \subseteq k$ be the smallest field such that all the coefficients of f(xy,y) are in k_1, over which all the singular points s_1, s_2, ..., s_{p^2} of F are defined, and such that f_{p+1} splits over k_1 into distinct linear factors. Then $\tilde{F}$ is defined over k_1 (see remark 4.9.0), and if q = cardinality of $k_1 = p^m$, the Weil zeta function of $\tilde{F}$ over k_1 is given by

$$Z(T) = \frac{1}{(1 - T)(1 - qT)^{p^3-p^2+p+1}(1 - q^2T)}$$

Proof: First, it is clear from the proof of theorem 4.1.0 that $\tilde{F}$ is defined over k_1. Let k_ν be a field with q^ν elements such that $k_1 \subseteq k_\nu \subseteq k$. Let us compute N_ν = the number of points of F rational over k_ν:

$$N_\nu = N_\nu^{(1)} + N_\nu^{(2)} + N_\nu^{(3)}$$

where $N_\nu^{(1)}$ = number of smooth points in F rational over k, $N_\nu^{(2)}$ = number of points of the curves

$$E_{ij}: \quad \begin{array}{l} 1 \le i \le p^2 \\ 1 \le j \le p - 1 \end{array}$$

rational over k_ν, and $N_\nu^{(3)}$ = number of points of $\bar{F}$ - F rational over k_ν.

$$N_\nu^{(1)} = q^{2\nu} - p^2$$

because for every value (x,y) in the affine plane over k we get exactly one point of F, and further, p^2 has to be subtracted to account for singular points. $N_\nu^{(2)} = p^2[(p - 1)q^\nu + 1]$ since the points entering into this sum arise from p^2 disjoint subschemes $E_i = \cup_{j=1}^{p-1} E_{ij}$, $i = 1, 2, \ldots, p^2$. E_i can be pictured as a union of projective lines over k_ν as follows:

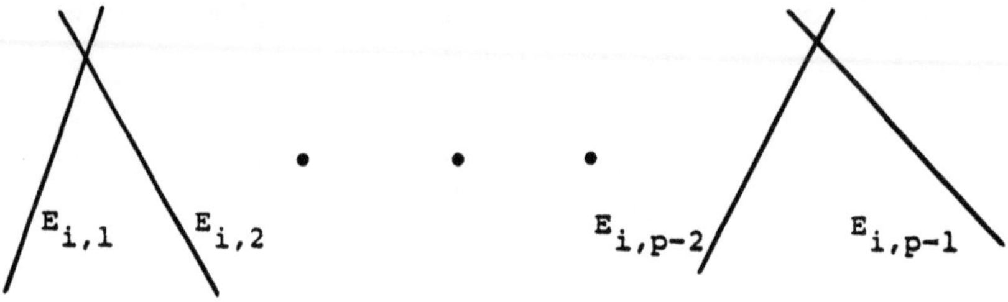

so E_i has $(p - 1)q^\nu + 1$ points. $N_\nu^{(3)} = (p + 1)q^\nu + 1$, since $\bar{F} - F$ consists of $p + 1$ projective lines all meeting in one point.

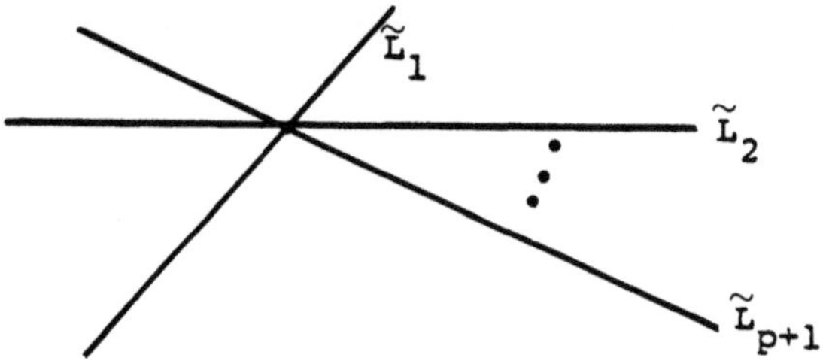

Thus

$$N_\nu = q^{2\nu} + (p^3 - p^2 + p + 1)q^\nu + 1$$

Therefore,

$$\sum_{\nu=1}^{\infty} N_\nu T^{\nu-1} = \frac{q^2}{1 - q^2T} + \frac{(p^3 - p^2 + p + 1)q}{1 - qT} + \frac{1}{1 - T}$$

$$\frac{d}{dT} \log \frac{1}{(1 - q^2T)(1 - qT)^{p^3-p^2+p+1}(1 - T)}$$

REMARK 4.10.1 The computation above provides another proof that $\rho = b_2 = p^3 - p^2 + p + 1$ in the case when k is the algebraic closure of a finite field. To see this, first note that the étale second

Betti number b_2 is equal to the exponent $p^3 - p^2 + p + 1$ in the zeta function. This fact follows from Deligne's proof of the Weil conjectures [see Deligne (1974) or the account in Hartshorne (1977, pp. 449-457, especially remarks following theorem 4.5)]. Then we can proceed exactly as in the proof of theorem 4.8.0, part (d).

5. RICHNESS OF THE CLASS OF ZARISKI SURFACES

Whereas in characteristic 0 all unirational surfaces are rational, the situation is far more interesting in characteristic $p > 0$.

We show in this section that there are infinitely many birationally distinct Zariski surfaces in characteristic $p \geq 5$ (corollary 5.1.1). We also prove a theorem (5.2.0) about the various possible types of surfaces that can be found among Zariski surfaces.† That theorem is meant to be merely a starting point toward the solution of the following general problem: For a given characteristic $p \geq 0$, what possible types of surfaces [from the Enriques classification table in Bombieri and Husemoller (1974)] can be found among Zariski surfaces over an algebraically closed field of characteristic p?

PROPOSITION 5.1.0 Let k be an algebraically closed field of characteristic $p \geq 5$. Let $m = p\ell + 1$, $\ell \geq 3$ an integer. Let F_ℓ: $z^p = x^m + y^m - x^2/2 - y^2/2$ be an affine surface, $\overline{F}_\ell$ its closure in projective space, $\tilde{F}_\ell$ a nonsingular model of $\overline{F}_\ell$. Then $p_g(\tilde{F}_\ell) \geq (p - 3)\ell$ and consequently, $p_g(\tilde{F}_\ell) \to +\infty$ as $\ell < +\infty$.

COROLLARY 5.1.1 There are infinitely many birationally distinct Zariski surfaces in characteristic $p \geq 5$.

Proof: The geometric genus is a birational invariant (Zariski, 1969, p. 33, proposition 9.6).

---

†This theorem was proved independently by Shioda (preprint).

COROLLARY 5.1.2 If $p \geq 5$, there are infinitely many fields between $k(x,y)$ and $k(x^{1/p}, y^{1/p})$ that are not isomorphic over k.

Proof: $k(x,y) \subsetneq k(\tilde{F}_\ell) \subsetneq k(x^{1/p}, y^{1/p})$ and $p_g(\tilde{F}_\ell)$ is a birational invariant (see Zariski, 1969, p. 33, proposition 9.6).

REMARK 5.1.3 If $k = \overline{Z/pZ}$, we can omit "over k" in the statement of corollary 5.1.2.

Proof of proposition 5.1.0: The Hessian with respect to x, y of the right-hand side of the equation of F_ℓ is identically equal to 1. Therefore, F has only rational double points by corollary 4.2.1.1. $\bar{F}$, the closure of F in projective space, is given by

$$\bar{F}: \quad z_0^{m-p} z^p = x^m + y^m - \frac{1}{2} z_0^{m-2}(x^2 + y^2)$$

The only singular point "at infinity" is $x = y = z_0 = 0$, $z = 1$. An affine neighborhood of that point is given by

$$F_0: \quad z_0^{m-p} = x^m + y^m - \frac{x^2 + y^2}{2} z_0^{m-2}$$

To find the adjoints in this case, instead of the usual resolution technique, we use another method which originated in a remark made by M. Hochster; namely, in the local ring at $p = (0,0,0)$ on F_0, the ideal of adjoints is integrally closed and it contains the three partials of the equation of F_0 (see 2.36.0 and 2.15.0).

This remark will enable us to find sufficiently many adjoints to prove that $p_g \to +\infty$ without studying explicitly the resolution of F_0.[†]

LEMMA 5.1.4 Let $p = (0,0,0)$ be the singular point of F_0. Let $\mathcal{O}_p$ be the corresponding local ring, $\mathcal{O}_p \subset k(F_0) = k(x,y,z_0)$. If a, b, c are nonnegative integers such that

[†]My original proof was by studying the resolution. That is substantially longer. It is important, for both proofs, to notice that the resolution of the singularity of F_0 at $p = (0,0,0)$ exists by Abhyankar (1956).

$$\frac{a}{m-1} + \frac{b}{m-1} + \frac{c}{m-p-1} \geq 1$$

the monomial $x^a y^b z_0^c$ is adjoint in $\mathcal{O}_p$.

Proof: We know that the partial derivatives of the equation of F_0 are adjoint in $\mathcal{O}_p$ and also that adjoints in $\mathcal{O}_p$ form an ideal. To compute the partials, let us write

$$E_0(x,y,z_0) = z_0^{m-p} - x^m - y^m + z_0^{m-2} \frac{x^2 + y^2}{2}$$

Now $\partial E_0/\partial z_0 = z_0^{m-p-1} \cdot$ (unit in $\mathcal{O}_p$). Therefore, z_0^{m-p-1} is adjoint. $\partial E_0/\partial x = x^{m-1} - xz_0^{m-2}$, so since z_0^{m-2} is a multiple of z_0^{m-p-1}, x^{m-1} is also an adjoint; similarly, y^{m-1} is an adjoint.

Now let $x^a y^b z^c$ be a monomial with

$$\frac{a}{m-1} + \frac{b}{m-1} + \frac{c}{m-p-1} \geq 1 \qquad (*)$$

Then we claim that $x^a y^b z^c$ is integrally dependent on the ideal $(x^{m-1}, y^{m-1}, z_0^{m-p-1})$, therefore also on the larger ideal of adjoints in $\mathcal{O}_p$.

To see the dependence, let us note that

$$(x^a y^b z^c)^{(m-1)(m-1)(m-p-1)}$$
$$= (x^{m-1})^{(m-1)(m-p-1)a} (y^{m-1})^{(m-1)(m-p-1)b} (z^{(m-p-1)})^{c(m-1)(m-1)}$$

Because of (*) we have

$$a(m-1)(m-p-1) + b(m-1)(m-p-1) + c(m-1)(m-1)$$
$$\geq (m-1)(m-1)(m-p-1)$$

which shows the integral dependence. Since $x^a y^b z_0^c$ is integrally dependent on the ideal of adjoints in $\mathcal{O}_p$, $x^a y^b z_0^c$ has to be actually an adjoint, by 2.36.0.

We return to the proof of the proposition: The monomial $x^a y^b z_0^c$ therefore defines an adjoint form to $\bar{F}_\ell$ of degree $a + b + c$. Let us take $a + b + c = m - 4 = \ell p - 3$ and a, b, c satisfying (*); then $x^a y^b z_0^c$ is an adjoint form of degree $m - 4$ (it is adjoint at all the singular points of F because of corollary 4.2.3) and by theorem 2.23.0 the k-vector space of such forms has dimension equal to $p_g(\tilde{F}_\ell)$.

Since monomials of degree m - 4 are distinct and linearly independent over k in the graded ring $k[\overline{F}_\ell]$, to prove our proposition it suffices to show that the number of distinct monomials of degree m - 4 satisfying (*) tends to infinity with ℓ. This can be shown by the following computation. Take $x^a y^b z_0^c$; assume that $a + b + c = m - 4 = p\ell - 3$, and take c to be any integer with $p\ell - 3 \geq c \geq 3\ell - 3$ (note that $p > 3$). It is clear that the number of such monomials is greater than $(p - 3)\ell$, and consequently it tends to $+\infty$ with ℓ. Let us therefore show that all such monomials satisfy (*).

We have

$$\begin{aligned}\frac{a}{m-1} + \frac{b}{m-1} + \frac{c}{m-p-1} &= \frac{a+b}{m-1} + \frac{c}{m-p-1}\\ &= \frac{m-4-c}{m-1} + \frac{c}{m-p-1}\\ &= 1 - \frac{3}{m-1} - \frac{c}{m-1} + \frac{c}{m-p-1}\\ &= 1 - \frac{3}{m-1} + c\left(\frac{1}{m-p-1} - \frac{1}{m-1}\right)\\ &\geq 1 - \frac{3}{m-1} + (3\ell - 3)\left(\frac{1}{m-p-1} - \frac{1}{m-1}\right)\\ &= 1 + \frac{3\ell - 3}{m-p-1} - \frac{3\ell}{m-1}\\ &= 1 + \frac{(3\ell - 3)(m-1) - 3\ell(m-p-1)}{(m-p-1)(m-1)} = 1\end{aligned}$$

since $m = p\ell + 1$.

REMARK 5.1.5 By computing the resolution explicitly, one can show that the theorem remains true for $p = 3$.

REMARK 5.1.6 We conjecture that p_g is also unbounded in the class of Zariski surfaces in the case of characteristic 2.

THEOREM 5.2.0[†] The following classes of surfaces are represented among Zariski surfaces.

[†]This fact was observed independently by Shioda (preprint).

(a) Surfaces of general type if the characteristic of the field is $p \geq 5$

(b) K3 surfaces if $p = 3$

(c) Rational surfaces for all p

Proof: (a) Take $p \geq 5$ and let F: $z^p = f(x,y)$ with f generic as in 4.4.0. Let $\tilde{F}$ be the nonsingular model of k(F) constructed in 4.5.0.

We have $(K_{\tilde{F}})^2 = (p - 3)(p + 1) \geq 1$ and P_2 of $\tilde{F}$, which is the number of biadjoint forms of degree $2(p - 3)$, satisfies $P_2 \geq 2$, since it follows from corollary 4.2.3 that every form of degree 2p - 6 is biadjoint. The fact that $\tilde{F}$ is of general type follows from Bombieri (1973, p. 176, theorem 1).

(b) Let k have characteristic $p = 3$. Consider F: $z^p = f_4(x,y)$ with f again generic, and let $\tilde{F}$ be the nonsingular model of k(F) as in 4.5.0. We have $\dim H^0(\tilde{F},\omega_{\tilde{F}}^{\otimes \ell})$ = [dimension of the space of ℓ-adjoint forms of degree $\ell(4 - 4) = 0$]. Because of corollary 4.2.3, $H^0(\tilde{F},\omega_{\tilde{F}}^{\otimes \ell}) \cong k$ for all ℓ. We see that the canonical ring of $\tilde{F}$ has transcendence degree 1 over k. We also have that the second étale Betti number b_2 of $\tilde{F}$ is 22 [see 4.8.0(c)] and F contains no exceptional curves of the first kind (see 4.8.1). Therefore, $\tilde{F}$ is a K3 surface by the table of Bombieri-Husemoller (1975, p. 373).

(c) In fact, it follows from definition 1.1.0 that every smooth rational surface in characteristic $p > 0$ is a ZS.

Acknowledgments: I owe a great debt of gratitude to my adviser, J. S. Milne, who helped me revive this thesis. I wish to thank M. Hochster for his hard work and hearty encouragement. O. Zariski's ideas, conversations with him, and his letters have been invaluable. I am grateful to H. Hironaka, who first suggested the problem, and Bill Haboush, who first taught me the language of schemes. Also, the following mathematicians have been particularly helpful through their conversation and letters: S. S. Abhyankar, A. Bauville, R. Berndt, S. Bloch, Y. Clifton, P. Deligne, J. Joel, D. J. Lewis, D. Mumford, R. Randall, F. Raymond, T. Shioda, F. Sullivan, T. Suwa, and J. Tate. In conclusion, I thank the referee for his suggestions and Jim Sturnfield for his careful reading.

REFERENCES

Abhyankar, S. S., Local uniformation on algebraic surfaces over ground fields of characteristic p ≠ 0, *Ann. Math.*, 63 (1956), 491-526.

Artin, M., On isolated rational singularities of surfaces, *Am. J. Math.*, 88 (1966), 129-136.

Blass, P., Zariski surfaces, thesis, University of Michigan, 1977.

Blass, P., and J. Lipman, Remarks on adjoints and arithmetic genera of algebraic varieties, *Am. J. Math.*, 101, No. 2 (1979), 331-336.

Bombieri, E., Canonical models of surfaces of general type, *Publ. Math. IHES*, No. 42 (1973).

Bombieri, E., and D. Husemoller, Classification and embeddings of surfaces. *Proc. Symp. Pure Math.*, 29 (Arcata), American Mathematical Society, Providence, R.I., 1975, p. 373, table.

Deligne, P., La conjecture de Weil, I, *Publ. Math. IHES,* 43 (1974), 273-307.

Enriques, F., *Le Supeficie Algebriche*, Nicola Zanichelli, Ed., Bologna, 1949.

Enriques, F., Sopra le superficie algebriche di Bigenere uno, Memorie Scelte Di Geometria, Vol. 2, p. 241.

Grauert, H., and O. Riemenschneider, Verschwindungssätze fur analytische Kohomologiegruppen auf komplexen Räumen, *Invent. Math.*, 11, Fasc. 4 (1970), 263-292.

Grothendieck, A., *SGA I*, Lecture Notes in Mathematics, No. 224, Springer-Verlag, New York, 1971.

Hartshorne, R., *Algebraic Geometry*, Graduate Texts in Mathematics, Springer-Verlag, New York, 1977.

Lang, S., *Abelian Varieties*, Interscience Tracts in Pure and Applied Mathematics, No. 7, Wiley, New York, 1959.

Lang, W. E., Quasi-elliptic surfaces in characteristic three, thesis, Harvard University, 1978.

Lipman, Joseph, Rational Singularities, *Publ. Math. IHES*, No. 36 (1969), 195-279.

Milne, J. S., Duality in the flat cohomology of a surface, *Ann. Sci. Ec. Norm. Sup.* ser. 4, vol. 9 (1976), 171-202, theorem 4.5.

Milne, J. S., *Étale Cohomology*, Princeton Mathematical Series, No. 33, Princeton University Press, Princeton, N.J., 1980.

Mumford, D., *Lectures on Curves on an Algebraic Surface*, Annals of Mathematics Studies, No. 59, Princeton University Press, Princeton, N.J., 1966.

Mumford, D., Introduction to algebraic geometry (preliminary version of the first three chapters), Harvard Lecture Notes, p. 335, Harvard University Press, Cambridge, Mass., 1966.

Nagata, Masayoshi, *Local Rings*, Interscience Tracts in Pure and Applied Mathematics, No. 13, Wiley, New York, 1962.

Nakai, Y., On the characteristic linear systems of algebraic families, *Ill. J. Math.*, 1 (1957), 552-561.

Šafarevič, I., Algebraic surfaces, *Proc. Steklov Inst. Math.*, 75, 1965a.

Šafarevič, I., *Proc. Steklov Inst. Math.*, 75, 1965b [English translation of Šafarevič (1965a)].

Šafarevič, I., *Minimal Models*, Tata Institute of Fundamental Research, Bombay, 1966.

Šafarevič, I., *Basic Algebraic Geometry*, Springer-Verlag, New York, 1974.

Serre, J. P., Faisceaux algebrique coherent, *Ann. Math.*, 61 (1955), 197-278.

Shioda, T., An example of unirational surfaces in characteristic p, *Math. Ann.*, 211 (1974), 233-236.

Zariski, Oscar, Introduction to the problem of minimal models in the theory of algebraic surfaces, *Publ. Math. Soc. Japan*, No. 4 (1958a).

Zariski, Oscar, The problem of minimal models in the theory of algebraic surfaces, *Am. J. Math.*, 80 (1958b), 153.

Zariski, Oscar, On Castelnuovo's criterion of rationality $p_a = g_g = 0$ of an algebraic surface, *Ill. J. Math.*, 2, No. 3 (1958c), 303 (especially p. 314).

Zariski, Oscar, *An Introduction to the Theory of Algebraic Surfaces*, Lecture Notes in Mathematics, No. 83, Springer-Verlag, New York, 1969.

APPENDIX 1: REMARKS ON ADJOINTS AND ARITHMETIC GENERA OF ALGEBRAIC VARIETIES†

Let $\mathbf{P}^{n+1}$ be projective (n + 1)-space over a fixed algebraically closed field k, let $F \subset \mathbf{P}^{n+1}$ be a (reduced) hypersurface of degree d, and let f: $\tilde{F} \to F$ be a desingularization. A classical characterization of the arithmetic genus of $\tilde{F}$:

$$p_a(\tilde{F}) = (-1)^n[\chi(\mathcal{O}_{\tilde{F}}) - 1] = h^n(\mathcal{O}_{\tilde{F}}) - h^{n-1}(\mathcal{O}_{\tilde{F}}) + \cdots + (-1)^{n-1}h^1(\mathcal{O}_{\tilde{F}})$$

---

†By Piotr Blass and Joseph Lipman (Purdue University, West Lafayette, Indiana). Reprinted with slight modifications from the *American Journal of Mathematics*, April 1979, 331-336.

is that

(#) $p_a(\tilde{F}) - 1$ is the virtual (or "postulated") dimension of the linear system of hypersurfaces in $\mathbb{P}^{n+1}$ having degree $d - n - 2$ and adjoint to F.

[For the case n = 2, cf. Zariski (1971, p. 73); for n = 3, see Zariski (1952, p. 590, footnote 13).]

Our main observation is that this characterization is an immediate corollary of duality theory and the Grauert-Riemenschneider vanishing theorem, the latter being valid when the characteristic of k is zero (Grauert and Riemenschneider, 1970), and if $n \geq 2$, also for characteristic > 0 [combine Wahl (1975, proposition 2.6) with Lipman (1978, theorem 2.3)].

To see this, we restate the characterization in a form which depends only on F (not its embedding in $\mathbb{P}^{n+1}$), and is in fact meaningful for any complete algebraic variety (reduced and pure n-dimensional).

Let $\omega_{\tilde{F}}$ be a dualizing sheaf on $\tilde{F}$, and let ω_F be a dualizing sheaf on F (ω_F is determined up to isomorphism by its property of representing the functor $\mathrm{Hom}_k(H^n(\mathcal{L}), k)$ of coherent $\mathcal{O}_F$-modules $\mathcal{L}$, cf. Hartshorne (1977, chap. III, sec. 7]. Also, $\omega_F = H^{-n}(\mathcal{R}_F)$, where $\mathcal{R}_F$ is a residual complex on F (Hartshorne, 1966, chap. VI). Since $\mathcal{O}_{\mathbb{P}^{n+1}}(-n - 2)$ is a dualizing sheaf on $\mathbb{P}^{n+1}$, the "adjunction formula" gives

$$\omega_F \cong \mathrm{Ext}^1_{\mathbb{P}^{n+1}}(\mathcal{O}_F, \mathcal{O}_{\mathbb{P}^{n+1}}(-n - 2)) \cong \mathcal{O}_F(d - n - 2)$$

There is a natural injective trace map†

$$\tau\colon\ f_*(\omega_{\tilde{F}}) \to \omega_F$$

†τ can be obtained from Hartshorne (1966, p. 369, theorem 2.1). In more down-to-earth terms, τ can be described (sketchily) as follows. Factor f as $\tilde{F} \xrightarrow{g} F_1 \xrightarrow{h} F$, where F_1 is the normalization of F. Over any affine open set $U \subseteq F$, the sections $\Gamma(U, \omega_F)$ can be identified with certain meromorphic n-differentials on F (see Kunz, 1975),

and the adjoint ideal $A \subseteq \mathcal{O}_F$ may be defined to be the annihilator of the cokernel C of τ, i.e.,

$$A = \mathrm{Hom}_{\mathcal{O}_F}(\omega_F, f_*(\omega_{\tilde{F}})) = \mathrm{Hom}_{\mathcal{O}_F}(\mathcal{O}_F(d - n - 2), f_*(\omega_{\tilde{F}}))$$
$$= (f_*(\omega_{\tilde{F}}))(-d + n + 2)$$

Let $\pi\colon \mathcal{O}_{\mathbb{P}^{n+1}} \to \mathcal{O}_F$ be the natural map, and let $A' = \pi^{-1}(A)$, so that we have an exact sequence

$$0 \to (\text{kernel of } \pi) = \mathcal{O}_{\mathbb{P}^{n+1}}(-d) \to A' \to A \to 0^{\dagger\dagger} \qquad (*)$$

What (#) says then is that

$$(-1)^n[\chi(\mathcal{O}_{\tilde{F}}) - 1] = \chi(A'(d - n - 2))$$
$$= \chi(A(d - n - 2)) + \chi(\mathcal{O}_{\mathbb{P}^{n+1}}(-n - 2))$$
$$= \chi(A(d - n - 2)) + (-1)^{n+1}$$
$$= \chi(f_*(\omega_{\tilde{F}})) + (-1)^{n+1}$$

and finally Serre duality gives

---

from which one also sees that

$$\Gamma(U, h_*\omega_F) \subseteq \Gamma(U, \omega_F)$$

Furthermore, $\Gamma(U, h_*\omega_F)$ consists of meromorphic n-forms having no polar divisors in $h^{-1}(U)$, while $\Gamma(U, f_*\omega_{\tilde{F}})$ consists of meromorphic n-forms without poles in $g^{-1}h^{-1}(U)$; thus

$$\Gamma(U, f_*\omega_{\tilde{F}}) \subseteq \Gamma(U, h_*\omega_{F_1}) \subseteq \Gamma(U, \omega_F)$$

and τ is just the resulting inclusion map.

††So the geometric genus $p_g(\tilde{F}) = H^0(f_*(\omega_{\tilde{F}}))$ satisfies

$$p_g(\tilde{F}) - 1 = H^0(A(d - n - 2)) - 1 = H^0(A'(d - n - 2)) - 1$$

[cf. (*)], which is the *actual* dimension of the linear system of hypersurfaces of degree $d - n - 2$ adjoint to F.

$$(-1)^n \chi(\mathcal{O}_{\tilde{F}}) = \chi(\omega_{\tilde{F}})$$

Thus, (#) simply says that

(##) $\chi(\omega_{\tilde{F}}) = \chi(f_*(\omega_{\tilde{F}}))$

But (##) holds for any reduced pure n-dimensional variety F proper over k, with $n \leq 2$ when k has characteristic > 0 [for such F a desingularization f: $\tilde{F} \to F$ always exists (Hironaka, 1964; Abhyankar, 1956)]. For the Leray spectral sequence for f gives

$$\chi(\omega_{\tilde{F}}) = \sum_{i=0}^{n-1} (-1)^i \chi(R^i f_*(\omega_{\tilde{F}}))$$

and the vanishing theorem says that $R^i f_*(\omega_{\tilde{F}}) = 0$ for $i > 0$. [In fact the spectral sequence degenerates,, so that $H^i(\tilde{F},\omega_{\tilde{F}}) = H^i(F, f_*\omega_{\tilde{F}})$ for all i]. Q.E.D.

REMARKS (1) The forms of degree j adjoint to F form a finite dimensional k-vector space

$$A_j = H^0(\mathbb{P}^{n+1}, A'(j)) \subsetneq H^0(\mathbb{P}^{n+1}, \mathcal{O}_{\mathbb{P}^{n+1}}(j)) = S_j \qquad (0 \leq j < \infty)$$

and $A = \oplus_{j=0}^{\infty} A_j$ is an ideal in the polynomial ring

$$k[X_0, X_1, \ldots, X_{n+1}] = \bigoplus_{j=0}^{\infty} S_j$$

The dimension A_j is given, for sufficiently large j, by the polynomial

$$H_A(j) = \chi(A'(j))$$

and so (#) says that

$$p_a(\tilde{F}) = H_A(d - n - 2)$$

All this begs the question: How to calculate A, or at least H_A?

In general, there is no easy answer. In case $\tilde{F}$ is the normalization of F (e.g., in the classical situation where F is a generic projection of $\tilde{F}$ into $\mathbb{P}^{n+1}$), we have

$$f_*(\omega_{\tilde{F}}) = \mathrm{Hom}_{\mathcal{O}_F}(f_*(\mathcal{O}_{\tilde{F}}),\omega_F)$$
$$= [\mathrm{Hom}_{\mathcal{O}_F}(f_*(\mathcal{O}_{\tilde{F}}),\mathcal{O}_F)](d - n - 2)$$

and τ is "evaluation at 1" [see Hartshorne, 1966, p. 319, (c); or use Kunz (1975) as in footnote †† on p. 103]. Hence

$$A = \mathrm{Hom}_{\mathcal{O}_F}(f_*(\mathcal{O}_{\tilde{F}}),\mathcal{O}_F)$$

which is the *conductor* of $\mathcal{O}_{\tilde{F}}$ in $\mathcal{O}_F$. Thus, in this case, adjoints coincide with *subadjoints*, which are more or less calculable (see Zariski, 1971, pp. 71-72).

(2) For any reduced pure m-dimensional Cohen-Macaulay variety F, proper over k, by Serre-Grothendieck duality $\chi(\mathcal{O}_F) = (-1)^m\chi(\omega_F)$. So given a desingularization f: $\tilde{F} \to F$, we have, using Grauert-Riemenschneider as above,

$$p_a(F) - p_a(\tilde{F}) = (-1)^m(\chi(\mathcal{O}_F) - \chi(\mathcal{O}_{\tilde{F}})) = \chi(\omega_F) - \chi(\omega_{\tilde{F}})$$
$$= \chi(\omega_F) - \chi(f_*\omega_{\tilde{F}})$$
$$= \chi(C)$$

(C is the cokernel of the injective map τ above). This generalizes formula (**) on p. 153 of Lipman (1978).

If, in particular, $F \subseteq \mathbb{P}^{n+1}$, and F is such that the dualizing sheaf ω_F is $\mathcal{O}_F(D - n - 2)$ for some D (e.g., if the vertex of the projecting cone over F is Gorenstein; in particular if F is a complete intersection of hypersurfaces of degrees $d_1, \ldots, d_{n+1-m}$, we can take $D = d_1 + \cdots + d_{n+1-m}$), then, proceeding as above, we have

$$C = \mathcal{O}_{\mathbb{P}^{n+1}}(D - n - 2)/A'(D - n - 2)$$

so that

$$(\#)' \quad p_a(F) - p_a(\tilde{F}) = \chi(C)$$
$$= \chi(\mathcal{O}_{\mathbb{P}^{n+1}}(D - n - 2)) - \chi(A'(D - n - 2))$$
$$= \binom{D - 1}{n + 1} - \chi(A'(D - n - 2))$$

(If F is a hypersurface of degree D then

$$p_a(F) = \binom{D-1}{n+1}$$

and (#)' reduces to (#).)

(3) Let F be as in (2), with $m \leq 2$ if k has characteristic > 0. If F is normal and has only isolated singularities, then $\mathcal{C}$ has zero-dimensional support (i.e., on the singular points) and

$$\chi(\mathcal{C}) = \sum_{x \text{ singular}} \dim_k(\mathcal{C}_x)$$

Here is another "dual" description of $\dim_k(\mathcal{C}_x)$. Let U be an affine neighborhood of a singular point x, containing no other singular point, let $\tilde{U} = f^{-1}(U)$, $E = f^{-1}(x)$, and assume (without real loss of generality) that f induces an isomorphism $\tilde{U} - E \xrightarrow{\sim} U - x$. We have an exact sequence

$$\begin{aligned} 0 \to H^0_E(\omega_{\tilde{F}}) &\to H^0(\tilde{U},\omega_{\tilde{F}}) \to H^0(U - x,\omega_{\tilde{F}}) \\ &\to H^1_E(\omega_{\tilde{F}}) \to H^1(\tilde{U},\omega_{\tilde{F}}) = 0 \end{aligned}$$

(the last equality by Grauert-Riemenschneider). Now

$$H^0(U - x,\omega_{\tilde{F}}) = H^0(U - x,\omega_F) = H^0(U,\omega_F)$$

because, F being normal, ω_F is the sheaf of meromorphic m-forms with no polar divisors on F.

We conclude that

$$\mathcal{C}_x \cong H^1_E(\omega_{\tilde{F}})$$

But by Lipman (1978, p. 188), $H^1_E(\omega_{\tilde{F}})$ is dual to the stalk $R^{m-1}f_*(\mathcal{O}_{\tilde{F}})_x$. Thus

$$\mathcal{C}_x \cong \operatorname{Hom}_k(R^{m-1}f_*(\mathcal{O}_{\tilde{F}})_x, k)$$

and so

$$\dim_k(\mathcal{C}_x) = \dim_k(R^{m-1}f_*(\mathcal{O}_{\tilde{F}})_x)$$

[This is essentially Theorem A in Yau (1977).]

Acknowledgment: Supported by NSF grant MCS76-8134.

REFERENCES

Abhyankar, S. S., Local uniformization on algebraic surfaces over ground fields of characteristic p $\neq$ 0, *Ann. Math.*, 63 (1956), 491-526.

Grauert, H., and O. Riemenschneider, Verschwindungssätze für analytische Kohomologiegruppen auf komplexen Räumen, *Invent. Math.*, 11, Fasc. 4 (1970), 263-292.

Hartshorne, R., *Residues and Duality*, Lecture Notes in Mathematics, No. 20, Springer-Verlag, New York, 1966.

Hartshorne, R., *Algebraic Geometry*, Graduate Texts in Mathematics, Springer-Verlag, New York, 1977.

Hironaka, H., Resolution of singularities of an algebraic variety over a field of characteristic zero, *Ann. Math.*, 79 (1964), 109-326.

Kunz, E., Holomorphe Differentialformen auf algebraischen Varietäten mit Singularitäten I, *Manuscripta Math.*, 15 (1975), 91-108.

Lipman, J., Desingularization of two-dimensional schemes, *Ann. Math.*, 107 (1978), 151-207.

Wahl, J., Vanishing theorems for resolutions of surface singularities, *Invent. Math.*, 31 (1975), 17-41.

Yau, S. S. T., Two theorems on higher dimensional singularities, *Math. Ann.*, 231 (1977), 55-59.

Zariski, O., *Algebraic Surfaces* (second supplemented edition), Springer-Verlag, New York, 1971.

Zariski, O., Complete linear systems on normal varieties and a generalization of a lemma of Enriques-Severi, *Ann. Math.*, 55 (1952), 552-592.

APPENDIX 2: COMPUTATION OF THE CONDUCTOR OF THE ALGEBRAIC SURFACE $Z^a = X^bY^c$†

Introduction

In this appendix we compute explicitly the ideal of adjoints in the coordinate ring of the algebraic surface V: $Z^a = X^bY^c$. The problem was proposed by O. Zariski (1971), who also solved the case

---

†By Piotr Blass and J. Blass (Bowling Green State University, Bowling Green, Ohio). Reprinted from Revue Roumaine de Mathematiques Pures et Appliquées, Tome XXII, No. 7 (1982), pp. 721-730.

$\max(\min(a,b),\min(a,c)) = 2$.

In the first two sections we find explicitly the normal model $\bar{V}$ of V. Then we compute the conductor of $\bar{V}$. Finally, we show that under our assumptions the conductor of $\bar{V}$ coincides with the ideal of adjoints in the coordinate ring of V (provided that the ground field has characteristic zero).

THEOREM 1 Let V: $Z^a = X^bY^c$ be an algebraic surface with $(a,b) = (a,c) = 1$. Assume also that a is not divisible by the characteristic of the ground field. Let K and L be the unique integers $1 \le K$, $L \le a - 1$ such that bK and cL are congruent to 1 modulo a. Define m and n by corresponding equations,

$$bK = ma + 1 \qquad cL = na + 1$$

and set

$$\alpha = \frac{Z^K}{X^m} \qquad \beta = \frac{Z^L}{Y^n}$$

For $1 \le i \le a - 1$ we can write $Z^i = (X^{m_i}/Y^{\bar{m}_i})\alpha^{K_i} = (Y^{n_i}/X^{\bar{n}_i})\beta^{L_i}$, where $1 \le K_i, L_i \le a - 1$, and $m_i, \bar{m}_i, n_i, \bar{n}_i$ are natural numbers. Then the integral closure of $\kappa[X,Y]$ in the function field $\kappa(V)$ of the surface V: $Z^a = X^bY^c$ is generated over $\kappa[X,Y]$ by

$$1, \frac{Z}{X^{m_1}Y^{n_1}}, \frac{Z^2}{X^{m_2}Y^{n_2}}, \ldots, \frac{Z^{a-1}}{X^{m_{a-1}}Y^{n_{a-1}}}$$

THEOREM 2 In the notation of theorem 1, if the characteristic of κ is zero, the ideal of adjoints in the coordinate ring of V is generated by

$$X^{m_{a-1}}Y^{n_{a-1}}, ZX^{m_{a-2}}Y^{n_{a-2}}, \ldots, Z^{a-2}X^{m_1}Y^{n_1}, Z^{a-1}$$

Preliminaries and Notation

Let V be the algebraic surface $Z^a = X^bY^c$. We assume that $(a,b) = (a,c) = 1$.

We will be using the following notation: κ is a field; $A = \kappa[X,Y]$; $\kappa[V] = \kappa[X,Y,Z]/(Z^a - X^bY^c)$; B will denote the integral closure of A (or of $\kappa[V]$) in the field of fractions of $\kappa[V]$ denoted $\kappa(V)$. We will also assume that $\kappa(V)$ is a separable extension of $\kappa(X,Y)$, so that $a \neq 0$ in κ. Let ρ: Spec B $\to$ Spec $\kappa[V]$ be the normalization map and let π: $\tilde{V} \to$ Spec B be a desingularization.

Following Artin (1966), we will say that Spec B has *only rational singularities* iff $R^1\pi_*\mathcal{O}_{\tilde{V}} = 0$. We will denote by σ the rational differential form $\alpha = dx\ dz/az^{a-1}$.

The conductor of B in $\kappa[V]$ is denoted C, $C = \{\alpha \in \kappa[V] : \alpha B \subset \kappa[V]\}$. Further, $f \in k(V)$ is called *subadjoint* if and only if $f\sigma$ has no polar curves (is regular) on Spec B, that is, if and only if f belongs to the conductor of B in $\kappa[V]$ (see Zariski, 1969, pp. 81-85); f is called *adjoint* if and only if f has no polar curves on $\tilde{V}$ (see Zariski, 1969, pp. 91-92). We also see easily that the set of adjoints denoted $\mathcal{A}$ forms an ideal in $\kappa[V]$ and that $\mathcal{A} \subset C$.

If $M \subset \kappa(V)$ is an A-module we will denote by D(M) the discriminant of M with respect to the trace function. In particular, if $M = \langle 1,\alpha,\ldots,\alpha^{n-1}\rangle$, we will use $D(\alpha)$ to denote D(M).

In the first part we show that B is a free A-module. In the second part we find a basis of B over A. The basic idea of calculations comes from Berwick (1929).

B is a Free A-Module

Since B is an integrally closed noetherian domain of dimension 2, B is a Macaulay ring (Kaplansky, 1970, p. 104, problem 25). Let $M \subset B$ be a maximal ideal. Set $m = M \cap A$. By arguments quoted above, B_M is a Macaulay ring. Since B_M is finitely generated over A_m, it is A_M-free (see Vascensalos, 1967). This implies that B is a projective A-module, and thus by the theorem of Seshadri (1958), it is a free A-module.

REMARK 1 By using similar arguments, one can prove that an integral closure A_L of a polynomial ring $A = \kappa[X_1,\ldots,X_n]$ in a finite separable extension L of $\kappa(X_1,\ldots,X_n)$ is a free A-module whenever A_L is Macaulay.

Computation of B

LEMMA 2.1 Let $h(X,Y)$ be an irreducible element of A. Assume that $g(u)$ is an element of a polynomial ring $A[u]$ such that

(i) $g(z)/h(X,Y) \in B$
(ii) $g(u) \not\equiv 0 \mod h(X,Y)$

Then there is a polynomial $g_1(u) \subset A[u]$ such that $g_1(z)/h(X,Y) \in B$, the highest coefficient of $g_1(u)$ is not divisible by h and degree $g_1 \leq$ degree g. Moreover, $g_1(u) \equiv g(u) \mod h(X,Y)$.

Proof: $g(u) = h_p u^p + \cdots + h_1 u + h_0$. Assume that $h|h_p$, $h|h_{p-1}$, ..., $h|h_{q+1}$, but $h \nmid h_q$. Then $g(u) = (h_p u^p + \cdots + h_{q+1}u^{q+1}) + g_1(u)$. Note that $g(z)/h = (\text{element of B}) + g_1(z)/h$.

LEMMA 2.2 If there is an element of B in the form $g(z)/h(X,Y)$ such that

(i) $g(u) \in A[u]$; degree $g < a$
(ii) $h(X,Y)$ is irreducible and monic in A
(iii) $g(u) \not\equiv 0 \mod h(X,Y)$

then $h(X,Y) = X$ or $h(X,Y) = Y$.

Proof: It is well known that B is contained in the free A-module generated by $1/z^{a-1}$, z/z^{a-1}, ..., z^{a-1}/z^{a-1} (see Bourbaki, 1972, chap. V, sec. 1, problem 19). Thus B is contained in the free A-module generated by 1, z/X^bY^c, ..., z^{a-1}/X^bY^c. If $g(z)/h(X,Y) \in B$, then by lemma 2.1 we may assume that $g(u) = h_q u^q + \cdots + h_0$ and $h \nmid h_q$. It follows that $g(z)/h(X,Y) = f(z)/X^bY^c$, deg $f(u) < a$. Thus $X^bY^c(h_q z^q + \cdots + h_0) = h(X,Y)f(z)$. Now z is an algebraic element of degree a over the field of fractions of A; thus the coefficients of corresponding powers of z in the preceding equation must be equal. In particular, $h|X^bY^c \cdot h_q$. Thus $h = X$ or $h = Y$.

LEMMA 2.3 Let $g(u) \in A[u]$ be a polynomial of the smallest degree satisfying

(i) $g(z)/X \in B$.
(ii) $g(u) \not\equiv 0 \mod X$

Then $g(u) = h(Y)u^q$ modulo X with $1 \le q < a$; $h(Y) \in K[Y]$.

Proof: Set $F(u) = u^a - X^bY^c$. By lemma 2.1, $g(u) = g_1(u) = h_q(X,Y)u + \cdots + h_c(X,Y)$ modulo X, where X $h_q(X,Y)$ and $g_1(z)/X \in B$. Set $H(Y) = h_q(0,Y)$. Note that under our assumptions $H(Y) \not\equiv 0$ and that $g_2(u) = H(Y)u^q + \cdots + h_0(X,Y)$ is congruent to $g_1(u)$ modulo X. Also, $g_2(z)/X \in B$. Applying Euclid's lemma in A[u] to F(u) and $g_2(u)$, we get

$$H(Y)^rF(u) = g_2(u)Q(u) + R(u) \qquad \deg R < \deg g_2 = \deg g \qquad (*)$$

Substituting u = z we get

$$0 = g_2(z)Q(z) + R(z)$$

Thus $R(z)/X \in B$ and this implies that $R(u) \equiv 0$ modulo X (since g has the smallest degree with the same property). Taking equation (*) modulo X, we get

$$H(Y)^ru^a \equiv g_2(u)Q(u) \text{ modulo } X$$

Hence $g_2(u) \equiv h(Y) \cdot u^q$ modulo X for some $q < a$.

If α is an algebraic element over a field *L* we will denote by $\mathrm{Irr}[\alpha,L]$ the minimal polynomial of α over *L*. In this notation we have:

LEMMA 2.4 Set $a_t = a/(a,t)$. Then

$$\mathrm{Irr}[z^t, (X,Y)] = u^{a_t} - X^{bt/(a,t)} \cdot Y^{ct/(a,t)}$$

for $1 \le t \le a - 1$.

REMARK 1 Lemma 2.3 shows that if there is an element in B with "X in the denominator," there is one of the form $h(Y)z^t/X$, where t is the smallest degree of g(u) admitting representation $g(z)/X \in B$.

REMARK 2 Lemma 2.3 implies that $h(Y)z^q/X \in B$. We can actually show that $z^q/X \in B$.

Proof: Using lemma 2.4, we obtain

$$\mathrm{Irr}\left[\frac{h(Y)z^q}{X},\kappa(X,Y)\right] = u^{a_q} - h(Y)^{a_q} X^{bq/(a,q)-a_q} Y^{cq/(a,q)}$$

Thus $bq/(a,q) - a_q \quad 0$.

$$\mathrm{Irr}\left[\frac{z^q}{X},\kappa(X,Y)\right] = \mathrm{Irr}\left[\frac{h(Y)z^q}{X}\cdot\frac{1}{h(Y)},\kappa(X,Y)\right]$$
$$= u^{a_q} - X^{bq/(a,q)-a_q}\cdot Y^{cq/(a,q)}$$

Thus $z^q/X \in B$.

LEMMA 2.5 If $b = 1$, there is no element in B of the form $g(z)/X$ with $g(u) \in A[u]$, $g(u) \not\equiv 0 \bmod X$, and $\deg g < a$.

Proof: By remark 2 we may assume that $z^q/X \in B$; $1 \le q < a$.

$$\mathrm{Irr}\left[\frac{z^q}{X},\kappa(X,Y)\right] = u^{a_q} - X^{q/(a,q)-a_q}\cdot Y^{cq/(a,q)-a_q}$$

But $q/(a,q) - a_q = q/(a,q) - a/(a,q) < 0$.

Let $V_X[I]$ be the highest power of X dividing an ideal I in A.

LEMMA 2.6 Assume that $Z^a = X \cdot Y^c$. Then $V_X[D(z)] = V_X[D(B)]$.

Proof: Assume that $\alpha_0, \alpha_1, \ldots, \alpha_{a-1}$ is a basis of B as an A-module. Let

$$z^t = \sum_{j=0}^{a-1} h_{t,j}\alpha_j \qquad \text{for } 0 \le t \le a-1;\ h_{t,j} \in A \qquad (*)$$

Then $D(z) = (\det[h_{t,j}])^2 \cdot D(B)$ (Samuel, 1970). If $V_X[D(z)] > V_X[D(B)]$, then $\det[h_{t,j}] \equiv 0 \bmod X$.

In this case we can find polynomials $g_0(Y), \ldots, g_{a-1}(Y)$ in [Y], not all equal to zero such that

$$[h_{t,j}]\begin{vmatrix} g_0(Y) \\ \vdots \\ g_{a-1}(Y) \end{vmatrix} = 0 \bmod X$$

Thus

$$g_0(Y) \cdot [h_{0,0}(0,Y);\ldots;h_{0,a-1}(0,Y)] + \cdots$$
$$+ g_{a-1}(Y) \cdot [h_{a-1,0}(0,Y);\ldots;h_{a-1,a-1}(0,Y)] = 0 \bmod X \qquad (**)$$

Set $\beta = g_0(Y)1 + g_1(Y)z + \cdots + g_{a-1}(Y)z^{a-1}$. Then $\beta \in B$ and $\beta \neq 0$. Using (*) we get

$$\beta = g_0(Y) \cdot (h_{0,0} \cdot \alpha_0 + h_{0,1} \cdot \alpha_1 + \cdots + h_{0,a-1}\alpha_{a-1})$$
$$+ g_1(Y) \cdot (h_{1,p} \cdot \alpha_0 + h_{1,1} \cdot \alpha_1 + \cdots + h_{1,a-1}\alpha_{a-1})$$
$$+ g_{a-1}(Y)(h_{a-1,0} \cdot \alpha_0 + h_{a-1,1} \cdot \alpha_1 + \cdots$$
$$+ h_{a-1,a-1} \cdot \alpha_{a-1}) \qquad (***)$$

By rearranging the terms in (***), we get

$$\beta = \alpha_0(g_0h_{0,0} + \cdots + g_{a-1}h_{a-1,0}) + \cdots$$
$$+ \alpha_{a-1}(g_0h_{0,a-1} + \cdots + g_{a-1}h_{a-1,a-1})$$

By (**), $g_0(Y)h_{0,t}(X,Y) + \cdots + g_{a-1}(Y)h_{a-1,t}(X,Y)$ is congruent to 0 modulo X, for $0 \leq t \leq a - 1$. Thus $\beta/X \in B$. But this contradicts lemma 2.5.

REMARK 3 The proof of lemma 2.6 can be used in a much more general situation. It is an analog of the well-known fact: A finite abelian p group contains an element of order p.

LEMMA 2.7 $Z^a = X^bY^c$. Let K and L be the unique integers as in theorem 1 satisfying:

$$1 \leq K \leq a - 1 \qquad Kb = ma + 1$$
$$1 \leq L \leq a - 1 \qquad Lc = na + 1$$

Set

$$\alpha = \frac{z^K}{X^m} \qquad \beta = \frac{z^L}{Y^n}$$

Then $V_X[D(\alpha)] = V_X[D(B)] = a - 1$ and $V_Y[D(\beta)] - V_Y[D(B)] = a - 1$.

Proof: Note that $\alpha^a = XY^{cK}$ and $\beta^a = YX^{bL}$. Thus α and β are elements of B. By lemma 2.6, $V_X[D(\alpha)] = V_X[D(B)]$ and $V_Y[D(\beta)] = V_Y[D(B)]$.

The second part of both equalities follows from the well-known fact that $D(\alpha) = N(\alpha^{a-1})$ (see Samuel, 1970).

NOTATION 2.8 In the notation of lemma 2.7, by rearranging terms according to the powers of z, we get

$$\langle 1,\alpha,\ldots,\alpha^{a-1}\rangle = \langle 1, zX^{-m_1}Y^{\bar{m}_1}, \ldots, z^{a-1}X^{-m_{a-1}}Y^{\bar{m}_{a-1}}\rangle$$

$$\langle 1,\beta,\ldots,\beta^{a-1}\rangle = \langle 1, zX^{n_1}Y^{-n_1}, \ldots, z^{a-1}X^{n_{a-1}}Y^{-n_{a-1}}\rangle$$

Note that $\bar{m}_i \geq 0$ and $\bar{n}_i \geq 0$ for $0 \leq i \leq a-1$. Also, lemma 2.5 implies that $\bar{m}_i \geq 0$ and $\bar{n}_i \geq 0$ for $1 \leq i \leq a-1$. We define $\gamma_i = z^i/X^{m_i}Y^{n_i}$, $1 \leq i \leq a-1$.

We denote by Γ the A-module generated by $1, \gamma_1, \ldots, \gamma_{a-1}$.

The following lemma is a restatement of theorem 1:

LEMMA 2.9 $V_X[D(\Gamma)] = V_X[D(B)]$; $V_Y[D(\Gamma)] = V_Y[D(B)]$. Thus $\Gamma = B$ as an A-module.

Proof: We will prove only the equality involving X. The matrix H of transition from $\Gamma = \langle 1,\gamma_1,\ldots,\gamma_{a-1}\rangle$ to $\langle 1,\alpha,\ldots,\alpha^{a-1}\rangle$ is given by

$$H = \begin{vmatrix} 1 & & & 0 \\ & Y^{-\bar{m}_1 n_1} & & \\ & & \ddots & \\ 0 & & & Y^{-\bar{m}_{a-1}n_{a-1}} \end{vmatrix}$$

Then $V_X[\det(H)] = 0$. Thus $V_X[D(\Gamma)] = V_X[D(\alpha)]$ and by lemma 2.7, $V_X[D(\Gamma)] = V_X[D(B)]$.

Let us look at the following example:

EXAMPLE 2.10 $Z^7 = X^3Y^5$.

(1) $K = 5$: $5 \cdot 3 = 7 \cdot 2 + 1$; $m = 2$

$L = 3$: $3 \cdot 5 = 7 \cdot 2 + 1$; $n = 2$

(2) $\alpha = z^5/X^2;\ \beta = z^3/Y^2$

(3) $<1,\alpha,\ldots,\alpha^6> = 1, ZY^{10}, z^2Y^{20}, \frac{z^3Y^5}{X}, \frac{z^4Y^{15}}{X}, \frac{z^5}{X^2}, \frac{z^6Y}{X^2}$

(4) $<1,\beta,\ldots,\beta^6> = 1, zX^6, \frac{z^2X^3}{Y}, \frac{z^3}{Y^2}, \frac{z^4X^6}{Y^2}, \frac{z^5X^3}{Y^3}, \frac{z^6}{Y^4}$

(5) $m_1 = 0,\ \bar{m}_1 = 10,\ n_1 = 0,\ \bar{n}_1 = 6$

$m_2 = 0,\ \bar{m}_2 = 20,\ n_2 = 1,\ \bar{n}_2 = 3$

$m_3 = 1,\ \bar{m}_3 = 5,\ n_3 = 2,\ \bar{n}_3 = 0$

$m_4 = 1,\ \bar{m}_4 = 15,\ n_4 = 2,\ \bar{n}_4 = 6$

$m_5 = 2,\ \bar{m}_5 = 0,\ n_5 = 3,\ \bar{n}_5 = 3$

$m_6 = 2,\ \bar{n}_6 = 10,\ n_6 = 4,\ \bar{n}_6 = 0$

(6) $\gamma_1 = z,\ \gamma_2 = \frac{z^2}{Y},\ \gamma_3 = \frac{z^3}{XY^2},\ \gamma_4 = \frac{z^4}{XY^2},\ \gamma_5 = \frac{z^5}{X^2Y^3},\ \gamma_6 = \frac{z^6}{X^2Y^4}$

(7) $B = \left\langle 1, z, \frac{z^2}{Y} \frac{z^3}{XY^2} \frac{z^4}{XY^2} \frac{z^5}{X^2Y^3} \frac{z^6}{X^2Y^4} \right\rangle$

Computation of the Conductor

We denote by C the conductor of the ring $\kappa[X,Y,z]$ in B. More precisely, $C = \{\alpha \in \kappa[X,Y,z];\ \alpha \cdot B \subset \kappa[X,Y,z]\}$.

LEMMA 3.1 In notation 2.8, C is generated over A by

$$<X^{m_{a-1}}Y^{n_{a-1}}, X^{m_{a-2}}Y^{n_{a-2}}z, \ldots, X^{m_1}Y^{n_1}z^{a-2}, z^{a-1}>$$

Proof: By lemma 2.9,

$$B = \left\langle 1, \frac{z}{X^{m_1}Y^{n_1}}, \ldots, \frac{z^{a-1}}{X^{m_{a-1}}Y^{n_{a-1}}} \right\rangle$$

Let $\alpha = f_0 + f_1z + \cdots + f_{a-1}z^{a-1} \in C;\ f_i \in \kappa[X,Y]$. Then

$$\alpha \frac{z^r}{X^{m_r}Y^{n_r}} = \frac{1}{X^{m_r}Y^{n_r}} (f_0z^r + f_1z^{r+1} + \cdots + f_{a-1-r}z^{a-1}) + h$$

with $h \in \kappa[X,Y,z]$. Thus $X^{m_r}Y^{n_r}|f_t$; $0 \le t \le a - 1 - r$. In particular, $X^{m_{a-1}} \cdot Y^{n_{a-1}}|f_0$; $X^{m_{a-2}} \cdot Y^{n_{a-2}}|f_1, \ldots, X^{m_1}Y^{n_1}|f_{a-2}$; this proves our assertion.

EXAMPLE 3.2 $Z^7 = X^3Y^5$ (see example 2.10).

$$C = \langle X^2Y^4, X^2Y^3z, XY^2z^2, XY^2z^3, Yz^4, z^5, z^6\rangle$$

Note that as an ideal C is generated by X^2Y^4, X^2Y^3z, XY^2t^2, Yz^4, z^5.

Conductor C Coincides with the Ideal of Adjoints in the Coordinate Ring of V

In this section κ denotes an algebraically closed field of characteristic zero.

LEMMA 4.1 Spec B has only rational singularities.

Proof: In the notation of 2.7, B is the integral closure of $\kappa[X,Y,\alpha]$ in the field $\kappa(X,Y,\alpha) = k(V)$. Now $\kappa[X,Y,\alpha] \approx \kappa[X,Y,Z]/(Z^a - XY^{ac})$. The integral closure of the latter ring has only rational singularities [see, e.g., Lipman (1975, pp. 207, 208) where characteristic zero is assumed].

PROPOSITION 4.2 The ideal of adjoints A in $\kappa[V]$ is equal to the conductor C.

Proof: Since Spec (B) has only rational singularities, it follows from Lipman (1978, p. 153, (i), and p. 157, remark) that every rational differential 2-form which is regular on Spec (B) remains regular on $\tilde{V}$. Therefore, $C \subset A$ and $A \subset C$ trivially.

Acknowledgments: We would like to acknowledge with pleasure useful conversations with F. Call, W. Heinzer, J. Lipman, and R. Pancotti [who independently solved the case max(min(a,c),min(a,b)) = 3].

REFERENCES

Artin, M., On isolated rational singularities of surfaces, *Am. J. Math.*, 88 (1966), 129-136.

Berwick, W. E. H., *Integral Bases*, Cambridge University Press, Cambridge, 19 , p. 127.

Bourbaki, N., *Commutative Algebra*, Addison-Wesley, Reading, Mass., 1972.

Kaplansky, I., *Commutative Rings*, Allyn and Bacon, Boston, 1970.

Lipman, J., Introduction to resolution of singularities, *Proc. Symp. Pure Math.*, Vol. XXIX (Arcata), American Mathematical Society, Providence, R.I., 1975, Algebraic Geometry.

Lipman, J., Desingularization of two-dimensional schemes, *Ann. Math.*, 107 (1978), 151-207.

Samuel, P., *Algebraic Theory of Numbers*, Houghton Mifflin, Boston, 1970.

Seshadri, C. S., Triviality of vector bundles over the affine space K^2, *Proc. Nat. Acad. Sci. USA*, 44 (1958), 456-458.

Vascensalos, W. V., Extensions of Macaulay rings, *An. Acad. Bras. Cienc.* (2), 39 (1967), 211-214.

Zariski, O., *An Introduction to the Theory of Algebraic Surfaces*, Springer Lecture Notes in Mathematics, No. 83, Springer-Verlag, New York, 1969.

Zariski, O., An unpublished note related to a problem suggested to P. Blass, Purdue University, June 11, 1971.

2

Links with Differential Equations in Characteristic $p > o$†

INTRODUCTION

The purpose of this chapter is to outline a program of work in the theory of the Zariski surfaces [see Blass (1977, 1980), Lang (1981), and the definition below]. We show how a method that was inspired by the work of Jacobson in 1937 and then continued in the 1950s by, among others, Barsotti, Hochschild, Cartier, and Samuel plays a decisive role in the geometry of Zariski surfaces.

Our approach is to treat a certain formula inspired by Jacobson (1937) and proven by Hochschild (1955) and Barsotti (1958) as a differential equation over a field of positive characteristic. We call this equation (JHBS); see proposition 1 below. It was amply illustrated by the work of my former student at Purdue, Jeffrey Lang (1981), that this differential equation can in many cases be effectively solved. We show here how this yields some very subtle numerical invariants of Zariski surfaces, such as its Artin invariant (provided that there is no torsion in Pic or that we know the torsion part of Pic). In a future paper we hope to apply this differential equation to the Artin period map and to the Chern class map in crystalline cohomology.

It is a pleasure to dedicate this paper to Nathan Jacobson. Since I am more a geometer than an algebraist, I had recently to discover Jacobson's deep work for myself through geometry.

---

†By Piotr Blass; reprinted with modifications from *Contemporary Mathematics*, AMS 13 (1982).

Some of the problems discussed in this paper belong properly in the framework of purely inseparable descent of Cartier and Grothendieck and the related notion of p-curvature introduced by Grothendieck and Katz. Jacobson's fundamental work on restricted Lie algebras of derivations laid the foundations for their work.

On a more technical level, I wish to mention that the differential equation that we study is mentioned as an exercise in Jacobson's textbook (1980) and also that it seems likely that the methods of semilinear algebra (Jacobson, 1943) may be very useful in solving that differential equation. This paper does not give complete proofs in some cases. We hope that the well-intentioned reader will have no difficulty in filling in the gaps. In any case, we hope to do so in a future publication. Our main purpose is to outline an emerging theory. The reader who wants to learn more about Zariski surfaces is referred to my thesis (Blass, 1977, 1980) and J. Lang's thesis (Lang, 1981).

The study of Zariski surfaces in Section 1 yields the following result in pure algebra: There are infinitely many *nonisomorphic* fields between $k(X_1 \cdots X_n)$ and $k(X_1^{1/p}, \ldots, X_n^{1/p})$. Such a result would be difficult if not impossible to prove by field theory alone. Thus the geometry of ZSs provides a useful complement to the "Galois theoretic" methods pioneered by Jacobson for the study of fields in characteristic $p > 0$.

Zariski surfaces form admittedly a fairly narrow class of algebraic surfaces. Their most important property is that they can be joined with the plane $\mathbf{P}^2$ by a dominant purely inseparable map. This naturally leads us to introduce a new concept of equivalence for surfaces in characteristic $p > 0$, which we call here β-equivalence. We hope that this concept will prove useful in a much broader context then Zariski surfaces. As some evidence that this indeed may be the fact, we summarize here our recent solution (Blass, 1982) of the problem of unrationality of Enriques surfaces in characteristic 2, which is best phrased in terms of β-equivalence.

Finally, I cannot resist the temptation, and I announce a number of open problems and conjectures in the theory of algebraic surfaces. I hope that they will provide an added stimulus for people who wish to work in this beautiful area of mathematics.

0. PRELIMINARIES AND NOTATION

k is an algebraically closed field of characteristic $p > 0$.

X, a smooth projective surface over k, is called a *Zariski surface* (or ZS) if there exists a purely inseparable rational dominant map of degree p, $\pi: X \to \mathbb{P}^2$. [Equivalently, there exists a purely inseparable extension of degree p, $k(X) \subsetneq k(\mathbb{P}^2) = k(t_1,t_2)$, where the t_i's are indeterminates.]

Next we outline a construction procedure for ZSs which is fundamental for this chapter. Let $f_{pe}(x_0,x,y)$ be a form of degree pe which is not a pth power of a form of degree e. We construct a two-dimensional variety $\bar{F}$ by gluing together the three possibly singular surfaces,

$$F_0: \quad z^p = f_{pe}(1,x,y)$$

$$F_1: \quad z_1^p = f_{pe}(x_0,1,\bar{y})$$

$$F_2: \quad z_2^p = f_{pe}(\bar{x}_0,\bar{x},1)$$

The gluing of F_0 and F_1 is given by $x \to 1/x_0$, $y \to \bar{y}/x_0$, $z \to z_1/x_0^e$, and similarly for the other cases.

We will denote by ρ the canonical map $\rho: \bar{F} \to \mathbb{P}^2$ and by $\pi: \tilde{F} \to \bar{F}$ a minimal desingularization of $\bar{F}$. Clearly, $\tilde{F}$ is a Zariski surface. If $\bar{F}$ is a normal surface, the map $\rho: \bar{F} \to \mathbb{P}^2$ is simply the normalization of $\mathbb{P}^2$ in the field $k(F_0) = k(F_1) = k(F_2)$. We will refer to $\bar{F}$ and $\tilde{F}$ as surfaces defined by $z^p = f_{pe}(x_0,x,y)$; in symbols, $\bar{F}, \tilde{F}: \; z^p = f_{pe}(x_0,x,y)$.

REMARK It is more natural to think of $\bar{F}$ as embedded in the weighted projective space $\mathbb{P}(1,1,1,e)$. We do not exploit this point of view in the present paper. We hope to do so in a subsequent one.

The singularities of $\bar{F}$ may be very difficult to understand and resolve in general. Therefore, in this chapter we discuss primarily the "generic case" of this construction, where the singularities of $\bar{F}$ are as mild as possible.

DEFINITION A form $f_{pe}(x_0,x,y)$ is called *generic*, or *of type* (B*) [using notation analogous to J. Lang (1981)], if the corresponding surface $\bar{F}$: $z^p = f_{pe}(x_0,x,y)$ has only finitely many singular points all of which are double points locally given by $z^p = uv$ + higher-order terms. The corresponding smooth surface $\tilde{F}$: $z^p = f_{pe}(x_0,x,y)$ will be referred to as *a generic* ZS *of degree* (p,pe) [or *a* ZS *of type* B* *and of degree* (p,pe)].

Let us denote by A_p^{pe} the affine space of all the forms of degree pe. Let GA_p^{pe} be the subset of generic forms (of type B*). It can be shown that GA_p^{pe} is a Zariski open dense subset of A_p^{pe}. We omit the proof. We will also need the corresponding projective spaces: $P_p^{pe} = A_e^{pe} - \{0\}/k^*$ and $GP_p^{pe} = GA_e^{pe}/k^*$. Let X be a ZS. We will denote by b_i the étale Betti numbers of X and by ρ the base number (see Milne, 1979, p. 216). Since X is unirational, we have $\rho = b_2$ and it can be shown that the discriminant of the intersection form on $N(X) = \text{Pic } X/\text{Pic}^n X$ is an even power of the prime p. We denote it by $p^{2\sigma_0}$ and σ_0 is called the *M. Artin invariant* of X.

Finally, we introduce some notation from Samuel (1964) and Lang (1981). Let F: $z^p = f(x,y)$ be an affine, *normal* surface, the ring $k[F] \approx k[x^p,y^p,f] \subset k[x,y]$. Consider the derivation of $k[x,y]$ given by

$$Dw = \frac{\partial f}{\partial x}\frac{\partial w}{\partial y} - \frac{\partial f}{\partial y}\frac{\partial w}{\partial x}$$

In our case, the units of $k[F]$ are just elements of k. We have the following proposition due to Samuel.

PROPOSITION 0 $C\ell\ k[F] \approx$ the additive group of polynomial solution of the differential equation

$$D^{p-1}t - at = -t^p \qquad \text{(JBHS)}$$

where a is defined by $D^p = aD$.

1. THE GEOMETRY OF GENERIC ZARISKI SURFACES

In this section we compute the usual numerical invariants of a generic ZS of degree (p,pe). This is joint work with J. Sturnfield. We then apply this knowledge to field theory. Then we introduce a basic exact sequence involving Pic $\tilde{F}$. Finally, we link it with the JBHS differential equation using Samuel (1964) and we quote some results of J. Lang. We also give a formula for the Artin invariant of a generic Zariski surface of degree (p,pe) in the case that there is no torsion in Pic. We conclude with some examples of the theory.

Let $f_{pe}(x_0,x,y) \in GA_e^{pe}$ be a generic form (see Section 0). Let $\tilde{F} \xrightarrow{\pi} \bar{F} \xrightarrow{\rho} \mathbb{P}^2$ be the corresponding surfaces. We denote by $K_{\tilde{F}}$ the canonical divisor and by $\tilde{\ell}_{\tilde{F}} = (\rho \circ \pi)^* \mathcal{O}_{\mathbb{P}^2}(1)$ as well as the corresponding divisor.

LEMMA 1 (S. Abhyankar) $K_{\tilde{F}} = (pe - e - 3)\tilde{\ell}_{\tilde{F}}$.

Proof: The proof consists in computing the divisor of a differential using the explicit equations for the F_0, F_1, F_2 and was carried out by Abhyankar. We omit the details.

LEMMA 2 (J. Sturnfield) $\bar{F}$ has $(pe)^2 - 3pe + 3$ singularities all of the form $z^p = uv$ + higher-order terms.

Idea of proof: Compute the intersection number of the curve $\partial f(1,x,y)/\partial x = 0$ and $\partial f_{pe}(1,x,y)/\partial y = 0$ at infinity.

THEOREM 3 (P. Blass and J. Sturnfield) (i) $p_g(\tilde{F}) = p_a(\tilde{F})$ = number of nonnegative integral solutions of the inequality $\alpha - \beta - e\gamma \leq pe - e - 3$ = number of integer points in the interior of the tetrahedron spanned by (0,0,0), (pe,0,0), (0,pe,0), and (0,0,p) = $(1/12)(p - 1)[e^2p(2p - 1) - 9pe + 12]$.

(ii) $b_1(\tilde{F}) = b_3(\tilde{F}) = 0$.

(iii) $b_2(\tilde{F}) = b(\tilde{F}) = [(pe)^2 - 3pe + 3](p - 1) + 1$.

(iv) $(\tilde{\ell}_{\tilde{F}})^2 = p$ and $K_{\tilde{F}}^2 = (pe - e - 3)^2 p$.

Proof: The proof is straightforward since we remember that each singularity of $\bar{F}$ contributes p - 1 exceptional curves on $\tilde{F}$ except perhaps for (i), which follows immediately from the technical lemma 8, which we put at the end of this section.

COROLLARY For every $n \leq 2$ there exist infinitely many nonisomorphic fields between $k(X_1, X_2, \ldots, X_n)$ and $k(X_1^{1/p}, X_2^{1/p}, \ldots, X_n^{1/p})$, where the X_i's are indeterminates.

Proof: By theorem 3, there exists a sequence of generic ZS, $\tilde{F}_1$, $\tilde{F}_2$, ..., $\tilde{F}_n$, ... with $p_g(\tilde{F}_n) \to +\infty$ with n. Obviously, we may assume that $k(X_1, X_2) \subsetneq k(\tilde{F}_n) \subsetneq k(X_1^{1/p}, X_2^{1/p})$. This proves the corollary for n = 2 since p_g is a birational invariant. To prove the general case, consider the varieties $V_i = \tilde{F}_i \times A^{n-2}$. Obviously, $h^{2,0}(V_i) = \dim H^0(V_i, \Omega^2_{V_i})$ satisfies $h^{2,0}(V_i) \to +\infty$ as $i \to \infty$ and the corollary follows from the birational invariance of the numbers $h^{2,0}$ (see Hartshorne, 1977, p. 190, exercise 8.7).

To study the Artin invariant of $\tilde{F}$, we first have to establish an exact sequence.

THEOREM 4 In the notation of theorem 3, let A be the free abelian group generated by the curve $\tilde{\ell}_{\tilde{F}}$ and by the exceptional curves for the desingularization π: $\tilde{F} \to \bar{F}$; then the discriminant of the intersection form on A is equal to $p^{(pe)^2 - 3pe + 4}$. Moreover, if all the singularities of $\bar{F}$ are contained in F_0, a condition that can always be realized by a linear change of the coordinates x_0, x, y, we have the exact sequence

$$0 \to A \to \text{Pic } \tilde{F} \to C\ell \; k[F_0] \to 0 \tag{4.1}$$

Outline of proof: The intersection matrix of the generators of A is given by $(pe)^2 - 3pe + 3$ blocks of the form

$$\underbrace{\begin{bmatrix} -2 & 1 & & & & & \\ 1 & -2 & \cdot & & & & \\ \cdot & \cdot & \cdot & \cdot & & & \\ & \cdot & \cdot & \cdot & \cdot & & \\ & & \cdot & \cdot & \cdot & \cdot & \\ & & & \cdot & \cdot & -2 & 1 \\ & & & & \cdot & 1 & -2 \end{bmatrix}}_{p-1} p - 1 \quad \text{with determinant} = p$$

and by one block [p] corresponding to $\tilde{\ell}^2 = p$. Thus the curves generating A are independent in Pic $\tilde{F}$, Pic $\tilde{F}$ = $C\ell$ $\tilde{F}$ and there is an obvious map $C\ell\ \tilde{F} \to C\ell\ k[F_i]$. We leave the remaining details to the reader.

Let us now assume that all the singularities of $\bar{F}$ are in F_0, and let us write $f(x,y)$ for $f_{pe}(1,x,y)$. Samuel's theory now applies to F_0: $z^p = f(x,y)$, and consequently $C\ell\ k[F_0]$ is isomorphic to the group of polynomial solutions of the (JBHS) equation

$$D^{p-1}t - at = -t^p \qquad \text{(JBHS)}$$

corresponding to

$$z^p = f(x,y) \qquad \text{(see Section 0, proposition 0)}$$

THEOREM 5 (J. Lang) $C\ell\ k[F_0]$ is a finite group isomorphic to the direct sum of ν copies of Z/pZ. Moreover, any polynomial $t(x,y)$ satisfying the (JBHS) equation has degree $\le \deg f - 2$ ($\le pe - 2$ in our case).

Proof: See Lang (1981).

Let us now assume in addition that Pic $\tilde{F}$ has no torsion.† We do not know whether this is always true for a generic Zariski surface—this remains to be investigated. However, we do know that it is true in certain cases. (See the examples below.)

PROPOSITION 6 If Pic $\tilde{F}$ has no torsion, Pic $\tilde{F}$ = NS $\tilde{F}$ = $N(\tilde{F})$ because the Picard scheme of $\tilde{F}$ is reduced and discrete (see Milne, 1979, pp. 213-216) and we have the following formula for the Artin invariant of $\tilde{F}$:

$$\sigma_0(\tilde{F}) = \frac{(pe - 1)(pe - 2)}{2} + 1 - \nu \qquad (6.1)$$

where $(Z/pZ)^\nu$ is the group of solutions of the differential equation (JBHS).

Proof: This follows from the exact sequence 4.1 and basic properties of quadratic forms.

---

†This has been proven by W. E. Lang in *Compositio Math.* 52(2) (1984), 197-202.

REMARK 7 A similar theory can be developed for the surface F: $z^p = f_{p+1}(x,y)$, where $f_{p+1}(x,y)$ is a generic polynomial of degree $p + 1$. If $\bar{F}$ is the closure of F in P^3 and $\tilde{F}$ is a minimal desingularization of $\bar{F}$, and if we assume that Pic $\tilde{F}$ has no torsion, then

$$\sigma_0(\tilde{F}) = \frac{p(p + 1)}{2} - \nu \tag{7.1}$$

where $C\ell\ k[F_1] \approx (\mathbf{Z}/p\mathbf{Z})^\nu$ and ν can be computed from the (JBHS) differential equation. The reader may wish to consult Blass (1977, chap. 4) for some details of the geometry of such an $\tilde{F}$.

Some Examples

1. $\tilde{F}$: $z^2 = f_6(x_0,x,y)$, $f_6 \in GA_2^6$. In this case $\bar{F}$ has 21 singularities. Thus $b_2(\tilde{F}) = 22$. $K_{\tilde{F}} \equiv 0$. Thus $\tilde{F}$ is a K3 surface and Pic $\tilde{F}$ has no torsion. J. Lang has shown (Lang, 1981) that $1 \leq \nu$ for a generic $\tilde{F}$; thus $\sigma_0 = 10 + 1 - \nu \leq 10$, as is well known (Šafarevič and Rudakov, 1979).

2. $\tilde{F}$: $z^3 = f_6(x_0,x,y)$, characteristic 3. Here J. Lang (1981) shows that for an open dense *subset* of GA_2^6, $\nu = 0$; thus $\sigma_0 = 11$, otherwise $\sigma_0 \leq 11$. We remark that Pic $\tilde{F}$ is automatically torsion free in this case.

3. F: $z^3 = f_4(x,y)$, characteristic 3. Let $\tilde{F}$ be the smooth projective model of k(F). For a generic choice of the fourth-degree polynomial $f_4(x,y)$, $\tilde{F}$ is a K3 surface. $\sigma_0(\tilde{F}) = 6 - \nu$; see (7.1) above. J. Lang has shown that $\nu = 0$ for a dense open set of such $f_4(x,y)$; he also shows that $z^3 = x^4 + y^4 - xy$ has $\nu = 5$, thus $\sigma_0 = 1$.

We end with a technical lemma.

LEMMA 8 (P. Blass and J. Sturnfield) Let $\tilde{F} \xrightarrow{\pi} P^2$ be a generic ZS of degree (p,pe), $e > 1$, $\tilde{\ell} = \pi^*\mathcal{O}_{P^2}(1)$. Then $\dim H^0(\tilde{\ell}) = 3$ if $\ell > 1$ and $\dim H^0((pe - e - 3)(\tilde{\ell}) = p_g(\tilde{F})$ = number of distinct monomials $x^\alpha y^\beta z^\gamma$ with $\alpha + \beta + \gamma e \leq pe - e - 3$.

Proof: By a suitable change of coordinates, we may assume that F: $z^p = f_{pe}(1,x,y)$ contains all the singularities and $\tilde{\ell}$ is the line at infinity. Then in order to find $H^0(\tilde{\ell})$, we only need to find polynomials

$$p(x,y,z) \in k[F_0]$$

which have a pole of order ≤ 1 on $\tilde{\ell}$. Clearly, 1, x, y, have that property. However, the function z has under our gluing a pole of of order e' along $\tilde{\ell}$. Thus if $e \ge 2$, $z \notin H^0(\tilde{\ell})$. Also, $p(x,y,z)$ has a unique representation $p = \Sigma\, c_{\alpha\beta\gamma} x^\alpha y^\beta z^\gamma$ with $0 \le \gamma \le p - 1$. To compute the order of p along $\tilde{\ell}$, we write the rational function p on the chart F_1: $z_1^p = f_{pe}(x_0, 1, \bar{y})$ as

$$p = \Sigma\, c_{\alpha\beta\gamma} x_0^{-\alpha-\beta-\gamma e} \bar{y}^\beta z_1^\gamma$$

If $p \in H^0(\tilde{\ell})$, then $x_0 p$ is regular on F_i since x_0 has a simple zero along $\tilde{\ell}$. Thus

$$x_0 p = \Sigma\, c_{\alpha\beta\gamma} x_0^{-\alpha-\beta-\gamma e+1} \bar{y}^\beta z_1^\gamma \in k[F_1]$$

This means that if $c_{\alpha\beta\gamma} \ne 0$, then $\alpha + \beta + \gamma e - 1 \le 0$, as can be seen by collecting all terms with a fixed γ and noting that monomials $x_0^m y^n$, where $m,n \in z$, are independent in $k[F_1]$. If $e \ge 2$, the only solutions are

$$\alpha = 0 \quad \beta = 0 \quad \gamma = 0$$
$$\alpha = 1 \quad \beta = 0 \quad \gamma = 0$$
$$\alpha = 0 \quad \beta = 1 \quad \gamma = 0$$

Therefore, p is a combination over k of 1, x, y. Similarly, multiplying by x_0^{pe-e-3}, we conclude that $H^0((pe - e - 3)\tilde{\ell}_{\tilde{F}})$ is spanned by the desired monomials in x_0, x, y.

2. MODULI INTRODUCTION: MODULI AND THE (CONJECTURAL) ARTIN STRATIFICATION BY σ_0 OR ν

Throughout this section we will assume for technical reasons that $e \ge 2$ and that $pe - e - 3 > 0$ and also $e + 3 \not\equiv 0(p)$. I hope that that last condition will eventually be removed; it is used only in the proof of lemma 10 below.†

†Added in proof: Indeed it may now be removed because of the result of W. E. Lang quoted in the previous footnote.

In this section we tackle the following problem: Let $\tilde{F}$, $\tilde{G}$ be generic ZS defined by $z^p = f_{pe}(x_0,x,y)$ and $z^p = g_{pe}(x_0,x,y)$, respectively. If $\tilde{F}$ is isomorphic to $\tilde{G}$ as surfaces over k, what can be said about the relationship of the forms $f_{pe}(x_0,x,y)$ and $g_{pe}(x_0,x,y)$? It turns out that the answer is quite simple. $\tilde{F} \approx \tilde{G}$ if and only if f_{pe} can be obtained from g_{pe} by a linear change of coordinates followed by adding the pth power of a form of degree e; see theorem 9 below.

This leads us naturally to the following construction of a moduli space for generic Zariski surfaces of degree (p,pe).[†] Consider the action of GL(3) on $A_p^{pe} \approx$ the space of homogeneous forms in x_0, x, y of degree pe. Let $B_p^{pe} \subseteq A_p^{pe}$ be the subspace generated by the monomials $x_0^{\alpha}y^{\beta}x^{\gamma}$ with $\alpha + \beta + \gamma = pe$ and $p|\alpha$, $p|\beta$, $p|\gamma$. Clearly, B_p^{pe} is stable under the action of GL(3). There is, therefore, an induced action on $C_p^{pe} = A_p^{pe}/B_p^{pe}$. Now the set of generic forms or forms of type B* (see Section 0), $GA_p^{pe} \subseteq A_p^{pe}$, is stable under the action of GL(3) and it is easy to see that the set of equivalence classes of elements of GA_p^{pe} forms an open and dense subset of C_p^{pe}.

We will denote this subset by GC_p^{pe}. Now GL(3) acts on GC_p^{pe} and consequently PGL(3) acts on GC_p^{pe}/k^*. In fact, PGL(3) acts on the projective space $C_p^{pe} - \{0\}/k^*$. We will show that every point of GC_p^{pe} is stable in the sense of geometric invariant theory and this will show that the universal categorical quotient $GC_p^{pe}/PGL(3)$ exists. As the next proposition will show, this quotient denoted GM_p^{pe} is a good candidate for a coarse moduli space for Zariski surfaces of type B* of degree (p,pe).

We use the same notions and notation as J. Shah (1980), who in turn follows Mumford (1965). By a suitable choice of coordinates, any given one-parameter subgroup λ of PGL_3 is diagonalized and has the form

---

[†]We are using the word "moduli" somewhat loosely here. I still have not investigated the question whether or not the space that we obtain is in fact a coarse or a fine moduli space in the sense of Mumford (1965).

$$\begin{bmatrix} t^{r_0} & 0 & 0 \\ 0 & t^{r_1} & 0 \\ 0 & 0 & t^{r_2} \end{bmatrix}$$

such that $\Sigma\, r_i = 0$ and $r_0 \geq r_1 \geq r_2$, $r_0 > 0$.

We wish to show that the class of any form $f_{pe}(x_0,x,y) \in GA_p^{pe}$ is stable. We consider three cases.

Case 1: The term $x_0^{pe-1}x$ or $x_0^{pe-1}y$ is present in $f_{pe}(x_0,x,y)$. Then $\mu(f,\lambda) \geq (pe - 1)r_0 + r_1 = (pe - 1)r_0 + r_1 + r_2 - r_2 = (pe - 2)r_0 - r_2 > 0$ or $(pe - 1)r_0 + r_2 = (pe - 2)r_0 + r_0 + r_1 + r_2 - r_1 = (pe - 2)r_0 - r_1 > 0$.

Case 2: $x_0^{pe-2}xy$ is present in f. $\mu \geq (pe - 2)r_0 + r_1 + r_2 = (pe - 3)r_0 + r_0 + r_1 + r_2 = (pe - 3)r_0 > 0$.

Case 3: The terms mentioned in cases 1 and 2 are missing, but then both terms $x_0^{pe-2}x^2$ and $x_0^{pe-2}y^2$ appear. Suppose that $(pe - 2)r_0 + 2r_1 \leq 0$ and $(pe - 2)r_0 + 2r_2 \leq 0$. Then $(pe - 2)r_0 + r_1 + r_2 \leq 0$, so that $(pe - 3)r_0 \leq 0$ is a contradiction. Thus $\mu([f],\lambda) > 0$, which shows stability by Mumford's theorem. We conclude that under the action of PGL(3) on C_p^{pe} the open set GC_p^{pe} is stable and it consists of stable points in the sense of Mumford. We conclude that the universal categorical quotient $GC_p^{pe}/PGL(3)$ exists and we denote it GM_p^{pe}. This is our "moduli space" for generic ZSs of degree (p,pe). As we said in the beginning of the section, this terminology is partially justified by the following theorem.

THEOREM 9 Assume that $pe - e - 3 \geq 1$ and that $p \nmid e + 3$. Let $\tilde{F}$: $z^p = f_{pe}(x_0,x,y)$ and $\tilde{G}$: $z^p = g_{pe}(x_0,x,y)$ be two generic Zariski surfaces defined by generic forms f_{pe}, g_{pe} (of type B*; see Section 0). Then $\tilde{F}$ is isomorphic to $\tilde{G}$ as k-schemes if and only if f_{pe} and g_{pe} define the same element of GM_p^{pe} (i.e., iff f_{pe} and g_{pe} can be obtained from each other by a linear change of coordinates followed by adding the pth power of a form of degree e).

Proof: The "if" part is easy. As to the "only if" part, we will only outline the proof in several lemmas.

Let $\alpha: \tilde{F} \to \tilde{G}$ be a k-isomorphism.

LEMMA 10 $\alpha^*\tilde{\ell}_{\tilde{G}} \approx \tilde{\ell}_{\tilde{F}}$.

Proof: Since $(pe - e - 3)\tilde{\ell}_{\tilde{G}}$ is the dualizing sheaf on $\tilde{G}$ and $(pe - e - 3)\tilde{\ell}_{\tilde{F}}$ is the dualizing sheaf on $\tilde{F}$, we have $(pe - e - 3)(\alpha^*\tilde{\ell}_{\tilde{G}} - \tilde{\ell}_{\tilde{F}}) = 0$. However, Pic $\tilde{F}$ has no torsion of order prime to p because of the exact sequence 4.1. Thus since $p \nmid e + 3$, $\alpha^*\tilde{\ell}_{\tilde{G}} = \tilde{\ell}_{\tilde{F}}$.

LEMMA 11 There exists a commutative diagram where $\bar{\alpha}$ is an isomorphism and γ is a linear isomorphism.

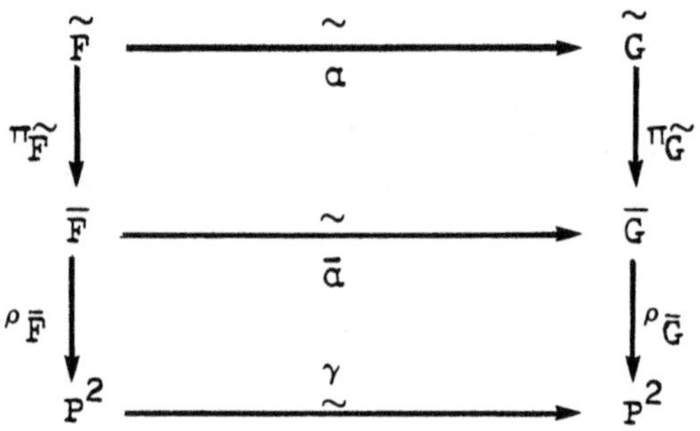

Proof: First we prove the existence of the linear map γ. Consider the two maps of $\tilde{F}$ to $\mathbf{P}^2$, $\rho_{\bar{G}} \circ \pi_{\tilde{G}} \circ \alpha$ and $\rho_{\bar{F}} \circ \pi_{\tilde{F}}$. Because of lemma 1 the pullback of $\mathcal{O}_{P^2}(1)$ is the same for both maps, namely, $\tilde{\ell}_{\tilde{F}}$. Also, $H^0(\tilde{\ell}_{\tilde{F}}) = 3$ by the technical lemma at the end of Section 1. Thus the two maps are both defined by the same complete linear system and the existence of γ, a linear isomorphism, is well known (see Hartshorne, 1977, chap. II, sec. 7). To prove the existence of the isomorphism $\bar{\alpha}$, note that an exceptional curve C on $\tilde{F}$ must be mapped to an exceptional curve of $\tilde{G}$ because of the existence of γ and from this the existence of the isomorphism $\bar{\alpha}$ follows.

Proof of theorem 9: Because of the above lemmas we may assume the following diagram:

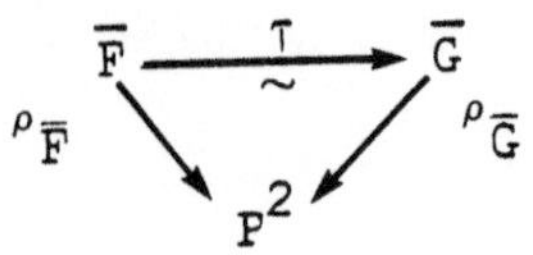

τ an isomorphism. Under such conditions we will show that f_{pe} and g_{pe} differ by the pth power of a form of degree e. Everything follows easily from the following lemma.

LEMMA 12 (P. Blass and W. Heinzer) Let F: $z^p = f(x,y)$ and G: $z_1^p = g(x,y)$ be normal surfaces in A^3 and such that there is a commutative diagram

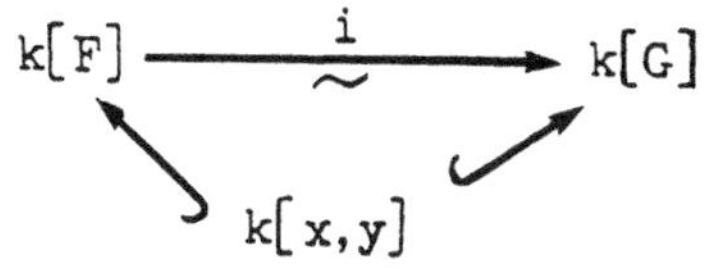

with i an isomorphism. Then $f(x,y) - cg(x,y) = [b(x,y)]^p$ for some polynomial $h(x,y) \in k[x,y]$ and some $c \in k$.

Proof: We have

$$g(x,y) = a_0^p + a_1^p f + \cdots + a_{p-1}^p f^{p-1}$$

for some $a_i \in k[x,y]$. Compute

$$\frac{\partial g}{\partial x} = \frac{\partial f}{\partial x}(a_1^p + 2a_2^p f + 3a_3^p f^2 + \cdots + (p-1)a_{p-1}^p f^{p-2})$$
$$\frac{\partial g}{\partial y} = \frac{\partial f}{\partial y}(a_1^p + 2a_2^p f + 3a_3^p f^2 + \cdots + (p-1)a_{p-1}^p f^{p-2})$$

Because G is a normal surface, the polynomial in parentheses must be a constant in k. Thus $a_1 + 2a_2 z + \cdots + (p-1)a_{p-1}z^{p-2} \in k$, but since $k[x,y,z]$ is a free $k[x,y]$ module on $1, z, z^2, \ldots, z^{p-1}$, we conclude that $a_2 = a_3 = \cdots = a_{p-1} = 0$ and that $a_1 \in k$. Thus $g = [a_0(x,y)]^p + a_1^p f$, $a_1 \in k$, and the lemma follows.

We leave to the reader the remaining details in the proof of the theorem.

Conjectural Stratification by σ_0 (or ν)

We state here two conjectures which are crystallized in the author's conversations with M. Artin and J. Lang. The idea is the following. As the form $f_{pe}(x_0,x,y)$ varies in GA_p^{pe}, the number of solutions

$(Z/pZ)^{\nu}$ of the differential equation (JBHS) will also change. This will result in a change of σ_0 (see formula 4.1, where we assume that Pic $\tilde{F}$ is torsion free). By analogy with Artin's study of supersingular K3 surfaces, we are led to the following conjectures. In what follows, we will assume that Pic $\tilde{F}$ is torsion free for all the surfaces in GM_p^{pe}, although this is probably not essential.

CONJECTURE 1 Let $M_\tau \subset GM_p^{pe}$ be the set of surfaces with the Artin invariant $\sigma_0 \leq \tau$ [or equivalently with $\nu \geq (pe - 1)(pe - 2)/2 + 1 - \tau$]. Then we have a filtration

$$\begin{array}{l} M_{\tau-1} \subset M_\tau \subset \cdots \subset M_{(\sigma_0)\max-1} \subset M_{(\sigma_0)\max} \subset GM_p^{pe} \\ \cup \\ \vdots \\ \cup \\ M_{(\sigma_0)\min} \end{array}$$

where $(\sigma_0)\max$, $(\sigma_0)\min$ are the largest and the smallest possible values of the Artin invariant and we conjecture that $M_{\tau-1}$ is closed in M_τ and that $M(\sigma_0)\max$ is dense in GM_p^{pe}. Of course, we could also state this conjecture in terms of ν (as long as Pic has no torsion).

CONJECTURE 2 (M. Artin) Codimension of $M_{\tau-1}$ in M_τ is $\leq p_g(\tilde{F})$.

This conjecture resembles Lefschetz's theorem in characteristic 0, which says that a cycle has to satisfy p_g conditions in order to be algebraic (vanishing of all the periods of abelian integrals).

These conjectures can also be stated in terms of the spaces GA_p^{pe}, as was pointed out to me by J. Lang.

3. β-EQUIVALENCE

Let S be the set of all (isomorphism classes) of smooth projective surfaces over $k = \bar{k}$, characteristic $k = p > 0$. We wish to introduce on S the weakest equivalence relation that identifies two surfaces X and Y provided that there exists a purely inseparable dominant rational map π: $X \to Y$. It is easy to see that the right definition is as follows.

DEFINITION X is said to be *β-equivalent* to Y, where $X,Y \in S$ iff there exists a k-inclusion of fields $k(X)^{p^n} \hookrightarrow k(Y)$ such that k(Y) becomes a finite purely inseparable extension of $k(X)^{p^n}$ for some $n \geq 0$.

It is straightforward to check that this is an equivalence relation. We write $X \sim_\beta Y$ and we denote by $[X]_\beta$ the equivalence class of X.

Here is another description of β-equivalence. Let $(X,\mathcal{O}_X) \in S$. Consider the schemes $(X,\mathcal{O}_X^{p^n}) = X_n$. Clearly, $X_n \in S$ and the function field of X_n is $k(X_n) = (k(X))^{p^n} \subset k(X)$. Consequently, there is a purely inseparable dominant rational map $X \to X_n$.

LEMMA 13 $X \sim_\beta Y$ iff for sufficiently large $n >> 0$ there exists a purely inseparable rational dominant map $Y \to X_n$.

Proof: Obvious.

Let us give one more equivalent condition.

LEMMA 14 $X \sim_\beta Y$ if and only if there exists a sequence $Z_0, Z_1, \ldots, Z_n$ with $Z_i \in S$, $Z_0 = X, \ldots, Z_n = Y$ and such that there exists a purely inseparable dominant map between Z_i and Z_{i+1} (in either direction). Thus, loosely speaking, $X \sim_\beta Y$ iff X and Y can be joined by a chain of purely inseparable dominant rational maps.

A number of étale cohomology invariants are preserved under β-equivalence:

PROPOSITION 15 If $X \sim_\beta Y$, then

(1) $b_1(X) = b_1(Y)$, i.e., dim Alb(X) = dim Alb(Y).

(2) $\lambda(X) = b_2(X) - \rho(X) = b_2(Y) - \rho(Y) = \lambda(Y)$.

(3) $\pi_1^{et}(X) \approx \pi_1^{et}(Y)$ (conjecture).

Proof: Omitted. [I have only checked (3) carefully in the simply connected case.]

As an application of β-equivalence, we sketch the proof of the following theorem [details appear elsewhere; see Blass (1982)].

THEOREM 16 A classical (or a supersingular) Enriques surface X in characteristic 2 is always β-equivalent to $\mathbb{P}^2$, thus certainly unirational.

Proof: Bombieri and Mumford (see Bombieri and Mumford, 1976, p. 220) have constructed a purely inseparable double cover $\tilde{X} \to X$. Let $X_1 \xrightarrow{\pi} \tilde{X}$ be a smooth model of $\tilde{X}$. Clearly, $X_1 \sim_\beta X$; thus $\lambda(X_1) = \lambda(X) = 0$. The last equality follows from Bombieri and Mumford (1976).

LEMMA 17 X_1 is either a ruled surface or a K3 surface.

Proof: This is proven in Blass (19) using the dualizing sheaf. Now we again use β-equivalence. If X_1 is ruled since $X_1 \sim_\beta X$ and $b_1(X) = 0$ by Bombieri and Mumford (1976); therefore also $b_1(X_1) = 0$, and thus X_1 is rational and we are done. Now if X_1 is K3, then, since $X_1 \sim_\beta X$, $\lambda(X_1) = \lambda(X) = 0$ [the last equality by Bombieri and Mumford (1976)]. Thus X_1 is a supersingular K3 surface in characteristic 2, and therefore, by the deep theorem of Šafarevič and Rudakov (1979), it is β-equivalent to $\mathbb{P}^2$. This completes our outline of the proof.

We have included the outline above to indicate the usefulness of β-equivalence. We propose to call any surface β-equivalent to $\mathbb{P}^2$ a *generalized Zariski surface* (GZS).

REMARK 18 In a recent note with M. Levine we have shown that the β-equivalence class of $\mathbb{P}^2$ is closed under specialization (Blass and Levine, 1982). We conjecture that this is true for β-equivalence classes of many other, perhaps all, surfaces in characteristic p.

PROBLEM Under what conditions on X is $[X]_\beta$ closed under specialization?

We state a number of open problems about β-equivalence in the following section.

4. OPEN PROBLEMS

1. Find a surface $X \in [\mathbb{P}^2]_\beta$ which is not a Zariski surface.†

REMARK For example, a generic surface $z^{25} = f_{26}(x,y)$ is a reasonable candidate in characteristic 5, but I do not know how to prove that it is *not* a *ZS*.

2. Does there exist a surface X of general type and such that if $Y \sim_\beta X$, then Y must also be of general type?
3. Zariski's problem: For $p \geq 5$ find a ZS with $p_g = 0$ but non-rational.

REMARK This was solved by P. Blass in characteristic 2 in 1977 (see Blass, 1977) and by W. E. Lang in characteristic 3 in 1978 (see Lang, 1979).

4. Problem (due to P. Blass and W. E. Lang): Let X be a Zariski surface with $p_g = 0$ and $\mathrm{Pic}\,^\tau X$ trivial. Does X have to be rational?
5. The same question as in problem 4 assuming that Pic X is torsion free instead of $\mathrm{Pic}\,^\tau X = 0$.
6. Is every simply connected and unirational surface β-equivalent to $\mathbb{P}^2$?
7. X a ZS with $p_g = 0$ and Pic X torsion free. Is X rational?
8. Is it possible to construct a sequence of surfaces $X_1, X_2, \ldots, X_n, \ldots$ over k with $b_2(X_i) - \rho(X_i) \to +\infty$ and such that every smooth surface over k is dominated by one of the surfaces X_i (via a rational map)?

Finally, three problems in characteristic 0:

---

†Solved by Katsura 1982.

9. Find a surface X that is not birationally equivalent to a *normal* surface in $\mathbb{P}^3$.
10. Let $X \subset \mathbb{P}^3$ be a normal surface with only one isolated singularity $g \in X$. Let $\tilde{X} \to X$ be a desingularization. Suppose that $H^1(\tilde{X}, \mathcal{O}_{\tilde{X}}) \neq 0$. Is $\tilde{X}$ necessarily ruled?
11. Mumford's question: Let $F \subseteq \phi\mathbb{P}^3$ be a surface of degree 5. Let $\tilde{F} \overset{\pi}{\to} F$ be a desingularization. Suppose that $H^1(\tilde{F}, \mathcal{O}_{\tilde{F}}) \neq 0$. Does $\tilde{F}$ have to be a (birationally) ruled surface?

Acknowledgments: Several theorems in this paper were proven jointly with James Sturnfield. I thank him for his cooperation. I also rely heavily on the extensive computations carried out by Jeffrey Lang in his 1981 Purdue thesis. I am also indebted to M. Artin, S. Abhyankar, W. Haboush, J. Lipman, W. Heinzer, and A. Fauntleroy for some very useful discussions of these topics.

REFERENCES

Artin, Michael, Supersingular K3 surfaces, *Ann. Sci. Norm. Sup.*, 4 serie, T7, fasc.4 (1974), pp. 543-568.

Barsotti, Iacopo, Repartitions on abelian varieties, *Ill. J. Math.*, 2 (1958), 43-69 (especially pp. 58-59).

Blass, Piotr, Zariski surfaces, thesis, University of Michigan, 1977, *Diss. Math.*, 200 (1983).

Blass, Piotr, Zariski surfaces, *C. R. Math. Rep. Acad. Sci. Canada*, II, No. 1 (1980).

Blass, Piotr, Unirationality of Enriques surfaces in characteristic two, *Compositio Mathematica*, Vol. 45, Fasc 3 (1982), 393-398.

Blass, Piotr, and Marc Levine, Families of Zariski surfaces, *Duke Math. J.*, 49 (1982), 129-136.

Bombieri, E., and David Mumford, Enriques' classification of surfaces in characteristic p > 0, III, *Invent. Math.*, 36 (1976), 197-232.

Hartshorne, R., *Algebraic Geometry*, Graduate Texts in Mathematics, Springer-Verlag, New York, 1977.

Hochschild, G., Simple algebras with purely inseparable splitting fields of exponent one, *Trans. Am. Math. Soc.*, 79 (1955), 447.

Jacobson, Nathan, Abstract derivations and Lie algebras, *Trans. Am. Math. Soc.*, 47 (1937), 206.

Jacobson, Nathan, *Basic Algebra*, Vol. II, W. H. Freeman, San Francisco, 1980 (especially p. 536, exercise 3).

Jacobson, Nathan, *The Theory of Rings*, A.M.S. Mathematical Surveys, No. II, American Mathematic Society, Providence, R.I., 1943.

Katsura, Toshiyuki, A note on Enriques surfaces in characteristic two, *Compositio Math.*, Vol. 47, Fasc 2 (1982), pp. 207-217.

Lang, Jeffrey, Ph.D. thesis, Purdue University, 1981.

Lang, W. E., Quasi-elliptic surfaces in characteristic three, *Ann. Sci. Ec. Norm. Sup.* (1979).

Milne, J. S., *Étale Cohomology*, Princeton Mathematical Series, No. 33, Princeton University Press, Princeton, N.J., 1979.

Mumford, David, *Geometric Invariant Theory*, Academic Press, New York, 1965.

Safarevic, Igor, and A. N. Rudakov, Supersingular K3 surfaces over fields of characteristic two, *Math. USSR Izv.*, 13 (1979), No. 1, 147-165.

Samuel, Pierre, Lectures on Unique Factorization Domains, Tata Lecture notes, 1964.

Shah, Jayant, A complete moduli space for K3 surfaces of degree 2, *Ann. Math.*, 112 (1980), 480-510 (especially pp. 489-490).

3

The Divisor Classes of the Surface $z^{p^n} = G(x,y)$†

INTRODUCTION

In this chapter we study the divisor class group of normal affine surfaces $F \subset \mathbf{A}_k^3$ defined by equations of the form $z^p = G(x,y)$, where the ground field k is assumed to be algebraically closed of characteristic $p > 0$.

Surfaces of this type are briefly considered by O. Zariski in his paper on Castelnuovo's criterion of rationality (see Zariski, 1958, p. 314) and investigations of their geometry have been made by P. Blass (see Blass, 1983) under whose advisement I began this project. P. Samuel in his 1964 Tata notes (Samuel, 1964a) describes the class group of several of these surfaces, such as $z^p = xy$ and $z^p = x^i + y^j$. Results from Samuel's notes form the foundation of this work and a brief discussion of these appears in Section 1. Facts concerning the order and type of the class group of F are proved in Section 2, together with "Ganong's formula (2.6), which plays a vital role throughout this exploration. Next, in Section 3, the $p = 2$ case is considered. An explicit description of $C\ell(F: z^2 = G)$ for G of degree ≤ 4 is given entirely in terms of the coefficients of G [see (3.15) and the discussion that follows]. Several questions are posed (3.5) that are investigated further in Section 4, where the surface $z^3 = G_4$ is studied (G_n is notation for a

---

†This chapter is reprinted from Jeffrey Lang's Ph.D. thesis, Purdue University, 1981.

polynomial of degree n). The section ends with some conjectures, among them a conjecture of M. Artin (4.25), which we confirm against the surfaces $z^5 = G_3$ and $z^2 = G_6$ in Section 5 and $z^3 = G^6$ and $z^p = G_3$ in Section 6.

We compute the class group of three concrete examples in Section 7. Perhaps the most interesting of these is the calculation of $C\ell(F)$ for the surface $z^p = x^{p+1} + y^{p+1} + (1/2)(x^2 - y^2)$, where p is assumed to be greater than 2. This surface is introduced by Zariski in the above-mentioned article.

Section 8 provides an example of a smooth nonrational surface with factorial coordinate ring, thus demonstrating that the separability condition in P. Russell's "affine theorem of Castelnuovo" is essential (see Russell, 1981, p. 2).

The scope of our research is broadened in Sections 9 and 10 to include normal hypersurfaces in $\mathbb{A}_k^{n+1}$ defined by equations of the form $z^{p^m} = G(x_1,x_2,\ldots,x_n)$. In Section 9 we develop an inductive method of studying $C\ell(F\colon\ z^{p^m} = G)$ and show that many results from earlier sections generalize nicely.

In Section 10, miscellaneous results are collected. The class group of the hypersurface $z^{p^m} = H(x_1,\ldots,x_n)$, where H is a form of degree not divisible by p, is calculated and the local behavior of $C\ell(F\colon\ z^p = G)$ is discussed. We end the section with a description of the class group of Krull rings A such that $k[x_1^{p^m},\ldots,x_n^{p^m}] \subset A \subset k[x_1,\ldots,x_n]$.

NOTATION

k algebraically closed field of characteristic $p > 0$, unless stated to the contrary

$\mathbb{A}_k^n$ affine n space over k

$\mathbb{P}_k^n$ projective n space over k

Surface irreducible, reduced, two-dimensional, quasi-projective variety over k

Hypersurface irreducible, reduced, n-dimensional, quasi-projective variety over k

For a smooth surface X we write p_g = geometric genus of X. The notation F: $f(x_1, x_2, \ldots, x_n) = 0$ means

$$F = \operatorname{Spec} \frac{k[x_1, \ldots, x_n]}{(f(x_1, \ldots, x_n))} \qquad F \subset \mathbf{A}^n$$

If A is a Krull ring, we denote by $C\ell(A)$ the divisor class group of A.

If X is a surface, we denote by $C\ell(X)$ the divisor class group of the coordinate ring of X.

1. PRELIMINARIES

The cornerstone of this chapter is P. Samuel's 1964 Tata notes (Samuel, 1964a). What follows is a brief discussion of some results from these notes.

DEFINITION Let A be a domain. A is a *Krull ring* if there exists a family $(v_i)_{i \in I}$ of discrete valuations of qt(A) such that

(1) $A = \bigcap_i R_{v_i}$, where R_{v_i} denotes the ring of v_i.
(2) For every $x \neq 0 \in A$, $v_i(x) = 0$ for almost all $i \in I$.

THEOREM 1.1 A noetherian integrally closed domain is a Krull ring.

DEFINITION Let A be a domain with quotient field K. A *fractionary ideal* a is an A-submodule of K for which there exists an element $d \in A (d \neq 0)$ such that $da \subset A$. A fractionary ideal is called a *principal ideal* if it is generated by one element. A is said to be *integral* if $a \subset A$. a is said to be *divisorial* if $a \neq (0)$ and if a is an intersection of principal ideals.

DEFINITION Let I(A) denote the set of nonzero fractionary ideals of the domain of A. On I(A) we define an equivalence relation by $a \sim b \Leftrightarrow A:a = A:b$ where $A:a = \{x \in k \mid xa \subset A\}$. The quotient set of I(A) by this equivalence relation is called the set of *divisors* of A, denoted by D(A). For each $a \in I(A)$, we denote by $\bar{a}$ the equivalence class of a in D(A).

DEFINITION Let A be a Krull domain. The composition law $(a,b) \rightsquigarrow ab$ on I(A) induces a well-defined operation on D(A), thus giving D(A) the structure of an abelian group with identity element $\bar{A}$ (see Samuel, 1964a, pp. 1-4). Hereafter we will write this composition law additively. Thus $\bar{a} + \bar{b} = \overline{ab}$ for $\bar{a},\bar{b} \in D(A)$. Let F(A) denote the subgroup of D(A) generated by the principal divisors (equivalence classes of principal ideals). We denote by $C\ell(A)$ the quotient group D(A)/F(A), called the *divisor class group* of A.

THEOREM 1.2 Let A be a Krull ring. Then
(1) $C\ell(A)$ is generated by the classes of the height 1 primes of A.
(2) A is factorial if and only if $C\ell(A) = 0$.

NOTATION Let $A \subset B$ be rings. Let $p \subset A$ and $q \subset B$ be prime ideals. We write q/p if $q \cap A = p$ and we say that q *lies over* p.

THEOREM 1.3 Let $A \subset B$ be Krull rings. Suppose that either B is integral over A or that B is a flat A algebra. Then there is a well-defined group homomorphism ϕ: $C\ell(A) \to C\ell(B)$ (Samuel, 1964a, pp. 19-20).

Let us now describe the homomorphism of theorem (1.3). If q and p are height 1 primes of B and A with q|p, we let e(q,p) denote the ramification index of q over p. Then for each height 1 prime p of A we define $\phi(p) = \Sigma_{q/p}\, e(q,p)q$, the sum taken over all height 1 primes in B lying over p. This sum is always finite since B is a Krull domain. We then extend ϕ by linearity.

THEOREM 1.4 Let A be a Krull ring, S a multiplicatively closed subset in A. Then $S^{-1}A$ is an A-flat Krull ring and
(1) ϕ: $C\ell(A) \to C\ell(S^{-1}A)$ is surjective.
(2) If S is generated by prime elements, then ϕ is bijective (Samuel, 1964a, p. 21).

REMARK 1.5 In theorem (1.4) $\ker \phi = H + F(A)/F(A)$, where $H \subset D(A)$ is the subgroup generated by those height 1 primes p of A such that $p \cap S \neq \emptyset$.

THEOREM 1.6 Let R be a Krull ring. Then R[X] is a Krull ring and ϕ: $C\ell(R) \to C\ell(R[X])$ is bijective (Samuel, 1964a, p. 22).

Let A be a noetherian ring and m an ideal contained in the Jacobsen radical of A. If we give A the m-adic topology, then (A,m) is called a *Zariski ring*. The completion $\hat{A}$ of A will also be a Zariski ring and is A flat with $A \subset \hat{A}$.

THEOREM 1.7 Let (A,m) be a Zariski ring. Then if $\hat{A}$ is a Krull ring, so is A. Also, ϕ: $C\ell(A) \to C\ell(\hat{A})$ is injective (Samuel, 1964a, p. 23).

Throughout this chapter we will concentrate for the most part on the case where qt(B)/qt(A) is a purely inseparable extension of degree $p > 0$. The following results, also from Samuel's notes, will be used extensively.

We let B be a Krull ring of characteristic $p \neq 0$. Let Δ be a derivation of qt(B) such that $\Delta(B) \subset B$. Let $K = \ker(\Delta)$ and $A = B \cap K$. Then A is a Krull ring with B integral over A. Thus we have a map ϕ: $C\ell(A) \to C\ell(B)$.

We set $L = \{t^{-1}\Delta t \mid t^{-1}\Delta t \in B,\ t \in qt(B)\}$. Note that L is an additive subgroup of B called the *group of logarithmic derivatives* of Δ. Set $L' = \{u^{-1}\Delta u \mid u \text{ is a unit in } B\}$. Then L' is a subgroup of L.

THEOREM 1.8 (a) There exists a canonical monomorphism $\bar{\phi}$: $\ker \phi \to L/L'$. (b) If $[qt(B) : K] = p$ and $\Delta(B)$ is not contained in any height 1 primes of B, then $\bar{\phi}$ is an isomorphism (Samuel, 1964a, p. 62).

THEOREM 1.9 If $[qt(B) : K] = p$, then (a) $\exists\, a \in A$ such that $\Delta^p = a\Delta$, and (b) an element $t \in B$ is in $L \Leftrightarrow \Delta^{p-1}(t) - at = -t^p$ (Samuel, 1964a, pp. 63-64).

REMARK 1.10 We take a moment to describe the monomorphism $\bar{\phi}$: $\ker \phi \to L/L'$. Let $\beta \in \ker \phi \subset C\ell(A)$. Then $\phi(\beta) = (t)$ for some $t \in qt(B)$. The map $\bar{\phi}$ takes β to $\Delta t/t$.

To see that $\Delta t/t$ is in B, we write β as a linear combination of height 1 primes of A. $\beta = n_1q_1 + \cdots + n_rq_r$, where the q_i are height 1 primes of A and the n_i are integers. For each i, there is a unique height 1 prime $\underline{p}_i$ in B lying over q_i. By definition $\phi(\beta) = n_1e_1\underline{p}_1 + \cdots + n_re_r\underline{p}_r$, where ℓ_i denotes the ramification index of $\underline{p}_i$, i = 1, ..., r, and $\phi(\beta) = (t)$ implies that $\underline{p}_1^{n_1e_1}\cdots\underline{p}_r^{n_re_r}B = tB$. Thus for each height 1 prime $\underline{p}$ of B the ramification index of $\underline{p}$ divides $v_{\underline{p}}(t)$, where $v_{\underline{p}}$ is the valuation corresponding to $\underline{p}$. It follows that there exists an $a \in K$ such that $v_{\underline{p}}(t) = v_{\underline{p}}(a)$, i.e., $t = au$, where u is a unit in $B_{\underline{p}}$. Thus $\Delta t/t = \Delta a/a + \Delta u/u = \Delta u/u$. Since $\Delta(B_{\underline{p}})$ is contained in $B_{\underline{p}}$, we conclude that $\Delta t/t \in B_{\underline{p}}$ for each height 1 prime $\underline{p}$ of B. Since B is a Krull ring, we have that $\Delta t/t \in$ B (see Fossum, 1973, p. 8).

2. PROPERTIES OF $C\ell(F)$ AND GANONG'S FORMULA

Let k be an algebraically closed field of characteristic $p > 0$. Most of our attention throughout this thesis will be given to surfaces $F \subsetneq \mathbf{A}_k^3$ defined by an equation of the form $z^p = G(x,y)$, where $G(x,y) \in k[x,y]$. Any such surface F is easily seen to be isomorphic to the surface F' defined by the equation $z^p = G(x,y) + [H(x,y)]^p$ for any $H(x,y) \in k[x,y]$ [just make the transformation $z \to z - H(x,y)$ in the equation $z^p = G(x,y)$]. Thus when we study surfaces F as above we will always assume that G(x,y) has no monomials that are elements of $k[x^p,y^p]$. Hereafter we will refer to this condition as condition (*) on G.

Furthermore, we will restrict our attention to surfaces F as above that are normal. By the jacobian criterion this is equivalent to assuming that $\partial G/\partial x$ and $\partial G/\partial y$ are relatively prime. We will refer to this as condition (**) on G.

LEMMA 2.1 Let $G(x,y) \in k[x,y]$ be a nonzero polynomial satisfying conditions (*) and (**). Let $F \subsetneq \mathbf{A}^3$ be the surface defined by the equation $z^p = G(x,y)$. Then the coordinate ring of F is isomorphic to $A = k[x^p,y^p,G]$.

Proof: The coordinate ring of F is $R = k[x,y,z]/I$, where I is the prime ideal in $k[x,y,z]$ generated by $z^p - G(x,y)$. Let $T: k[x,y,z] \to A$ be the mapping that sends an element $\Sigma\, \alpha_{ijk} x^i y^j z^k \in k[x,y,z]$ to the element $\Sigma\, \alpha^p_{ijk} x^{pi} y^{pj} G^k$. This is a surjective homomorphism since k is algebraically closed (hence perfect). Thus the kernel of T is a height 1 prime containing I. Since I is height 1 we have that $I = \ker(T)$. Therefore, R is isomorphic to A.

Let $D: k(x,y) \to k(x,y)$ be the k-derivation defined by

$$D = \frac{\partial G}{\partial y}\frac{\partial}{\partial x} - \frac{\partial G}{\partial x}\frac{\partial}{\partial y}$$

D is called the *jacobian derivation* of G.

LEMMA 2.2 With D as above and A as in (2.1) we have that $D^{-1}(0) \cap k[x,y] = A$.

Proof: Let $B = D^{-1}(0) \cap k[x,y]$. Then $k[x^p,y^p] \subseteq A \subseteq B \subseteq k[x,y]$. Thus B is integral over A. Since (*) is in force we know that either $G_x \neq 0$ or $G_y \neq 0$ (i.e., $Dx \neq 0$ or $Dy \neq 0$). Therefore, we have that $k(x^p,y^p) \subsetneq Q_A \subsetneq Q_B \subsetneq k(x,y)$. We conclude that $Q_A = Q_B$. Since A is integrally closed [condition (**)] we have that $A = B$.

LEMMA 2.3 Let D and A be as in (2.2). Let L be the group of logarithmic derivatives of D, $L = \{Df/f \mid f \in k(x,y) \text{ and } Df/f \in k[x,y]\}$. Then $C\ell(A) \approx L$.

Proof: $Dx = Gy$ and $Dy = -G_x$. Since G_x and G_y are relatively prime, the image of D is not contained in any height 1 prime of $k[x,y]$.

By lemma (2.2) and theorem (1.8) we have that $C\ell(A) \approx L/L'$, where $L' = \{Du/u \mid u \text{ is a unit in } k[x,y]\}$. Therefore, $L' = 0$.

LEMMA 2.4 Let D, A, and L be as in (2.3). If an element $t \in k[x,y]$ is in L, then the degree of $t \leq$ (degree of G) - 2.

Proof: Suppose that $t \in L$. Then $t \in k[x,y]$ and there exists $f \in k(x,y)$ such that $Df/f = t$. $\exists\, h,g \in k[x,y]$ such that $f = h/g^p$. Thus $Dh/h = t$. We have that $Dh = h_x G_y - k_y G_x$ is of degree $\leq \deg h + \deg G - 2$. This shows that $\deg t \leq \deg G - 2$.

REMARK 2.5 By bounding the degree of any logarithmic derivative we obtain a method for computing the divisor class group of a normal surface of the form $z^p = G(x,y)$ over an algebraically closed field of characteristic $p > 0$. We write t as a polynomial in x and y of degree equal to deg G - 2 with undetermined coefficients. We then substitute t into our differential equation (1.9b) and compare coefficients. We then see that the coefficients of t must satisfy a certain system of equations. The number of solutions of this system will be the order of the class group. Since $L \subsetneq k[x,y]$ each element in the class group will have p-torsion.

We will show shortly that these class groups are always of finite order. First we prove a formula inspired by R. Ganong which will prove useful in studying the differential equation (1.9b).

In attempting to understand the class groups of these surfaces it would seem that a formula for a in (1.9a) might be of value. R. Ganong conjectured that a formula for a is given by $a = \Sigma_{i=0}^{p-1} G^i \nabla \times (G^{p-(i+1)})$, where $\nabla = \partial^{2(p-1)}/\partial x^{p-1}\partial y^{p-1}$. While trying to prove this formula for a, we discovered the following formula, which will streamline many calculations.

THEOREM 2.6 Let $G \in k[x,y]$ satisfy condition (**) and D be the jacobian derivation of G. Then $\forall\ \alpha \in k(x,y)$, $D^{p-1}\alpha - a\alpha = -\Sigma_{i=0}^{p-1} G^i \nabla(G^{p-(i+1)}\alpha)$.

Note that Ganong's formula for a is obtained from theorem (2.6) by making the substitution $\alpha = 1$. Before we prove (2.6) we will need the following lemma.

LEMMA 2.7 Let D be as in (2.6) and E be the derivation on k(x,y) defined by $E = (1/G_x)(\partial/\partial x)$. Then $\forall\ \alpha \in k(x,y)$, $E^{p-1}(D^{p-1}(\alpha) - a\alpha) = \nabla(\alpha)$.

Proof: By (1.9a), $a \in A$ [i.e., $D(a) = 0$]. Given $\alpha \in k(x,y)$ we have that $D(D^{p-1}(\alpha) - a\alpha) = D^p\alpha - aD\alpha = 0$. Thus

$$D^{p-1}(\alpha) - a\alpha \in qt(A) \quad \forall\ \alpha \in k(x,y) \tag{2.7.1}$$

In the proof of (2.4) we saw that for $t \in k[x,y]$, $\deg(Dt) \leq \deg t + \deg G - 2$. Thus it follows that $\deg(a) \leq (p-1)[\deg(G)-2]$. It therefore follows that for every $t \in k[x,y]$,

$$\deg(D^{p-1}t - at) \leq \deg t + (p-1)(\deg G - 2) \tag{2.7.2}$$

Let $h \in A = k[x^p, y^p, G]$. Then $\exists$ unique $\beta_i \in k[x,y]$ such that $h = \Sigma_{i=0}^{p-1} \beta_i^p G^i$. Then $E(h) = \Sigma_{i=1}^{p-1} i\beta_i^p G^{i-1}$. In particular, we see that $E(h) \in A$ and $\deg E(h) \leq \deg h - \deg G$. If we combine this statement with (2.7.2), we obtain that $\forall\ t \in k[x,y]$ either

$$\begin{aligned} &(1)\quad E^{p-1}(D^{p-1}t - at) = 0, \text{ or} \\ &(2)\quad \deg(E^{p-1}(D^{p-1}t - at)) \leq \deg t - 2(p-1) \end{aligned} \tag{2.7.3}$$

Any differential operator on $k(x,y)$ can be written uniquely as a linear combination of $\alpha^{i+j}/\alpha x^i \alpha y^j$, $0 \leq i, j \leq p-1$, with coefficients in $k(x,y)$ (see Jacobson, 1964, p. 187). Thus there exists a unique $b_{ij} \in k(x,y)$, $0 \leq i, j \leq p-1$, such that

$$E^{p-1}(D^{p-1}\alpha - a\alpha) = \sum_{0 \leq i,j \leq p-1} b_{ij} \frac{\partial^{i+j}(\alpha)}{\partial x^i \partial y^j} \tag{2.7.4}$$

for all $\alpha \in k(x,y)$.

We now show that each $b_{ij} = 0$ for $0 \leq i + j \leq 2p - 3$. By (2.7.3), $E^{p-1}(D^{p-1}(1) - a) = 0$. If we substitute $\alpha = 1$ into the right side of (2.7.4), we see that $b_{00} = 0$. We proceed now by induction on $i + j$. Suppose that $b_{ij} = 0$ for all i,j such that $i + j < k$, where $k \leq 2p - 3$. Let $0 \leq i_0, j_0 \leq p - 1$ with $i_0 + j_0 = k$. Substituting $\alpha = x^{i_0} y^{j_0}$ into (2.7.4), we obtain $i_0! j_0! b_{i_0 j_0}$ on the right-hand side and 0 on the left by (2.7.3). Therefore, $b_{i_0 j_0} = 0$. We conclude that each $b_{ij} = 0$ for $i + j \leq 2p - 3$, which shows that

$$E^{p-1}(D^{p-1}\alpha - a\alpha) = b_{(p-1)(p-1)} \nabla(\alpha) \qquad \text{for all } \alpha \in k(x,y) \tag{2.7.5}$$

That $b_{(p-1)(p-1)} = 1$ is not difficult to see. The reasoning goes like this. We have the following two facts.

2.7.6 The highest-order differential form of the operator D^r is

$$\sum_{i=0}^{r} \binom{r}{i}(-1)^i G_y^{r-i} G_x^i \frac{\partial^r}{\partial x^{r-i}\partial y^i} \qquad \text{for } 0 < r < p$$

2.7.7 The highest-order differential form of the operator $E^s D^{p-1}$ is

$$\sum_{i=s}^{p-1} G_y^{p-(i+1)} G_x^{i-s} \frac{\partial^{p+s-1}}{\partial x^{p+s-(i+1)}\partial y^i} \qquad 0 \le s \le p-1$$

The proof of (2.7.6) is by simple induction on r. The proof of (2.7.7) uses (2.7.6) with $r = p - 1$ and induction on s.

If we substitute $s = p - 1$ into (2.7.7), we obtain that the form of order $2(p - 1)$ in $E^{p-1}D^{p-1}$ is $\partial^{2(p-1)}/\partial x^{p-1}\partial y^{p-1} = \nabla$. Since the differential operator $\alpha \to E^{p-1}(a\alpha)$ is of order at most $p - 1$, we conclude that the differential form of order $2(p - 1)$ in the operator $\alpha \to E^{p-1}(D^{p-1}\alpha - a\alpha)$ is ∇, i.e., $b_{(p-1)(p-1)} = 1$. Thus the lemma is proved.

Proof of theorem 2.6: Let $\alpha \in k(x,y)$ and j be a positive integer with $0 \le j \le p - 1$. We know by (2.7.1) that there is a unique $\alpha_i \in k(x,y)$ such that $D^{p-1}\alpha - a\alpha = \Sigma_{i=0}^{p-1} \alpha_i^p G^i$. On the one hand, we know from lemma (2.7) that $E^{p-1}(D^{p-1}(G^j\alpha) - aG^j\alpha) = \nabla(G^j\alpha)$. On the other hand, we see directly that $E^{p-1}(D^{p-1}(G^j\alpha) - aG^j\alpha) = E^{p-1}(G^j(D^{p-1}\alpha - a\alpha)) = E^{p-1}(\Sigma_{i=0}^{p-1} \alpha_i^p G^{i+j}) = -\alpha_{p-(j+1)}^p$. We conclude that

$$D^{p-1}\alpha - a\alpha = -\sum_{i=0}^{p-1} G^i \nabla(G^{p-(i+1)}\alpha) \qquad \text{Q.E.D.}$$

Hereafter we will refer to the formula in theorem (2.6) as *Ganong's formula.*

LEMMA 2.8 Let $G \in k[x,y]$ be of degree g. Let $H = \{t \in k[x,y] \mid (G^{p-1}t) = t^p\}$. Then H is a finite p-group of type $(p,\ldots,p)$ of order p^M, where $M \le g(g - 1)/2$.

Proof: Given $t \in [x,y]$, we have that $\deg\{\nabla(G^{p-1}t)\} \le (p - 1)g + \deg t - 2(p - 1)$. It follows that if $t \in H$, then $p \deg t \le (p - 1)g + \deg t - 2(p - 1)$ and hence $\deg t \le g - 2$. Thus $t = \Sigma_{i+j\le g-2} \alpha_{ij} x^i y^j$

for some $\alpha_{ij} \in k$. Substituting t into the equation

$$\nabla(G^{p-1}t) = t^p \tag{2.8.1}$$

we obtain on the left side of this equation a polynomial in x and y whose coefficients are linear expressions in the α_{ij} with coefficients in k. Comparing the coefficients of $x^{rp}y^{sp}$ on both sides of (2.8.1), we see that for each pair of nonnegative integers (r,s) with $r + s \leq g - 2$, α_{rs} must satisfy an equation of the form

$$L_{rs} = \alpha_{rs}^p \tag{2.8.2}$$

where L_{rs} is a linear expression in the α_{ij} with coefficients in k. There are a total of $g(g - 1)/2$ such equations.

Note that the ring $R = k[\ldots,\alpha_{rs},\ldots]$ with the relations $L_{rs} = \alpha_{rs}^p$ is a finite-dimensional k-vector space spanned by all monomials in the α_{rs} of degree less than or equal to $(p - 1)g(g - 1)/2$. Thus R is artinian and has a finite number of maximal ideals (see Atiyah and Macdonald, 1969, p. 89).

Therefore, the $g(g - 1)/2$ equations $L_{rs} = \alpha_{rs}^p$ intersect in a finite number of points. By Bezout's theorem this number is at most $p^{g(g-1)/2}$ (see Griffiths and Harris (1978), p. 670).

This shows that the order of $H \leq p^{g(g-1)/2}$. H is a p-group of type (p,...,p) since $H \subset k[x,y]$.

PROPOSITION 2.9 Let $G \in k[x,y]$ satisfy conditions (*) and (**), $F \subsetneq \mathbf{A}^3$ be the surface defined by $z^p = G(x,y)$. Then $C\ell(F)$ is a finite p group of type (p,p,...,p) of order p^M where $M \leq g(g - 1)/2$. [By $C\ell(F)$ we mean the class group of the coordinate ring of F.]

Proof: Let D: $k(x,y) \to k(x,y)$ be the jacobian derivation of G, and L be the corresponding group of logarithmic derivatives of D. By (2.3), $C\ell(F) \approx L$. By (1.9b) an element $t \in k[x,y]$ is in $L \Leftrightarrow D^{p-1}t - at = -t^p$. Using Ganong's formula, we see that $t \in k[x,y]$ is in $L \Leftrightarrow$

$$\nabla(G^{p-1}t) = t^p \text{ and } \nabla(G^jt) = 0 \quad \text{for } j = 0, 1, \ldots, p - 2 \tag{2.9.1}$$

Thus we see that $L \subsetneq H$, where H is the group defined in (2.8) and the proposition follows immediately.

We can use the other equations in (2.9.1) to reduce this upper bound somewhat. For example, if $t \in L$, then $\nabla t = 0$. So we see that t can have no monomials of the form $x^i y^i$, where $i \equiv j \equiv -1 \pmod p$. Thus we see that the order of $C\ell(F)$ is p^M, where $M \leq g(g-1)/2 - N$, where N = the number of ordered pairs (i,j) such that $0 \leq i + j \leq g - 2$ and $i \equiv j \equiv -1 \pmod p$. We take up the matter of refining this upper bound in Section 10.

In addition to imposing conditions (*) and (**) on G(x,y) we will often make the assumption that the polynomials G_x, G_y, and $G_{xx}G_{yy} - G_{xy}^2$ do not have a point in common (a generic assumption on G).

P. Blass (1983) has shown that under these conditions all singularities on F: $z^p = G(x,y)$ are rational. He has also shown that the local equation of any singularity will be of the form $z^p = xy$ + (higher-degree terms). Hereafter we will refer to this additional condition on G(x,y) as condition (B).

2.10 We finish this section with the following remark. Given a logarithmic derivative $t \in L$, how do we find an $f \in k(x,y)$ such that $Df/f = t$? Samuel gives such a procedure in his proof of theorem (3.2) in his Tata notes (Samuel, 1964a).

We let Δ: $k(x,y) \to k(x,y)$ be the derivation $\Delta = (1/t)D$. Then choose $s \in k(x,y)$ such that $Ds \neq 0$. Let $y_1 = \Delta s$ and $y_i = \Delta y_{i-1} - (i-1)y_{i-1}$, $i = 2, \ldots, p$. Samuel shows that $y_p = 0$. Thus there exists an i such that $y_i \neq 0$ and $y_{i+1} = 0$. From this element we obtain t, i.e., $Dy_i/y_i = t$.

3. THE SURFACE $z^2 = G$

Let k be an algebraically closed field of characteristic 2. Let $G(x,y) \in k[x,y]$ satisfy conditions (*) and (**). Let D be the jacobian derivation of G and $L \subseteq k[x,y]$ be the corresponding group of logarithmic derivatives. Then our A in lemma (2.1) becomes $k[x^2,y^2,G]$ and our differential equation (1.9b) becomes

3.1 An element $t \in k[x,y]$ is in $L \Leftrightarrow Dt + at = t^2$, where $D^2 = aD$.

We begin by calculating a in (3.1). We can assume that $G_y \neq 0$. Then $a = D^2(x)/Dx = DG_y/G_y = G_{xy}$.

3.2 Thus $a = G_{xy}$.

3.3 Note that this shows that $G_{xy} \in L$, since $G_{xy} = DG_y/G_y$. Therefore if $G_{xy} \neq 0$, then $L \neq 0$. We conclude that if $G_{xy} \neq 0$, the surface F defined by $z^2 = G(x,y)$ has nonfactorial coordinate ring. In other words, for a generic G, $C\ell(F) \neq 0$.

REMARK 3.4 The same argument shows that the surface F defined by $z^p = G(x,y)$ over an algebraically closed field of characteristic $p > 0$ has nontrivial class group whenever $G_{yy} = 0$ (or $G_{xx} = 0$) and $G_{xy} \neq 0$.

3.5 Two questions immediately arise.

(1) We know that when $p = 2$, $C\ell(F) \neq 0$ for a generic G. Then the obvious question is: What is $C\ell(F)$ for a generic choice of G when $p = 2$? Also,

(2) For $p > 2$, does the surface F have factorial coordinate ring for a generic G?

We attempt to shed some light on these questions in this section and in Sections 4, 5, 6, and 10.

Our approach toward answering these questions is to study these problems locally. By this we mean to study the divisor class group of the local ring of F at a singular point. This we do in Section 10.

Another attack might be to bound the degree of G and study the corresponding system of equations (see remark 2.5). This we do below and in Sections 4, 5, and 6.

Let us begin with the "simplest" case. Let $F \subseteq \mathbf{A}^3$ be the surface defined by the equation $z^2 = G_4(x,y)$, where G_4 stands for a polynomial of degree ≤ 4 satisfying conditions (*) and (**). Thus G_4 is of the form

$$G_4 = \alpha_{10}x + \alpha_{01}y + \alpha_{11}xy + \alpha_{30}x^3 + \alpha_{21}x^2y + \alpha_{12}xy^2 + \alpha_{03}y^3 + \alpha_{31}x^3y + \alpha_{13}xy^3$$

By (2.3), $C\ell(F) \approx L$, where L is the group of logarithmic derivatives of the jacobian derivation, D, of G_4.

By (2.4), $\deg(t) \le 2$. Thus t is of the form

$$t = c_{00} + c_{10}x + c_{01}y + c_{20}x^2 + c_{11}xy + c_{02}y^2$$

As in the proof of (2.9), $t \in L \Leftrightarrow \nabla(Gt) = t^2$ and $\nabla(t) = 0$. $\nabla(t) = 0 \Rightarrow c_{11} = 0$. We can quickly calculate $\nabla(Gt)$, obtaining

$$\begin{aligned}\nabla(Gt) = {} & (c_{00} + c_{20}x^2 + c_{02}y^2)(\alpha_{11} + \alpha_{31}x^2 + \alpha_{13}y^2) \\ & + c_{10}(\alpha_{01} + \alpha_{21}x^2 + \alpha_{03}y^2) \\ & + c_{01}(\alpha_{10} + \alpha_{30}x^2 + \alpha_{12}y^2)\end{aligned} \tag{3.6}$$

If we compare the coefficients of (3.6) with those of t^2, we see that the c_{ij} must satisfy the following system of equations:

$$\begin{aligned}
&(1)\quad \alpha_{11}c_{00} + \alpha_{01}c_{10} + \alpha_{10}c_{01} = c_{00}^2 \\
&(2)\quad \alpha_{31}c_{00} + \alpha_{21}c_{10} + \alpha_{30}c_{01} + \alpha_{11}c_{20} = c_{10}^2 \\
&(3)\quad \alpha_{13}c_{00} + \alpha_{03}c_{10} + \alpha_{12}c_{01} + \alpha_{11}c_{02} = c_{01}^2 \\
&(4)\quad \alpha_{13}c_{02} = c_{02}^2 \\
&(5)\quad \alpha_{13}c_{02} = c_{02}^2 \\
&(6)\quad \alpha_{13}c_{20} + \alpha_{31}c_{02} = 0
\end{aligned} \tag{3.7}$$

We have that the order of $C\ell(F)$ = the number of solutions to the system (3.7). Let us analyze this system, with the eventual goal of finding the order of $C\ell(F)$ for a generic G_4.

To perform this analysis we will employ the following three lemmas.

LEMMA 3.8 Let k be an algebraically closed field of characteristic $p \ne 0$. Let S_1 be a system of equations of the form

$$
\begin{aligned}
S_1:\quad &(1)\quad \alpha_{11}x_1 + \cdots + \alpha_{1n}x_n = x_1^p\\
&(2)\quad \alpha_{21}x_1 + \cdots + \alpha_{2n}x_n = x_2^p\\
&\quad\vdots\\
&(n)\quad \alpha_{n1}x_1 + \cdots + \alpha_{nn}x_n = x_n^p\\
&(\ell_1)\quad \alpha_1x_1 + \cdots + \alpha_nx_n = 0 \qquad \text{where } \alpha_n \neq 0
\end{aligned}
$$

Then S_1 is equivalent (in the sense that it has the same number of solutions) to the system S_2.

$$
\begin{aligned}
S_2:\quad &(1)\quad \beta_{11}x_1 + \cdots + \beta_{1(n-1)}x_{n-1} = x_1^p\\
&(2)\quad \beta_{21}x_1 + \cdots + \beta_{2(n-1)}x_{n-1} = x_2^p\\
&\quad\vdots\\
&(n-1)\quad \beta_{(n-1)1}x_1 + \cdots + \beta_{(n-1)(n-1)}x_{n-1} = x_{n-1}^p\\
&(\ell_1)\quad \alpha_1x_1 + \cdots + \alpha_nx_n = 0\\
&(\ell_2)\quad \beta_1x_1 + \cdots + \beta_{n-1}x_{n-1} = 0
\end{aligned}
$$

where

$$\beta_{ij} = \alpha_{ij}\alpha_n - \alpha_{in}\alpha_j$$

$$\beta_j = \alpha_1^p\beta_{ij} + \alpha_2^p\beta_{2j} + \cdots + \alpha_{n-1}^p\beta_{(n-1)j} + \alpha_n^p(\alpha_{nj}\alpha_n - \alpha_{nn}\alpha_j)$$

Proof: Multiply equation (i) in S_1 by α_n and subtract equation (ℓ_1) multiplied by α_{in}. We obtain the equation

$$(\bar{i})\quad \beta_{i1}x_1 + \cdots + \beta_{i(n-1)}x_{n-1} = \alpha_n x_i^p$$

where $\beta_{ij} = \alpha_{ij}\alpha_n - \alpha_{in}\alpha_j$ for $i = 1, \ldots, n$, $j = 1, 2, \ldots, n-1$. Now multiply each equation $(\bar{i})$ by α_i^p and sum these together. We obtain the equation

$$(3.8.1)\quad \beta_1x_1 + \cdots + \beta_{n-1}x_{n-1} = \alpha_n(\alpha_1x_1 + \cdots + \alpha_nx_n)^p \qquad (3.8.1)$$

where

$$\beta_j = \alpha_1^p\beta_{ij} + \cdots + \alpha_{n-1}^p\beta_{(n-1)j} + \alpha_n^p\beta_{nj}$$

Combining (3.8.1) with ℓ_1 we obtain the linear equation

$$(\ell_2)\quad \beta_1 x_1 + \cdots + \beta_{n-1} x_{n-1} = 0$$

We conclude that our original system is equivalent to the system

$$(3.8.2)\qquad \begin{array}{rl} (\bar{1}) & \beta_{11} x_1 + \cdots + \beta_{1(n-1)} x_{n-1} = \alpha_n x_1^p \\ (\overline{n-1}) & \beta_{(n-1)1} x_1 + \cdots + \beta_{(n-1)(n-1)} x_{n-1} = \alpha_n x_{n-1}^p \\ (\ell_1) & \alpha_1 x_1 + \cdots + \alpha_n x_n = 0 \\ (\ell_2) & \beta_1 x_1 + \cdots + \beta_{n-1} x_{n-1} = 0 \end{array}\qquad (3.8.2)$$

Note that if we look at the points of intersection of these hypersurfaces [in (3.8.2)] in projective space $\mathbb{P}_k^n$, we see that all intersections occur at finite distance.

But the system

$$\begin{array}{l} u^{p-1}\beta_{11} x_1 + \cdots + u^{p-1}\beta_{1(n-1)} x_{n-1} = \alpha_n x_1^p \\ \vdots \\ u^{p-1}\beta_{(n-1)1} x_1 + \cdots + u^{p-1}\beta_{(n-1)(n-1)} x_{n-1} = \alpha_n x_{n-1}^p \\ (\ell_1) \text{ and } (\ell_2) \end{array}$$

is clearly isomorphic to the system

$$\begin{array}{l} u^{p-1}\beta_{11} x_1 + \cdots + u^{p-1}\beta_{1(n-1)} x_{n-1} = x_1^p \\ \vdots \\ u^{p-1}\beta_{(n-1)1} x_1 + \cdots + u^{p-1}\beta_{(n-1)(n-1)} x_{n-1} = x_{n-1}^p \\ (\ell_1) \text{ and } (\ell_2) \end{array}$$

Thus we see that the system (3.8.2) is equivalent to S_2.

LEMMA 3.9 Let k be an algebraically closed field of characteristic $p \neq 0$. Consider the system of equations

$$\begin{bmatrix} a_{11} & a_{12} & \cdots & a_{1n} & c_{11} & c_{12} & \cdots & c_{1m} \\ a_{21} & a_{22} & \cdots & a_{2n} & c_{21} & c_{22} & \cdots & c_{2m} \\ \cdots & & & & \cdots & & & \\ a_{n1} & a_{n2} & \cdots & a_{nn} & c_{n1} & c_{n2} & \cdots & c_{nm} \\ b_{11} & b_{12} & \cdots & b_{1n} & d_{11} & d_{12} & \cdots & d_{1m} \\ b_{21} & b_{22} & \cdots & b_{2n} & d_{21} & d_{22} & \cdots & d_{2m} \\ \cdots & & & & \cdots & & & \\ b_{m1} & b_{m2} & \cdots & b_{mn} & d_{m1} & d_{m2} & \cdots & d_{mm} \end{bmatrix} \begin{bmatrix} x_1 \\ x_2 \\ \vdots \\ x_n \\ y_1 \\ y_2 \\ \vdots \\ y_m \end{bmatrix} = \begin{bmatrix} x_1^p \\ x_2^p \\ \vdots \\ x_n^p \\ y_1^p \\ y_2^p \\ \vdots \\ y_m^p \end{bmatrix} \qquad (3.9.1)$$

Suppose that the rank of the matrix above is m and $\det[d_{rs}] \neq 0$. Then the system (3.9.1) is equivalent to the system

$$\begin{bmatrix} A & 0 & \cdots & 0 & E_{11} & E_{12} & \cdots & E_{1m} \\ 0 & A & \cdots & 0 & E_{21} & E_{22} & \cdots & E_{2m} \\ \cdots & & & & \cdots & & & \\ 0 & 0 & \cdots & A & E_{n1} & E_{n2} & \cdots & E_{nm} \\ b_{11} & b_{12} & \cdots & b_{1n} & d_{11} & d_{12} & \cdots & d_{1m} \\ b_{21} & b_{22} & \cdots & b_{2n} & d_{21} & d_{22} & \cdots & d_{2m} \\ \cdots & & & & \cdots & & & \\ b_{m1} & b_{m2} & \cdots & b_{mn} & d_{m1} & d_{m2} & \cdots & d_{mm} \end{bmatrix} \begin{bmatrix} x_1 \\ x_2 \\ \vdots \\ x_n \\ y_1 \\ y_2 \\ \vdots \\ y_m \end{bmatrix} = \begin{bmatrix} 0 \\ 0 \\ \vdots \\ 0 \\ y_1^p \\ y_2^p \\ \vdots \\ y_n^p \end{bmatrix} \qquad (3.9.2)$$

where $A^p = \det[d_{rs}]$ and where E_{ij}^p is the cofactor of b_{j1} in the $(m + 1) \times (m + 1)$ matrix

$$\begin{bmatrix} a_{i1} & c_{i1} & c_{i2} & \cdots & c_{im} \\ b_{11} & d_{11} & d_{12} & \cdots & d_{1m} \\ b_{21} & d_{21} & d_{22} & \cdots & d_{2m} \\ b_{m1} & d_{m1} & d_{m2} & \cdots & d_{mm} \end{bmatrix}$$

Proof: Let i = 1, ..., n. Since the rank of the matrix in (3.9.1) is m and $\det[d_{rs}] \neq 0$, there is a nontrivial linear combination of row i and rows n + j, j = 1, ..., m, in (3.9.1) that gives 0. As a matter of fact, we know that A^p row(i) + E_{i1}^p row(n + 1) + ⋯ + E_{im}^p row(n + m) will give 0 on the left of the "=" in (3.9.1), while the right side of the "=" will be

$$A^p x_i^p + E_{i1}^p y_1^p + \cdots + E_{im}^p y_m^p$$

Thus we see that we can replace row (i) in system (3.9.1) by the equation

$$Ax_i + E_{i1}y_1 + \cdots + E_{im}y_m = 0$$

LEMMA 3.10 Let k be as in (3.9). Let A be the $(n + m) \times (n + m)$ matrix of (3.9.1). Then the system $A\vec{x} = \vec{x}^p$ has p^m distinct solutions.
Proof: Using (3.9), we reduce to the case where our system is as in (3.9.2). If we homogenize this system, we see by Bézout's theorem that there are p^m points of intersection. But it is easy to see that there are no points of intersection at infinity. Since the determinant of (3.9.2) is not zero, each of these intersection points has multiplicity 1.

3.11 Lemmas (3.8) and (3.10) demonstrate that $C\ell(F)$ can always be determined for any normal surface F: $z^p = G(x,y)$ over an algebraically closed field k of characteristic $p \neq 0$. For we have seen (2.8) that the order of $C\ell(F)$ is the number of distinct solutions to a system of equations of the form $\ell_{ij} = x_{ij}^p$ and $\ell_r = 0$, where ℓ_{ij} and ℓ_r are linear expressions in the x_{ij} with coefficients in k. Using (3.8), we can reduce this system to a system of n linear equations in n unknowns having only the trivial solution or else to a system of the type discussed in (3.10).

We now resume our analysis of system (3.7). We will consider two cases, $\alpha_{13}\alpha_{31} \neq 0$ and $\alpha_{13}\alpha_{31} = 0$.
Case 1: $\alpha_{31}\alpha_{13} \neq 0$. From equation (5) we know that $c_{02} = 0$ or α_{13}. If $c_{02} = 0$, then from (6) we have that $c_{20} = 0$, and system 3.7 reduces to the system

$$\begin{aligned}
&(1)\quad \alpha_{11}c_{00} + \alpha_{01}c_{10} + \alpha_{10}c_{01} = c_{00}^2 \\
&(2)\quad \alpha_{31}c_{00} + \alpha_{21}c_{10} + \alpha_{30}c_{01} = c_{10}^2 \qquad (3.12) \\
&(3)\quad \alpha_{13}c_{00} + \alpha_{03}c_{10} + \alpha_{12}c_{01} = c_{01}^2
\end{aligned}$$

The number of solutions to (3.12) is 2^3 if and only if

$$\begin{bmatrix} \alpha_{11} & \alpha_{01} & \alpha_{10} \\ \alpha_{31} & \alpha_{21} & \alpha_{30} \\ \alpha_{13} & \alpha_{03} & \alpha_{12} \end{bmatrix} \neq 0 \qquad \text{by (3.10)} \tag{3.13}$$

If $c_{03} = \alpha_{13}$, then from (6) we have that $c_{20} = \alpha_{31}$ and our system (3.7) reduces to the system

$$\begin{aligned} &(1) \quad \alpha_{11}c_{00} + \alpha_{01}c_{10} + \alpha_{10}c_{01} = c_{00}^2 \\ &(2) \quad \alpha_{31}c_{00} + \alpha_{21}c_{10} + \alpha_{30}c_{01} = c_{10}^2 + \alpha_{11}\alpha_{31} \\ &(3) \quad \alpha_{13}c_{00} + \alpha_{03}c_{10} + \alpha_{12}c_{01} = c_{01}^2 + \alpha_{11}\alpha_{13} \end{aligned} \tag{3.14}$$

If in this system we make the substitutions $c_{00} \to c_{00} + \alpha_{11}$, we obtain the system (3.12).

Thus we conclude that if $\alpha_{13}\alpha_{31} \neq 0$, then (3.7) has 2^4 solutions $\Leftrightarrow$ the determinant in (3.13) is not zero.

Case 2: $\alpha_{13}\alpha_{31} = 0$. Assume that $\alpha_{13} = 0$. Then from (5) of (3.7) we have that $c_{02} = 0$. Equations (4) and (6) imply that $c_{20} = 0$. And our system again reduces to the system (3.12), which has 2^3 solutions if (3.13) holds.

We review the discussion above in the following proposition.

PROPOSITION 3.15 The surface F: $z^2 = G_4$ has class group of order 2^4 $\Leftrightarrow$

$$\alpha_{31}\alpha_{13}\begin{vmatrix} \alpha_{11} & \alpha_{01} & \alpha_{10} \\ \alpha_{31} & \alpha_{21} & \alpha_{30} \\ \alpha_{13} & \alpha_{03} & \alpha_{12} \end{vmatrix} \neq 0$$

In other words, for a generic G_4, $C\ell(F)$ consists of four copies of $\mathbb{Z}/2\mathbb{Z}$.

From case 2 we see that if $\alpha_{13}\alpha_{31} = 0$, then the order of $C\ell(F) \leq 2^3$ and we get equality if (3.13) holds.

Let us examine what other possibilities for $C\ell(F)$ can occur when F is given by the equation $z^2 = G_4$, where we assume that the degree of G is 4. By (3.10) we have that if $\alpha_{31}\alpha_{13} = 0$,

$$\begin{vmatrix} \alpha_{11} & \alpha_{01} & \alpha_{10} \\ \alpha_{31} & \alpha_{21} & \alpha_{30} \\ \alpha_{13} & \alpha_{03} & \alpha_{12} \end{vmatrix} = 0 \quad \text{and} \quad \begin{vmatrix} \alpha_{31} & \alpha_{21} \\ \alpha_{13} & \alpha_{03} \end{vmatrix} \neq 0$$

then $C\ell(F)$ has order 2^2.

Applying (3.10) again, we have that if

$$\alpha_{13}\alpha_{31} = \begin{vmatrix} \alpha_{11} & \alpha_{01} & \alpha_{10} \\ \alpha_{31} & \alpha_{21} & \alpha_{30} \\ \alpha_{13} & \alpha_{03} & \alpha_{12} \end{vmatrix} = \begin{vmatrix} \alpha_{31} & \alpha_{21} \\ \alpha_{13} & \alpha_{03} \end{vmatrix} = 0$$

then $C\ell(F)$ has order 2.

Of course, this is the smallest that the order of $C\ell(F)$ can be. For if deg $G_4 = 4$, then either α_{13} or $\alpha_{31} \neq 0$. Then (3.3) implies that $C\ell(F) \neq 0$.

So we see that we have a kind of stratification on the set of coefficients of G_4. If we view the coefficients of G_4 as being coordinate functions of projective space $\mathbb{P}_k^8$, we see that on a dense open subset, U say, of $\mathbb{P}_k^8$, $C\ell(F)$ has order 2^4. On a subset $C_1 \subsetneq U$ of codimension 1, the class group is of order 2^3. On a subset $C_2 \subsetneq C_1$ of codimension 2, the class group is of order 2^2, and $C\ell(F)$ is $\mathbf{Z}/2\mathbf{Z}$ on a subset $C_3 \subsetneq C_2$ of codimension 3.

REMARK 3.16 In the examples to be considered in Sections 4, 5, and 6, we will see that a quite opposite type of stratification occurs (see 4.18, 4.25, 5.15, 5.16, 5.26, 6.14, 6.25). In particular, we will show that in each of these examples $C\ell(F\colon\ z^p = G_n)$ has minimum possible order for a generic choice of G_n.

Let us backtrack for a moment. We just observed that if deg $G_4 = 4$, then $C\ell(F) \neq 0$, where F is defined by $z^2 = G_4$ and the characteristic of k is 2.

What if we assume that the degree of G_3 is 3; is $C\ell(F) = 0$ a possibility, where F is defined by the equation $z^2 = G_3$? This case can quickly be studied by just making the substitution $\alpha_{13} = \alpha_{13} = 0$ into the expression for G_4 on page 152. We are then considering the surface $z^2 = G_3$ where G_3 is of the form $G = \alpha_{10}x + \alpha_{01}y + \alpha_{11}xy + \alpha_{30}x^3 + \alpha_{21}x^2y + \alpha_{12}xy^2 + \alpha_{03}y^3$. Substituting $\alpha_{13} = \alpha_{31} = 0$ into system (3.7), we see that the order of $C\ell(F\colon\ z^2 = G_3)$ is equal to the number of distinct solutions to the system of equations

$$\begin{aligned} \alpha_{11}c_{00} + \alpha_{01}c_{10} + \alpha_{10}c_{01} &= c_{00}^2 \\ \alpha_{21}c_{10} + \alpha_{30}c_{01} &= c_{10}^2 \\ \alpha_{03}c_{10} + \alpha_{12}c_{01} &= c_{01}^2 \end{aligned} \tag{3.17}$$

Using (3.10), we see that $C\ell(F\colon\ z^2 = G_3)$ has order 2^3 if and only if

$$\alpha_{11}\begin{vmatrix} \alpha_{21} & \alpha_{30} \\ \alpha_{03} & \alpha_{12} \end{vmatrix} \neq 0$$

Also, we observe that the class group has order 2^2 when

$$\alpha_{11} = 0 \qquad \text{and} \qquad \begin{vmatrix} \alpha_{21} & \alpha_{30} \\ \alpha_{03} & \alpha_{12} \end{vmatrix} \neq 0$$

and has order 2 when

$$\alpha_{11} = \begin{vmatrix} \alpha_{21} & \alpha_{30} \\ \alpha_{03} & \alpha_{12} \end{vmatrix} = 0 \qquad \text{and} \qquad \alpha_{12} \neq 0$$

Finally, we observe that $C\ell(F) = 0$ if and only if

$$\alpha_{11} = \begin{vmatrix} \alpha_{21} & \alpha_{30} \\ \alpha_{03} & \alpha_{12} \end{vmatrix} = \alpha_{21} = \alpha_{12} = 0 \tag{3.18}$$

The "only if" implication follows from (3.3), (3.10), and the discussion above. The "if" implication is obvious.

Note that we have imposed on G_3, as we do throughout this paper, that G_{3x} and G_{3y} are relatively prime. Thus we obtain the following proposition.

PROPOSITION 3.19 Let $F \subsetneq \mathbf{A}^3$ be a normal surface defined by the equation $z^2 = G(x,y)$, where $G \in k[x,y]$ has degree ≤ 4 and k is an algebraically closed field of characteristic 2. Then $C\ell(F) = 0$ if and only if F is a plane.

Proof: We can assume that G satisfies condition $(*)$. We have seen that if $\deg G = 4$, then $C (F) \neq 0$. If $\deg G = 3$ and $C\ell(F) = 0$, then

$$\alpha_{11} = \alpha_{21} = \alpha_{12} = \begin{vmatrix} \alpha_{21} & \alpha_{30} \\ \alpha_{03} & \alpha_{12} \end{vmatrix} = 0 \qquad \text{by (3.18)}$$

This leaves only two possibilities for G, either $G = \alpha_{10}x + \alpha_{01}y + \alpha_{30}x^3$ or $G = \alpha_{10}x + \alpha_{01}y + \alpha_{03}y^3$. If $G = \alpha_{10}x + \alpha_{01}y + \alpha_{30}x^3$, then to maintain normality we must have that $\alpha_{01} \neq 0$. If we make the substitution $y \to \alpha_{10}x + \alpha_{01}y + \alpha_{30}x^3$, we see that F is isomorphic to the surface $z^2 = y$. The case $G = \alpha_{10}x + \alpha_{01}y + \alpha_{03}y^3$ is handled similarly.

By (3.3), the degree of G cannot be 2. Thus if $C\ell(F) = 0$, then F is a plane. The reverse implication is obvious.

We finish this section with the following discussion. We saw in (3.3) that if k is an algebraically closed field of characteristic 2 and $F \subsetneq \mathbf{A}^3$ is a surface defined by $z^2 = G(x,y)$, then $G_{xy} \neq 0$ implies that $C\ell(F) \neq 0$. We then should be able to produce a nonprincipal height 1 prime in $A = k[x^2,y^2,G]$, which we know is isomorphic to the coordinate ring of F. We do this with the aid of the next lemma.

LEMMA 3.20 Let $f \in k[x,y]$ be such that $Df/f \in k[x,y]$. Suppose that $f = g^r h$, where $g \in k[x,y]$ is irreducible, r is a positive integer not divisible by p (the characteristic of k), and that $h \in k[x,y]$ is relatively prime to g. Then $Dg/g \in k[x,y]$.

Proof: Let $t = Df/f$. Then $ft = Df = D(g^r h) = rg^{r-1}(Dg)h + g^r Dh$. Thus we see that g divides $rhDg \Rightarrow g$ divides Dg.

We now continue the search for the nonprincipal height 1 prime. Let $G_x = G_1^{r_1} \cdots G_n^{r_n}$ be a factorization of G_x into irreducible factors in $k[x,y]$. Since $G_{xy} \neq 0$, one of the r_i is not divisible by 2, say r_1. By (3.20),

$$DG_1/G_1 \in k[x,y]$$

Let $I = G_1k[x,y] \cap A$. I is clearly a height 1 prime. We now show that I is not principal. Note first that $D(G_xG_y) = 0$, i.e., $G_xG_y \in A$. We have that $r_1 = 2s_1 + 1$ for some nonnegative integer s_1. Also, $G_xG_y/G_1^{2s_1} = G_1G_2^{r_2}\cdots G_n^{r_n}G_y \in k[x,y] \cap qt(A) = A$. We conclude that $G_xG_y/G_1^{2s_1}$ is an element of I of value 1 in the valuation on $k(x,y)$ induced by $G_1k[x,y]$. It follows that the ramification index of $G_1k[x,y]$ over I is 1. Thus θ: $C\ell(A) \to L$ maps I to DG_1/G_1. Since θ is well defined, I must be nonprincipal.

4. THE SURFACE $z^3 = G_4$ AND ARTIN'S CONJECTURE

We continue our investigation of questions (1) and (2) of (3.5). Throughout this section we let k be an algebraically closed field of characteristic 3 and $G_4 \in k[x,y]$ be of degree 4 satisfying conditions (*) and (**). Then G_4 has the form $G_4 = \alpha_{10}x + \alpha_{01}y + \alpha_{20}x^2 + \alpha_{11}xy + \alpha_{02}y^2 + \alpha_{21}x^2y + \alpha_{12}xy^2 + \alpha_{40}x^4 + \alpha_{31}x^3y + \alpha_{22}x^2y^2 + \alpha_{13}xy^3 + \alpha_{04}y^4$. Let $F \subseteq \mathbf{A}_k^3$ be the surface defined by $z^3 = G_4$. Let D be the jacobian derivation of G_4 and $L \subseteq k[x,y]$ the group of logarithmic derivatives. Then $C\ell(F) \approx L$.

Let $t \in L$. By (2.4), deg $t \leq 2$. By (1.9) and Ganong's formula, we have that

$$\nabla(G^2t) = t^3 \qquad \text{and} \qquad \nabla(Gt) = \nabla(t) = 0 \tag{4.1}$$

t must be of the form $t = c_{00} + c_{10}x + c_{01}y + c_{20}x^2 + c_{11}xy + c_{02}y^2$. Substituting this expression for t into (4.1), we see that the order of $C\ell(F\colon\ z^3 = G_4)$ is equal to the number of distinct solutions to the system

$$
\begin{aligned}
(1)\quad & (\alpha_{11}^2 - \alpha_{10}\alpha_{12} - \alpha_{01}\alpha_{21} - \alpha_{20}\alpha_{02})c_{00} \\
& - (\alpha_{10}\alpha_{02} + \alpha_{01}\alpha_{11})c_{10} - (\alpha_{10}\alpha_{11} + \alpha_{01}\alpha_{20})c_{01} \\
& + \alpha_{01}^2 c_{20} - \alpha_{10}\alpha_{01}c_{11} + \alpha_{10}^2 c_{02} = c_{00}^3 \\
(2)\quad & -(\alpha_{21}\alpha_{31} + \alpha_{12}\alpha_{40})c_{00} \\
& + (\alpha_{21}^2 - \alpha_{20}\alpha_{22} - \alpha_{11}\alpha_{31} - \alpha_{02}\alpha_{40})c_{10} \\
& - (\alpha_{20}\alpha_{31} + \alpha_{11}\alpha_{40})c_{01} \\
& - (\alpha_{10}\alpha_{22} + \alpha_{01}\alpha_{31} + \alpha_{20}\alpha_{12} + \alpha_{11}\alpha_{21})c_{20} \\
& - (\alpha_{10}\alpha_{31} + \alpha_{01}\alpha_{40} + \alpha_{20}\alpha_{21})c_{11} - \alpha_{10}\alpha_{40}\alpha_{02} = c_{10}^3 \\
(3)\quad & -(\alpha_{21}\alpha_{04} + \alpha_{12}\alpha_{13})c_{00} - (\alpha_{11}\alpha_{04} + \alpha_{02}\alpha_{13})c_{10} \qquad (4.2) \\
& + (\alpha_{12}^2 - \alpha_{20}\alpha_{04} - \alpha_{11}\alpha_{13} - \alpha_{02}\alpha_{22})c_{01} - \alpha_{01}\alpha_{04}c_{20} \\
& - (\alpha_{04}\alpha_{10} + \alpha_{01}\alpha_{13} + \alpha_{02}\alpha_{12})c_{11} \\
& - (\alpha_{13}\alpha_{10} + \alpha_{01}\alpha_{22} + \alpha_{11}\alpha_{12} + \alpha_{02}\alpha_{21})c_{02} = c_{01}^3 \\
(4)\quad & (\alpha_{31}^2 - \alpha_{40\ 22})c_{20} - \alpha_{40}\alpha_{31}c_{11} + \alpha_{40}^2 c_{02} = c_{20}^3 \\
(5)\quad & -(\alpha_{31}\alpha_{04} + \alpha_{22}\alpha_{13})c_{20} + (\alpha_{22}^2 - \alpha_{04}\alpha_{40} - \alpha_{31}\alpha_{13})c_{11} \\
& - (\alpha_{40}\alpha_{13} + \alpha_{22}\alpha_{31})c_{02} = c_{11}^3 \\
(6)\quad & \alpha_{04}^2 c_{20} - \alpha_{13}\alpha_{04}c_{11} + (\alpha_{13}^2 - \alpha_{22}\alpha_{04})c_{02} = c_{02}^3 \\
(\ell_1)\quad & \alpha_{22}c_{00} + \alpha_{12}c_{10} + \alpha_{21}c_{01} + \alpha_{02}c_{20} + \alpha_{11}c_{11} + \alpha_{20}c_{02} = 0
\end{aligned}
$$

We will now argue that for a generic choice of the α_{ij}, this system has only the trivial solution. Let $A = (\alpha_{ij})$, $0 \le i,j \le 6$, be the coefficient matrix we obtain from system (4.2), equations (1)-(6), where $a_{ij}, a_{2j}, a_{3j}, \ldots, a_{6j}$ stands for the coefficient of $c_{00}, c_{10}, c_{01}, \ldots, c_{02}$, respectively, in equation (j).

If we now assume that $\alpha_{20} \neq 0$, we can apply lemma (3.8) to obtain an equivalent system of equations

$$\begin{bmatrix} b_{11} & b_{12} & \cdots & b_{15} \\ b_{21} & b_{22} & \cdots & b_{25} \\ \cdots & & & \\ b_{51} & b_{52} & \cdots & b_{55} \end{bmatrix} \begin{bmatrix} c_{00} \\ c_{10} \\ c_{01} \\ c_{20} \\ c_{11} \end{bmatrix} = \begin{bmatrix} c_{00}^3 \\ c_{10}^3 \\ c_{01}^3 \\ c_{20}^3 \\ c_{11}^3 \end{bmatrix} \qquad (4.3)$$

$$(\ell_2) \quad b_1c_{00} + b_2c_{10} + b_3c_{01} + b_4c_{20} + b_5c_{11} = 0$$

and (ℓ_1), where (ℓ_1) is the linear equation appearing in (4.2) and where we employ lemma (3.8) to calculate the b_{ij} and b_j. We list these calculations in Table 1.

Table 1

$b_{11} = \alpha_{20}\alpha_{11}^2 - \alpha_{20}\alpha_{10}\alpha_{12} - \alpha_{20}^2\alpha_{02} - \alpha_{20}\alpha_{01}\alpha_{21} - \alpha_{10}^2\alpha_{22}$,

$b_{12} = -(\alpha_{10}\alpha_{02}\alpha_{20} + \alpha_{01}\alpha_{11}\alpha_{20} + \alpha_{10}^2\alpha_{12})$, $b_{13} = -(\alpha_{10}\alpha_{11}\alpha_{20} + \alpha_{01}\alpha_{20}^2 + \alpha_{10}^2\alpha_{21})$,

$b_{14} = \alpha_{01}^2\alpha_{20} - \alpha_{10}^2\alpha_{02}$, $b_{15} = (\alpha_{10}\alpha_{01}\alpha_{20} + \alpha_{10}^2\alpha_{11})$

$b_{21} = -(\alpha_{21}\alpha_{31}\alpha_{20} + \alpha_{12}\alpha_{40}\alpha_{20} - \alpha_{10}\alpha_{40}\alpha_{22})$, $b_{22} = \alpha_{21}^2\alpha_{20} - \alpha_{20}^2\alpha_{22} - \alpha_{11}\alpha_{31}\alpha_{20} -$

$\alpha_{20}\alpha_{02}\alpha_{40} + \alpha_{10}\alpha_{40}\alpha_{12}$, $b_{23} = \alpha_{10}\alpha_{40}\alpha_{21} - \alpha_{20}^2\alpha_{31} - \alpha_{20}\alpha_{11}\alpha_{40}$,

$b_{24} = \alpha_{10}\alpha_{40}\alpha_{02} - \alpha_{10}\alpha_{22}\alpha_{20} - \alpha_{20}\alpha_{01}\alpha_{31} - \alpha_{20}^2\alpha_{12} - \alpha_{20}\alpha_{11}\alpha_{21}$,

$b_{25} = \alpha_{10}\alpha_{11}\alpha_{40} - \alpha_{10}\alpha_{20}\alpha_{31} - \alpha_{01}\alpha_{20}\alpha_{40} - \alpha_{20}^2\alpha_{21}$

$b_{31} = \alpha_{13}\alpha_{10}\alpha_{22} + \alpha_{01}\alpha_{22}^2 + \alpha_{11}\alpha_{12}\alpha_{22} + \alpha_{02}\alpha_{21}\alpha_{22} - \alpha_{20}\alpha_{21}\alpha_{04} - \alpha_{12}\alpha_{13}\alpha_{20}$,

$b_{32} = \alpha_{13}\alpha_{10}\alpha_{12} + \alpha_{01}\alpha_{22}\alpha_{12} + \alpha_{11}\alpha_{12}^2 + \alpha_{02}\alpha_{21}\alpha_{12} - \alpha_{11}\alpha_{04}\alpha_{20} - \alpha_{20}\alpha_{02}\alpha_{13}$,

$b_{33} = \alpha_{21}\alpha_{13}\alpha_{10} + \alpha_{21}\alpha_{01}\alpha_{22} + \alpha_{11}\alpha_{12}\alpha_{21} + \alpha_{21}^2\alpha_{02} + \alpha_{20}\alpha_{12}^2 - \alpha_{20}^2\alpha_{04} -$

$\alpha_{20}\alpha_{02}\alpha_{22} - \alpha_{20}\alpha_{11}\alpha_{13}$, $b_{34} = \alpha_{02}\alpha_{13}\alpha_{10} + \alpha_{01}\alpha_{22}\alpha_{02} + \alpha_{11}\alpha_{12}\alpha_{02} -$

Table 1 (continued)

$\alpha_{20}\alpha_{01}\alpha_{04} + \alpha_{02}^2\alpha_{21}$, $b_{35} = \alpha_{11}\alpha_{13}\alpha_{10} + \alpha_{11}\alpha_{01}\alpha_{22} + \alpha_{11}\alpha_{02}\alpha_{21} -$

$\alpha_{20}\alpha_{04}\alpha_{10} - \alpha_{20}\alpha_{01}\alpha_{13} - \alpha_{20}\alpha_{02}\alpha_{12}$

$b_{41} = -\alpha_{40}^2\alpha_{22}$, $b_{42} = -\alpha_{40}^2\alpha_{12}$, $b_{43} = -\alpha_{40}^2\alpha_{21}$, $b_{44} = -\alpha_{40}\alpha_{22}\alpha_{20} +$
$\alpha_{31}^2\alpha_{20} - \alpha_{40}^2\alpha_{02}$, $b_{45} = -\alpha_{40}\alpha_{31}\alpha_{20} - \alpha_{40}^2\alpha_{11}$

$b_{51} = \alpha_{40}\alpha_{13}\alpha_{22} + \alpha_{22}^2\alpha_{31}$, $b_{52} = \alpha_{12}\alpha_{40}\alpha_{13} + \alpha_{12}\alpha_{22}\alpha_{31}$,

$b_{53} = \alpha_{21}\alpha_{40}\alpha_{13} + \alpha_{22}\alpha_{31}\alpha_{21}$, $b_{54} = \alpha_{40}\alpha_{02}\alpha_{13} + \alpha_{02}\alpha_{22}\alpha_{31} - \alpha_{20}\alpha_{31}\alpha_{04} -$

$\alpha_{20}\alpha_{22}\alpha_{13}$, $b_{55} = \alpha_{11}\alpha_{40}\alpha_{13} + \alpha_{11}\alpha_{22}\alpha_{31} + \alpha_{22}^2\alpha_{20} - \alpha_{40}\alpha_{04}\alpha_{20} - \alpha_{13}\alpha_{31}\alpha_{20}$.

$b_1 = \alpha_{22}^3 b_{11} + \alpha_{12}^3 b_{21} + \alpha_{21}^3 b_{31} + \alpha_{02}^3 b_{41} + \alpha_{11}^3 b_{51} + \alpha_{20}^3(\alpha_{22}^2\alpha_{04} - \alpha_{22}\alpha_{13}^2)$

$b_2 = \alpha_{22}^3 b_{12} + \alpha_{12}^3 b_{22} + \alpha_{21}^3 b_{32} + \alpha_{02}^3 b_{42} + \alpha_{11}^3 b_{52} + \alpha_{20}^3(\alpha_{12}\alpha_{22}\alpha_{04} - \alpha_{12}\alpha_{13}^2)$

$b_3 = \alpha_{22}^3 b_{13} + \alpha_{12}^3 b_{23} + \alpha_{21}^3 b_{33} + \alpha_{02}^3 b_{43} + \alpha_{11}^3 b_{53} + \alpha_{20}^3(\alpha_{21}\alpha_{22}\alpha_{04} - \alpha_{21}\alpha_{13}^2)$

$b_4 = \alpha_{22}^3 b_{14} + \alpha_{12}^3 b_{24} + \alpha_{21}^3 b_{34} + \alpha_{02}^3 b_{44} + \alpha_{11}^3 b_{54} + \alpha_{20}^3(\alpha_{04}^2\alpha_{20} + \alpha_{02}\alpha_{22}\alpha_{04} -$
$\alpha_{02}\alpha_{12}^2)$

$b_5 = \alpha_{22}^3 b_{15} + \alpha_{12}^3 b_{25} + \alpha_{21}^3 b_{35} + \alpha_{02}^3 b_{45} + \alpha_{11}^3 b_{55} + \alpha_{20}^3(\alpha_{11}\alpha_{22}\alpha_{04} - \alpha_{11}\alpha_{13}^2 -$
$\alpha_{04}\alpha_{13}\alpha_{20})$.

We assume that $b_5 \neq 0$ and apply lemma (3.8). Then we have that system (4.3) is equivalent to the system of equations

$$\begin{bmatrix} c_{11} & c_{12} & c_{13} & c_{14} \\ c_{21} & c_{22} & c_{23} & c_{24} \\ c_{31} & c_{32} & c_{33} & c_{34} \\ c_{41} & c_{42} & c_{43} & c_{44} \end{bmatrix} \begin{bmatrix} c_{00} \\ c_{10} \\ c_{01} \\ c_{20} \end{bmatrix} = \begin{bmatrix} c_{00}^3 \\ c_{10}^3 \\ c_{01}^3 \\ c_{20}^3 \end{bmatrix} \tag{4.4}$$

$$(\ell_3) \quad c_1 c_{00} + c_2 c_{10} + c_3 c_{01} + c_4 c_{20} = 0$$

(ℓ_2) and (ℓ_1), where (ℓ_2) and (ℓ_1) are as above and where we list the formulas for the c_{ij} and c_j below.

c_{ij}	1	2	3	4
1	$b_{11}b_5 - b_{15}b_1$	$b_{12}b_5 - b_{15}b_2$	$b_{13}b_5 - b_{15}b_3$	$b_{14}b_5 - b_{15}b_4$
2	$b_{21}b_5 - b_{25}b_1$	$b_{22}b_5 - b_{25}b_2$	$b_{23}b_5 - b_{25}b_3$	$b_{24}b_5 - b_{25}b_4$
3	$b_{31}b_5 - b_{35}b_1$	$b_{32}b_5 - b_{35}b_2$	$b_{33}b_5 - b_{35}b_3$	$b_{34}b_5 - b_{35}b_4$
4	$b_{41}b_5 - b_{45}b_1$	$b_{42}b_5 - b_{45}b_2$	$b_{43}b_5 - b_{45}b_3$	$b_{44}b_5 - b_{45}b_4$

$$c_1 = b_1^3 c_{11} + b_2^3 c_{21} + b_3^3 c_{31} + b_4^3 c_{41} + b_5^3 (b_{51}b_5 - b_{55}b_1) \tag{4.5}$$

$$c_2 = b_1^3 c_{12} + b_2^3 c_{22} + b_3^3 c_{32} + b_4^3 c_{42} + b_5^3 (b_{52}b_5 - b_{55}b_2)$$

$$c_3 = b_1^3 c_{13} + b_2^3 c_{23} + b_3^3 c_{33} + b_4^3 c_{43} + b_5^3 (b_{53}b_5 - b_{55}b_3)$$

$$c_4 = b_1^3 c_{14} + b_2^3 c_{24} + b_3^3 c_{34} + b_4^3 c_{44} + b_5^3 (b_{54}b_5 - b_{55}b_4)$$

We will now show that c_4 is a nonzero polynomial in the α_{rs}. If we make the substitution

$$\alpha_{21} = i \qquad \text{all other } \alpha_{rs} = 1 \tag{4.6}$$

then the matrix appearing on the left side of "=" in (4.3), which we will hereafter refer to as B, becomes

$$\begin{bmatrix} 1-i & 0 & 1-i & 0 & 1 \\ -i & 0 & i+1 & 1-i & i-1 \\ -1 & i+1 & 0 & i-1 & i \\ -1 & -1 & -i & -1 & 1 \\ -1 & -1 & -i & 0 & 1 \end{bmatrix} \tag{4.7}$$

and the b_j become

$$b_1 = -i - 1 \qquad b_2 = -i - 1 \qquad b_3 = i - 1 \qquad b_4 = -1$$
$$b_5 = -i - 1$$

So if we let C be the matrix in system (4.4), we see that under the substitution (4.6), C becomes

$$\begin{bmatrix} i-1 & i+1 & -i-1 & 1 \\ -i-1 & i & i+1 & -i \\ -i & -i-1 & i+1 & i-1 \\ -i-1 & -i-1 & 0 & i-1 \end{bmatrix} \tag{4.8}$$

and $c_1 = 1$, $c_2 = -i$, $c_3 = i$, $c_4 = -i - 1$.

Thus we see that not only is c_4 a nonzero polynomial in the α_{rs}, but so are the polynomials c_1, c_2, c_3, and det C.

Assume that $c_4 \neq 0$. By (3.8) we obtain the equivalent system of equations

$$\begin{bmatrix} d_{11} & d_{12} & d_{13} \\ d_{21} & d_{22} & d_{23} \\ d_{31} & d_{32} & d_{33} \end{bmatrix} \begin{bmatrix} c_{00} \\ c_{10} \\ c_{01} \end{bmatrix} = \begin{bmatrix} c_{00}^3 \\ c_{10}^3 \\ c_{01}^3 \end{bmatrix} \tag{4.9}$$

$$(\ell_4) \quad d_1c_{00} + d_2c_{10} + d_3c_{01} = 0$$

and ℓ_3, ℓ_2, ℓ_1 as above.

A table of formulas appears below.

$$\begin{aligned} d_1 &= c_1^3d_{11} + c_2^3d_{21} + c_3^3d_{31} + c_4^3(c_{41}c_4 - c_{44}c_1) \\ d_2 &= c_1^3d_{12} + c_2^3d_{22} + c_3^3d_{32} + c_4^3(c_{42}c_4 - c_{44}c_2) \\ d_3 &= c_1^3d_{13} + c_2^3d_{23} + c_3^3d_{33} + c_4^3(c_{43}c_4 - c_{44}c_3) \end{aligned} \tag{4.10}$$

d_{ij}	1	2	3
1	$c_{11}c_4 - c_{14}c_1$	$c_{12}c_4 - c_{14}c_2$	$c_{13}c_4 - c_{14}c_3$
2	$c_{21}c_4 - c_{24}c_1$	$c_{22}c_4 - c_{24}c_2$	$c_{23}c_4 - c_{24}c_3$
3	$c_{31}c_4 - c_{34}c_1$	$c_{32}c_4 - c_{34}c_2$	$c_{33}c_4 - c_{34}c_3$

(4.10) cont.

To see that d_3 is a nonzero homogeneous polynomial in the α_{rs}, we continue with our substitution (4.6). If we let D be the matrix in (4.10), then under (4.6) D becomes

$$\begin{bmatrix} 1 & -i & i \\ 0 & -i - 1 & i - 1 \\ 0 & i - 1 & -i + 1 \end{bmatrix} \tag{4.11}$$

and we obtain $d_1 = -1$, $d_2 = -1$, $d_3 = -i - 1$.

We assume now that $d_3 \neq 0$, thus obtaining the equivalent system

$$\begin{bmatrix} e_{11} & e_{12} \\ e_{21} & e_{22} \end{bmatrix} \begin{bmatrix} c_{00} \\ c_{10} \end{bmatrix} = \begin{bmatrix} c_{00}^3 \\ c_{10}^3 \end{bmatrix} \tag{4.12}$$

$$(\ell_5) \quad e_1 c_{00} + e_2 c_{01} = 0$$

and ℓ_1, ℓ_2, ℓ_3, and ℓ_4, where we list the corresponding formulas

e_{ij}	1	2
1	$d_{11}d_3 - d_{13}d_1$	$d_{12}d_3 - d_{13}d_2$
2	$d_{21}d_3 - d_{23}d_1$	$d_{22}d_3 - d_{23}d_2$

(4.13)

$$e_1 = d_1^3 e_{11} + d_2^3 e_{21} + d_3^3 (d_{31}d_3 - d_{33}d_1)$$

$$e_2 = d_1^3 e_{12} + d_2^3 e_{22} + d_3^3 (d_{32}d_3 - d_{33}d_2)$$

Let E be the matrix of (4.13). Under the substitution (4.6) we find that E becomes

$$\begin{bmatrix} -1 & -i - 1 \\ i - 1 & -1 \end{bmatrix} \tag{4.14}$$

and we have that $e_1 = i - 1$, $e_2 = -i$. Hence e_2 is a nonzero polynomial in the α_{rs}.

So we may assume that $e_2 \neq 0$ and system (4.12) is then equivalent to the system

$$f_{11}c_{00} = c_{00}^3 \quad \text{and} \quad (\ell_6) \quad f_1 c_{00} = 0 \tag{4.15}$$

together with equations ℓ_1, ℓ_2, ℓ_3, ℓ_4, ℓ_5, and ℓ_6, and where $f_{11} = e_{11}e_2 - e_{12}e_1$ and

$$f_1 = e_1^3 f_{11} + e_2^3(e_{21}e_2 - e_{22}e_1)$$

Under (4.6) these become

$$f_{11} = i + 1 \quad \text{and } f_1 = i + 1 \tag{4.16}$$

Thus f_1 is a nonzero polynomial.

From the discussion above, we draw some immediate conclusions.

4.17 1. If we choose the α_{rs} so that $a_6b_5c_4d_3e_2f_1 \neq 0$, then our original system (4.2) is equivalent to the system (4.15). Then the linear equations ℓ_1, ℓ_2, ..., ℓ_6 of (4.15) have only the trivial solution (i.e., $c_{00} = c_{10} = c_{01} = c_{20} = c_{11} = c_{02} = 0$). Thus we conclude that generically the surface F: $z^3 = G_4$ has trivial class group. The surface $z^3 = x + y + x^2 + xy + y^2 + x^4 + x^3y + x^2y^2 + xy^3 + y^4$ provides an example of a surface with $C\ell(F) = 0$.

2. Note that this is the opposite of the behavior that we have when $p = 2$. There we saw that the surface $z^2 = G_4$ generically had divisor class group of maximum possible order 2^4.

3. If we study system (4.2) more carefully, we find that the maximum number of solutions to this system is 3^5. For if one of α_{22}, α_{12}, α_{21}, α_{02}, α_{11}, α_{20} is not 0, then by (3.8) and (3.10) we have that system (4.2) has at most 3^5 solutions. If $\alpha_{22} = \alpha_{12} = \alpha_{21} = \alpha_{02} = \alpha_{11} = \alpha_{20} = 0$, the matrix A will have 0's in the first three columns. By (3.10) the order of the solution set of system

(4.2) will be at most 3^3. We conclude that the order of $C\ell(F: z^3 = G_4) \leq 3^5$. We will see shortly that this upper bound is reached.

4.18 Let us now examine, as we did in Section 3, what sort of "stratification" does $C\ell(F)$ induce on the coefficient space of G_4. From (4.3) we have that the order of $C\ell(F:\ z^3 = G_4)$ is e^5 if $b_1 = b_2 = b_3 = b_4 = b_5 = 0$ and α_{20} det $B \neq 0$. An example of a surface whose coefficients satisfy these conditions is $z^3 = -x^4 + x^3y + y^4 + x^2 + xy$. If we let C_5 be the subset of the coefficient space of G_4 corresponding to those surfaces F such that the order of $C\ell(F) = 3^5$ we see that the codimension of C_5 is at most 5.

From (4.4) we have that the order of $C\ell(F)$ is 3^4 if $b_2\alpha_{20} \times \det(C) \neq 0$ and $c_1 = c_2 = c_3 = c_4 = 0$. An example of such a surface is given by $z^3 = xy + x^2 + x^4 - y^4$. We let C_4 be the subset of the coefficient space of G_4 consisting of those coefficients for which $C\ell(F)$ has order 3^4. It follows that C_4 is of codimension ≤ 4.

If $c_4b_5\alpha_{20}$ det $D \neq 0$ and $d_1 = d_2 = d_3 = 0$, then $C\ell(F)$ has order 3^3. Thus C_3, the subset on which $C\ell(F)$ has order 3^3, has codimension ≤ 3. The surface $z^3 = -x^4 + x^3y + xy^3 + y^4 + x^2 - xy$ provides an example with $c_4b_5\alpha_{20}$ det $D \neq 0$ and $d_1 = d_2 = d_3 = 0$.

From (4.12) we have that the order of $C\ell(F)$ is 3^2 whenever $d_3c_4b_5\alpha_{20}$ det $E \neq 0$ and $e_1 = e_2 = 0$. An example of such a surface is given by $z^3 = xy + x^2 + ix^2y + xy^2 + x^4 + x^3y + x^2y^2 + xy^3$. The codimension of C_2, the subset of the coefficient space on which $C\ell(F)$ has order 3^2, is of codimension ≤ 2.

Using (4.15), we have that coefficients satisfying $e_2d_3c_4b_5 \times \alpha_{20}f_{11} \neq 0$ and $f_1 = 0$ describe surfaces with class groups of order 3. $z^2 = x^2 + xy + x^2y + xy^2 + x^4 + x^3y + ix^2y^2 + xy^3$ is such a surface. Thus C_1, the subset of the coefficient space of G_4 corresponding to those coefficients for which $C\ell\ F = \mathbb{Z}/3\mathbb{Z}$, has codimension 1.

We conclude that the coefficient space of F: $z^3 = G_4$ behaves very differently from that of F^1: $z^2 = G_4$, where the ground fields of F and F^1 are of characteristic 3 and 2, respectively.

The coefficient space of F^1 contains subsets $U' \supset C_1^1 \supset C_2^1 \supset C_3^1$ of codimension 0, 1, 2, 3 on which the order of $C\ell(F^1)$ is 2^4, 2^3,

2^2, 2, respectively. On the coefficient space of F we have subsets U, C_1, C_2, C_3, C_4, C_5 of codimension 0, 1, 2, 3, 4, and 5 on which the order of $C\ell(F)$ is 0, 3, 3^2, 3^3, 3^4, and 3^5.

We are in a position, admittedly a rather tenuous one, to produce some questions and conjectures.

4.19 We have now seen that the surface $z^3 = G_4$ generically has trivial class group. Is this the case when $p > 2$? That is, does the surface F: $z^p = G$ have trivial class group when $p > 2$ for generic G?

4.20 If $p > 2$, is there then always a stratification of the coefficient space of $z^p = G_n$ of the type described for the surface F: $z^3 = G_4$?

4.21 If the characteristic of k is 2, is the maximum order of $C\ell(F\colon\ z^2 = G_n)$ reached on a dense open subset of the coefficient space, as was the case when $n = 4$? We will see in Section 5 that this is not true in general.

4.22 P. Blass (1982) has shown that for generic G_4 the surface F^1; $z^2 = G_4$ is rational and that the surface F: $z^3 = G_4$ is K3. Is there some relationship between the "generic genus" of the surface $z^p = G_n$ and the behavior of $C\ell(z^p = G_n)$ with respect to the coefficient space?

In an effort to attack these questions we begin with some perhaps fragile conjectures.

4.23 When $p > 2$, the surface $z^p = G$ has trivial class group for a generic G.

4.24 When $p = 2$, the surface $z^2 = G$ has class group of order 2 for a generic G if deg $G > 4$. Motivation for this conjecture is given in Sections 5 and 10.

4.25 That a relationship exists between the generic genus of $z^p = G_n$ and the behavior of $C\ell(z^p = G_n)$ with respect to the coefficient space of G_n was first conjectured by M. Artin. Artin's conjecture, as I understand it, is this: If the surface $z^p = G_n$ has geometric genus $p_g = m > 0$ for a generic G_n and if the order of $C\ell(z^p = G_n)$ is p^s for a generic G_n, then the subset of the coefficient space corresponding to those coefficients of G_n for which $C\ell(F)$ has order p^{s+i} has codimension $\leq im$.

At this point we do not possess the tools to attack these questions completely. But the need for more examples is immediate.

To shed some light on (4.19), (4.20), and (4.22) we will study the surface $z^5 = G_3$ where the characteristic of k is 5. P. Blass (1982) has shown that this surface is K3 generically. We conduct this study in Section 5.

In the same section, the surface $z^2 = G_6$ over k of characteristic 2, which from Blass we know has generic genus 1, will also be studied, thus providing input into questions (4.20), (4.21), and (4.22).

In Section 6, the surface $z^3 = G_6$, having generic genus 3, is considered. Evidence supporting conjectures (4.23), (4.24), and (4.25) is obtained.

5. THE SURFACES $z^5 = G_3$ AND $z^2 = G_6$

Let k be an algebraically closed field of characteristic 5, and let $G_3 \in k[x,y]$ be of degree 3 satisfying conditions (*) and (**). Then G_3 is of the form

$$G_3 = \alpha_{10}x + \alpha_{01}y + \alpha_{20}x^2 + \alpha_{11}xy + \alpha_{02}y^2 + \alpha_{30}x^3 + \alpha_{21}x^2y + \alpha_{12}xy^2 + \alpha_{03}y^3$$

where the $\alpha_{rs} \in k$. Let $F \subseteq \mathbf{A}^3_k$ be the surface defined by $z^5 = G_3$. Let D be the jacobian derivation of G_3 and $L \subseteq k[x,y]$ the group of logarithmic derivatives. Then $C\ell(F) \approx L$.

Given $t \in L$, we have that $\deg t \leq 1$ by (2.4). By (1.9) and Ganong's formula, we have that

$$\nabla(G_3^4 t) = t^5 \quad \text{and} \quad \nabla(G_3^3 t) = \nabla(G_3^2 t) = \nabla(G_3 t) = \nabla(t) = 0 \qquad (5.1)$$

where $\nabla = \partial^8/\partial x^4 \partial y^4$. Since $\deg t \leq 1$, t must be of the form

$t = c_{00} + c_{10}x + c_{01}y$. Such a t automatically satisfies $\nabla(G_3^2 t) = \nabla(G_3 t) = \nabla(t) = 0$ since $\deg(G_3^i t) \le 7$ for i = 0, 1, 2 and ∇ when applied to any polynomial just gives the coefficients appearing in front of monomials $x^k y^\ell$ with $k \equiv \ell \equiv 4 \pmod 5$.

$$\begin{aligned}
\nabla G_3^3 &= \alpha_{20}(\alpha_{21}\alpha_{03} + 3\alpha_{12}^2) + \alpha_{11}(\alpha_{30}\alpha_{03} + \alpha_{12}\alpha_{21}) \\
&\quad + \alpha_{02}(\alpha_{30}\alpha_{12} + 3\alpha_{21}^2) \\
\nabla G_3^3 x) &= \alpha_{10}(3\alpha_{12}^2 + \alpha_{21}\alpha_{03}) + \alpha_{01}(\alpha_{30}\alpha_{03} + \alpha_{21}\alpha_{12}) + 3\alpha_{30}\alpha_{02}^2 \\
&\quad + \alpha_{21}\alpha_{11}\alpha_{02} + \alpha_{12}(\alpha_{20}\alpha_{02} + 3\alpha_{11}^2) + \alpha_{03}\alpha_{20}\alpha_{11} \qquad (5.2) \\
\nabla(G_3^3 y) &= \alpha_{10}(\alpha_{30}\alpha_{03} + \alpha_{21}\alpha_{12}) + \alpha_{01}(\alpha_{30}\alpha_{12} + 3\alpha_{21}^2) \\
&\quad + \alpha_{30}\alpha_{11}\alpha_{02} + \alpha_{21}(3\alpha_{11}^2 + \alpha_{20}\alpha_{02}) + \alpha_{12}\alpha_{20}\alpha_{11} \\
&\quad + 3\alpha_{03}\alpha_{20}^2
\end{aligned}$$

Let $a_1 = \nabla G_3^3$, $a_2 = \nabla(G_3^3 x)$, $a_3 = \nabla(G_3^3 y)$. Then $\nabla(G_3^3 t) = 0$ implies that (c_{00}, c_{10}, c_{01}) must satisfy the linear equation

$$(\ell_1) \quad a_1 c_{00} + a_2 c_{10} + a_3 c_{01} = 0$$

Also, we have that

$$\begin{aligned}
\nabla(G_3^4) &= \alpha_{10}^2(\alpha_{12}^2 + 2\alpha_{21}\alpha_{03}) + 4\alpha_{10}\alpha_{01}(\alpha_{30}\alpha_{03} + \alpha_{21}\alpha_{12}) \\
&\quad + \alpha_{01}^2(\alpha_{21}^2 + 2\alpha_{30}\alpha_{12}) + 2\alpha_{20}^2\alpha_{01}\alpha_{03} \\
&\quad + 2\alpha_{11}^2(\alpha_{12}\alpha_{10} + \alpha_{21}\alpha_{01}) + 2\alpha_{02}^2\alpha_{10}\alpha_{30} \\
&\quad + 4\alpha_{20}\alpha_{11}(\alpha_{10}\alpha_{03} + \alpha_{01}\alpha_{12}) + 4\alpha_{20}\alpha_{02}(\alpha_{10}\alpha_{12} + \alpha_{01}\alpha_{21}) \\
&\quad + 4\alpha_{11}\alpha_{02}(\alpha_{10}\alpha_{21} + \alpha_{01}\alpha_{30}) + \alpha_{20}^2\alpha_{02}^2 + 2\alpha_{20}\alpha_{11}^2\alpha_{02} + \alpha_{11}^4 \\
\nabla(G_3^4 x) &= [2\alpha_{10}^2(\alpha_{11}\alpha_{03} + \alpha_{02}\alpha_{12}) \\
&\quad + 4\alpha_{10}\alpha_{01}(\alpha_{20}\alpha_{03} + \alpha_{11}\alpha_{12} + \alpha_{02}\alpha_{21}) \qquad (5.3) \\
&\quad + 2\alpha_{01}^2(\alpha_{20}\alpha_{12} + \alpha_{11}\alpha_{21} + \alpha_{02}\alpha_{30}) + 2\alpha_{10}(\alpha_{20}\alpha_{02}^2 + \alpha_{11}^2\alpha_{02}) \\
&\quad + 4\alpha_{01}(\alpha_{20}\alpha_{11}\alpha_{02} + \alpha_{11}^3)] + [\alpha_{30}^2(\alpha_{12}^2 + 2\alpha_{21}\alpha_{03}) \\
&\quad + 2\alpha_{30}\alpha_{21}^2\alpha_{12} + \alpha_{21}^4]\,x^5 \\
&\quad + (4\alpha_{30}\alpha_{03}^3 + 2\alpha_{21}\alpha_{12}^2\alpha_{03} + 4\alpha_{12}^3\alpha_{03})y^5
\end{aligned}$$

$$
\begin{aligned}
\nabla(G_3^4 y) = {} & [2\alpha_{01}^2(\alpha_{11}\alpha_{30} + \alpha_{20}\alpha_{21}) \\
& + 4\alpha_{01}\alpha_{10}(\alpha_{02}\alpha_{30} + \alpha_{11}\alpha_{21} + \alpha_{20}\alpha_{12}) \\
& + 2\alpha_{10}^2(\alpha_{02}\alpha_{21} + \alpha_{11}\alpha_{12} + \alpha_{20}\alpha_{02}) \\
& + 2\alpha_{01}(\alpha_{02}\alpha_{20}^2 + \alpha_{11}^2\alpha_{20}) + 2\alpha_{01}(\alpha_{02}\alpha_{20}^2 + \alpha_{11}^2\alpha_{20}) \\
& + 4\alpha_{10}(\alpha_{02}\alpha_{11}\alpha_{20} + \alpha_{11}^3)] + (4\alpha_{03}\alpha_{30}^3 + 2\alpha_{12}\alpha_{21}\alpha_{30}^2 \\
& + 4\alpha_{21}^3\alpha_{30})x^5 \\
& + [\alpha_{03}^2(\alpha_{21}^2 + 2\alpha_{12}^2\alpha_{30}) + 2\alpha_{03}\alpha_{12}\alpha_{21} + \alpha_{12}^4]y^5
\end{aligned}
\tag{5.3 cont.}
$$

5.4 Let $a_{11} = \nabla(G_3^4)$, a_{12} be the constant term in $\nabla(G_3^4 x)$, a_{13} be the constant coefficient in $\nabla(G_3^4 y)$, a_{22} be the coefficient of x^5 in $\nabla(G_3^4 x)$, and a_{32}, a_{33} be the coefficients of y^5 in $\nabla(G_3^4 x)$, $\nabla(G_3^4 y)$, respectively. Then $\nabla(G_3^4 t) = t^5$ implies that (c_{00}, c_{10}, c_{01}) must satisfy the system of equations

$$
\begin{aligned}
a_{11}c_{00} + a_{12}c_{10} + a_{13}c_{01} &= c_{00}^5 \\
a_{22}c_{10} + a_{23}c_{01} &= c_{10}^5 \\
a_{32}c_{10} + a_{33}c_{01} &= c_{01}^5
\end{aligned}
\tag{5.5}
$$

Combining (5.5) and (ℓ_1), we see that $t \in L$ if and only if (c_{00}, c_{10}, c_{01}) satisfies the system of equations

$$
\begin{bmatrix} a_{11} & a_{12} & a_{13} \\ 0 & a_{22} & a_{23} \\ 0 & a_{32} & a_{33} \end{bmatrix}
\begin{bmatrix} c_{00} \\ c_{10} \\ c_{01} \end{bmatrix}
=
\begin{bmatrix} c_{00}^5 \\ c_{10}^5 \\ c_{01}^5 \end{bmatrix}
\tag{5.6}
$$

$$(\ell_1) \quad a_1c_{00} + a_2c_{10} + a_3c_{01} = 0$$

We now analyze this system with the aid of lemmas (3.8) and (3.10).

If $a_3 \neq 0$, then by (3.8) system (5.6) is equivalent to the system of equations

$$
\begin{bmatrix} b_{11} & b_{12} \\ b_{21} & b_{22} \end{bmatrix}
\begin{bmatrix} c_{00} \\ c_{10} \end{bmatrix}
=
\begin{bmatrix} c_{00}^5 \\ c_{10}^5 \end{bmatrix}
\tag{5.7}
$$

(ℓ_2) $b_1c_{00} + b_2c_{10} = 0$ and equation (ℓ_1) that appears in (5.6).

The formulas for the b_{ij} and b_j are as follows:

b_{ij}	1	2
1	$a_3a_{11} - a_1a_{13}$	$a_3a_{12} - a_2a_{13}$
2	$-a_1a_{23}$	$a_3a_{22} - a_2a_{23}$

$$b_1 = a_1^5b_{11} + a_2^5b_{21} + a_3^5(-a_1a_{33}) \tag{5.8}$$

$$b_2 = a_1^5b_{12} + a_2^5b_{22} + a_3^5(a_3a_{32} - a_2a_{33})$$

We will see in a moment that b_2 is a nonzero polynomial expression in the α_{rs} with coefficients in the prime subfield of k.

Assuming that $b_2 \neq 0$, we see from (3.8) that system (5.7) is equivalent to the system of equations

$d_{11}c_{00} = c_{00}^5$ $\quad(\ell_3)$ $d_1c_{00} = 0$

and equations ℓ_1 and ℓ_2 that appear in (5.7), where (5.9)

$d_{11} = b_2b_{11} - b_1b_{12}$ and $d_1 = b_1^5d_{11} - b_2^5b_1b_{22}$

5.10 a_3, b_2, and d_1 are all nonzero polynomials in the α_{rs} with coefficients in the prime subfield of k.

Proof: If we make the substitution $\alpha_{30} = \xi$, where $\xi^2 = 2$, and let all other $\alpha_{rs} = 1$, then the matrix in (5.6), which we will call A, becomes

$$A = \begin{bmatrix} 1 & 0 & 0 \\ 0 & 2\xi + 2 & 1 - \xi \\ 0 & 2\xi - 1 & 2\xi - 1 \end{bmatrix} \tag{5.10.1}$$

The a_j become $a_1 = 2$, $a_2 = 3$, $a_3 = 3$.

If we let B be the matrix in (5.7), we can use the formulas in (5.8) to see what B and b_1, b_2 become under the substitution above. These are

$$B = \begin{bmatrix} 3 & 0 \\ 2\xi - 2 & -\xi - 2 \end{bmatrix} \qquad (5.10.2)$$

$$b_1 = 1 - \xi \qquad b_2 = 2\xi - 1$$

Finally, we see from the formulas in (5.9) that under this substitution we obtain

$$d_{11} = 2 + \xi \qquad d_1 = 2$$

Hence a_3, b_2, and d_1 are nonzero polynomials in the α_{rs}.

5.11 If we choose that α_{rs} such that $a_3 b_2 d_1 \neq 0$, then our original system of equations (5.6) is equivalent to the system (5.9). Remember, by "equivalent" we mean that they have the same number of distinct solutions. In system (5.9) equations ℓ_1, ℓ_2, and ℓ_3 will then have only the trivial solution ($c_{00} = c_{10} = c_{01} = 0$).

We see, therefore, that generically the surface F: $z^5 = G_3$ has trivial divisor class group. For if we let U be the open subset of the coefficient space of G_3 defined by $a_3 b_2 d_1 \neq 0$, then $C\ell(F: z^5 = G_3) = 0$ whenever the coefficients of G_3 are in U. We have just seen that the surface F: $z^5 = x + y + x^2 + xy + y^2 + x^3 + ix^2 y + xy^2 + y^3$ is such an example.

5.12 Studying (5.6), we see by lemma (3.10) that this system of equations has at most 5^3 solutions, and that this happens only if $a_1 = a_2 = a_3 = 0$ and $\det A \neq 0$.

We will now show that there is no surface F: $z^5 = G_3$ whose coefficients satisfy these conditions. Let g be the homogeneous form of degree 2 of G_3. g is either 0, a perfect square, or a product of distinct factors. Thus after a change of coordinates we have two cases to consider: case 1, $\alpha_{11} = \alpha_{02} = 0$, and case 2, $\alpha_{20} = \alpha_{02} = 0$, $\alpha_{11} = 1$.

Case 1: We have that $a_1 = \alpha_{20}(\alpha_{21}\alpha_{03} + 3\alpha_{12}^2)$, $a_2 = \alpha_{10}(3\alpha_{12}^2 + \alpha_{21}\alpha_{03}) + \alpha_{01}(\alpha_{30}\alpha_{03} + \alpha_{12}\alpha_{21})$, $a_3 = \alpha_{10}(\alpha_{30}\alpha_{03} + \alpha_{21}\alpha_{12}) +$

$\alpha_{01}(\alpha_{30}\alpha_{12} + 3\alpha_{21}^2) + 3\alpha_{03}\alpha_{20}^2$, and $a_{11} = \alpha_{10}^2(\alpha_{12}^2 + 2\alpha_{21}\alpha_{03}) +$ $4\alpha_{10}\alpha_{01}(\alpha_{30}\alpha_{03} + \alpha_{21}\alpha_{12}) + \alpha_{01}^2(\alpha_{21}^2 + 2\alpha_{30}\alpha_{12}) + 2\alpha_{20}^2\alpha_{01}\alpha_{03}$. Thus $a_{11} = 2\alpha_{10}a_2 + 2\alpha_{01}a_3 + \alpha_{20}^2\alpha_{01}\alpha_{03}$.

If $a_1 = a_2 = a_3 = 0$, then $a_{11} = \alpha_{20}^2\alpha_{01}\alpha_{03}$. If $\det A \neq 0$, then $a_{11} \neq 0$. Hence

$$\alpha_{21}\alpha_{03} + 3\alpha_{12}^2 = 0 \Rightarrow \alpha_{30}\alpha_{03} + \alpha_{21}\alpha_{12} = 0 \Rightarrow a_3 = \alpha_{01}(\alpha_{30}\alpha_{12}$$

$$+ 3\alpha_{21}^2) + 3\alpha_{03}\alpha_{20}^2 = \alpha_{01}\left(\frac{-\alpha_{21}\alpha_{12}^2}{\alpha_{03}} + \frac{\alpha_{21}\alpha_{12}^2}{\alpha_{03}}\right) + 3\alpha_{03}\alpha_{20}^2 = \alpha_{03}\alpha_{20}^2$$

Thus $\alpha_{03}\alpha_{20}^2 = 0$, which implies that $a_{11} = 0$, contradiction.

Case 2: Assuming that $a_1 = a_2 = a_3 = 0$, we have that $a_1 = \alpha_{30}\alpha_{03} +$ $\alpha_{21}\alpha_{12}$, $a_2 = \alpha_{10}(3\alpha_{12}^2 + \alpha_{21}\alpha_{03}) + 3\alpha_{12}$, $a_3 = \alpha_{01}(3\alpha_{21}^2 + \alpha_{12}\alpha_{30}) +$ $3\alpha_{21}$. Then $a_{11} = 2\alpha_{10}(-3\alpha_{12}) + 2\alpha_{01}(-3\alpha_{21}) + 2(\alpha_{12}\alpha_{10} + \alpha_{21}\alpha_{01}) +$ $1 = \alpha_{12}\alpha_{10} + \alpha_{21}\alpha_{01} + 1$.

From a_2 and a_3 we obtain $2\alpha_{10}\alpha_{21}\alpha_{03} = -(\alpha_{01}\alpha_{12} + 1)\alpha_{12}$ and $2\alpha_{01}\alpha_{12}\alpha_{30} = -(\alpha_{10}\alpha_{21} + 1)\alpha_{21}$. Multiplying these two expressions, we obtain

$$\alpha_{12}\alpha_{21}\alpha_{10}\alpha_{01}(\alpha_{30}\alpha_{03} + \alpha_{21}\alpha_{12}) + \alpha_{12}\alpha_{21}(\alpha_{21}\alpha_{01} + \alpha_{12}\alpha_{10} + 1) = 0$$

$$\Rightarrow \alpha_{12}\alpha_{21}(\alpha_{21}\alpha_{01} + \alpha_{12}\alpha_{10} + 1) = \alpha_{12}\alpha_{21}a_{11} = 0$$

If $a_{11} = 0$, then $\det A = 0$. If $\alpha_{21} = 0$, then $a_2 = 3\alpha_{12}a_{11} = 0$. Thus if $a_{11} \neq 0$, then $\alpha_{12} = 0 \Rightarrow \alpha_{30}\alpha_{03} = 0 \Rightarrow a_{22} = a_{23} = 0 \Rightarrow \det A = 0$.

Thus we conclude that the order of $C\ell(F\colon\ z^5 = G_3)$ is at most 5^2.

5.13 Let C_2 be the subset of the coefficient space corresponding to those coefficients of G_3 for which $C\ell(F\colon\ z^5 = G_3)$ has order 5^2. If we let B be the matrix of (5.8), then those surfaces whose coefficients satisfy $a_3 \det(B) \neq 0$, $b_1 = b_2 = 0$, define a subset of C_2 by (5.7) and (3.10). $z^5 = xy^2 + yx^2 + xy$ is such a surface. Thus C_2 has codimension ≤ 2.

5.14 Let C_1 be the subset of the coefficient space defined by those coefficients of G_3 for which $C\ell(F\colon\ z^5 = G_3) \equiv \mathbb{Z}/5\mathbb{Z}$. From (5.9) and (3.10), a surface whose coefficients satisfy $a_3 b_2 c_{11} \neq 0$, $d_1 = 0$ has coefficients in C_1. An example of such a surface is provided by $z^5 = x + y + x^2 + xy + y^2 + x^3 + 2x^2y + xy^2 + y^3$. We conclude that C_1 has codimension 1.

What conclusions can be drawn?

5.15 From P. Blass (see Blass, 1982) we have that the surface F: $z^5 = G_3$ has geometric genus 1 for a generic G_3. In (5.11) we saw that for a generic G_3, $C\ell(F\colon\ z^5 = G_3) = 0$. This gives further support to conjecture (4.23).

5.16 We also have that the subsets of the coefficient space of G_3, C_1, and C_2, defined by those coefficients for which $C\ell(F)$ has order 5 and 5^2 have codimension ≤ 1 and 2, respectively. This agrees with Artin's conjecture (4.25).

In the remainder of this section we perform the same type of analysis on the surface F: $z^2 = G_6$ over a field of characteristic 2. As already mentioned, these surfaces are of genus 1 for a generic G_6 (see Blass, 1982).

We have seen that if $\partial^2 G_6/\partial x \partial y \neq 0$, then $C\ell(F\colon\ z^2 = G_6)$ is not trivial. We will now show that this group is $\mathbb{Z}/2\mathbb{Z}$ for a generic G_6.

Let k be an algebraically closed field of characteristic 2, $G_6 \in k[x,y]$ be of degree 6 satisfying conditions (*) and (**). Then $G_6 = \alpha_{10}x + \alpha_{01}y + \alpha_{11}xy + \alpha_{30}x^3 + \alpha_{21}x^2y + \alpha_{12}xy^2 + \alpha_{03}y^3 + \alpha_{31}x^3y + \alpha_{13}xy^3 + \alpha_{50}x^5 + \alpha_{41}x^4y + \alpha_{32}x^3y^2 + \alpha_{23} + \alpha_{14}xy^4 + \alpha_{05}y^5 + \alpha_{51}x^5y + \alpha_{33}x^3y^3 + \alpha_{15}xy^5$. As usual, we let $F \subseteq \mathbb{A}_k^3$ be the surface defined by $z^2 = G_6$, D be the jacobian derivation of G_6, and $L \subseteq k[x,y]$ the group of logarithmic derivatives of D.

Given $t \in L$, we have from (2.4) that $\deg t \leq 4$. We also have from Ganong's formula and (1.9) that $t \in L$ iff

$$\nabla(G_6 t) = t^2 \qquad \text{and} \qquad \nabla t = 0 \tag{5.17}$$

From (5.17) we have that the order of $C\ell(F)$ is equal to the number of distinct solutions to the system of equations

$$
\begin{aligned}
&(1) && \alpha_{11}c_{00} + \alpha_{01}c_{10} + \alpha_{10}c_{01} = c_{00}^2 \\
&(2) && \alpha_{31}c_{00} + \alpha_{21}c_{10} + \alpha_{30}c_{01} + \alpha_{11}c_{20} + \alpha_{01}c_{30} + \alpha_{10}c_{21} = c_{10}^2 \\
&(3) && \alpha_{13}c_{00} + \alpha_{03}c_{10} + \alpha_{12}c_{01} + \alpha_{11}c_{02} + \alpha_{01}c_{12} + \alpha_{10}c_{03} = c_{01}^2 \\
&(4) && \alpha_{51}c_{00} + \alpha_{41}c_{10} + \alpha_{50}c_{01} + \alpha_{31}c_{20} + \alpha_{21}c_{30} + \alpha_{30}c_{21} \\
& && \quad + \alpha_{11}c_{40} = c_{20}^2 \\
&(5) && \alpha_{15}c_{00} + \alpha_{05}c_{10} + \alpha_{14}c_{01} + \alpha_{13}c_{02} + \alpha_{03}c_{12} + \alpha_{12}c_{03} \\
& && \quad + \alpha_{11}c_{04} = c_{02}^2 \\
&(6) && \alpha_{51}c_{20} + \alpha_{50}c_{21} + \alpha_{41}c_{30} + \alpha_{31}c_{40} = c_{30}^2 \\
&(7) && \alpha_{33}c_{20} + \alpha_{51}c_{02} + \alpha_{23}c_{30} + \alpha_{32}c_{21} + \alpha_{41}c_{12} + \alpha_{50}c_{03} \\
& && \quad + \alpha_{13}c_{40} + \alpha_{31}c_{22} = c_{21}^2 \qquad\qquad (5.18) \\
&(8) && \alpha_{15}c_{20} + \alpha_{33}c_{02} + \alpha_{05}c_{30} + \alpha_{14}c_{21} + \alpha_{23}c_{12} + \alpha_{32}c_{03} \\
& && \quad + \alpha_{13}c_{22} + \alpha_{31}c_{04} = c_{12}^2 \\
&(9) && \alpha_{15}c_{02} + \alpha_{05}c_{12} + \alpha_{14}c_{03} + \alpha_{13}c_{04} = c_{03}^2 \\
&(\tilde{\ell}_1) && \alpha_{33}c_{00} + \alpha_{23}c_{10} + \alpha_{32}c_{01} + \alpha_{13}c_{20} + \alpha_{31}c_{02} + \alpha_{03}c_{30} \\
& && \quad + \alpha_{12}c_{21} + \alpha_{21}c_{12} + \alpha_{30}c_{03} + \alpha_{11}c_{22} = 0 \\
&(10) && \alpha_{51}c_{40} = c_{40}^2 \\
&(\tilde{\ell}_2) && \alpha_{51}c_{22} + \alpha_{33}c_{40} = 0 \\
&(11) && \alpha_{15}c_{40} + \alpha_{33}c_{22} + \alpha_{51}c_{04} = c_{22}^2 \\
&(\tilde{\ell}_3) && \alpha_{15}c_{22} + \alpha_{33}c_{04} = 0 \\
&(12) && \alpha_{15}c_{04} = c_{04}^2
\end{aligned}
$$

5.19 We have that one of $\alpha_{51}, \alpha_{33}, \alpha_{15} \neq 0$ since we are assuming that $\deg G_6 = 6$. Then if we restrict our attention to the subsystem of (4.18) consisting of equations (10), (11), (12), $(\tilde{\ell}_2)$, and $(\tilde{\ell}_3)$ we see that this subsystem has only two possible solutions.

They are the trivial solution, $c_{40} = c_{22} = c_{04} = 0$, and a nontrivial solution, $c_{40} = \alpha_{51}$, $c_{22} = \alpha_{33}$, $c_{04} = \alpha_{15}$.

If we suppose that $c_{40} = \alpha_{51}$, $c_{22} = \alpha_{33}$, $c_{04} = \alpha_{15}$ and make the change of coordinates $c_{00} \to c_{00} + \alpha_{11}$, $c_{20} \to c_{20} + \alpha_{31}$, $c_{02} \to c_{02} + \alpha_{13}$, the system (4.18) just becomes the same system of equations (5.18) along with the substitutions $c_{40} = c_{22} = c_{04} = 0$.

More explicitly, system (5.18) becomes

$$
\begin{aligned}
&(1)\quad \alpha_{11}c_{00} + \alpha_{01}c_{10} + \alpha_{10}c_{01} = c_{00}^2\\
&(2)\quad \alpha_{31}c_{00} + \alpha_{21}c_{10} + \alpha_{30}c_{01} + \alpha_{11}c_{20} + \alpha_{01}c_{30} + \alpha_{10}c_{21} = c_{10}^2\\
&(3)\quad \alpha_{13}c_{00} + \alpha_{03}c_{10} + \alpha_{12}c_{01} + \alpha_{11}c_{02} + \alpha_{01}c_{12} + \alpha_{10}c_{03} = c_{01}^2\\
&(4)\quad \alpha_{51}c_{00} + \alpha_{41}c_{10} + \alpha_{50}c_{01} + \alpha_{31}c_{20} + \alpha_{21}c_{30} + \alpha_{30}c_{21} = c_{20}^2\\
&(5)\quad \alpha_{15}c_{00} + \alpha_{05}c_{10} + \alpha_{14}c_{01} + \alpha_{13}c_{02} + \alpha_{03}c_{12} + \alpha_{12}c_{03} = c_{02}^2\\
&(6)\quad \alpha_{51}c_{20} + \alpha_{50}c_{21} + \alpha_{41}c_{30} = c_{30}^2 \qquad\qquad (5.20)\\
&(7)\quad \alpha_{33}c_{20} + \alpha_{51}c_{02} + \alpha_{23}c_{30} + \alpha_{32}c_{21} + \alpha_{41}c_{12} + \alpha_{50}c_{03} = c_{21}^2\\
&(8)\quad \alpha_{15}c_{20} + \alpha_{33}c_{02} + \alpha_{05}c_{30} + \alpha_{14}c_{21} + \alpha_{23}c_{12} + \alpha_{32}c_{03} = c_{12}^2\\
&(9)\quad \alpha_{15}c_{02} + \alpha_{05}c_{12} + \alpha_{14}c_{03} = c_{03}^2\\
&(\ell_1)\quad \alpha_{33}c_{00} + \alpha_{23}c_{10} + \alpha_{32}c_{01} + \alpha_{13}c_{20} + \alpha_{31}c_{02} + \alpha_{03}c_{30} + \alpha_{12}c_{21}\\
&\qquad\qquad + \alpha_{21}c_{12} + \alpha_{30}c_{03} = 0
\end{aligned}
$$

We conclude that the number of solutions to (5.18) is twice the number of solutions to system (5.20), which we will now show has only the trivial solution for a generic choice of the α_{rs}. The argument will be the same as that used in the two preceding sections and the first part of this one.

To organize notation a bit, let $x_1 = c_{10}$, $x_2 = c_{01}$, $x_3 = c_{00}$, $x_4 = c_{20}$, $x_5 = c_{02}$, $x_6 = c_{30}$, $x_7 = c_{21}$, $x_8 = c_{12}$, and $x_9 = c_{03}$. For each i,j = 1, 2, ..., 9, let $a_{ij}^{(1)}$ be the coefficient of x_i in equation (j) of (5.20), and let $a_j^{(1)}$ be the coefficient of x_j in equa- $(\tilde{\ell}_1)$ of (5.20).

For k = 2, ..., 9 we inductively define $a_{ij}^{(k)}$, $a_j^{(k)}$ for i,j = 1, 2, ..., 10 - k by the formulas

$$a_{ij}^{(k)} = a_{ij}^{(k-1)} a_{10-k}^{(k-1)} - a_{i(10-k)}^{(k-1)} a_j^{(k-1)}$$

$$a_j^{(k)} = a_1^{(k-1)2} a_{1j}^{(k)} + a_2^{(k-1)2} a_{2j}^{(k)} + \cdots + a_{10-(k+1)}^{(k-1)2} a_{10-k-1}^{(k)}$$
$$+ a_{10-k}^{(k-1)2} a_{(10-k)j}^{(k-1)} a_{10-k}^{(k-1)} - a_{(10-k)(10-k)}^{(k-1)} a_j^{(k-1)} \qquad (5.21)$$

For each k = 1, 2, ..., 9, let S_k be the system of equations defined by the matrix equation

$$S_k: \quad A_k \begin{bmatrix} x_1 \\ \vdots \\ x_{10-k} \end{bmatrix} = \begin{bmatrix} x_1^2 \\ \vdots \\ x_{10-k}^2 \end{bmatrix} \qquad (5.22)$$

and the linear equations $\ell_1, \ell_2, \ldots, \ell_k$, where A_k is the (10 - k) by (10 - k) matrix with entries $a_{ij}^{(k)}$ and ℓ_m is the equation

$$a_1^{(m)} x_1 + \cdots + a_{10-m}^{(m)} x_{10-m} = 0$$

5.23 We see that the $a_{ij}^{(k)}$ and $a_j^{(k)}$ are polynomials in the α_{rs} with coefficients in the prime subfield.

Furthermore, by (3.10) we have that system (5.20) is equivalent to system S_k for a choice of α_{rs} provided that $a_9^{(1)} a_8^{(2)} \cdots a_{10-k+1}^{(k-1)} \neq 0$.

As always, we give an example to show that the polynomials $a_9^{(1)}, a_8^{(2)}, \ldots, a_1^{(9)}$, are not zero.

Let ξ be a primitive cube root of unity. Then $\xi^3 = 1$ and $\xi^2 + \xi + 1 = 0$. If we make the substitution $\alpha_{10} = \alpha_{13} = \alpha_{15} = \xi$, all other $\alpha_{rs} = 1$, we obtain

$$a_9^{(1)} = 1 \quad a_8^{(2)} = \xi^2 \quad a_1^{(3)} = 1 \quad a_6^{(4)} = \xi^2 \quad a_5^{(5)} = 1$$
$$a_4^{(6)} = \xi \quad a_3^{(7)} = \xi^2 \quad a_2^{(8)} = \xi \quad a_1^{(9)} = \xi^2 \qquad (5.24)$$

Thus from (5.23) we conclude that system (5.20) has the trivial solution whenever $a_9^{(1)} a_8^{(2)} \cdots a_1^{(9)} \neq 0$

Since our original system (5.18) has twice the number of distinct solutions that system (5.20) has, we see that system (5.18) has exactly two solutions if $a_9^{(1)}a_8^{(2)}\cdots a_1^{(9)} \neq 0$. These are the trivial solutions, all $c_{ij} = 0$, and the nontrivial solution, $c_{00} = \alpha_{11}$, $c_{20} = \alpha_{31}$, $c_{02} = \alpha_{13}$, $c_{40} = \alpha_{51}$, $c_{22} = \alpha_{33}$, $c_{04} = \alpha_{15}$. Therefore, we have that the order of $C\ell(F\colon\ z^2 = G_6)$ is 2 for a generic choice of the α_{rs}.

Also, we see that the subsets E_i of the coefficient space of G_6 corresponding to those coefficients for which $C\ell(F\colon\ z^2 = G_6)$ has order 2^{i+1} have codimension $\leq i$ for $i = 1, \ldots, 9$.

The reasoning is the same as that used in previous discussions. For we have that system (5.20) is equivalent to system S_k whenever $a_9^{(1)}a_8^{(2)}\cdots a_{10-k+1}^{(k-1)} \neq 0$. If we add the conditions that $\det(A_k) \neq 0$ and $a_1^{(k)} = a_2^{(k)} = \cdots = a_{10-k}^{(k)} = 0$, we see that system (5.20) has 2^{10-k} distinct solutions by (3.10), and hence system (5.18) has 2^{10-k+1} solutions. Thus the subset of the coefficient space corresponding to those α_{rs} for which $C\ell(F\colon\ z^2 = G_6)$ has order 2^{10-k+1} is of codimension at most $10 - k$.

In the following table we give a list of surfaces $F\colon\ z^2 = G_6$ with their corresponding orders of $C\ell(F\colon\ z^2 = G_6)$.

Table 5.25

F	Order of $C\ell(F)$
$z^2 = xy^5 + y$	2
$z^2 = xy^4(y + \theta_1) + y$	2^2
$z^2 = xy^2(y + \theta_1)^2(y + \theta_2)^2 + y$	2^3
$z^2 = xy^2(y + \theta_1)(y + \theta_2)(y + \theta_3) + y$	2^4
$z^2 = xy(y + \theta_1)(y + \theta_2)(y + \theta_3)(y + \theta_4) + y$	2^5
$z^2 = x^5 + y^5 + x^3y^3 + x^3y + xy^3$	2^6
$z^2 = x^2y + xy^2 + x^3y^2 + x^2y^3 + x^5y + xy^5$	2^7
$z^2 = y + x^3 + x^5y + xy^4$	2^8
$z^2 = y + x^5y + xy^4$	2^9
$z^2 = xy + x^4y + y^4x + x^5y + y^5x$	2^{10}

where $\theta_1, \theta_2, \theta_3, \theta_4 \in k$ are pairwise distinct [see (7.15), p. 215].

REMARK 5.26 In the study above we saw that for generic G_6, the class group of the surface F: $z^2 = G_6$ is $\mathbb{Z}/2\mathbb{Z}$. This is the first bit of evidence we give supporting conjecture (4.24). We have also seen that Artin's conjecture (4.25) holds for this example. Observe also that since $\deg(G_6) = 6$, then $(\partial^2/\partial x \partial y)(G_6) \neq 0$ and hence $C\ell(F: z^2 = G_6)$ is not trivial, by (3.3).

In Section 3, proposition (3.19), we showed that if $\deg G \leq 4$, the surface F: $z^2 = G$ has trivial class group if and only if F is a plane. In Section 8 we give an example to show that this proposition cannot be improved. That is, there exists G of degree 5 such that the surface F: $z^2 = G_5$ has trivial class group and such that F is not a plane.

Of course, one might ask why we ignored the surfaces defined by equations of the form $z^2 = G^5$, where $\deg G_5$ is 5, and proceeded to the degree 6 case instead. There are several reasons.

Classically, geometers would study the degree 6 case rather than the degree 5 case because any $\tilde{F}$: $z^2 = G_5$ is birational to some F as above of degree 6. To see this, we let $G_5(x,y,u)$ be a homogenization of $G_5(x,y)$. Then one sees readily that the three surfaces $z^2 = uG_5(x,1,u)$, $z^2 = uG_5(1,y,u)$, and $z^2 = G_5(x,y)$ are all birational and one of these has degree 6.

Another reason is that the study of the class groups in the degree 6 case is no more difficult than the degree 5 case. Both of these studies depend on solving a system of 10 equations in nine unknowns.

Also, once we know the system of equations (5.18) corresponding to the sixth-degre case, we can quickly obtain those corresponding to the class group of surfaces $\tilde{F}$: $z^2 = G_5$ by just making the substitution $\alpha_{51} = \alpha_{33} = \alpha_{15} = 0$ in system (5.20). Then we can analyze this new system of equations in the usual way. One obtains the following results.

5.27 For a generic choice of G_5, the surfaces $\tilde{F}$: $z^2 = G_5$ have divisor class group $\mathbb{Z}/2\mathbb{Z}$.

5.28 The subsets $\tilde{E}_1, \tilde{E}_2, \ldots, \tilde{E}_9$ of the coefficient space of G_5, corresponding to those α_{rs} for which the order of $C\ell(\tilde{F}: z^2 = G_5)$ is $2^2, 2^3, \ldots, 2^9$ have codimension $\leq 1, 2, \ldots, 9$, respectively.

5.29 The order of $C\ell(\tilde{F}: z^2 = G_5)$ is bounded above by 2^9.

5.30 Let T be the closed subset of the coefficient space of G_5 defined by $\alpha_{11} = \alpha_{31} = \alpha_{13} = 0$. Then there exists an open subset $V \subseteq T$ such that $C\ell(\tilde{F}: z^2 = G_5) = 0$ if the coefficients of G_5 belong to V.

5.31 A list of surfaces, $z^2 = G_5$, and their corresponding class groups appears in the following table.

$\tilde{F}$: $z^2 = G_5$	Order of $C\ell(\tilde{F}: z^2 = G_5)$
$z^2 = xy^4 + y$	2^1
$z^2 = xy^3(y + \theta_1) + y$	2^2
$z^2 = xy^2(y + \theta_1)(y + \theta_2) + y$	2^3
$z^2 = xy(y + \theta_1)(y + \theta_2)(y + \theta_3) + y$	2^4
$z^2 = x^3y + y^3x + x^5 + y^5$	2^5
$z^2 = xy + x^4y + xy^4 + x^3y^2 + x^2y^3$	2^6
$z^2 = xy + x^4y + xy^4 + x^3y^2$	2^7
$z^2 = xy + x^3 + x^4y + xy^4$	2^8
$z^2 = xy + x^4y + y^4x$	2^9

It is interesting to note that both tables (5.25) and (5.31) have several examples of regular surfaces with nontrivial class group.

We will show in Section 7 that if k is an algebraically closed field of characteristic $p \neq 0$ and n is a positive integer, there exist a regular surface F: $z^p = G(x,y)$ with divisor class group of order p^n.

Miyanishi showed that this cannot happen when k has characteristic 0. Miyanishi proved the following.

5.32 Let k be an algebraically closed field characteristic 0, let A be a regular k-subalgebra of k[x,y] such that k[x,y] is a finitely generated A-module. Then A is a polynomial ring over k.

Also, P. Russell proved the following in the characteristic $p > 0$ case.

5.33 (Affine Theorem of Castelnuovo) Suppose that k is perfect. Let $A \subset k^{[2]}$ be a finitely generated, regular k-algebra of dimension 2 such that $\bar{k} \otimes_k A$ is factorial ($\bar{k}$ an algebraic closure of k) and $qt(k^{[2]})/qt(A)$ is a separable extension. Then $A \approx k^{[2]}$.

In Section 8 we show that Russell's theorem does not hold if we drop the condition that $qt(k^{[2]})/qt(A)$ is a separable extension.

6. THE SURFACES $z^3 = G_6$ AND $z^p = G_3$, $p \geq 5$

In this section we continue with our investigation of the relationship between the coefficient space of G_n ($G_n \in k[x,y]$ of degree n) and the divisor class group of the surface $z^p = G_n$. We first consider the case $p = 3$, $n = 6$ and finish the section with the case $p > 3$, $n = 3$. The $p = 3$, $n = 6$ case provides an example of surfaces with "generic genus" 3, while the $p > 3$, $n = 3$ case gives examples of surfaces with positive generic genus that increases without bound with increasing p.

Let k be an algebraically closed field of characteristic 3, $G_6 \in k[x,y]$ be of degree 6 satisfying conditions (*) and (**). Let $F \subseteq \mathbf{A}_k^3$ be the surface defined by $z^3 = G_6$.

Then, as in preceding discussions, we can show with the aid of (1.9), (2.4), and Ganong's formula that the order of $C\ell(F)$ is equal to the number of distinct solutions to the system of equations

$$A\begin{bmatrix} x_1 \\ \vdots \\ x_{14} \end{bmatrix} = \begin{bmatrix} x_1^3 \\ \vdots \\ x_{14}^3 \end{bmatrix} \qquad B\begin{bmatrix} x_1 \\ \vdots \\ x_{14} \end{bmatrix} = \vec{0} \tag{6.1}$$

where A is a 14 × 14 matrix with components a_{ij} and B is a 7 × 14 matrix with components b_{kl} with the formulas for these given below (the α_{rs} are the coefficients of G_6).

(6.2)

$a_{11} = \alpha_{11}^2 + 2(\alpha_{20}\alpha_{02}+\alpha_{10}\alpha_{12}+\alpha_{01}\alpha_{21})$, $a_{21} = 2(\alpha_{10}\alpha_{42}+\alpha_{01}\alpha_{51}+\alpha_{20}\alpha_{32}+\alpha_{11}\alpha_{41}+\alpha_{02}\alpha_{50}+\alpha_{21}\alpha_{31}+\alpha_{12}\alpha_{40})$, $a_{31} = 2(\alpha_{10}\alpha_{15}+\alpha_{01}\alpha_{24}+\alpha_{20}\alpha_{05}+\alpha_{11}\alpha_{14}+\alpha_{02}\alpha_{23}+\alpha_{21}\alpha_{04}+\alpha_{12}\alpha_{13})$, $a_{41} = \alpha_{41}^2 + 2(\alpha_{50}\alpha_{32}+\alpha_{40}\alpha_{42}+\alpha_{31}\alpha_{51})$, $a_{51} = 2(\alpha_{50}\alpha_{05}+\alpha_{41}\alpha_{14}+\alpha_{32}\alpha_{23}+\alpha_{40}\alpha_{15}+\alpha_{31}\alpha_{24}+\alpha_{13}\alpha_{42}+\alpha_{04}\alpha_{51})$, $a_{61} = \alpha_{14}^2 + 2(\alpha_{23}\alpha_{05}+\alpha_{13}\alpha_{15}+\alpha_{04}\alpha_{24})$,

$a_{71} = a_{81} = a_{91} = a_{101} = a_{111} = a_{121} = a_{131} = a_{141} = 0.$

$a_{12} = 2(\alpha_{10}\alpha_{02}+\alpha_{01}\alpha_{11})$, $a_{22} = \alpha_{21}^2 + 2(\alpha_{10}\alpha_{32}+\alpha_{01}\alpha_{41}+\alpha_{20}\alpha_{22}+\alpha_{11}\alpha_{31}+\alpha_{02}\alpha_{40})$,

$a_{32} = 2(\alpha_{10}\alpha_{05}+\alpha_{01}\alpha_{14}+\alpha_{11}\alpha_{04}+\alpha_{02}\alpha_{13})$, $a_{42} = 2(\alpha_{21}\alpha_{51}+\alpha_{40}\alpha_{32}+\alpha_{31}\alpha_{41}+\alpha_{22}\alpha_{50})$.

$a_{52} = 2(\alpha_{21}\alpha_{24}+\alpha_{40}\alpha_{05}+\alpha_{31}\alpha_{14}+\alpha_{22}\alpha_{23}+\alpha_{13}\alpha_{32}+\alpha_{04}\alpha_{41})$, $a_{62} = 2(\alpha_{13}\alpha_{05}+\alpha_{04}\alpha_{14})$,

$a_{72} = \alpha_{51}^2$, $a_{82} = 2\alpha_{51}\alpha_{24}$, $a_{92} = \alpha_{24}^2$, $a_{102} = a_{112} = a_{122} = a_{132} = a_{142} = 0.$

$a_{13} = 2(\alpha_{01}\alpha_{20}+\alpha_{10}\alpha_{11})$, $a_{23} = 2(\alpha_{01}\alpha_{50}+\alpha_{10}\alpha_{41}+\alpha_{11}\alpha_{40}+\alpha_{20}\alpha_{31})$,

$a_{33} = \alpha_{12}^2 + 2(\alpha_{01}\alpha_{23}+\alpha_{10}\alpha_{14}+\alpha_{02}\alpha_{22}+\alpha_{11}\alpha_{13}+\alpha_{20}\alpha_{04})$, $a_{43} = 2(\alpha_{31}\alpha_{50}+\alpha_{40}\alpha_{41})$,

$a_{53} = 2(\alpha_{12}\alpha_{42}+\alpha_{04}\alpha_{50}+\alpha_{13}\alpha_{41}+\alpha_{22}\alpha_{32}+\alpha_{31}\alpha_{23}+\alpha_{40}\alpha_{14})$, $a_{63} = 2(\alpha_{12}\alpha_{15}+\alpha_{04}\alpha_{23}+\alpha_{13}\alpha_{14}+\alpha_{22}\alpha_{05})$, $a_{73} = 0$, $a_{83} = \alpha_{42}^2$, $a_{93} = 2\alpha_{15}\alpha_{42}$, $a_{103} = \alpha_{15}^2$,

$a_{113} = a_{123} = a_{133} = a_{143} = 0.$

$a_{14} = \alpha_{01}^2$, $a_{24} = 2(\alpha_{10}\alpha_{22}+\alpha_{01}\alpha_{31}+\alpha_{20}\alpha_{12}+\alpha_{11}\alpha_{21})$, $a_{34} = 2\alpha_{01}\alpha_{04}$,

$a_{44} = \alpha_{31}^2 + 2(\alpha_{40}\alpha_{22}+\alpha_{20}\alpha_{42}+\alpha_{11}\alpha_{51}+\alpha_{21}\alpha_{41}+\alpha_{12}\alpha_{50})$, $a_{54} = 2(\alpha_{31}\alpha_{04}+\alpha_{22}\alpha_{13}+\alpha_{31}\alpha_{04}+\alpha_{22}\alpha_{13}+\alpha_{20}\alpha_{15}+\alpha_{11}\alpha_{24}+\alpha_{21}\alpha_{14}+\alpha_{12}\alpha_{23})$, $a_{64} = \alpha_{04}^2$, $a_{74} = 2(\alpha_{50}\alpha_{42}+\alpha_{51}\alpha_{41})$,

$a_{84} = 2(\alpha_{50}\alpha_{15}+\alpha_{41}\alpha_{24}+\alpha_{23}\alpha_{42}+\alpha_{14}\alpha_{51})$, $a_{94} = 2(\alpha_{23}\alpha_{15}+\alpha_{14}\alpha_{24})$,

$a_{104} = a_{114} = a_{124} = a_{134} = a_{144} = 0$

$a_{15} = 2\alpha_{10}\alpha_{01}$, $a_{25} = 2(\alpha_{10}\alpha_{31}+\alpha_{01}\alpha_{40}+\alpha_{20}\alpha_{21})$, $a_{35} = 2(\alpha_{10}\alpha_{04}+\alpha_{01}\alpha_{13}+\alpha_{02}\alpha_{12})$,

$a_{45} = 2(\alpha_{40}\alpha_{31}+\alpha_{20}\alpha_{51}+\alpha_{21}\alpha_{50})$, $a_{55} = \alpha_{22}^2 + 2(\alpha_{40}\alpha_{04}+\alpha_{31}\alpha_{13}+\alpha_{20}\alpha_{24}+\alpha_{02}\alpha_{42}+$

$\alpha_{21}\alpha_{23}+\alpha_{12}\alpha_{32})$, $a_{65} = 2(\alpha_{13}\alpha_{04}+\alpha_{02}\alpha_{15}+\alpha_{12}\alpha_{05})$, $a_{75} = 2\alpha_{50}\alpha_{51}$,

$a_{85} = 2(\alpha_{50}\alpha_{24}+\alpha_{32}\alpha_{42}+\alpha_{23}\alpha_{51})$, $a_{95} = 2(\alpha_{32}\alpha_{15}+\alpha_{23}\alpha_{24}+\alpha_{05}\alpha_{42})$,

$a_{105} = 2\alpha_{05}\alpha_{15}$, $a_{115} = a_{125} = a_{135} = a_{145} = 0$.

$a_{16} = \alpha_{10}^2$, $a_{26} = 2\alpha_{10}\alpha_{40}$, $a_{36} = 2(\alpha_{01}\alpha_{22}+\alpha_{10}\alpha_{13}+\alpha_{02}\alpha_{21}+\alpha_{11}\alpha_{12})$, $a_{46} = \alpha_{40}^2$,

$a_{56} = 2(\alpha_{13}\alpha_{40}+\alpha_{22}\alpha_{31}+\alpha_{02}\alpha_{51}+\alpha_{11}\alpha_{42}+\alpha_{12}\alpha_{41}+\alpha_{21}\alpha_{32})$, $a_{66} = \alpha_{13}^2 + 2(\alpha_{04}\alpha_{22}+$

$\alpha_{02}\alpha_{24}+\alpha_{11}\alpha_{15}+\alpha_{12}\alpha_{14}+\alpha_{21}\alpha_{05})$, $a_{76} = 0$, $a_{86} = 2(\alpha_{32}\alpha_{51}+\alpha_{41}\alpha_{42})$,

$a_{96} = 2(\alpha_{05}\alpha_{51}+\alpha_{14}\alpha_{42}+\alpha_{32}\alpha_{24}+\alpha_{41}\alpha_{15})$, $a_{106} = 2(\alpha_{05}\alpha_{24}+\alpha_{15}\alpha_{14})$,

$a_{116} = a_{126} = a_{136} = a_{146} = 0$.

$a_{17} = 0$, $a_{27} = a_{11}$, $a_{37} = 0$, $a_{47} = a_{21}$, $a_{57} = a_{31}$, $a_{67} = 0$, $a_{77} = a_{41}$,

$a_{87} = a_{51}$, $a_{97} = a_{61}$, $a_{107} = a_{117} = a_{127} = a_{137} = a_{147} = 0$.

$a_{18} = 0$, $a_{28} = 2(\alpha_{20}\alpha_{11}+\alpha_{10}\alpha_{21})$, $a_{38} = \alpha_{02}^2$, $a_{48} = 2(\alpha_{10}\alpha_{51}+\alpha_{20}\alpha_{41}+\alpha_{11}\alpha_{50}+$

$\alpha_{21}\alpha_{40})$, $a_{58} = 2(\alpha_{10}\alpha_{24}+\alpha_{20}\alpha_{14}+\alpha_{11}\alpha_{23}+\alpha_{02}\alpha_{32}+\alpha_{21}\alpha_{13}+\alpha_{12}\alpha_{22})$, $a_{68} = 2\alpha_{02}\alpha_{05}$,

$a_{78} = 2(\alpha_{50}\alpha_{41}+\alpha_{40}\alpha_{51})$, $a_{88} = \alpha_{32}^2 + 2(\alpha_{50}\alpha_{14}+\alpha_{41}\alpha_{23}+\alpha_{40}\alpha_{24}+\alpha_{22}\alpha_{42}+\alpha_{13}\alpha_{51})$,

$a_{98} = 2(\alpha_{23}\alpha_{14}+\alpha_{22}\alpha_{15}+\alpha_{13}\alpha_{24})$, $a_{108} = \alpha_{05}^2$, $a_{118} = a_{128} = a_{138} = a_{148} = 0$.

$a_{19} = 0$, $a_{29} = \alpha_{20}^2$, $a_{39} = 2(\alpha_{02}\alpha_{11}+\alpha_{01}\alpha_{12})$, $a_{49} = 2\alpha_{20}\alpha_{50}$,

$a_{59} = 2(\alpha_{01}\alpha_{42}+\alpha_{02}\alpha_{41}+\alpha_{11}\alpha_{32}+\alpha_{20}\alpha_{23}+\alpha_{12}\alpha_{31}+\alpha_{21}\alpha_{22})$, $a_{69} = 2(\alpha_{01}\alpha_{15}+\alpha_{02}\alpha_{14}+$

$\alpha_{11}\alpha_{05}+\alpha_{12}\alpha_{04})$, $a_{79} = \alpha_{50}^2$, $a_{89} = 2(\alpha_{32}\alpha_{41}+\alpha_{22}\alpha_{51}+\alpha_{31}\alpha_{42})x^6y^3$,

$a_{99} = \alpha_{23}^2 + 2(\alpha_{05}\alpha_{41}+\alpha_{14}\alpha_{32}+\alpha_{04}\alpha_{42}+\alpha_{22}\alpha_{24}+\alpha_{31}\alpha_{15})$, $a_{109} = 2(\alpha_{05}\alpha_{14}+\alpha_{04}\alpha_{15})$,

$a_{119} = a_{129} = a_{139} = a_{149} = 0$, $a_{110} = 0$, $a_{210} = 0$, $a_{310} = a_{11}$, $a_{410} = 0$,

$a_{510} = a_{21}$, $a_{610} = a_{31}$, $a_{710} = 0$, $a_{810} = a_{41}$, $a_{910} = a_{51}$, $a_{1010} = a_{61}$,

$a_{1110} = a_{1210} = a_{1310} = a_{1410} = 0$.

$a_{111} = 0$, $a_{211} = a_{12}$, $a_{311} = 0$, $a_{411} = a_{22}$, $a_{511} = a_{32}$, $a_{611} = 0$,

$a_{711} = a_{42}$, $a_{811} = a_{52}$, $a_{911} = a_{62}$, $a_{1011} = 0$, $a_{1111} = a_{72}$,

$a_{1211} = a_{82}$, $a_{1311} = 0$, $a_{1411} = 0$.

$a_{112} = 0$, $a_{212} = a_{13}$, $a_{312} = 0$, $a_{412} = a_{23}$, $a_{512} = a_{33}$, $a_{612} = 0$,

$a_{712} = a_{43}$, $a_{812} = a_{53}$, $a_{912} = a_{63}$, $a_{1012} = 0$, $a_{1112} = 0$, $a_{1212} = a_{83}$,

$a_{1312} = a_{103}$, $a_{1412} = 0$.

$a_{113} = 0$, $a_{213} = 0$, $a_{313} = a_{12}$, $a_{413} = 0$, $a_{513} = a_{22}$, $a_{613} = a_{32}$,

$a_{713} = 0$, $a_{813} = a_{42}$, $a_{913} = a_{52}$, $a_{1013} = a_{62}$, $a_{1113} = 0$,

$a_{1213} = a_{72}$, $a_{1313} = a_{92}$, $a_{1413} = 0$.

$a_{114} = 0$, $a_{214} = 0$, $a_{314} = a_{13}$, $a_{414} = 0$, $a_{514} = a_{23}$, $a_{614} = a_{33}$.

$a_{714} = 0$, $a_{814} = a_{43}$, $a_{914} = a_{53}$, $a_{1014} = a_{63}$, $a_{1114} = 0$, $a_{1214} = 0$,

$a_{1314} = a_{93}$, $a_{1414} = a_{103}$.

$b_{11} = \alpha_{22}$, $b_{21} = 0$, $b_{31} = 0$, $b_{41} = b_{51} = b_{61} = b_{71} = 0$, $b_{12} = \alpha_{12}$,

$b_{22} = \alpha_{42}$, $b_{32} = \alpha_{15}$, $b_{42} = b_{52} = b_{62} = b_{72} = 0$, $b_{13} = \alpha_{21}$, $b_{23} = \alpha_{51}$,

$b_{33} = \alpha_{24}$, $b_{43} = b_{53} = b_{63} = b_{73} = 0$, $b_{14} = \alpha_{02}$, $b_{24} = \alpha_{32}$, $b_{34} = \alpha_{05}$,

$b_{44} = b_{54} = b_{64} = b_{74} = 0$, $b_{15} = \alpha_{11}$, $b_{25} = \alpha_{41}$, $b_{35} = \alpha_{14}$,

$b_{45} = b_{55} = b_{65} = b_{75} = 0$, $b_{16} = \alpha_{20}$, $b_{26} = \alpha_{30}$, $b_{36} = \alpha_{23}$,

$b_{46} = b_{56} = b_{66} = b_{76} = 0$, $b_{17} = 0$, $b_{27} = \alpha_{22}$, $b_{37} = 0$,

$b_{47} = b_{57} = b_{67} = b_{77} = 0$, $b_{18} = \alpha_{01}$, $b_{28} = \alpha_{31}$, $b_{38} = \alpha_{04}$,

$b_{48} = b_{58} = b_{68} = b_{78} = 0$, $b_{19} = \alpha_{10}$, $b_{29} = \alpha_{40}$, $b_{39} = \alpha_{13}$,

$b_{49} = b_{59} = b_{69} = b_{79} = 0$

$b_{1\,10} = 0$, $b_{2\,10} = 0$, $b_{3\,10} = \alpha_{22}$, $b_{4\,10} = 0$, $b_{5\,10} = 0$, $b_{6\,10} = 0$,

$b_{7\,10} = 0$, $b_{1\,11} = 0$, $b_{2\,11} = \alpha_{12}$, $b_{3\,11} = 0$, $b_{4\,11} = \alpha_{42}$, $b_{5\,11} = \alpha_{15}$,

$b_{6\,11} = 0$, $b_{7\,11} = \alpha_{24}^2$, $b_{1\,12} = 0$, $b_{2\,12} = \alpha_{21}$, $b_{3\,12} = 0$, $b_{4\,12} = \alpha_{51}$,

$b_{5\,12} = \alpha_{24}$, $b_{6\,12} = 0$, $b_{7\,12} = 2\alpha_{15}\alpha_{42}$, $b_{1\,13} = 0$, $b_{2\,13} = 0$, $b_{3\,13} = \alpha_{12}$,

$b_{4\,13} = 0$, $b_{5\,13} = \alpha_{42}$, $b_{6\,13} = \alpha_{15}$, $b_{8\,13} = 2\alpha_{51}\alpha_{24}$, $b_{1\,14} = 0$, $b_{2\,14} = 0$,

$b_{3\,14} = \alpha_{21}$, $b_{4\,14} = 0$, $b_{5\,14} = \alpha_{51}$, $b_{6\,14} = \alpha_{24}$, $b_{7\,14} = \alpha_{42}^2$.

Let us begin our analysis of system (6.1) by restricting our attention for the moment to the subsystem of (6.1) consisting of the last four cubic equations of the system $A\vec{x} = \vec{x}^3$ together with the last four linear equations of the system $B\vec{x} = 0$. We write this subsystem below.

$$\begin{bmatrix} \alpha_{51}^2 & 0 & 0 & \\ 2\alpha_{51}\alpha_{24} & \alpha_{42}^2 & \alpha_{51}^2 & \\ 0 & \alpha_{15}^2 & \alpha_{24}^2 & 2\alpha_{15}\alpha_{42} \\ 0 & 0 & 0 & \alpha_{15}^2 \end{bmatrix} \begin{bmatrix} x_{11} \\ x_{12} \\ x_{13} \\ x_{14} \end{bmatrix} = \begin{bmatrix} x_{11}^3 \\ x_{12}^3 \\ x_{13}^3 \\ x_{14}^3 \end{bmatrix} \begin{bmatrix} \alpha_{42} & \alpha_{51} & 0 & 0 \\ \alpha_{15} & \alpha_{24} & \alpha_{42} & \alpha_{51} \\ 0 & 0 & \alpha_{15} & \alpha_{24} \\ \alpha_{24}^2 & (2\alpha_{15}\alpha_{42}) & (2\alpha_{51}\alpha_{24}) & \alpha_{42}^2 \end{bmatrix} \begin{bmatrix} x_{11} \\ x_{12} \\ x_{13} \\ x_{14} \end{bmatrix} \begin{bmatrix} 0 \\ 0 \\ 0 \\ 0 \end{bmatrix} \tag{6.3}$$

By explicitly solving this subsystem we find that (6.3) always has exactly three solutions. These are

$$\begin{bmatrix} x_{11} \\ x_{12} \\ x_{13} \\ x_{14} \end{bmatrix} = \begin{bmatrix} 0 \\ 0 \\ 0 \\ 0 \end{bmatrix}, \begin{bmatrix} \alpha_{51} \\ -\alpha_{42} \\ \alpha_{24} \\ -\alpha_{15} \end{bmatrix}, \begin{bmatrix} -\alpha_{51} \\ \alpha_{42} \\ -\alpha_{24} \\ \alpha_{15} \end{bmatrix} \tag{6.4}$$

If we now shift our concentration to the subsystem of (6.1) consisting of the first 10 equations of the cubic system $A\vec{x} = \vec{x}^3$, together with the first linear equation of $B\vec{x} = 0$, we obtain the following system:

$$\begin{bmatrix} a_{11} & a_{12} & \cdots & a_{110} \\ a_{21} & a_{22} & \cdots & a_{210} \\ \cdots & \cdots & \cdots & \cdots \\ a_{101} & a_{102} & \cdots & a_{1010} \end{bmatrix} \begin{bmatrix} x_1 \\ x_2 \\ \cdots \\ x_{10} \end{bmatrix} = \begin{bmatrix} x_1^3 \\ x_2^3 \\ \cdots \\ x_{10}^3 \end{bmatrix} \tag{6.5}$$

(ℓ_1) $\alpha_{22}x_1 + \alpha_{12}x_2 + \alpha_{21}x_3 + \alpha_{02}x_4 + \alpha_{11}x_5 + \alpha_{20}x_6 + \alpha_{01}x_8 + \alpha_{10}x_9 = 0$

Just as in previous discussions, one shows that for a generic choice of the α_{rs} (the coefficients of G_6), system (6.5) has only the trivial solution. Also, if we let $E_1, E_2, \ldots, E_9$ be the subsets of the coefficient space of G_6 corresponding to those choices of α_{rs} for which system (6.5) has $3, 3^2, \ldots, 3^9$ solutions, respectively, we have that the codimension of each E_i is at most i, i = 1, 2, ..., 9. It is easy to see that 3^9 is the maximum number of solutions to (6.5).

We now combine systems (6.3) and (6.5) to obtain the subsystem of (6.1) described below.

$$\begin{bmatrix} \alpha_{42} & \alpha_{51} & 0 & 0 \\ \alpha_{15} & \alpha_{24} & \alpha_{42} & \alpha_{51} \\ 0 & 0 & \alpha_{15} & \alpha_{24} \\ \alpha_{24}^2 & (2\alpha_{15}\alpha_{42}) & (2\alpha_{51}\alpha_{24}) & \alpha_{42}^2 \end{bmatrix} \begin{bmatrix} x_{11} \\ x_{12} \\ x_{13} \\ x_{14} \end{bmatrix} = \begin{bmatrix} 0 \\ 0 \\ 0 \\ 0 \end{bmatrix}$$

$$\begin{bmatrix} a_{11} & a_{12} & \cdots & a_{110} & & & & \\ a_{21} & a_{22} & \cdots & a_{210} & & & & \\ \cdots & & \cdots & & & & & \\ a_{101} & a_{102} & \cdots & a_{1010} & & & & \\ & & & & \alpha_{51}^2 & 0 & 0 & 0 \\ & & & & 2\alpha_{51}\alpha_{24} & \alpha_{42}^2 & \alpha_{51}^2 & 0 \\ & & & & 0 & \alpha_{15}^2 & \alpha_{24}^2 & 2\alpha_{15}\alpha_{42} \\ & & & & 0 & 0 & 0 & \alpha_{15}^2 \end{bmatrix} \begin{bmatrix} x_1 \\ x_2 \\ \vdots \\ x_{10} \\ x_{11} \\ x_{12} \\ x_{13} \\ x_{14} \end{bmatrix} \quad \begin{bmatrix} x_1^3 \\ x_2^3 \\ \vdots \\ x_{10}^3 \\ x_{11}^3 \\ x_{12}^3 \\ x_{13}^3 \\ x_{14}^3 \end{bmatrix}$$

$$\alpha_{22}x_1 + \alpha_{12}x_2 + \alpha_{21}x_3 + \alpha_{02}x_4 + \alpha_{11}x_5 + \alpha_{20}x_6 + \alpha_{01}x_8 + \alpha_{10}x_9 = 0 \tag{6.6}$$

From the discussion above it follows that generically this system has exactly three distinct solutions. They are

$$x_1 = \cdots = x_{10} = 0 \qquad \begin{bmatrix} x_{11} \\ x_{12} \\ x_{13} \\ x_{14} \end{bmatrix} = \begin{bmatrix} 0 \\ 0 \\ 0 \\ 0 \end{bmatrix}, \begin{bmatrix} \alpha_{51} \\ -\alpha_{42} \\ \alpha_{24} \\ -\alpha_{15} \end{bmatrix}, \begin{bmatrix} -\alpha_{51} \\ \alpha_{42} \\ -\alpha_{24} \\ \alpha_{15} \end{bmatrix} \tag{6.7}$$

To obtain our original system (6.1) from (6.6) we must add the two linear equations

$$(\ell_2) \quad \alpha_{42}x_2 + \alpha_{51}x_3 + \alpha_{32}x_4 + \alpha_{41}x_5 + \alpha_{50}x_6 + \alpha_{22}x_7 + \alpha_{31}x_8 + \alpha_{40}x_9 + \alpha_{12}x_{11} + \alpha_{21}x_{12} = 0$$

$$(\ell_3) \quad \alpha_{15}x_2 + \alpha_{24}x_3 + \alpha_{05}x_4 + \alpha_{14}x_5 + \alpha_{23}x_6 + \alpha_{04}x_8 + \alpha_{13}x_9 + \alpha_{22}x_{10} + \alpha_{12}x_{13} + \alpha_{21}x_{14} = 0$$

It then follows that the solutions in (6.7) satisfy system (6.1) only if $\alpha_{12}\alpha_{51} = \alpha_{21}\alpha_{42}$ and $\alpha_{12}\alpha_{24} = \alpha_{21}\alpha_{15}$. We conclude that for a generic choice of the α_{rs}, the system (6.1) has only the trivial solution.

As seen previously, system (6.3) always has exactly three solutions and system (6.5) has at most 3^9 solutions. Thus system (6.6) and hence system (6.1) has at most 3^{10} solutions.

6.8 We therefore have that the order of $C\ell(F\colon\ z^3 = G_6)$ is bounded above by 3^{10} and is 1 for a generic choice of G_6. We will now show that this upper bound can be refined, that the order of $C\ell(F)$ is $\le 3^8$.

As mentioned previously, there are only three possibilities for the four-tuple $(x_{11}, x_{12}, x_{13}, x_{14})$. They are $(0,0,0,0)$, $(\alpha_{51}, -\alpha_{42}, \alpha_{24}, -\alpha_{15})$, and $(-\alpha_{51}, \alpha_{42}, -\alpha_{24}, \alpha_{15})$.

We will demonstrate that after substituting any one of these into system (6.1) there will remain at most 3^7 possibilities for the 10-tuple $(x_1, \ldots, x_{10})$. It will then follow that system (6.1) has at most 3^8 solutions.

If we substitute $(x_{11}, x_{12}, x_{13}, x_{14}) = (0,0,0,0)$ into equation (6.1), the resulting system we obtain is

$$\begin{bmatrix} a_{11} & a_{12} & \cdots & a_{110} \\ a_{21} & a_{22} & \cdots & a_{210} \\ \cdots & \cdots & \cdots & \cdots \\ a_{101} & a_{102} & \cdots & a_{1010} \end{bmatrix} \begin{bmatrix} x_1 \\ x_2 \\ \vdots \\ x_{10} \end{bmatrix} = \begin{bmatrix} x_1^3 \\ x_2^3 \\ \vdots \\ x_{10}^3 \end{bmatrix} \tag{6.9}$$

$$(\ell_1)\quad \alpha_{15}x_2 + \alpha_{24}x_3 + \alpha_{05}x_4 + \alpha_{14}x_5 + \alpha_{23}x_6 + \alpha_{04}x_8 + \alpha_{13}x_9 + \alpha_{22}x_{10} = 0$$

$$(\ell_2)\quad \alpha_{42}x_2 + \alpha_{51}x_3 + \alpha_{32}x_4 + \alpha_{41}x_5 + \alpha_{50}x_6 + \alpha_{22}x_7 + \alpha_{31}x_8 + \alpha_{40}x_9 = 0$$

$$(\ell_3)\quad \alpha_{22}x_1 + \alpha_{12}x_2 + \alpha_{21}x_3 + \alpha_{02}x_4 + \alpha_{11}x_5 + \alpha_{20}x_6 + \alpha_{01}x_8 + \alpha_{10}x_9 = 0$$

We will consider several cases.

Case 1: Equations ℓ_1 and ℓ_2 are linearly dependent. That is, $\ell_1 = a\ell_2$ for some $a \in k$. Let $\tilde{a} = \sqrt[3]{a}$. If we make the replacement x by $x - \tilde{a}$, the monomials x^2y^4, xy^5, x^2y^3, xy^4, y^5, xy^3, y^4, and x^2y^2

will vanish in G_6. Thus we can assume that the y-degree of G_6 is less than 3. It then follows that any logarithmic derivative has y-degree less than 2. In terms of the system of equations above, this translates into the statement $x_6 = x_9 = x_{10} = 0$, which shows that system (6.9) has at most 3^7 solutions.

Case 2: Equations ℓ_1, ℓ_2, and ℓ_3 are linearly dependent. From case 1 we may assume that equations ℓ_1 and ℓ_2 are independent over k. Then $\exists$ a,b $\in$ k such that $\ell_3 = a\ell_1 + b\ell_2$. If we now replace x by $x - \tilde{b}$ and y by $y - \tilde{a}$, where $\tilde{a} = \sqrt[3]{a}$, $\tilde{b} = \sqrt[3]{b}$, then each of the monomials $x^i y^j$ with $i + j \leq 3$ vanishes in G_6. That is, G_6 has leading form of degree at least 4. Thus any logarithmic derivative of G_6 will have leading form of degree = 2. In terms of system (6.9), this just says that $x_0 = x_1 = x_2 = 0$.

Case 3: Suppose now that ℓ_1, ℓ_2, ℓ_3 are linearly independent over k. Then after performing elementary row operations we can assume ℓ_1, ℓ_2, and ℓ_3 have the form

$$(\tilde{\ell}_1) \quad x_{10} = \beta_{11}x_1 + \beta_{12}x_2 + \cdots + \beta_{17}x_7$$

$$(\tilde{\ell}_2) \quad x_9 = \beta_{21}x_1 + \beta_{22}x_2 + \cdots + \beta_{27}x_7$$

$$(\tilde{\ell}_3) \quad x_8 = \beta_{31}x_1 + \beta_{32}x_2 + \cdots + \beta_{37}x_7$$

If we combine these three equations with equations (1) through (7) of the cubic equations of (6.9) we obtain the following subsystem of (6.9):

$$\begin{bmatrix} a_{11} & a_{12} & \cdots & a_{17} \\ \cdot & \cdot & \cdots & \cdot \\ \cdot & \cdot & \cdots & \cdot \\ \cdot & \cdot & \cdots & \cdot \\ a_{71} & a_{72} & \cdots & a_{77} \end{bmatrix} \begin{bmatrix} x_1 \\ \cdot \\ \cdot \\ \cdot \\ x_7 \end{bmatrix} = \begin{bmatrix} x_1^3 \\ \cdot \\ \cdot \\ \cdot \\ x_7^3 \end{bmatrix} \tag{6.10}$$

together with equations $\tilde{\ell}_1$, $\tilde{\ell}_2$, and $\tilde{\ell}_3$ above.

Note that $k[x_1,x_2,\ldots,x_{10}]$ with the relations given in system (6.10) is a finite-dimensional k-vector space spanned by the monomials $\{x_i^n x_j^m \mid i,j = 1, 2, \ldots, 7, n,m = 0, 1, 2\}$. Thus $k[x_1,\ldots,x_{10}]$

with these relations is an artinian ring and thus has only finitely many maximal ideals. This shows, by Bezout's theorem, that system (6.10) has at most 3^7 distinct solutions, hence so does system (6.9).

Exactly the same argument works, practically word for word, if we began with either of the other possibilities for the four-tuple $(x_{11}, x_{12}, x_{13}, x_{14})$.

At this point I have not been able to produce an example of a surface F: $z^3 = G_6$ with class group of order 3^8. The best that I have been able to come up with so far are examples of surfaces with class group of order 3^6. The surface $z^3 = x^5y + xy^3 + y$ is such a surface. It is my feeling that 3^7 may be the actual "reachable" upper bound on the order of $C\ell(F\colon\ z^3 = G_6)$.

We know that for a generic G_6, $C\ell(F\colon\ z^3 = G_6)$ is trivial. Let us continue this exploration of the relationship of $C\ell(F\colon\ z^3 = G_6)$ and the coefficient space of G_6.

We have the following lemma, which is an immediate consequence of lemma (3.8) and elementary linear algebra.

LEMMA 6.11 Let k be an algebraically closed field of characteristic $p \neq 0$. Let S_1 be a system of equations of the form

$$S_1\colon\quad \begin{array}{ll} (1) & \alpha_{11}x_1 + \cdots + \alpha_{1n}x_n = x_1^p \\ (2) & \alpha_{21}x_1 + \cdots + \alpha_{2n}x_n = x_2^p \\ & \vdots \\ (n) & \alpha_{n1}x_1 + \cdots + \alpha_{nn}x_n = x_n^p \\ (\ell_1) & \beta_{11}x_1 + \cdots + \beta_{1n}x_n = 0 \\ (\ell_2) & \beta_{21}x_1 + \cdots + \beta_{2n}x_n = 0 \\ & \vdots \\ (\ell_m) & \beta_{m1}x_1 + \cdots + \beta_{mn}x_n = 0 \end{array}$$

where $\beta_{1n} \neq 0$.

Then S_1 is equivalent (in the sense that it has the same number of distinct solutions) to the system S_2 given below.

$$\begin{aligned}
S_2: \quad (1)\quad & a_{11}x_1 + \cdots + a_{1(n-1)}x_{n-1} = x_1^p \\
(2)\quad & a_{21}x_1 + \cdots + a_{2(n-1)}x_{n-1} = x_2^p \\
& \vdots \\
(n-1)\quad & a_{(n-1)1}x_1 + \cdots + a_{(n-1)(n-1)}x_{n-1} = x_{n-1}^p \\
(\ell_1')\quad & b_{11}x_1 + \cdots + b_{1(n-1)}x_{n-1} = 0 \\
(\ell_2')\quad & b_{21}x_1 + \cdots + b_{2(n-1)}x_{n-1} = 0 \\
& \vdots \\
(\ell_m')\quad & b_{m1}x_1 + \cdots + b_{m(n-1)}x_{n-1} = 0 \\
(\ell_1)\quad & \beta_{11}x_1 + \cdots + \beta_{1n}x_n = 0
\end{aligned}$$

where

$$\begin{aligned}
a_{ij} &= \alpha_{ij}\beta_{1n} - \alpha_{in}\beta_{1j} \\
b_{1j} &= \beta_{11}^p a_{1j} + \beta_{12}^p a_{2j} + \cdots + \beta_{1(n-1)}^p a_{(n-1)j} \\
&\quad + \beta_{1n}^p(\alpha_{nj}\beta_{1n} - \alpha_{nn}\beta_{1j}) \\
b_{ij} &= \beta_{ij}\beta_{1n} - \beta_{in}\beta_{1j} \qquad \text{for } i = 2, \ldots, m,\ j = 1, \ldots, n-1
\end{aligned}$$

Using lemma (6.11) we further analyze system (6.1). Recall that there are exactly three possibilities for the four-tuple $(x_{11}, x_{12}, x_{13}, x_{14})$, namely, $(0,0,0,0)$, $(\alpha_{51}, -\alpha_{42}, \alpha_{24}, -\alpha_{15})$, and $(-\alpha_{51}, \alpha_{42}, -\alpha_{24}, \alpha_{15})$. Let us make the assumption that

$$\alpha_{51}\alpha_{12} = \alpha_{21}\alpha_{42} \qquad \text{and} \qquad \alpha_{12}\alpha_{24} = \alpha_{15}\alpha_{21} \tag{6.12}$$

Then if we substitute any one of the above four-tuples into system (6.1), system (6.1) will reduce to system (6.9). Thus, under assumption (6.12), our original system has exactly three times the number of distinct solutions that system (6.9) has.

If we suppose that $\alpha_{22} \neq 0$, then we can apply lemma (6.11) to system (6.9) to obtain the equivalent system of equations

$$\begin{aligned}
\ell_1^{(2)}:\quad & b_{11}x_1 + b_{12}x_2 + \cdots + b_{19}x_9 = 0 \\
\ell_2^{(2)}:\quad & b_{21}x_1 + b_{22}x_2 + \cdots + b_{29}x_9 = 0
\end{aligned}$$

$$\ell_3^{(2)}: \quad b_{31}x_1 + b_{32}x_2 + \cdots + b_{39}x_9 = 0$$

$$\begin{bmatrix} a_{11}^{(2)} & a_{12}^{(2)} & \cdots & a_{19}^{(2)} \\ a_{21}^{(2)} & a_{22}^{(2)} & \cdots & a_{29}^{(2)} \\ \cdots & & \cdots & \\ a_{91}^{(2)} & a_{92}^{(2)} & \cdots & a_{99}^{(2)} \end{bmatrix} \begin{bmatrix} x_1 \\ x_2 \\ \\ x_9 \end{bmatrix} = \begin{bmatrix} x_1^3 \\ x_2^3 \\ \vdots \\ x_9^3 \end{bmatrix} \tag{6.13}$$

together with ℓ_1, where the $a_{ij}^{(2)}$ and $b_{ij}^{(2)}$ are polynomials in the α_{rs} with coefficients in the prime subfield of k and where the formulas for these are gotten from lemma (6.9).

Note that if each $b_{ij}^{(2)} = 0$ and $\det(a_{ij}^{(2)}) \neq 0$, then from (3.10) system (6.13) has 3^9 solutions. We conclude that $C\ell(F\colon\ z^3 = G_6)$ has order 3^{10} on a subset W_{10} of the coefficient space of codimension ≤ 29. Of course, this is no great discovery since the coefficient space of G_6 has dimension 22 and we have just shown that the order of $C\ell(F)$ is $\leq 3^8$.

If we now assume that $b_{19}^{(2)} \neq 0$, then system (6.13) is equivalent [by lemma (6.11)] to the system

$$\begin{bmatrix} a_{11}^{(3)} & a_{12}^{(3)} & \cdots & a_{18}^{(3)} \\ a_{21}^{(3)} & a_{22}^{(3)} & \cdots & a_{28}^{(3)} \\ \cdots & & \cdots & \\ a_{31}^{(3)} & a_{32}^{(3)} & \cdots & a_{88}^{(3)} \end{bmatrix} \begin{bmatrix} x_1 \\ x_2 \\ \vdots \\ x_8 \end{bmatrix} = \begin{bmatrix} x_1^3 \\ x_2^3 \\ \vdots \\ x_8^3 \end{bmatrix} \tag{6.14}$$

$$\ell_1^{(3)}: \quad b_{11}^{(3)}x_1 + b_{12}^{(3)}x_2 + \cdots + b_{18}^{(3)}x_8 = 0$$

$$\ell_2^{(3)}: \quad b_{21}^{(3)}x_1 + b_{22}^{(3)}x_2 + \cdots + b_{28}^{(3)}x_8 = 0$$

$$\ell_3^{(3)}: \quad b_{31}^{(3)}x_1 + b_{32}^{(3)}x_2 + \cdots + b_{38}^{(3)}x_8 = 0$$

together with ℓ_1, $\ell_1^{(2)}$, where the formulas for the $a_{ij}^{(3)}$ and $b_{ij}^{(3)}$ are as in lemma (6.11). If each $b_{ij}^{(3)} = 0$ and $\det(a_{ij}^{(3)}) \neq 0$, then (6.14) has 3^8 solutions, from which it follows that the order of $C\ell(F)$ is 3^9 on a subset W_9 of the coefficient space of codimension 26. Again, this is a vacuous discovery.

But we do see that by continuing in this manner we will obtain the following.

6.15 Let W_i, $i = 1, 2, 3, \ldots, 8$ be the subset of the coefficient space consisting of those α_{rs} for which $C\ell(F\colon\ z^3 = G_6)$ has order 3^i. Then the codimension of W_i is less than or equal to $3i - 1$. Also, generically $C\ell(F\colon\ z^3 = G_6) = 0$. Following is a list of some surfaces and their corresponding divisor class group orders.

G_6	Order of $C\ell(F\colon\ z^3 = G_6)$
$x + y^2 + y^2x + y^4 + y^5 + x^5y$	1
$xy + y^4$	3
$xy(y + \theta_1) + y^4$	3^2
$xy(y + \theta_1)(y + \theta_2) + y^4$	3^3
$xy(y + \theta_1)(y + \theta_2)(y + \theta_3) + y^4$	3^4
$xy(y + \theta_1)(y + \theta_2)(y + \theta_3)(y + \theta_4) + y^4$	3^5
$x^5y + xy^3 + y$	3^6

Let us summarize the results that we have obtained. We showed that for a generic G_6 the surface $F\colon\ z^3 = G_6$ has trivial class group. Also, the order of $C\ell(F)$ is bounded above by 3^8, and if we let W_i be the subset of the coefficient space corresponding to those α_{rs} for which the order of $C\ell(F)$ is 3^i, then W_i has codimension at most $3i - 1$. Comparing these results with conjectures (4.23) and (4.25), we see that both of these are supported by this example.

Recall that the generic genus of $F\colon\ z^3 = G_6$ is 3. Artin's conjecture states that since $C\ell(F) = 0$ for a generic G_6, the order of $C\ell(F)$ will be 3^i on a subset of codimension $\leq 3i$. So not only does the surface $F\colon\ z^3 = G_6$ provide another example supporting Artin's conjecture, but it also provides an example where equality in the codimension relationship does not hold. The surface $F\colon\ z^p = G_3$ to be considered shortly will provide another.

REMARK 6.16 To show that the surface $z^p = G_n$ has trivial class group for generic G_n [where n = deg G and char(k) = p], we first consider the corresponding system of equations we obtain via (1.9), (2.4), and Ganong's formula.

This system will have the form

$$S_1: \quad A_1 \begin{bmatrix} x_1 \\ \cdot \\ \cdot \\ \cdot \\ x_m \end{bmatrix} = \begin{bmatrix} x_1^p \\ \cdot \\ \cdot \\ \cdot \\ x_m^p \end{bmatrix} \qquad B_1 \begin{bmatrix} x_1 \\ \cdot \\ \cdot \\ \cdot \\ x_m \end{bmatrix} = \vec{0}$$

where A_1 is an $m \times m$ matrix and B_1 an $m \times r$ matrix with entries in $F_p[\alpha_{rs}]$ (F_p is the prime subfield of k and α_{rs} are the coefficients of G_n), and where $r \le m$ are positive integers.

By repeated application of lemma (6.11), we successively obtain systems of equations

$$S_j: \quad A_j \begin{bmatrix} x_1 \\ \cdot \\ \cdot \\ \cdot \\ x_{m+1-j} \end{bmatrix} = \begin{bmatrix} x_1^p \\ \cdot \\ \cdot \\ \cdot \\ x_{m+1-j}^p \end{bmatrix} \qquad B_j \begin{bmatrix} x_1 \\ \cdot \\ \cdot \\ \cdot \\ x_{m+1-j} \end{bmatrix} = \vec{0}$$

for $0 < j \le m$, of the type just described and whose solution set has order equal to the order of $C\ell(F)$ provided that at the j - 1 stage, B_{j-1} was not identically zero.

Thus one way to show that $C\ell(F) = 0$ generically is to treat the α_{rs} as indeterminates and show explicitly that the matrices B_1, B_2, ..., B_{m-1} do not vanish in this process. Of course, another attack is to produce an actual example $\tilde{F}$: $z^p = \breve{G}_n$ with the resulting matrices $\tilde{B}_1$, $\tilde{B}_2$, ..., $\breve{B}_{m-1}$ all nonzero. This is exactly what we have been doing all along.

For the $z^3 = G_6$ case such an example is provided by the surface $z^3 = x + y^2 + y^2x + y^4 + y^5 + x^5y$. If we substitute $\alpha_{10} = \alpha_{02} = \alpha_{04} = \alpha_{05} = \alpha_{51} = 1$, all other $\alpha_{ij} = 0$, into system (6.9) and successively apply lemma (6.11) to eliminate (in order) x_9, x_{10}, x_8, x_6, x_5, x_7, x_4, x_3, x_2, we obtain the following for B_i.

$$B_1 = \begin{bmatrix} 0&0&0&1&0&0&0&1&0&0 \\ 0&0&1&0&0&0&0&0&0&0 \\ 0&1&0&1&0&0&0&0&1&0 \end{bmatrix} \qquad B_2 = \begin{bmatrix} 0&0&0&1&0&0&0&1&0 \\ 0&0&1&0&0&0&0&0&0 \\ 0&0&0&0&0&2&2&2&2 \end{bmatrix}$$

$$B_3 = \begin{bmatrix} 0&0&0&1&0&0&0&1 \\ 0&0&1&0&0&0&0&0 \\ 0&1&0&1&1&1&0&0 \end{bmatrix} \qquad B_4 = \begin{bmatrix} 0&0&0&1&0&2&0 \\ 0&0&1&0&0&0&0 \\ 0&1&0&1&1&1&0 \end{bmatrix}$$

$$B_5 = \begin{bmatrix} 0&2&0&1&1&0 \\ 0&0&1&0&0&0 \\ 0&1&0&2&1&0 \end{bmatrix} \qquad B_6 = \begin{bmatrix} 2&0&0&0&1 \\ 0&2&0&1&0 \\ 0&2&0&1&0 \end{bmatrix} \qquad B_7 = \begin{bmatrix} 1&2&0&2 \\ 0&0&1&0 \\ 0&2&0&1 \end{bmatrix} \tag{6.17}$$

$$B_8 = \begin{bmatrix} 0&2&0 \\ 0&0&1 \\ 1&1&0 \end{bmatrix} \qquad B_9 = \begin{bmatrix} 0&2 \\ 2&2 \\ 1&1 \end{bmatrix} \qquad B_{10} = \begin{bmatrix} 1 \\ 2 \\ 1 \end{bmatrix}$$

Thus, as already stated, system (6.9) has only the trivial solution for a generic G_6, from which it follows (see p. 192) that system (6.1) has only the trivial solution for a generic G_6. We repeat this type of argument below when we show that $C\ell(F\colon\ z^p = G_3) = 0$ for a generic G_3 and $p > 5$.

The Surface F: $z^p = G_3$ for $p > 5$

Let k be an algebraically closed field of characteristic $p > 5$, $G_3 \in k[x,y]$ be of degree 3 satisfying (*) and (**). Let F be the surface defined by the equation $z^p = G_3$.

We begin by showing that for a generic G_3, $C\ell(F) = 0$. We employ the method described above.

Proof: Choose $\alpha \in k$ such that α is algebraic of degree greater than 2p over the prime subfield of k. Let $\breve{G}_3 = xy + (x + y)(x - y)(x + \alpha y)$. Then $\tilde{G}_3$ satisfies (*) and (**).

We calculate the following:

$$\nabla(G_3^{p-3}) = 4(1 + 9\alpha^2 - \alpha^4) \qquad \nabla(G_3^{p-3}x) = 3\alpha(3 - \alpha^2)$$

$$\nabla(G_3^{p-3}y) = 3(1 - 3\alpha^2) \qquad \nabla(G_3^{p-2}) = -12\alpha \qquad \nabla(G_3^{p-2}x) = 2$$

$$\nabla(G_3^{p-2}y) = -2\alpha \qquad \nabla(G_3^{p-1}) = 1$$

$$\nabla(G_3^{p-1}x) = \left(\alpha^{p+1} - \frac{1}{\alpha^2} - 1\right)x^p + \left(\alpha - \frac{\alpha^p}{\alpha^2} - 1\right)y^p \tag{6.18}$$

$$\nabla(G_3^{p-1}y) = \left(\alpha - \frac{\alpha^p}{\alpha^2} - 1\right)x^p + \left(\alpha^{p+1} - \frac{1}{\alpha^2} - 1\right)y^p$$

It then follows from (1.9), (2.4), and Ganong's formula that the order of $C\ell(F)$ is less than or equal to the number of distinct solutions to the system of equations

$$\tilde{A}_1\begin{bmatrix}x_1\\x_2\\x_3\end{bmatrix} = \begin{bmatrix}x_1^p\\x_2^p\\x_3^p\end{bmatrix} \qquad B_1\begin{bmatrix}x_1\\x_2\\x_3\end{bmatrix} = \begin{bmatrix}0\\0\\0\end{bmatrix} \tag{6.19}$$

where

$$\tilde{A}_1 = \begin{bmatrix}1 & 0 & 0\\ 0 & \dfrac{\alpha^{p+1} - 1}{\alpha^2 - 1} & \dfrac{\alpha - \alpha^p}{\alpha^2 - 1}\\ 0 & \dfrac{\alpha - \alpha^p}{\alpha^2 - 1} & \dfrac{\alpha^{p+1} - 1}{\alpha^2 - 1}\end{bmatrix}$$

$$\tilde{B}_1 = \begin{bmatrix}6 & -1 & \alpha\\ 4(1 + 9\alpha^2 - \alpha^4) & 3\alpha(3 - \alpha^2) & 3(1 - \alpha^2)\end{bmatrix}$$

We apply (6.9) to this system, eliminating x_3 to obtain the equivalent system

$$\tilde{A}_2\begin{bmatrix}x_1\\x_2\end{bmatrix} \quad \begin{bmatrix}x_1^p\\x_2^p\end{bmatrix} \qquad \tilde{B}_2\begin{bmatrix}x_1\\x_2\end{bmatrix} = 0 \tag{6.20}$$

where

$$\tilde{A}_2 = \begin{bmatrix}\alpha & 0\\ \dfrac{-6\alpha(\alpha - \alpha^p)}{\alpha^2 - 1} & \alpha^p\end{bmatrix}$$

$$\tilde{B}_2 = \begin{bmatrix} 6\alpha^{p+1} - 6\alpha^2(\alpha^2(\alpha^2 - 1))^{p-1} & 0 \\ 4\alpha^5 - 54\alpha^3 + 14\alpha & 3\alpha^4 - 6\alpha^2 - 3 \end{bmatrix}$$

We have by our choice of α that det $B_2 \neq 0$ and hence systems (6.19) and (6.20) have only the trivial solution. As noted on page 197, since we were able to apply (6.9) in such a way so that the sequence of matrices $\tilde{B}_1$, $\tilde{B}_2$ are nonzero, it follows that for a generic G_3, system (6.19) has only the trivial solution. Of course, this implies that $C\ell(F) = 0$ for a generic G_3.

At this point we would normally study the relationship between $C\ell(F\colon\ z^3 = G_3)$ and the coefficient space of G_3.

But rather than look at the entire family of third-degre polynomials, G_3, satisfying conditions (*) and (**), we will restrict our attention to the subfamily of those G_3 satisfying condition (B) (see the discussion on p. 150). We justify this in two ways. One justification is that condition (B) is a generic one. Thus we will be deleting a relatively small number of surfaces from our discussion. Another is that computations and arguments will be greatly simplified.

Assume then that $\bar{G}_3$ satisfies condition (B). Then (see p. 150) after a linear change of coordinates we can assume that $\bar{G}_3$ has the form $\bar{G}_3 = xy + \beta_{30}x^3 + \beta_{21}x^2y + \beta_{12}xy^2 + \beta_{03}y^3$. Furthermore, we have that

$$\begin{aligned} &\nabla(\bar{G}_3^{p-2}) = 6(\beta_{30}\beta_{03} + \beta_{12}\beta_{21}) \\ &\nabla(\bar{G}_3^{p-2}x) = -2\beta_{12} \qquad \nabla(\bar{G}_3^{p-2}y) = -2\beta_{21} \end{aligned} \tag{6.21}$$

Let $\bar{L}$ be the group of logarithmic derivatives that corresponds to $\bar{G}_3$. From lemma (2.4) any logarithmic derivative $t \in \bar{L}$ has degree at most 1.

Then using Ganong's formula and theorem (1.9), we have that the order of $C\ell(\bar{F}\colon\ z^3 = \bar{G}_3)$ is equal to the number of distinct solutions to a system of equations of the form

$$A\begin{bmatrix}x_1\\x_2\\x_3\end{bmatrix} = \begin{bmatrix}x_1^3\\x_2^3\\x_3^3\end{bmatrix} \qquad B\begin{bmatrix}x_1\\x_2\\x_3\end{bmatrix} = \begin{bmatrix}0\\0\\0\end{bmatrix} \tag{6.22}$$

where A is a 3×3 matrix and B is a $n \times 3$ matrix for some n.

From the fact that $\nabla(\bar{G}^{p-2}t) = 0$, we see from (6.21) that one of the linear equations in system (6.22) is given by

$$3(\beta_{30}\beta_{03} + \beta_{12}\beta_{21})x_1 - \beta_{12}x_2 - \beta_{21}x_3 = 0 \tag{6.23}$$

If any one of the coefficients in (6.23) is nonzero, then from (3.8) and (3.10) we know that (6.22) has at most p^2 solutions.

If each of these in (6.23) is zero, then $\bar{G}_3 = x(y + \beta_{30}x^2)$ or $\bar{G}_3 = y(x + \beta_{03}y^2)$. In either case the surface $z^p = \bar{G}_3$ will be isomorphic to the surface $z^p = xy$, which has class group $\mathbb{Z}/p\mathbb{Z}$ [see (7.1), p. 203]. Thus we conclude that the order of $C\ell(F\colon z^p = \bar{G}_3)$ is at most p^2.

We now show that on the coefficient space of $\bar{F}\colon\ z^3 = \bar{G}_3$ the order of $C\ell(\bar{F})$ is p on a subset of codimension 1 and p^2 on a subset of codimension 2, where we here are viewing the coefficient space of $\bar{F}$ as contained in A^4.

We assume that $\beta_{30} = 0$ and $\beta_{21}\beta_{12}\beta_{03} \neq 0$. Then if we let D be the jacobian derivation corresponding to $\bar{G}_3$ and L the logarithmic derivatives of D, we see that $Dy/y = -(y + 2\beta_{21}xy + \beta_{12}y^2)/y = -(1 + 2\beta_{21}x + \beta_{12}y) \in L$. Hence we see that the order of $C\ell\ F \geq p$ when $\beta_{30} = 0$.

We have already seen that the order of $C\ell(\bar{F})$ is equal to the number of solutions to a system of equations of the form given in (6.22). From (6.23) we know that one of the linear equations in this system is given by

$$3\beta_{12}\beta_{21}x_1 - \beta_{12}x_2 - \beta_{21}x_3 = 0 \tag{6.24}$$

From the fact that $\nabla(G^{p-3}t) = 0$ for $t \in L$, we also obtain the equation

$$(3\beta_{21}^2\beta_{12}^2 + 2\beta_{21}^3\beta_{03})x_1 - (\beta_{21}\beta_{12}^2 + \beta_{21}^2\beta_{03})x_2$$
$$- \beta_{21}^2\beta_{12}x_3 = 0 \qquad (6.25)$$

as another of these linear equations in system (6.22). If $\beta_{21\ 03} \neq 0$, these two equations are independent over k and system (6.22) will have at most p solutions.

We conclude that with $\beta_{30} = 0$ and $\beta_{21}\beta_{12}\beta_{03} = 0$, the order of $C\ell(\bar{F}: z^p = \bar{G}_3)$ is p. We thereby have shown that the subspace of the coefficient space of $\bar{G}_3$ corresponding to those β_{rs} for which $C\ell(\bar{F}: z^p = \bar{G}_3)$ has order p has codimension 1.

To show that $C\ell(\bar{F})$ has order p^2 on a subset of the coefficient space of $\bar{G}_3$ of codimension ≤ 2, we assume that $\beta_{30} = \beta_{03} = 0$. Then $\bar{G}_3 = xy + \beta_{21}x^2y + \beta_{12}xy^2$, $Dx/x = 1 + \beta_{21}x + 2\beta_{12}y$, and $Dy/y = -(1 + 2\beta_{21}x + \beta_{12}y)$. Thus these elements, Dx/x and Dy/y, generate a subgroup of L of order p^2. Since the order of $\bar{L}$ is at most p^2, we have that the order of $C\ell(\bar{F}: z^3 = \bar{G}_3)$ is p^2 on a subset of the coefficient space of $\bar{G}_3$ of codimension ≤ 2. Also we note that for a generic $\bar{G}_3$, $C\ell(\bar{F}) = 0$. For we know that condition (B) is a generic condition on G_3 and we have shown that for a generic G_3 the coordinate ring of G_3 is factorial.

6.26 Let us summarize our findings. We have shown that for a generic choice of G_3, the surface F: $z^p = G_3$ has trivial class group if $p > 5$, thus giving further support to conjecture (4.23).

Even though the coefficient space of $\bar{G}_3 = xy + \beta_{30}x^3 + \beta_{21}x^2y + \beta_{12}xy^2 + \beta_{03}y^3$ is a proper subset of the coefficient space of G_3, we justified its study since for a generic choice of G_3, the surface F: $z^p = G_3$ can be written in the form $\bar{F}$: $z^p = \bar{G}_3$ after a linear change of coordinates.

We saw that Artin's conjecture held on the coefficient space of $\bar{G}_3$, for the "generic genus" of $\bar{F}$ is positive and the order of $C\ell(\bar{F})$ is p^i on subsets of codimension i for i = 0, 1, 2. We have also shown that p^2 is the maximum order of $C\ell(\bar{F}: z^p = \bar{G}_3)$.

This concludes our attack on conjectures (4.23), (4.24), and (4.25), although we give more evidence to support (4.24) in Section 10.

7. EXAMPLES

Throughout this section we let k be an algebraically closed field of characteristic $p > 0$. We calculate $C\ell(F)$ for some concrete surfaces.

When calculating $C\ell(F)$ for surfaces of the form $z^p = G$, probably the most natural place to start is with the case where G is homogeneous.

We begin this section with a study of normal surfaces $z^p = G$, where G is a product of distinct linear factors.

PROPOSITION 7.1 Let k be an algebraically closed field of characteristic $p > 0$. Let $G = (x - \alpha_1 y)(x - \alpha_2 y)\cdots(x - \alpha_n y)$, where the $\alpha_i \in k$ are pairwise distinct. Assume that G satisfies condition (**). Then $C\ell(F\colon\ z^p = G)$ is a direct sum of $n - 1$ copies of $\mathbf{Z}/p\mathbf{Z}$.

REMARK 7.2 The case $n = 2$ was proven by P. Samuel in his Tata notes (see Samuel, 1964a, pp. 65, 66).

Proof: Let D be the jacobian derivation of G and L the group of logarithmic derivatives of D. Let $t \in L$.

7.2.1 By Ganong's formula we have that $\nabla(G^{p-1}t) = t^p$. Let t_ℓ and t_m be the forms of t of lowest and highest degree. Let $\text{degree}(t_\ell) = \ell$ and $\text{degree}(t_m) = m$. Comparing the lowest and highest degree forms on both sides of the equality in (7.2.1), we have that

$$p\ell \geq (p - 1)n + \ell - 2(p - 1)$$
$$pm \leq (p - 1)n + m - 2(p - 1)$$

Thus we obtain $\ell \geq n - 2 \geq m$, which shows that $\ell = m = n - 2$. Therefore, t is homogeneous of degree $(n - 2)$.

Duplicating the argument used in the proof of (2.8), we have that the order of $C\ell(F)$ is at most p^{n-1}.

We will show that this order equals p^{n-1} by demonstrating that

$$\frac{D(x - \alpha_1 y)}{x - \alpha_1 y}, \quad \frac{D(x - \alpha_2 y)}{x - \alpha_2 y}, \quad \ldots, \quad \frac{D(x - \alpha_{n-1} y)}{x - \alpha_{n-1} y}$$

are in L and linearly independent over $\mathbb{Z}/p\mathbb{Z}$.

For suppose that

$$a_1 \frac{D(x - \alpha_1 y)}{x - \alpha_1 y} + a_2 \frac{D(x - \alpha_2 y)}{x - \alpha_2 y} + \cdots + a_n \frac{D(x - \alpha_n y)}{x - \alpha_n y} = 0$$

where the $a_i \in \mathbb{Z}/p\mathbb{Z}$ not all zero.

After clearing denominators and replacing Dx with G_y and Dy with $-G_x$, we obtain

$$\left[\sum_{j=1}^{n-1} a_j (x - \alpha_1 y)\cdots(x - \alpha_{j-1} y)(x - \alpha_{j+1} y)\cdots(x - \alpha_{n-1} y)\right] G_y$$

$$= \left[\sum_{j=1}^{n-1} a_j\, {}_j (x - \alpha_1 y)\cdots(x - \alpha_{j-1} y)(x - \alpha_{j+1} y)\cdots(x - \alpha_{n-1} y)\right] G_x$$

Note that for each j, the degree of $(x - \alpha_1 y)\cdots(x - \alpha_{j-1} y) \times (x - \alpha_{j+1} y)\cdots(x - \alpha_{n-1} y)$ is less than $n - 1$, and the degree of G_y is $n - 1$. Thus G_y and G_x must have a factor in common, contradicting condition (**).

REMARK 7.3 Let F and G be as in (7.1). Recall (2.1) that the coordinate ring of G is isomorphic to $A = k[X^p, Y^p, G]$. It follows from (1.10) that $C\ell(A)$ is generated by the nonprincipal height 1 primes $q_i = (x - \alpha_i y)k[x,y] \cap A$, $i = 1, 2, \ldots, n - 1$. Note that q_i is generated by the elements $x^p - \alpha_i^p y^p$ and G. From (2.1) we see that the primes q_i correspond to the curves $z = x - \alpha_i y = 0$ on F.

Another surface whose class group we will be computing is given by the equation $z^p = x^{p+1} + y^{p+1} + xy$. An isomorphic form of this surface was first studied by O. Zariski in his paper on Castelnuovo's criterion of rationality (Zariski, 1958).

The surface Zariski considered is defined by the equation

$$z^p = x^{p+1} + y^{p+1} + \frac{1}{2}(x^2 - y^2) \tag{7.4}$$

To see that these two surfaces are isomorphic we replace x with x + y and y with x - y in the equation

$$z^p = x^{p+1} + y^{p+1} + xy$$

from which we obtain the isomorphic surface

$$z^p = 2x^{p+1} + 2y^{p+1} + x^2 - y^2 \tag{7.5}$$

If in (7.5) we replace z by 2z and reverse the roles of x and y, we obtain equation (7.4).

Zariski used this surface (7.4) to show that the separability condition in Castelnuovo's theorem on the rationality of plane involution is essential. This theorem states:

Let $k(x,y)$ be a purely transcendental extension of an algebraically closed field k, of transcendence degree 2, and let Σ be a field between k and $k(x,y)$, also of degree 2 over k. If $k(x,y)$ is a separable extension of Σ, then Σ is a pure transcendental extension of k.

Zariski shows that the geometric genus p_g of this surface is > 0.

We prove the following formula.

LEMMA 7.6 Let $G = ax + by + cxy$, where $a,b,c \quad k[x^p,y^p]$. Let D be the jacobian derivation of G. Then for each $r = 1, 2, \ldots, p - 1$,

$$D^{p-1}(x^r) = [(cx)^r - (-b)^r]c^{p-r-1}$$
$$D^{p-1}(y^r) = [(cy)^r - (-a)^r]c^{p-r-1}$$

Proof: The proof involves several steps.

Step 1: For each positive integer m and nonnegative integer n, define

$$A(n,m) = \sum_{i=0}^{m-1} (-1)^i \binom{m-1}{i} (m - i)^n$$

We will show that for all pairs n and m as above,

$$mA(n,m) + (m + 1)A(n, m + 1) = A(n + 1, m + 1)$$

We have that

$$mA(n,m) + (m + 1)A(n, m + 1)$$

$$= m \sum_{i=0}^{m-1} (-1)^i \binom{m-1}{i} (m - i)^n + (m + 1) \sum_{i=0}^{m} (-1)^i \binom{m}{i} (m + 1 - i)^n$$

$$= m \sum_{i=0}^{m-1} (-1)^i \binom{m - 1}{i} (m - i)^n$$

$$+ \sum_{i=1}^{m} (-1^i \binom{m}{i} (m + 1 - i)^n (m + 1) + (m + 1)^{n+1}$$

$$= m \sum_{i=0}^{m-1} (-1)^i \binom{m - 1}{i} (m - i)^n$$

$$+ \sum_{i=0}^{m-1} (-1)^{i+1} \binom{m}{i + 1} (m - i)^n (m + 1) + (m + 1)^{n+1}$$

$$= \sum_{i=0}^{m-1} (m - i)^n \left[(-1)^i m \binom{m - 1}{i} + (-1)^{i+1} (m + 1) \binom{m}{i + 1}\right] + (m + 1)^{n+1}$$

$$= \sum_{i=0}^{m-1} (-1)^i (m - i)^n \left[m \binom{m - 1}{i} - (m + 1) \binom{m}{i + 1}\right] + (m + 1)^{n+1}$$

$$= \sum_{i=0}^{m-1} (-1)^i (m - i)^n \left[\frac{m!}{i!(m - i - 1)!} - \frac{(m + 1)!}{(i + 1)!(m - i - 1)!}\right]$$
$$+ (m + 1)^{n+1}$$

$$= \sum_{i=0}^{m-1} (-1)^i (m - i)^n \left[1 - \frac{m + 1}{i + 1}\right] \frac{m!}{i!(m - i - 1)!} + (m + 1)^{n+1}$$

$$= \sum_{i=0}^{m-1} (-1)^{i+1} (m - i)^{n+1} \frac{m!}{(i + 1)!(m - i - 1)!} + (m + 1)^{n+1}$$

$$= \sum_{i=0}^{m-1} (-1)^{i+1} (m - i)^{n+1} \binom{m}{i + 1} + (m + 1)^{n+1}$$

$$= \sum_{i=1}^{m} (-1)^i (m + 1 - i)^{n+1} \binom{m}{i} + (m + 1)^{n+1}$$

$$= \sum_{i=0}^{m} (-1)^i (m + 1 - i)^{n+1} \binom{m}{i}$$

$$= A(n + 1, m + 1)$$

Step 2: For any positive integer t, $A(p - 2, t) = (-1)^{t+1}/t$. We have that

$$A(p - 2, t) = \sum_{i=0}^{t-1} (-1)^i \binom{t - 1}{i} (t - i)^{p-2}$$

$$= \sum_{i=0}^{t-1} (-1)^i \binom{t - 1}{i} \frac{1}{t - i} \quad [\text{since } (t - i)^{p-1} = 1 \text{ for } i < t]$$

$$= \sum_{i=0}^{t-1} (-1)^i \binom{t - 1}{t - i - 1} \frac{1}{t - i}$$

$$= (-1)^{t-1} \sum_{i=0}^{t-1} (-1)^i \binom{t - 1}{i} \frac{1}{i + 1}$$

$$= (-1)^{t-1} \frac{1}{t} \quad \text{(see Gould, 1959-1960, identity 1.37)}$$

Step 3: We prove the following for n and r positive integers.

(a) $$D^n(x^r) = r \sum_{i=0}^{r-1} \binom{r - 1}{i} A(n - 1, r - i) c^{n-r+i} x^i (Dx)^{r-i}$$

(b) $$D^n(y^r) = (-1)^{r+n} r \sum_{i=0}^{r-1} (-1)^i \binom{r - 1}{i} A(n - 1, r - i) c^{n-r+i} y^i (Dy)^{r-i}$$

To prove (b) we proceed by induction on n. When n = 1, the right side of (b) is

$$(-1)^{r+1} r \sum_{i=0}^{r-1} (-1)^i \binom{r - 1}{i} A(0, r - i) c^{1-r+i} y^i (Dy)^{r-i} = y^{r-1} Dy$$

since

$$A(0,t) = \begin{cases} 0 & \text{if } t > 1 \\ 1 & \text{if } t = 1 \end{cases}$$

So we assume that the formula holds for integers less than n + 1. Then

$$D^{n+1}(y^r) = D(D^n(y^n)) \quad \text{which by the induction hypothesis}$$

$$= (-1)^{r+n} r \sum_{i=0}^{r-1} (-1)^i \binom{r-1}{i} A(n-1, r-i) c^{n-r+i}$$

$$\times [iy^{i-1}(Dy)^{r-i+1} + y^i(r-i)(Dy)^{r-i-1} D^2y]$$

$$= (-1)^{r+n} r \sum_{i=0}^{r-1} (-1)^i \binom{r-1}{i} A(n-1, r-i) c^{n-r+i}$$

$$\times [iy^{i-1}(Dy)^{r-i+1} + (r-i)y^i(Dy)^{r-i}(-c)]$$

$$= (-1)^{r+n} r \left[\sum_{i=1}^{r-1} (-1)^i i \binom{r-1}{i} A(n-1, r-i) c^{n-r+i} y^{i-1}\right.$$

$$\times (Dy)^{r-i+1} - \sum_{i=0}^{r-1} (-1)^i (r-i) \binom{r-1}{i}$$

$$\left.\times A(n-1, r-i) c^{n-r+i+1} y^i (Dy)^{r-i}\right]$$

$$= (-1)^{r+n} r \left[\sum_{i=0}^{r-2} (-1)^{i+1} (i+1) \binom{r-1}{i+1} A(n-1, r-i-1)\right.$$

$$\times c^{(n-r+i+1)} y^i (Dy)^{r-i} - \sum_{i=0}^{r-2} (-1)^i (r-i) \binom{r-1}{i}$$

$$\left.\times A(n-1), r-i) c^{(n-r+i+1)} y^i (Dy)^{r-i}\right]$$

$$+ (-1)^n r c^n y^{r-1} Dy$$

$$= (-1)^{r+n+1} r \sum_{i=0}^{r-2} (-1)^i c^{n-r+i+1} y^i (Dy)^{r-i}$$

$$\times \left[(i+1) \binom{r-1}{i+1} A(n-1, r-i+1)\right.$$

$$\left.+ (r-i) \binom{r-1}{i} A(n-1, r-i)\right] + (-1)^n r c^n y^{r-1} Dy$$

$$= (-1)^{r+n+1} r \sum_{i=0}^{r-2} (-1)^i c^{n-r+i+1} y^i (Dy)^{r-i} \binom{r-1}{i}$$

$$\times [(r-i-1) A(n-1, r-i-1)$$

$$+ (r-i) A(n-1, r-i)] + (-1)^n r c^n y^{r-1} Dy$$

$$= (-1)^{r+n+1} r \sum_{i=0}^{r-2} (-1)^i \binom{r-1}{i} A(n, r-i) c^{n-r+i+1}$$

$$\times (Dy)^{r-i} + (-1)^n r c^n y^{r-1} Dy \quad \text{by step 1}$$

$$= (-1)^{r+n+1} r \sum_{i=0}^{r-1} (-1)^i \binom{r-1}{i}$$

$$\times A(n, r-i) c^{n-r+i+1} y^i (Dy)^{r-i}$$

which is the desired expression. The proof of (a) is entirely symmetric.

Step 4: From step 3 we have that

$$D^{p-1}(x^n) = r \sum_{i=0}^{r-1} \binom{r-1}{i} A(p-2, r-i) c^{p-r+i-1} x^i (Dx)^{r-i}$$

$$= r \sum_{i=0}^{r-1} \binom{r-1}{i} \frac{(-1)^{r-i+1}}{r-i} c^{p-r+i-1} x^i (Dx)^{r-i} \quad \text{by step 2}$$

$$= -c^{p-1} \sum_{i=0}^{r-1} \binom{r}{i} x^i \left(\frac{-Dx}{c}\right)^{r-i}$$

$$= -c^{p-1} \left[\left(x - \frac{Dx}{c}\right)^r - x^r\right]$$

$$= [(cx)^r - (-b)^r] c^{p-r-1}$$

Similarly,

$$D^{p-1}(y^r) = [(cy)^r - (-a)^r] c^{p-r-1}$$

LEMMA 7.7 Let G and D be as in (7.6). Then

$$D^{p-1}(x^r) - ax^r = -(-b)^r c^{p-r-1}$$

$$D^{p-1}(y^r) - ay^r = -(-a)^r c^{p-r-1}$$

where $a \in k[x,y]$ is such that $D^p = aD$ [see theorem (1.9)].

Proof: From lemma (7.6), $D^{p-1}(x) = (cx + b)c^{p-2}$. Thus $D^p(x) = c^{p-1} Dx$, which implies that $a = c^{p-1}$. Now use lemma (7.6) again.

LEMMA 7.8 Let $u,v \in k(x,y)$. Then

$$\nabla(uv) = \sum_{0\le i+j\le p-1} (-1)^{i+j} \frac{\partial^{i+j}}{\partial x^i \partial y^j}(u) \frac{\partial^{2p-(i+j+2)}}{\partial x^{p(i+1)} \partial y^{p-(j+1)}}(v)$$

Proof: Follows from calculus and the fact that $\binom{p-1}{i} = (-1)^i$ for $0 \le i < p$.

LEMMA 7.9 Let $G = xu + yv + xyw$, where u, v, and $w \in k[x^p, y^p]$ are homogeneous such that deg u = deg v → deg w + 2. Let $0 \le i, j \le p - 1$. Then the highest-degree form of $(\partial^{i+j}/\partial x^i \partial y^j)(G^{p-2})$ is a nonzero scalar multiple of

$$(xu + yv)^{p-(i+j+2)} u^i v^j \qquad \text{if } 0 \le i + j \le p - 2$$
$$u^{p-(j+2)} v^{p-(i+2)} w^{i+j+2-p} \qquad \text{if } p - 2 < i + j \le 2p - 4$$

Proof: We have that $G^{p-2} = \Sigma_{n=0}^{p-2} \binom{p-2}{n} (xu + yv)^n (xyw)^{p-(2+n)}$. For each $n = 0, 1, \ldots, p - 2$, $\deg[(xu + yv)^n (xyw)^{p-(2+n)}] = n(\deg u + 1) + [p - 2 + n)](\deg w + 2)$. Thus the form of highest degree in G^{p-2} occurs when $n = p - 2$ in the sum above, the next when $n = p - 3$, the next when $n = p - 4, \ldots.$

Thus if $0 \le i + j \le p - 2$, the form of highest degree in $(\partial^{i+j}(G^{p-2})/\partial x^i \partial y^j)$ is $(\partial^{i+j}(xu + yv)^{p-2}/\partial x^i \partial y^j)$, which is a nonzero constant multiple of $(xu + yv)^{p-(w+i+j)} u^i v^j$.

If $p - 2 < i + j \le 2p - 4$, the largest value of n such that $(xu + yv)^n (xy)^{p-(2+n)}$ is not zeroed by $\partial^{i+j}/\partial x^i \partial y^j$ is $n = 2p - 4 - (i + j)$. Therefore, the highest-degree form of $\partial^{i+j}(G^{p-2})/\partial x^i \partial y^j$ is a multiple of

$$\frac{\partial^{i+j}}{\partial x^i \partial y^j} (xu + yv)^{2p-(4+i+j)} (xyw)^{i+j+2-p}$$

i.e., a multiple of

$$\frac{\partial^{i+j}}{\partial x^i \partial y^j} (x^i y^j u^{p-j-2} v^{p-i-2} w^{i+j+2-p})$$

which is a multiple of $u^{p-j-2} v^{p-i-2} w^{i+j+2-p}$.

LEMMA 7.10 Let u and v be pth powers of distinct linear homogeneous polynomials. Let $G = xu + yv + xy$, D the jacobian derivation of G, and L the group of logarithmic derivatives of D. Then if $t \in L$, $(\partial^2(t)/\partial x\partial y) = 0$.

Proof: Note that $\partial G/\partial x = u + y$ and $\partial G/\partial y = v + x$ are relatively prime. We have by Ganong's formula and (7.8) that

$$0 = \nabla(G^{p-2}t) = \sum_{0\le i,j\le p-1} (-1)^{i+j} \frac{\partial^{i+j}(G^{p-2})}{\partial x^i \partial y^j} \frac{\partial^{2p-(i+j+2)}(t)}{\partial x^{p-(i+1)}\partial y^{p-(j+2)}} \tag{7.10.1}$$

By (2.4), $\deg t \le p - 1$. Thus

$$\frac{\partial^{2p-(i+j+2)}(t)}{\partial x^{p-(i+1)}\partial y^{p-(j+1)}} = 0$$

for $i = j \le p - 2$ and (7.10.1) becomes

$$\sum_{\substack{p-2<i+j\le 2p-4\\ 0\le i,j\le p-1}} (-1)^{i+j} \frac{\partial^{i+j}(G^{p-2})}{\partial x^i \partial y^j} \frac{\partial^{2p-(i+j+2)}(t)}{\partial x^{p-(i+1)}\partial y^{p-(j+1)}} = 0 \tag{7.10.2}$$

Let $\tilde{t}$ be the form of highest degree of t such that $\partial^2(\tilde{t})/\partial x\partial y \ne 0$. From (7.9) it follows that if $p - 2 < i + j \le 2p - 4$, either

$$\deg \frac{\partial^{i+j}}{\partial x^i \partial y^j}(G^{p-2}) \frac{\partial^{2p-(i+j+2)}(t)}{\partial x^{p-(i+2)}\partial y^{p-(j+1)}}$$

$$= \deg(\tilde{t}) + 2(p^2 - 3p + 1) + (1 - p)i + (1 - p)j \tag{7.10.3a}$$

with the leading form of

$$\frac{\partial^{i+j}}{\partial x^i \partial y^j}(G^{p-2}) \frac{\partial^{2p-(i+j+2)}(t)}{\partial x^{p-(i+1)}\partial y^{p-(j+1)}}$$

being

$$u^{p-(j+2)}v^{p-(i+2)} \frac{\partial^{2p-(i+j+2)}(\tilde{t})}{\partial x^{p-(i+1)}\partial y^{p-(j+1)}}$$

or

$$\deg \frac{\partial^{i+j}(G^{p-2})}{\partial x^i \partial y^j} \frac{\partial^{2p-(i+j+2)}(t)}{\partial x^{p-(i+1)} \partial y^{p-(j+1)}} < \deg(\tilde{t}) + 2(p^2 - 3p + 1) \tag{7.10.3b}$$

$$+ (1-p)i + (1-p)j \quad \text{and} \quad \frac{\partial^{2p-(i+j+2)}(\tilde{t})}{\partial x^{p-(i+1)} \partial y^{p-(j+1)}} = 0$$

Let (i_o, j_o) be the smallest (in the lexiographic order) ordered pair such that

$$\frac{\partial^{2p-(i+j+2)}(\tilde{t})}{\partial x^{p-(i+1)} \partial y^{p-(j+1)}} \neq 0$$

By (7.10.3) we must have that

$$\sum_{i+j=i_o+j_o} u^{p-(j+2)} v^{p-(i+2)} \frac{\partial^{2p-(i+j+2)}(\tilde{t})}{\partial x^{p-(i+1)} \partial y^{p-(j+1)}} = 0 \tag{7.10.4}$$

for this will be the highest-degree form of the expression on the left side of (7.10.2). But this (7.10.4) is clearly impossible since u and v are distinct primes and u (respectively, v) cannot possibly divide

$$\frac{\partial^{2p-(i+j+2)}(\tilde{t})}{\partial x^{p-(i+1)} \partial y^{p-(j+1)}}$$

for any i and j because $\deg(\tilde{t}) < p$.

PROPOSITION 7.11 Let G be as in (7.10). Then the surface F: $z^p = G$ has divisor class group of order p^M where $0 < M \le 2p - 1$.

Proof: By (7.10), if a polynomial $t \in L$, then t has the form

$$t = \sum_{i=1}^{p-1} \alpha_{i0} x^i + \alpha_{0i} y^i + \alpha_{00} \tag{7.11.1}$$

By (7.7), we have that $D^{p-1}(x^i) - ax^i = -(-v)^i$ and $D^{p-1}(y^i) - ay^i = -(-u)^i$ for $i = 0, 1, \ldots, p-1$.

Applying theorem (1.9b) to t, we see that

$$\alpha_{00} + \sum_{i=1}^{p-1} \alpha_{i0}(-v)^i + \alpha_{0i}(-u)^i = \alpha_{00}^p + \sum_{i=1}^{p-1} \alpha_{i0}^p x^{ip} + \alpha_{0i}^p y^{ip} \tag{7.11.2}$$

A comparison of homogeneous forms of the same degree in (7.11.2) yields

$$\alpha_{00} = \alpha_{00}^p \quad \text{and} \quad \alpha_{i0}v^i + \alpha_{0i}u^i = (-1)^i(\alpha_{i0}^p x^{ip} + \alpha_{0i}^p y^{ip})$$

$$i = 1, 2, \ldots, p - 1 \qquad (7.11.3)$$

By hypothesis, $\exists$ a,b,c,d $\in$ k such that $u = ax^p + by^p$ and $v = cx^p + dy^p$. Then (7.11.3) is equivalent to

$$\alpha_{00} = \alpha_{oo}^p \qquad (7.11.4)$$

$$\alpha_{io}c^i + \alpha_{0i}a^i = (-1)^i\alpha_{i0}^p, \quad \alpha_{i0}d^i + \alpha_{0i}b^i = (-1)^i\alpha_{0i}^p$$

$$\alpha_{i0}c^{i-j}d^j + \alpha_{0i}a^{i-j}b^j = 0$$

for $i = 1, 2, \ldots, p - 1$ and $0 < j \leq i - 1$. Clearly, system (7.11.4) has at most p^{2p-1} solutions. Thus the order of L and hence $C\ell(F) \leq p^{2p-1}$.

We can make still other observations from the proof of (7.11). Note that if $abcd \neq 0$ and $ad - bc \neq 0$, then $\alpha_{i0} = \alpha_{0i} = 0$ for $i = 3, 4, \ldots, p - 1$. This is because from (7.11.4) we have that

$$\alpha_{i0}c^{i-1}d + \alpha_{0i}a^{i-1}b = 0$$

$$\alpha_{i0}c^{i-2}d^2 + \alpha_{0i}a^{i-2}b^2 = 0 \qquad \text{for } i = 3, \ldots, p - 1$$

This system has only the trivial solution when

$$(ac)^{i-2}bd(bc - ad) \neq 0$$

Also, if $p > 2$, we have from system (7.11.4) the equations

$$\alpha_{20}cd + \alpha_{02}ab = 0$$

$$\alpha_{20}c^2 + \alpha_{02}a^2 = \alpha_{20}^p$$

$$\alpha_{20}d^2 + \alpha_{02}b^2 = \alpha_{02}^p$$

If abcd(ad - bc) $\neq 0$, this system has p solutions if $a^{p-1}b^{p+1} - c^{p+1}d^{p-1} = 0$ and only the trivial solution if $a^{p-1}b^{p+1} - c^{p+1}d^{p-1} \neq 0$.

The equations involving α_{10} and α_{01} in (7.11.4) are

$$c\alpha_{10} + d\alpha_{01} = -\alpha_{10}^p$$

$$a\alpha_{10} + b\alpha_{01} = -\alpha_{01}^p$$

which have p^2 solutions when ad - bc $\neq 0$.

Finally, we have only one equation involving α_{00}, namely,

$$\alpha_{00} = \alpha_{00}^p$$

which always has p solutions.

We therefore have proved the following.

PROPOSITION 7.12 Let $u = ax^p + by^p$, $v = cx^p + dy^p$ with abcd(ad - bc) $\neq 0$. Let $G = xy + xu + yv$. Then the order of $C\ell(F:\ z^p = G)$ is p^3 if $p = 2$ or if $a^{p-1}b^{p+1} - c^{p+1}d^{p-1} \neq 0$. $C\ell(F)$ has order p^4 if $p > 2$ and $a^{p-1}b^{p+1} - c^{p+1}d^{p-1} = 0$.

We return to our original project, the determination of C (F: $z^p = x^{p+1} + xy$). Then in the notation of proposition (7.11), we have that a = 1, b = 0, c = 0, d = 1. Thus system (7.11.4) becomes

$$\alpha_{00} = \alpha_{00}^p \quad \text{and} \quad \alpha_{i0} = (-1)^i\alpha_{0i}^p,\ \alpha_{0i} = (-1)^i\alpha_{i0}^p$$
$$\text{for } i = 1, 2, \ldots, p - 1$$

This system has exactly p^{2p-1} solutions. We therefore have the following.

PROPOSITION 7.13 The divisor class group of the surface F: $z^p = x^{p+1} + y^{p+1} + xy$ is a direct sum of 2p - 1 copies of $\mathbb{Z}/p\mathbb{Z}$.

REMARK 7.14 With the same ease we can calculate the divisor class group of the surface $z^p = xy^p + yx^p + xy$. Again this group will be of order p^{2p-1}.

We come to the last example of this section. Let $f(y^p)$ and $g(y^p) \in k[y^p]$ be relatively prime and assume f has r distinct roots. Miyanishi and Russell in a recent article showed that the divisor class group of the surface $z^p = xf(y^p) + yg(y^p)$ has order p^r (see Miyanishi and Russell, 1983).

We modify this example a bit and calculate the class group of the surface $z^p = xf(y) + yg(y)$, where we assume that f(y) and $xf'(y) + yg'(y) + g(y)$ are relatively prime.

PROPOSITION 7.15 Let $f(y), g(y) \in k[y]$, and $G = xf(y) + yg(y)$. Assume that $\partial G/\partial x$ and $\partial G/\partial y$ are relatively prime and that f has r distinct roots. Then the surface $z^p = G$ has class group of order p^r.

Proof: Let D be the jacobian derivation of G and L the corresponding group of logarithmic derivatives. Let $\theta_1, \theta_2, \ldots, \theta_r$ be the roots of f and $t \in L$.

Note that for all $h \in k[x,y]$, $\deg_x(Dh) \le \deg_x(h)$. It follows that $\deg_x t = 0$, that is, $t \in k[y]$.

Let Δ be the k-derivation defined by $\Delta = (1/t)D$. Let $y_1 = \Delta(y)$ and $y_i = \Sigma(y_{i-1}) - y_{i-1}$ for $j = 2, \ldots, p - 1$. Then each $y_i \in k(y)$.

From (2.10) we have that for some $i = 1, 2, \ldots, p - 1$, $Dy_j/y_j = t$. After multiplying y_j by a suitable pth we can assume that $y_j \in k[y]$. Let $y_j = \alpha_0(y - \alpha_1)^{s_1}(y - \alpha_2)^{x_2}\cdots(y - \alpha_n)^{s_n}$ be a factorization of y_j into linear factors. Dividing y_j by a pth power yields an element that still gives t as a logarithmic derivative. Thus we may assume that $1 \le s_i \le p - 1$ for each s_i above. We then have by (3.19) that for each $i = 1, 2, \ldots, n$, $D(y - \alpha_i)/(y - \alpha_i) \in L$. But for each i, $D(y - \alpha_i) = -f(y)$. Thus $\alpha_i \in \{\theta_1, \theta_2, \ldots, \theta_r\}$ for each i.

We conclude that $t = \Sigma_{i=1}^{n} D(y - \alpha_i)/(y - \alpha_i)$ and thus the logarithmic derivatives $D(y - \theta_1)/(y - \theta_1), \ldots, D(y - \theta_r)/(y - \theta_r)$ span L. These polynomials are easily seen to be independent. Thus L has order p^r.

REMARK 7.16 From this proposition we obtain many examples of smooth, rational surfaces with class groups of arbitrarily large order. For instance, if $f(y) \in k[y]$ has r distinct roots and $g(y) \in yk[y]$ is such that g(y) and f(y) have no common roots, the surface

$$z^p = x(f(y))^n + g(y) \qquad \text{where } n > 1$$

is a smooth rational surface with class group of order p^r.

This provides an answer to a question posed by Samuel (see Samuel, 1964b, p. 83). Samuel asks: "If A is an integral domain of characteristic $p \neq 0$, D a derivation of A, and t a logarithmic derivative belonging to the ideal D(A)A in A, then must t be the logarithmic derivative of a unit in A?"

Clearly the answer is "no." For if we choose f(y) and g(y) so that $k[x^p, y^p, xf(y) + yg(y)]$ is regular and not factorial, then the derivation $D = G_y(\partial/\partial x) - G_x(\partial/\partial y)$ on k[x,y], where $G = xf(y) + yg(y)$, is such that $D(k[x,y])k[x,y] = k[x,y]$ but $L \neq 0$. Thus there exists a logarithmic derivative in the ideal generated by D(k[x,y]) that is not a logarithmic derivative of a unit.

Other counterexamples have been found by S. Yuan (see Yuan, 1968, p. 50) and B. Singh (see Singh, 1968, p. 159).

Samuel's question for the case where A is a local ring has to my knowledge remained open. That is:

7.17 If A is a local domain of characteristic $p \neq 0$, D a derivation of A, and t a logarithmic derivative belonging to the ideal $D(A) \cdot A$ in A, then must t be the logarithmic derivative of a unit in A?

Some special cases of (7.17) have been verified by N. Hallier (see Hallier, 1965a) and Samuel (see Samuel, 1964b, p. 88). But we now show that in general (7.17) is not true.

7.18 Let k be a field of characteristic $p \neq 0$. Let $A \in k[x,y]$, D the k-derivation of A defined by $Dx = x$ and $Dy = xy$. Then $Dy/y = x \in D(A) \cdot A$. But x is not the logarithmic derivative of a unit in A.

Proof: Suppose that u is a unit in A with $u^{-1}Du = x$. We can assume that u is a power series of the form $u = 1 + \Sigma_{i+j>0}\, \alpha_{ij} x^i y^j$, where each $\alpha_{ij} \in k$. Since $Du = xu$, we obtain

$$\sum_{i+j>0} \alpha_{ij}(ix^i y^j + jx^{i+1} y^j) = x + \sum_{i+j>0} \alpha_{ij} x^{i+1} y^j \tag{7.18.1}$$

This implies that

$$\sum_{i+j>0} i\alpha_{ij} x^i y^j = x + \sum_{i+j>0} (1 - j)\alpha_{ij} x^{i+1} y^j \tag{7.18.2}$$

Comparing coefficients we have that

$$\alpha_{10} = 1 \quad \text{and} \quad (i + 1)\alpha_{(i+1)j} = (1 - j)\alpha_{ij} \tag{7.18.3}$$

for $i + j > 0$. In particular we see that

$$(i + 1)\alpha_{(i+1)0} = \alpha_{i0} \qquad \text{for } i > 0 \tag{7.18.4}$$

Inductively, we have that $\alpha_{10} = (i + 1)!\alpha_{(i+1)0}$ for all $i \geq 1$. Thus $\alpha_{10} = p!\alpha_{p0} = 0$, contradicting (7.18.3).

8. AN EXAMPLE RELATED TO THE AFFINE THEOREM OF CASTELNUOVO

In a recent article, P. Russell (1981, p. 2) proved the following theorem.

AFFINE THEOREM OF CASTELNUOVO Suppose that k is perfect. Let $A \subset k^{[2]}$ be a finitely generated, regular k-algebra of dimension 2 such that $\bar{k} \otimes_k A$ is factorial ($\bar{k}$ an algebraic closure of k) and $qt(k^{[2]})/qt(A)$ is a separable extension. Then $A = k^{[2]}$.

REMARK 8.1 In the theorem above, $k^{[2]}$ stands for a polynomial ring in two variables and qt(A) is notation for the quotient field of A.

A question that naturally arises is: Does the conclusion of the theorem still hold if we drop the condition that $qt(k^{[2]})/qt(A)$ is a separable extension? We answer in the negative by considering the surface X_1: $z^2 = x(xy + 1)^2 + y^3$ over an algebraically closed field k of characteristic 2.

From (2.1) we have that the coordinate ring of X_1 is isomorphic to $A = k[x^2, y^2, x(xy + 1)^2 + y^3]$. We will first show that A is factorial. All of the other conditions of the theorem are easily seen to be met except that $qt(k^{[2]})/qt(A)$ is a purely inseparable extension. Second, we show that the geometric genus $p_g(\tilde{X})$ is positive, where $\tilde{X}$ is a smooth projective model of qt(A). This will imply that X is not rational and hence A is not isomorphic to $k^{[2]}$.

Factoriality

With X_1 as above, let us calculate $C\ell(X_1)$. Let $G = x(xy + 1)^2 + y^3$. Let D be the jacobian derivation of G, L the group of logarithmic derivatives of D, and t be an element of L. By (2.4), the degree of $t \le 3$ and by (1.9) and Ganong's formula we have that $\nabla(t) = 0$ and $\nabla(Gt) = t^2$. Thus t must be of the form

$$t = (c_{00} + c_{20}x^2 + c_{02}y^2) + (c_{10} + c_{30}x^2 + c_{12}y^2)x + (c_{01} + c_{21}x^2 + c_{03}y^2)y \tag{8.1}$$

for some $c_{ij} \in k$.

Substituting this expression for t into the equation $(Gt) = t^2$, we obtain

$$(c_{10} + c_{30}x^2 + c_{12}y^2)y^2 + (c_{01} + c_{12}x^2 + c_{03}y^2)(xy + 1)^2 = t^2 \tag{8.2}$$

Comparing coefficients in (8.2) we see that each $c_{ij} = 0$. This implies that $t = 0$. Hence $L = 0$. Therefore, A is factorial.

Nonrationality

To show that A is not isomorphic to $k^{[2]}$, we will accomplish this by showing that the surface X_1 is not rational. We do this in the following steps.

1. We make X_1 an affine piece of a projective k-scheme x.
2. We define a double differential σ on X.
3. We resolve X to obtain a smooth projective surface $\tilde{X}$, birational to X_1, and show that σ lifts to a nonzero regular differential $\tilde{\sigma}$ on $\tilde{X}$.

It then follows that $p_g(\tilde{X}) > 0$ and that X_1 is not rational.

Step 1: Let X_2 be the surface in $\mathbf{A}^2_k$ defined by the equation $w^2 = u^3v + uv^5 + v^{32}$ and X_3 be the surface defined by the equation $t^2 = r^2s + s^5 + r^3s^3$. We glue X_1, X_2, and X_3 together in the following way. Let

U_{12} (resp., U_{13}) be the open subset of X_1 defined by $y \neq 0$ (resp., $x \neq 0$).

V_{12} (resp., V_{23}) be the open subset of X_2 defined by $v \neq 0$ (resp., $u \neq 0$).

W_{13} (resp., W_{23}) be the open subset of X_3 defined by $s \neq 0$ (resp., $r \neq 0$).

Let

ϕ_{12}: $U_{12} \to V_{12}$ be the isomorphism defined by $x \to \frac{u}{v}$, $y \to \frac{1}{v}$, $z \to \frac{w}{v^3}$.

ϕ_{13}: $U_{13} \to W_{13}$ be the isomorphism defined by $x \to \frac{1}{s}$, $y \to \frac{r}{s}$, $z \to \frac{t}{s^3}$.

ϕ_{23}: $V_{23} \to W_{23}$ be the isomorphism defined by $u \to \frac{1}{r}$, $v \to \frac{s}{r}$, $w \to \frac{t}{r^3}$.

We glue X_1, X_2, and X_3 together via these isomorphisms to obtain a scheme X. We note that the coordinate ring of X_1 (resp., X_2, X_3) is the integral closure of $k[x,y]$ (resp., $k[u,v], k[r,s]$) in its quotient field. Thus we have a finite morphism $X \to P^2$. Since a finite morphism is projective (see Grothendieck, 1961, p. 113) and a composition of projective morphisms is projective, it follows that X is a projective k-scheme.

Step 2: For each $i = 1, 2, 3$, define σ_i, a differential on X_i as follows:

$$\text{On } X_1: \quad \sigma_1 = \frac{dx\ dz}{y^2} = \frac{dy\ dz}{(xy + 1)^2}$$

$$\text{On } X_2: \quad \sigma_2 = \frac{du\ dw}{u^3 + uv^4 + v^2} = \frac{dy\ dw}{u^2v + w^5}$$

$$\text{On } X_3: \quad \sigma_3 = \frac{dr\ dt}{r^2 + s^4 + r^3s^2} = \frac{ds\ dt}{r^2s^3}$$

We check that these differentials agree on the foregoing overlaps.

Under ϕ_{12}, σ_1 becomes

$$\frac{d\left(\frac{1}{v}\right)\, d\left(\frac{w}{v^3}\right)}{(u/v)^2(1/v)^2 + 1} = \frac{\left(\frac{1}{v^2}\right)\, dv\left(\frac{1}{v^2}\,\frac{v\ dw + w\ dv}{v^2}\right)}{u^2/v^4 + 1} = \frac{dv\ dw}{vu^2 + v^5} = \sigma_2$$

Similarly, σ_2 maps to σ_3 under ϕ_{23} and σ_1 maps to σ_3 under ϕ_{13}.

Thus these differentials glue together to give a differential σ on X. We now resolve X to obtain a smooth projective scheme $\tilde{X}$ and show that σ lifts to a regular differential $\tilde{\sigma}$ on $\tilde{X}$.

Step 3: $\tilde{\sigma}$, the lifting of σ to $\tilde{X}$, will be a regular differential on $\tilde{X}$ if we show X has only rational singularities (see Lipman, 1978, p. 153).

Since X_1 is smooth, X can only have singularities on $X_2 \cup X_3$. On X_2: $w^2 = u^3v + uv^5 + v^3$ singularities can occur only when $v = 0$. Otherwise, we would be considering points on $X_1 \cap X_2$, which we know is smooth. So we see that X_2 has only an isolated singularity at $(u,v,w) = (0,0,0)$.

Similarly, we see that X_3: $t^2 = r^2s + s^5 + r^3s^3$ has only an isolated singularity at $(r,s,t) = (0,0,0)$.

These double point singularities will be rational if we show that they can be resolved by quadratic transformations along (see Lipman, 1969, p. 255).

J. Lipman has shown that if an isolated singularity on a normal affine surface has local equation of the form

$$z^2 = xy^2 + x^2g(x,y),\ g(x,y) \in k[x,y] \tag{8.3}$$

then the singularity is rational (see Lipman, 1969, p. 266). Thus we see immediately that the singularity on X_3 is rational. This leaves only the singularity on X_2.

We begin by blowing up the origin on X_2. Since w is integrally dependent on the ideal generated by u and v, the blowup of (0,0,0) is covered by two charts (see Blass, 1983, p. 96). Namely,

$$F_1:\ w^2 = u^2v + u^4v^5 + uv^3$$

F_2: $w^2 = u^3v^2 + uv^4 + v$

F_1 has only an isolated singularity at the origin which is a rational singularity by (8.3). F_2 is smooth since $\partial/\partial v = 1$. Thus the singularity on X_2 can be resolved by quadratic transformations alone and is a rational singularity. Therefore, $\tilde{\sigma}$ is a regular differential on $\tilde{X}$, which shows that X_1 is not rational.

REMARK 8.4 After the circulation of a preliminary version of this paper, M. Miyanishi and P. Russell have shown that the ring $A = k[x^p, y^p, x(xy + 1)^p + y^{p+1}]$ gives an example of a regular, factorial, nonrational ring for all primes $p > 0$ (see Miyanishi and Russell, 1983).

Also, Miyanishi and Russell have observed that a theorem of Ganong (1982) yields a more concise proof that A is not isomorphic to $k^{[2]}$.

9. THE SURFACE $z^{p^n} = G(x,y)$

In this section we study the divisor class group of surfaces of the form $z^{p^n} = G(x,y)$. Throughout this section, as always, we assume that k is an algebraically closed field of characteristic $p > 0$. Let $G(x,y) \in k[x,y]$ satisfy condition (**). For each $n = 1, 2, \ldots$ we let $F_n \subset \mathbb{A}_k^3$ be the surface defined by the equation $z^{p^n} = G(x,y)$. By the jacobian criterion F_n is a normal surface for each n.

LEMMA 9.1 For each n, the coordinate ring of F_n is isomorphic to $A_n = k[x^{p^n}, y^{p^n}, G]$.

Proof: Similar to the proof of (2.1).

We will develop an inductive procedure for studying $C\ell(F_n)$. For each positive integer n, let $B_n = k[x^{p^{n+1}}, y^{p^{n+1}}, G^p]$. Clearly, B_n is isomorphic to A_n and we have that $B_n \subset A_{n+1} \subset A_n$. Also, A_n is integral over A_{n+1}, A_{n+1} is integral over B_n, and $[qt(A_n) : qt(A_{n+1})] = [qt(A_{n+1} : qt(B_n)] = p$.

Then by theorem (1.3) there exist group homomorphisms θ_n: $C\ell(B_n) \to C\ell(A_{n+1})$ and ϕ_n: $C\ell(A_{n+1}) \to C\ell(A_n)$. We use derivations to study θ_n and ϕ_n. We start with θ_n.

Let E_n be the restriction of the $k(x,y)$ derivation $(1/G_x)(\partial/\partial x)$ to A_{n+1}.

LEMMA 9.2 E_n maps A_{n+1} into A_{n+1} and has kernel B_n.

Proof: Let $\alpha = \Sigma_{i=0}^{p-1} \beta_i G^i$ for unique $\beta_i \in B_n$. We have that $E_n(\alpha) = \Sigma_{i=1}^{p-1} i\beta_i G^{i-1}$. Thus $E_n(\alpha) \in A_{n+1}$ and $E_n(\alpha) = 0$ if and only if $\beta_1 = \beta_2 = \cdots = \beta_{p-1} = 0$, that is, if and only if $\alpha \in B_n$.

PROPOSITION 9.3 For each positive integer n, $C\ell(F_n)$ injects into $C\ell(F_{n+1})$.

Proof: With E_n: $A_{n+1} \to A_{n+1}$ as above, let $\tilde{\tilde{L}}_n \subset A_{n+1}$ be the group of logarithmic derivatives of E_n. Let $\tilde{\tilde{L}}'_n \subset A_{n+1}$ be the group of logarithmic derivatives of units of A_{n+1}. The units of A_{n+1} are exactly the elements of k. Thus $\tilde{\tilde{L}}'_n = 0$. By (1.8) we have that $\ker(\theta_n) \approx \tilde{\tilde{L}}_n$.

Let $\alpha \in A_{n+1}$. One sees easily that $\deg(E_n(\alpha)) \leq \deg(\alpha) - \deg G$. It follows that if $E_n(\alpha)/\alpha \in A_{n+1}$, then $E_n(\alpha) = 0$. This shows that $\tilde{\tilde{L}}_n = 0$. Therefore, $\ker(\theta_n) = 0$.

COROLLARY 9.4 The divisor class group of the surface $z^{2^n} = G(x,y)$ is nontrivial whenever $\partial^2 G/\partial x \partial y \neq 0$.

Proof: Use (3.3) and (9.3).

To gain some understanding of ϕ_n: $C\ell(A_{n+1}) \to C\ell(A_n)$ we define a derivation D_n: $A_n \to A_n$. Given $\alpha \in A_n$, there exists unique $\alpha_i \in k[x,y]$ such that $\alpha = \Sigma_{i=0}^{p^n-1} \alpha_i^{p^n} G^i$.

Define $D_n(\alpha) = \Sigma_{i=0}^{p^n-1} (D\alpha_i)^{p^n} G^i$, where $D = (\partial G/\partial x)(\partial/\partial y) - (\partial G/\partial y)(\partial/\partial x)$ is the jacobian derivation of G.

Of course, we must show that D_n is indeed a derivation.

LEMMA 9.5 D_n, as defined above, is a derivation.

Proof: Clearly, D_n is additive. We show that the multiplicative property holds. That is, $D_n(uv) = uD_n(v) + vD_n(u)$ for $u,v \in A_n$.

We have that $u = \Sigma_{i=0}^{p^n-1} \alpha_i^{p^n} G^i$ and $v = \Sigma_{i=0}^{p^n-1} \beta_i^{p^n} G^i$ for unique $\alpha_i, \beta_i \in k[x,y]$. We proceed by induction on the number of nonzero coefficients appearing in u plus the number of nonzero coefficients appearing in v.

Suppose that this sum is 2. Then $u = \alpha^{p^n} G^i$ and $v = \beta^{p^n} G^j$ for some $\alpha, \beta \in k[x,y]$ and positive integers i and j.

$$D_n(uv) = D_n((\alpha\beta)^{p^n} G^{i+j}) = (D(\alpha\beta))^{p^n} G^{i+j} = (\alpha D\beta + \beta D\alpha)^{p^n} G^{i+j}$$
$$= \alpha^{p^n} G^i D_n(\beta^{p^n} G^j) + \beta^{p^n} G^j D_n(\alpha^{p^n} G^i) = uD_n(v) + vD_n(u)$$

Now assume that the sum of nonzero coefficients appearing in u and v is greater than 2. Say that v has more than 2. Let $0 < j_0 < p^n$ be the highest power of G with nonzero coefficient appearing in v. Let this coefficient be γ^{p^n}. Then $D_n(uv) = D_n(u(v - \gamma^{p^n} G^{j_0})) + D_n(u\gamma^{p^n} G^{j_0})$, which by the inductive hypothesis $= uD_n(v - \gamma^{p^n} G^{j_0}) + (v - \gamma^{p^n} G^{j_0})D_n(u) + uD_n(\gamma^{p^n} G^{j_0}) + \gamma^{p^n} G^{j_0} D_n(u) = uD_n(v) + vD_n(u)$.

LEMMA 9.6 Let $D_n: A_n \to A_n$ be the derivation described above.

(i) $A_{n+1} = \ker D_n$.

(ii) Let $L_n \subseteq A_n$ be the logarithmic derivatives of D_n. Then $\ker \phi_n$ is isomorphic to L_n.

(iii) Let $a \in A_1$ be such that $D^p = aD$. Then $D_n^p = a^{p^n} D_n$.

Proof: (i) Obviously, $A_{n+1} \subset \ker(D_n)$. If $\alpha = \Sigma_{i=0}^{p^n-1} \alpha_i^{p^n} G^i \in \ker D_n$, then $D\alpha_i = 0$, for each i, which implies that each $\alpha_i \in A$ and $\alpha \in A_{n+1}$.

(ii) Let $L_n' \subset L_n$ be the subgroup of L_n of logarithmic units of D_n. Since the units of A_n are precisely the elements of k, we have that $L_n' = 0$. By theorem (1.8) $\ker \phi_n = L_n$.

(iii) By theorem (1.9) there exists $a_n \in A_{n+1}$ such that $D_n^p = a_n D_n$. Assume that $D(x) \neq 0$. Then $D_n^p(x^{p^n}) = (D^p(x))^{p^n} = (aD(x))^{p^n} = a^{p^n} D_n(x^{p^n})$. Thus $a_n = a^{p^n}$.

PROPOSITION 9.7 Let $t = \alpha_0^{p^n} + \alpha_1^{p^n} G + \cdots + \alpha_{p^n-1}^{p^n} G^{p^n-1} \in A_n$. Then $t \in L_n$ if and only if

(1) $\nabla(G^i\alpha_j) = 0$ for $i = 0, 1, \ldots, p-1$, $0 \leq j \leq p^n - 1$ and $j \neq 0 \pmod p$.

(2) $\nabla(G^i\alpha_{ep}) = \alpha^p_{e+(p-(i+1))p^{n-1}}$ for $0, 1, \ldots, p^{n-1} - 1$.

Proof: Using the notation above and theorem (1.9), we have that $t \in L_n \Leftrightarrow D_n^{p-1}(t) - a^{p^n} t = -t^p$. Thus

$$t \in L_n \Leftrightarrow \sum_{j=0}^{p^n-1} (D^{p-1}\alpha_j - a\alpha_j)^{p^n} G^j = -\sum_{j=0}^{p^n-1} \alpha_j^{p^{n+1}} G^{jp} \tag{9.7.1}$$

Comparing coefficients in (9.7.1), we obtain

$$\begin{aligned} t \in L_n \Leftrightarrow (1) \quad & D^{p-1}\alpha_j - a\alpha_j = 0 \qquad \text{for } j \neq 0 \pmod p, \\ & 0 \leq j \leq p^{n-1} \\ (2) \quad & \sum_{e=0}^{p^{(n-1)}-1} (D^{p-1}\alpha_{ep} - a\alpha_{ep})^{p^n} G^{ep} = -\sum_{r=0}^{p^n-1} \alpha_r^{p^{n+1}} G^{rp} \end{aligned} \tag{9.7.2}$$

Taking pth roots we see that (2) of (9.7.2) becomes

$$\sum_{e=0}^{p^{(n-1)}-1} (D^{p-1}\alpha_{ep} - a\alpha_{ep})^{p^{(n-1)}} G^e = -\sum_{r=0}^{p^n-1} \alpha_r^{p^n} G^r \tag{9.7.3}$$

Comparing both sides of (9.7.3), we have that (2) is equivalent to

$$(D^{p-1}\alpha_{ep} - a\alpha_{ep})^{p^{(n-1)}} = -\sum_{i=0}^{p-1} \alpha^{p^n}_{e+ip^{(n-1)}} G^{ip^{(n-1)}} \tag{9.7.4}$$

for $0 \leq e \leq p^{(n-1)}) - 1$, which is equivalent to

$$D^{p-1}\alpha_{ep} - a\alpha_{ep} = -\sum_{i=0}^{p-1} \alpha^p_{e+ip^{(n-1)}} G^i \qquad \text{for } 0 \le e \le p^{(n-1)} - 1 \tag{9.7.5}$$

If we apply Ganong's formula to the left side of (1) of (9.7.2) and to the left side of (9.7.5) and compare coefficients of G^i, we obtain the desired result.

LEMMA 9.8 Let $t = \alpha_0^{p^n} + \alpha_1^{p^n} G + \cdots + \alpha_{p^n-1} G^{(p^{n-1})} \in A_n$. Assume that $t \in L_n$. Then each α_j is determined by α_0 for $j = 0, 1, \ldots, p^n - 1$.

Proof: We proceed by reverse induction on $v(j)$, where $v(j)$ = the highest power of p that divides j.

Note that if $v(j) \ge n$, then j must be 0. Now assume that $v(j) = m < n$. We can write $j = e + (p - (i + 1))p^{(n-1)}$ for unique $e = 0, 1, \ldots, p^{(n-1)} - 1$ and $i = 0, 1, \ldots, p - 1$. Since $v(j) = m < n$ we have that $j = sp^m$ for some $s = 0, 1, \ldots, p - 1$.

By (9.7),

$$\alpha^p_{e+(p-(i+1))p^{(n-1)}} = \nabla(G^i\alpha_{ep}) = \nabla(G^i\alpha_{sp^{(m+1)}})$$

By the induction hypothesis $\alpha_{sp^{(m+1)}}$ is determined by α_0, hence so is α_j.

THEOREM 9.9 For each n, ker ϕ_n is a p-group of type $(p,p,\ldots,p)$ and of order p^f, where $f \le \deg G(\deg G - 1)/2$.

Proof: By lemma (9.6ii), $\ker \phi_n \approx L_n$. Thus ker ϕ_n is a p-group of type $(p,p,\ldots)$. From (9.7) and (9.8) the order of L_n is less than or equal to the order of the additive group

$$H = \{\alpha_0 \in k[x,y] \mid \nabla(G^{p-1}\alpha_0) = \alpha_0^p\}$$

In lemma (2.8) we saw that the order of H is p^f, where $f \le \deg G \times (\deg G - 1)/2$.

THEOREM 9.10 For each n, $C\ell(F_n)$ is a finite p-group of type $(p^{i_1}, p^{i_2}, \ldots p^{i_r})$, where each $i_j \le n$. The order of $C\ell(F_n) \le p^{ng(g-1)/2}$, where $g = \deg G$.

Proof (by induction on n): For n = 1, this is just proposition (2.9). For each $n \ge 1$, we have the exact sequence

$$0 \to \ker \phi_n \to C\ell(F_{n+1}) \xrightarrow{\phi_n} C\ell(F_n) \to 0 \tag{9.10.1}$$

Now just use the induction hypothesis and (6.10).

REMARK 9.11 Using the mappings θ_n and ϕ_n, we obtain an inductive procedure for studying $C\ell(F_n)$. For we have the following diagram for every n,

$$C\ell(B_n) \xrightarrow{\theta_n} C\ell(A_{n+1}) \xrightarrow{\phi_n} C\ell(A_n)$$

$$D_n\colon\ A_n \to A_n \tag{9.12}$$

$$\ker \phi_n \simeq L_n = \{D_n(f)/f \mid f \in qt(A_n),\ D_n(f)/f \in A_n\}$$

We finish this section by demonstrating, with some examples, how this procedure is conducted.

EXAMPLE 1 A natural question at this point is: If $F_n\colon\ z^{p^n} = G(x,y)$ has trivial class group, then does $F_{n+1}\colon\ z^{p^{n+1}} = G(x,y)$ have trivial class group also?

By (9.3) we have that $C\ell(F_m) = 0$ for each $m < n$, but what if $m > n$? The answer to this question is "no." We consider the surfaces $F_n\colon\ z^{2^n} = x(xy + 1)^2 + y^3$ over an algebraically closed field k of characteristic 2.

In Section 8 we saw that $C\ell(F_1) = 0$. We now show that $C\ell(F_2) = \mathbb{Z}/2\mathbb{Z}$.

By (9.6) we have that $C\ell(F_2) \approx L_1$. Let $t = \alpha_0^2 + \alpha_1^2 G \in A_1$, where $G = x(xy + 1)^2 + y^3$.

By (9.7) we have that $t \in L_1$ if and only if

$$\begin{aligned}&(1)\quad \nabla(G\alpha_1) = \nabla(\alpha_1) = 0\\&(2)\quad \nabla(G\alpha_0) = \alpha_0^2,\ \nabla(\alpha_0) = \alpha_1^2\end{aligned} \tag{9.13}$$

From (2.8) we have that $\deg \alpha_0 \le 3$.

We write $\alpha_0 = (a_{00} + a_{20}x^2 + a_{02}y^2) + (a_{10} + a_{30}x^2 + a_{12}y^2)x + (a_{01} + a_{21}x^2 + a_{03}y^2)y + a_{11}xy$, where $a_{ij} \in k$. Substituting α_0 into (2), we obtain

$$\begin{aligned}&(a_{10} + a_{11}y + a_{30}x^2 + a_{12}y^2)y^2 + (a_{01} + a_{11}x + a_{21}x^2\\&+ a_{03}y^2)(xy + 1)^2 + a_{11}(x(xy + 1)^2 + y^3)\\&= a_{00}^2 + a_{10}^2x^2 + a_{01}^2y^2 + a_{20}^2x^4 + a_{11}^2x^2y^2 + a_{02}^2y^4 + a_{30}^2x^6\\&\quad + a_{21}^2x^4y^2 + a_{12}^2x^2y^4 + a_{03}^2y^6\end{aligned} \tag{9.14}$$

and $a_{11} = \alpha_1^2$.

Comparing coefficients in the first equation of (9.14), we obtain

$$\begin{aligned}&a_{20} = a_{02} = a_{30} = a_{12} = a_{03} = 0 \quad \text{and} \quad a_{01} = a_{00}^2\\&a_{21} = a_{10}^2 \quad a_{10} = a_{01}^2 \quad a_{01} = a_{11}^2 \quad a_{21} = a_{21}^2\end{aligned} \tag{9.15}$$

This system has two solutions, the zero zolution and $a_{00} = a_{10} = a_{01} = a_{11} = a_{21} = 1$, all other $a_{ij} = 0$. Thus there is only one non-zero possibility for α_0. This is

$$\alpha_0 = 1 + x + y + xy + x^2y \tag{9.16}$$

Also, from the second equation of (9.13), we see that when $\alpha_0 = 0$, so is α_1 and when $\alpha_0 \ne 0$, then $\alpha_1 = 1$. Note that either $\alpha_1 = 0$ or $\alpha_1 = 1$ satisfies (1) of (9.13).

We conclude that L_1 has order 2 generated by

$$t = (1 + x + y + xy + x^2y)^2 + G$$

EXAMPLE 2 In Section 7 we calculated the divisor class group of the surface $z^p = G$, where G is a product of distinct linear factors. In the next proposition we calculate the divisor class group of the surface $z^{p^n} = G$ with the same assumptions on G.

PROPOSITION 9.17 Let k be an algebraically closed field of characteristic $p > 0$. Let $G = (x - \alpha_1 y)(x - \alpha_2 y)\cdots(x - \alpha_n y)$, where the $\alpha_i \in k$ are pairwise distinct. Assume that G satisfies condition (**). Then $C\ell(F_m\colon\ z^{p^m} = G)$ is a direct sum of $(n - 1)$ copies of $\mathbf{Z}/p^m\mathbf{Z}$, generated by the curves $z = x - \alpha_i y = 0$ for $i = 1, 2, \ldots, n - 1$.

REMARK 9.18 If n is not divisible by p, then condition (**) on G is implicit in the assumption that the α_i are pairwise distinct. For by Euler's formula we have that $G = (1/n)(xG_x + yG_y)$. Thus if G_x and G_y have a factor in common, they would have to have a linear factor in common which would also be a factor of G, hence a multiple factor of G.

Proof: The case $m = 1$ is covered in proposition (7.1) and remark (7.3). Suppose that the proposition holds for integers less than $m + 1$. Let $D_m\colon\ A_m \to A_m$ be the derivation described on page 222 and L_m the group of logarithmic derivatives of D_m. As in the proof of theorem (9.9), we have that the order of L_m is less than or equal to the order of $H = \{\alpha \in k[x,y] \mid \nabla(G^{p-1}\alpha) = \alpha^p\}$. In the proof of proposition (7.1) we showed that the order of $H \le p^{n-1}$ and that

$$\frac{D(x - \alpha_1 y)}{x - \alpha_1 y}, \frac{D(x - \alpha_2 y)}{x - \alpha_2 y}, \ldots, \frac{D(x - \alpha_{n-1} y)}{x - \alpha_{n-1} y} \in L$$

and are independent over $\mathbb{Z}/p\mathbf{Z}$. It then follows that the elements

$$\frac{D_m(x - \alpha_i y)^p}{(x - \alpha_i y)^p} = \left[\frac{D(x - \alpha_i y)}{x - \alpha_i y}\right]^p \qquad i = 1, \ldots, n - 1$$

belong to L_m and are independent over $\mathbb{Z}/p\mathbf{Z}$. Therefore, these elements generate L_m. We conclude that the order of $\ker \phi_m\colon C\ell(A_{m+1}) \to C\ell(A_m)$ is p^{n-1}.

Let B_m be as defined on page 221 and θ_m: $C\ell(B_m) \to C\ell(A_{m+1})$ the injection described in proposition (9.3). Let p_i, $i = 1, \ldots, n - 1$, be the height 1 primes of A_m generated by $(x - \alpha_i y)^{p^m}$ and G. Let $q_i = p_i \cap A_{m+1}$ and $\bar{p}_i = q_i \cap B_m$ for each i. By the induction hypothesis the p_i (respectively, $\bar{p}_i$) are of order p^m and generate $C\ell(A_m)$ [respectively, $C\ell(B_m)$].

Note that for $i = 1, \ldots, n - 1$, $G \in q_i$ and has value 1 in the valuation on A_m defined by p_i. Thus:

9.19 (The ramification index of p_i over q_i is 1.) It follows that the ramification index of q_i over $\bar{p}_i$ is p because the ramification index of p_i over $\bar{p}_i$ is obviously p.

From statement (9.19) we derive two conclusions:

(1) ϕ_m is a surjection with kernel containing the elements $p^m q_1, \ldots, p^m q_{n-1}$.

(2) $p^m q_1, \ldots, p^m a_{n-1}$ are linearly independent over $\mathbb{Z}/p\mathbb{Z}$ since θ_m is an injection and the elements $p^{m-1}\bar{p}_1, \ldots, p^{m-1}\bar{p}_{n-1}$ of $C\ell(B_m)$ are linearly independent over $\mathbb{Z}/p\mathbb{Z}$.

Since ker ϕ_m has order p^{n-1} we see that $p^m q_1, \ldots, p^m q_{n-1}$ are a basis for ker ϕ_m. It then follows that $C\ell(A_{m+1})$ is a direct sum of $n - 1$ copies of $\mathbb{Z}/p^{m+1}\mathbb{Z}$ generated by $q_1, \ldots, q_{n-1}$, where the q_i are of order p^{m+1}.

Clearly, q_i corresponds to the curve $z = (x - \alpha_i y) = 0$ on F_{m+1}.

EXAMPLE 3 In Section 8 we showed that the surface $z^2 = x(xy + 1)^2 + y^3$ has trivial class group and we remarked that Miyanishi and Russell extended this result to characteristic $p \geq 2$. That is, the surface $z^p = x(xy + 1)^p + y^{p+1}$ has trivial class group where we assume that the ground field has characteristic $p > 2$. In this section we showed that the class group of the surface $z^4 = x(xy + 1)^2 + y^3$ is not trivial and is in fact $\mathbb{Z}/2\mathbb{Z}$. One might expect then that the class group of the surface $z^{p^2} = x(xy + 1)^p + y^{p+1}$ is nontrivial also for

$p > 2$. But we will see that this is not the case. In fact, we will show that for all $p > 3$, the surface $z^{p^2} = x(xy + 1) + y^{p+1}$ has zero class group.

LEMMA 9.20 Let k be an algebraically closed field of characteristic $p > 2$, $G = x(xy + 1)^p + y^{p+1}$ and $H = \{\alpha \in k[x,y] \mid \nabla(G^{p-1}\alpha) = \alpha^p\}$. Then a polynomial $\alpha \in H$ if and only if α has the form $\alpha = \Sigma_{e=0}^{p-1} a_e(xy + 1)^e y^{p-e-1} + dx(xy + 1)^{p-1}$, where $d \in Z/pZ$, $a_e = 0$ for $0 \leq e < (p - 1)/2$, $a_{p-n-1}^p = (-1)^n\binom{2n}{n}a_{2n}$ for $0 \leq n \leq (p - 1)/2$.

Proof: Let $\alpha \in H$. From the proof of (2.9) we have that $\deg(\alpha) \leq 2p - 1$. Thus we can write $\alpha = \Sigma_{0\leq i,j\leq p-1} \alpha_{ij}^p x^i y^j$, where each $\alpha_{ij} \in k[x,y]$ has degree ≤ 1. We have that $G^{p-1} = \Sigma_{e=0}^{p-1} (-1)^e(xy + 1)^{ep} \times y^{(p-e-1)p}x^e y^{(p-e-1)}$ and

$$\nabla(G^{p-1}\alpha) = \sum_{e=0}^{p-1} (-1)^e(xy + 1)^{ep}y^{(p-e-1)p}\alpha^p_{(p-e-1)} \tag{9.20.1}$$

Therefore,

$$\nabla(G^{p-1}\alpha) = \alpha^p \Leftrightarrow \sum_{e=0}^{p-1} (-1)^e(xy + 1)^e y^{(p-e-1)}\alpha_{(p-e-1)e} = \alpha \tag{9.20.2}$$

From (9.20.2) we conclude that $\deg_x(\alpha) \leq p$ and $\deg_y(\alpha) \leq p$, from which it follows that $\deg_x(\alpha_{ij}) = 0$ for $0 < i \leq p - 1$ and $\deg_y(\alpha_{ij}) = 0$ for $0 < j \leq p - 1$. Thus we see that $\alpha_{(p-e-1)e} \in k$ for $0 < e < p - 1$ and $\alpha_{(p-1)0} = a + by$, $\alpha_{0(p-1)} = c + dx$ for some a, b, c, $d \in K$. So we can rewrite α in (9.20.2) as

$$\alpha = \sum_{e=0}^{p-1} a_e(xy + 1)^e y^{p-e-1} + by^p + dx(xy + 1)^{p-1} \tag{9.20.3}$$

With α in this form (9.20.3) we have that

$$\nabla(G^{p-1}\alpha) = \sum_{n=0}^{p-1/2} (-1)^n\binom{2n}{n}a_{2n}(xy + 1)^{p(p-n-1)}y^{pn} + dx^p(xy + 1)^{p(p-1)} \tag{9.20.4}$$

If we let y = 0 in (9.20.3), we obtain $\alpha(y = 0) = a_{p-1} + dx$ and in (9.20.4) $\nabla(G^{p-1}\alpha) = a_0 + dx^p$. We conclude that $a_0 = a_{p-1}^p$ and $d = d^p$. If we let x = 0, then (9.20.3) becomes $\alpha(x = 0) = \Sigma_{e=0}^{p-1} ay^{p-e-1} + by^p$ and (9.20.4) becomes $\nabla(G^{p-1}\alpha) = \Sigma_{n=0}^{p-1/2} (-1)^n\binom{2n}{n} \times a_{2n}y^{pn}$.

Comparing these two expressions we obtain b = 0 and

$$\begin{aligned} a_e &= 0 \qquad \text{for } 0 \le e < (p - 1)/2 \\ a_{p-n-1}^p &= (-1)^n\binom{2n}{n}a_{2n} \qquad \text{for } 0 \le n \le (p - 1)/2 \end{aligned} \tag{9.20.5}$$

Summarizing, we have that $\alpha \in H \Rightarrow \alpha$ has the form $\alpha = \Sigma_{e=0}^{p-1} a_e(xy + 1)^e y^{p-e-1} + dx(xy + 1)^{p-1}$, where d and the a are as in the statement of the lemma. The reverse implication is easily proved with a straightforward computation.

The next lemma is a refinement of (9.20).

LEMMA 9.21 Let G, *H*, and k be as in (9.20).

(1) If $p \not\equiv 1 \pmod 3$, then $\alpha \in H$ if and only if $\alpha = dx(xy + 1)^{p-1}$ for some $d \in \mathbf{Z}/p\mathbf{Z}$.

(2) If $p \equiv 1 \pmod 3$, then $\alpha \in H$ if and only if $\alpha = c(xy + 1)^{n_0}y^{m_0} + dx(xy + 1)^{p-1}$, where $n_0 = 2(p - 1)/3$, $m_0 = (p - 1)/3$, $c \in k$ with $c^p = \binom{2n_0}{n_0}c$, and $d \in \mathbf{Z}/p\mathbf{Z}$.

Proof: (1) $p \not\equiv 1 \pmod 3$. Let $\alpha \in H$. By (9.20), $\alpha = \Sigma_{e=0}^{p-1} a_e(xy + 1)^e y^{p-e-1} + dx(xy + 1)^{p-1}$, where the a and d satisfy the conditions of (9.20). We have that $a_e = 0$ for $0 \le e < (p - 1)/2$. We proceed by induction on e to show that each $a_e = 0$.

Let $(p - 1)/2 \le e_0 \le p - 1$. Then $e_0 = (p - n - 1)$ for some $0 \le n \le (p - 1)/2$. Since $p \not\equiv 1 \pmod 3$, either $p - n - 1 > 2n$ or $p - n - 1 < 2n$.

If $p - n - 1 > 2n$, then by (9.20), $a_{e_0}^p = a_{p-n-1}^p = (-1)^n\binom{2n}{n}a_{2n}$. By induction $a_{2n} = 0$, hence $a_{e_0} = 0$.

If $p - n - 1 < 2n$, then $2p - 4n - 2 < p - n - 1$. By (9.20), $a_{e_0}^p = a_{p-n-1}^p = (-1)^n\binom{2n}{n}a_{2n}$ and $a_{2n} = (-1)^{p-2n-1}\binom{2p - 4n - 2}{p - 2n - 1}a_{2p-4n-2}$.

By induction $a_{2p-4n-2} = 0$. Hence $a_{2n} = a_{e_0} = 0$. Therefore, $\alpha = dx(xy + 1)^{p-1}$.

(2) $p \equiv 1 \pmod 3$. Let $\alpha = \Sigma\, a_e(xy + 1)^e y^{p-e-1} \in H$. We show that $a_e = 0$ for $e \neq 2(p - 1)/3$. From (9.20), $a_e = 0$ for $0 \leq e < (p - 1)/2$. We proceed by induction on e. Let $e_0 \neq 2(p - 1)/3$ be such that $(p - 1)/2 \leq e_0 \leq p - 1$. Then $e_0 = p - n - 1$ for $0 \leq n \leq (p - 1)/2$. Then $p - n - 1 \neq 2n$. Now repeat the argument used in (1).

Let $n_0 = 2(p - 1)/3$ and $m_0 = (p - 1)/3$. Since each $a_e = 0$ for $e \neq n_0$, we have by (9.20) that $\alpha = c(xy + 1)^{n_0} y^{m_0} + dx(xy + 1)^{p-1}$, where $c \in k$ is such that $c^p = \binom{2n_0}{n_0} c$ and $d \in \mathbb{Z}/p\mathbb{Z}$.

PROPOSITION 9.22 Let k, H, and G be as in (9.20). Let $F_2 \subset \mathbf{A}_k^3$ be defined by $z^{p^2} = G$. If $p = 3$, then $C\ell(F_2) \approx \mathbb{Z}/2\mathbb{Z}$. If $p > 3$, then $C\ell(F_2) = 0$.

Proof: Let $A_1 = k[x^p, y^p, G]$, $A_2 = k[x^{p^2}, y^{p^2}, G]$, and $L_1 \subset A_1$ be as described in (9.6). As mentioned previously, Miyanishi and Russell (1983) have shown that $C\ell(A_1) = 0$. By (9.6), $C\ell(A_2) \approx L_1$.

We again consider two cases.

Case 1: $p \not\equiv 1 \pmod 3$. By (9.21), H is generated by $f_1 = x(xy + 1)^{p-1}$. From (9.7) we have that if $t \in L_1$, then

$$t = \sum_{j=0}^{p-1} \alpha_j^p G^j \tag{9.22.1}$$

where $\alpha_0 = a_1 f_1$ for some $a_1 \in \mathbb{Z}/p\mathbb{Z}$ and where the rest of the α_j are given by (2) of (9.7).

We have that $\nabla(G^{p-2} f_1) = \ell_1(xy + 1)^{p(p-3)/2} y^{p(p-1)/2}$, where $\ell_1 \neq 0 \in \mathbb{Z}/p\mathbb{Z}$ and $\nabla(G^{p-3} f_1) = 0$. Thus by (9.7),

$$\begin{aligned} \alpha_1^p &= \nabla(G^{p-2}\alpha_0) = a_1 \nabla(G^{p-2} f_1) \\ \alpha_2^p &= a_1 \nabla(G^{p-3} f_1) \end{aligned} \tag{9.22.2}$$

We obtain

$$\alpha_1 = a_1\ell_1(xy + 1)^{(p-3)/2}y^{(p-1)/2} \quad \text{and} \quad \alpha_2 = 0 \tag{9.22.3}$$

We conclude that if $p = 3$, then $t = \alpha_0^3 + \alpha_1^3 G$, where $\alpha_1 = a_1\ell_1 y$, and $\nabla(G^2\alpha_1) = \nabla(G\alpha_1) = \nabla(\alpha_1) = 0$. All of the conditions of (9.7) are met, from which it follows that $L_1 = \mathbb{Z}/p\mathbb{Z}$.

If $p > 3$, then in order for t to be in L_1 we must have by (9.7) that

$$0 = \nabla(G^{(p+3)/2}\alpha_1) = a_1\ell_1(xy + 1)^{p(p+1)/2}y^p \tag{9.22.4}$$

Thus if $p > 3$, $a_1 = 0$, which implies that α_0 and hence t are both 0. Therefore, $L_1 = 0$ if $p > 3$ and $p \neq 1 \pmod 3$.

Case 2: $p = 1 \pmod 3$. By (9.21), $H \simeq \mathbb{Z}/p\mathbb{Z} \oplus \mathbb{Z}/p\mathbb{Z}$, generated by the polynomials $f_1 = x(xy + 1)^{p-1}$ and $f_2 = c(xy + 1)^{n_0}y^{m_0}$, where $c \neq 0 \in k$ is such that

$$c^p = \binom{2n_0}{n_0} c$$

If $t \in L_1$, then by (9.7) we have that

$$t = \sum_{j=0}^{p-1} \alpha_j^p G^j \tag{9.22.5}$$

where $\alpha_0 = a_1 f_1 + a_2 f_2$ for some $a_1, a_2 \in \mathbb{Z}/p\mathbb{Z}$ and the rest of the α_j are given by (2) of (9.7).

We have that $\nabla(G^{p-3}f_2) = \ell_2 c(xy + 1)^{p(n_0-1)}y^{p(m_0-1)}$ for some $\ell_2 \neq 0 \in \mathbb{Z}/p\mathbb{Z}$, and $\nabla(G^{p-3}f_1) = 0$. Therefore, by (9.7),

$$\alpha_2^p = \nabla(G^{p-3}\alpha_0) = a_2\ell_2 c(xy + 1)^{p(n_0-1)}y^{p(m_0-1)} \tag{9.22.6}$$

We then have that

$$\alpha_2 = a_2\ell_2 c^*(xy + 1)^{(n_0-1)}y^{(m_0-1)} \quad \text{where } (c^*)^p = c \tag{9.22.7}$$

By (9.7), $\nabla(G^{p-2}\alpha_2) = 0$. But $\nabla(G^{p-2}(xy + 1)^{n_0-1}y^{m_0-1})$ is a nonzero integral multiple of $(xy + 1)^{pm_0}y^{p(n_0-1)}$. Thus $a_2 = 0$ and $\alpha_0 = a_1 f_1$.

Now just repeat the argument used in case 1 to show that $a_1 = 0$. We conclude that if $p \equiv 1 \pmod 3$, then $L_1 = 0$.

REMARK 9.23 For $p > 3$, the surface $z^{p^2} = x(xy + 1)^p + y^{p+1}$ has trivial class group and is regular. In light of remark (8.4) we are led to the question:

9.24 Is the surface $z^{p^2} = x(xy + 1)^p + y^{p+1}$ a plane for $p > 3$?

As of this writing this matter is undecided.

10. MISCELLANEOUS RESULTS

In this section we resume an exploration begun in Section 3. In that section we asked: What is the class group of the surface F: $z^p = G(x,y)$ for a generic choice of G?

We attempt to shed some light on the local version of this question. We ask: For a generic G, what is the class group of F: $z^p = G(x,y)$ at a singular point of F? That is: Does there exist a group M and a generic way of choosing G such that for each singular point $Q \in F$, $C\ell(F_Q) \cong M$ (by F_Q we mean the local ring of F at Q)?

LEMMA 10.1 Let k be an algebraically closed field of characteristic $p > 0$. Let $F \subsetneq \mathbb{A}_k^3$ be the surface defined by the equation $z^{p^n} = (x - \alpha_1 y)(x - \alpha_2 y)\cdots(x - \alpha_m y)$. Let Q be the origin of F, $x = y = z = 0$. Then $C\ell(F_Q)$ is a direct sum of $m - 1$ copies of $\mathbb{Z}/p^n\mathbb{Z}$.

Proof: Let $A = k[x^{p^n}, y^{p^n}, (x - \alpha_1 y)\cdots(x - \alpha_m y)]$. By proposition (7.1), $C\ell(A) \cong \mathbb{Z}/p^n\mathbb{Z} \oplus \cdots \oplus \mathbb{Z}/p^n\mathbb{Z}$ ($m - 1$ copies). By (1.5) we have a surjective morphism ϕ: $C\ell\ A \to C\ell(F_Q)$ with $\ker \phi = H + F(A)/F(A)$, where H is the subgroup of $C\ell(A)$ generated by those height 1 primes q of A with $q \cap m \neq \phi$ with $m = (x^{p^n}, y^{p^n}, (x - \alpha_1 y)\cdots(x - \alpha_m y))$ and F(A) is the subgroup of $C\ell(A)$ generated by the principal divisorial ideals of A. From (7.1) we have that $C\ell(A)$ is generated by the non-principal height 1 primes $I_i = (x - \alpha_1 y)\cdots(x - \alpha_n y)A + (x - \alpha_i y)^{p^n}A$, $i = 1, 2, \ldots, m - 1$.

We have that $I_i \subseteq m$ for each i. Thus $\ker \phi = 0$.

PROPOSITION 10.2 The divisor class group of $R_n = k[x^{p^n}, y^{p^n}, xy]$ is isomorphic to $\mathbb{Z}/p^n\mathbb{Z}$.

We give two very different and interesting proofs of (10.2). The first of these, which makes use of a proposition proved by N. Hallier (1964), involves logarithmic derivatives. The second, a geometric argument, uses results of J. Lipman (see Lipman, 1969, pp. 224-240) and P. Blass (see Blass, 1983, pp. 107-121).

The following proposition, whose proof we provide, is due to N. Hallier (1964, p. 2).

PROPOSITION 10.2.1 Let A be a local Krull domain with maximal ideal m such that A and A/m are of equal characteristic $p > 0$. Let D: $A \to A$ be a derivation such that the ideal $I - D(A) \cdot (A)$ in A is contained in m. Let $a \in A$ be such that $D^p = aD$. If a is a unit in A, then each $t \in m$ that is the logarithmic derivative of an element $f \in qt(A)$ is the logarithmic derivative of a unit in A.

Proof: Replacing f by an element of $A^p f$, we can assume that $f \in A$. If f is a unit in A, we are done. If $f \in m$, we have by induction that $D^n f/af \in m$ for all positive integers n. Let $u = -1 + (D^{p-1}f/af)$. Then $D(u^{-1})/u^{-1} = Df/f = t$.

PROPOSITION 10.2.2 Let A, m, D, I, and a be as in (10.2.1). Let U be the multiplicative group of units of A. Let $L = \{Df/f \mid f \in qt(A)$ and $Df/f \in A\}$ be the group of logarithmic derivatives in A and $L' = \{Du/u \mid u \in U\}$. Then either L/L' is trivial or isomorphic to $\mathbb{Z}/p\mathbb{Z}$.

Proof: Let θ: $L \to A/m$ be the additive group homomorphism, mapping each $t \in L$ to its image $\bar{t}$ in A/m. By (10.2.1), $\ker \theta = L'$. Thus θ induces an injection $\bar{\theta}$: $L/L' \to A/m$.

By (1.9), an element $t \in L \Leftrightarrow D^{p-1}(t) - at = -t^p$. Thus if $t \in L$, we have that $\overline{at} = \bar{t}^p$. Since $\bar{a} \neq 0$, the polynomial $X^p - \bar{a}X$ has p distinct roots in an algebraic closure of A/m. It follows that the order of $\bar{\theta}(L/L')$ is at most p. Since L/L' is a p group, the result follows.

First proof of proposition 10.2: Let D: $k[x,y] \to k[x,y]$ be the derivation defined by $D = x(\partial/\partial x) - y(\partial/\partial y)$. Then, as in (9.5), we can use D to define a derivation D_n: $R_n \to R_n$ with ker $D_n = R_{n+1}$ and with $D_n^p = D_n$. If we let ϕ_n: $C\ell(R_{n+1}) \to C\ell(R_n)$ be the homomorphism of theorem (1.3), then by (1.8) ker $\phi_n \simeq L_n/L_n'$, where $L_n = \{D_n(f)/f \mid f \in qt(R_n), Df/f \in R_n\}$ and $L_n' = \{D_n(u)/u \mid u$ is a unit in $R_n\}$.

By (10.2.2), the order of ker$(\phi_n) \le p$. Thus, by induction, we see that the order of $C\ell(R_n) \le p^n$. By (1.7) and (10.1) we have the desired result.

Since the second proof of (10.2) uses techniques not developed in this paper, we give only an outline. For more details, see the articles mentioned above.

Second proof of proposition 10.2: Let m be the maximal ideal of R_n. There exists a desingularization f: $X \to \mathrm{Spec}(R_n)$ such that the closed fiber has distinct components $E_1, E_2, \ldots, E_{p^n-1}$ with intersection numbers as follows:

$$E_i \cdot E_j = \begin{cases} 1 & \text{if } |i - j| = 1 \\ 0 & \text{if } 0 \ne |i - j| \ne 1 \end{cases}$$

$$E_i^2 = -2$$

[see Blass (1983, pp. 107-121) for the case n = 1, p > 2].

The intersection matrix $((E_i \cdot E_j))$ is given by the $(p^n - 1) \times (p^n - 1)$ square matrix

$$\begin{bmatrix} -2 & 1 & & & \\ 1 & -2 & \ddots & & \\ & \ddots & \ddots & \ddots & \\ & & \ddots & -2 & 1 \\ & & & 1 & -2 \end{bmatrix} \quad \text{with determinant } p^n$$

From proposition (17.1) of Lipman (1969) and the discussion on page 225 of Lipman (1969), we have that the order of $C\ell(R_n)$ is equal to

$$\frac{1}{d_1 d_2 \cdots d_{p^n-1}} \det((E_i \cdot E_j)) \qquad \text{where } d_i = \text{the degree of } E_i$$

Applying (1.7) and (10.1), we conclude that

$$\frac{1}{d_1 d_2 \cdots d_{p^n-1}} \det((E_i \cdot E_j)) = p^n \qquad \text{and} \qquad C\ell(R_n) \approx \mathbb{Z}/p^n\mathbb{Z}$$

PROPOSITION 10.3 Let $G \in k[x,y]$ satisfy condition (B) of Section 2. Let Q be a singular point of the surface $F \subsetneq \mathbf{A}_k^3$ defined by the equation $z^{p^n} = G$. Then $C\ell(F_Q) \subsetneq \mathbb{Z}/p^n\mathbb{Z}$.

Proof: After a linear change of coordinates we may assume that Q is the origin $(x = y = z = 0)$ of F and that G has the form $G = xy +$ higher-degree terms (see the discussion following proposition (2.9).

Let $A = k[x^{p^n}, y^{p^n}, G]$ and m be the maximal ideal of A corresponding to Q. By (1.7) there exists an injection $C\ell(A_m) \to C\ell(\hat{A})$, where $\hat{A}$ is the completion A_m at m. We have that $\hat{A} = k[x^{p^n}, y^{p^n}, G]$. In $k[x,y]$ G factors into a product $G = uv$, where u and v are of the form $u = x +$ higher-degree terms, $v = y +$ higher-degree terms. Clearly, $k[x,y] = k[u,v]$. Thus $\hat{A} = k[u^{p^n}, v^{p^n}, uv]$. By (10.2), $C\ell(\hat{A}) \cong \mathbb{Z}/p^n\mathbb{Z}$.

Thus the question posed at the beginning of this section can be answered when $p = 2$.

COROLLARY 10.4 Let k be an algebraically closed field of characteristic 2 and $G \in k[x,y]$ satisfy condition (B). Let Q be a singular point of the surface F: $z^2 = G(x,y)$. Then $C\ell(F_Q) \simeq \mathbb{Z}/2\mathbb{Z}$.

Proof: Note that $\partial^2(G)/\partial x \partial y \neq 0$ since G satisfies condition (b) and F has a singularity. Thus by (3.4) we have that $C\ell(F) \neq 0$ and $G_x k[x,y] \cap k[x^2, y^2, G]$ is a nonprincipal height 1 prime on $k[x^2, y^2, G]$. Since Q is a singularity, $G_x k[x,y] \cap k[x^2, y^2, G]$ is contained in the maximal ideal of $k[x^2, y^2, G]$ corresponding to Q. Therefore, the mapping $C\ell(F) \to C\ell(F_Q)$ of (1.4) is not the zero mapping. By (10.3), $C\ell(F_Q) \subsetneq \mathbb{Z}/2\mathbb{Z}$, from which it follows that $C\ell(F_Q) = \mathbb{Z}/2\mathbb{Z}$.

REMARK 10.5 This corollary provides us with further hope that conjecture (4.24) is correct. As for the $p > 2$ case, (10.3) tells us that if Q is a singularity of the surface F: $z^p = G$, where G satisfies condition (B), then $C\ell(F_Q) = 0$ or $\mathbb{Z}/p\mathbb{Z}$. The question as to which if either of these groups is $C\ell(F_Q)$ for a generic G is an open one.

Finally, we note that the surfaces $z^p = xu + yv + xy$ of (7.10) satisfy condition (B). Thus these surfaces have local class group $\mathbb{Z}/p\mathbb{Z}$ at each singular point by an argument identical to that used in (10.4).

In Section 2 we showed that the order of the class group of the surface F: $z^p = G(x,y)$ is a pth power p^M where $M \le \deg G(\deg G - 1)/2$.

Actually, we showed that this upper bound could be refined slightly to p^{M-N}, where N = the number of ordered pairs (i,j) of positive integers such that $i + j \le \deg G - 2$ and $i \equiv j \equiv -1 \pmod p$. In the next proposition we refine this bound a bit further.

PROPOSITION 10.6 Let k be an algebraically closed field of characteristic $p > 0$ and $G \in k[x,y]$ satisfy conditions (*) and (**). Let n be a positive integer such that $2np \le \deg G < 2(n + 1)p$. Let $F \subset \mathbf{A}^3$ be the surface defined by the equation $z^p = G$. Then the order of $C\ell(F) = p^f$, where $f \le [\deg G(\deg G - 1) - (p - 1)n(n + 1)]/2$.

Proof: Let $g = \deg G$, D be the jacobian derivation of G, and L the group of logarithmic derivatives. Suppose that $t \in L$.

By (2.4), $\deg t \le g - 2$. Thus $t = \Sigma_{0\le i+j\le g-2}\, c_{ij}x^iy^j$ for some $c_{ij} \in k$. From (1.9) and Ganong's formula we have that

$$\nabla(G^q t) = \begin{cases} 0 & \text{for } 0 \le q \le p - 2 \\ t^p & \text{for } q = p - 1 \end{cases} \tag{10.6.1}$$

Since $\nabla(G^{p-1}t) = t^p$ we obtain for each of the c_{ij} an equation of the form

$$\ell_{ij} = c_{ij}^p \quad \text{where } \ell_{ij} \text{ is a linear expression in the } c_{im} \text{ with coefficients in k} \tag{10.6.2}$$

If we regroup terms we can write $t = \Sigma_{0 \le u, v \le p-1} \alpha_{uv} x^u y^v$ with $\alpha_{uv} = \Sigma\, c_{(u+rp)(v+sp)} x^{rp} y^{sp}$, where this sum is taken over all pairs (r,s) such that $0 \le u + rp + v + sp \le g - 2$.

Since $\nabla(G^q t) = 0$, we have for $q = 0, 1, \ldots, p - 2$, the equation

$$\ell_q: \sum_{0 \le u, v \le p-1} \alpha_{uv} \nabla(G^q x^u y^v) = 0 \qquad \text{for } 0 \le q \le p - 2 \quad (10.6.3)$$

We have that these $p - 1$ equations in the α_{uv} with coefficients $\nabla(G^q x^u y^v)$ are independent over $k(x^p, y^p)$. For suppose that $\beta_q \in k(x^p, y^p)$ such that $\beta_0 \ell_0 + \cdots + \beta_{p-2} \ell_{p-2} = 0$. Then it follows that

$$\nabla((\beta_0 + \beta_1 G + \cdots + \beta_{p-2} G^{p-2}) x^u y^v) = 0 \qquad (10.6.4)$$
$$\text{for all } 0 \le u, v \le p - 1$$

which implies that $\beta_0 + \beta_1 G + \cdots + \beta_{p-2} G^{p-2} = 0$ and hence $\beta_0 = \beta_1 = \cdots = \beta_{p-2} = 0$.

We conclude that $(p - 1)$ of the α_{uv}'s are determined by the remaining ones. Note that each α_{uv} involves at least $n(n + 1)/2$ of the c_{ij}'s. Thus we have that among the c_{ij}'s there are $(p - 1)n \times (n + 1)/2$ of them that are determined by the choice of the remaining $[g(g - 1) - (p - 1)n(n + 1)]/2$ ones. Each of these remaining ones has to satisfy an equation of the form (10.6.2). By an argument used in the proof of (2.9) and (6.7), we have that there are p^s possibilities for the $g(g - 1)/2$-tuple $(c_{00}, c_{10}, c_{01}, \ldots, c_{g-20}, \ldots, c_{0g-2})$, where $s \le \{g(g - 1) - (p - 1)n(n + 1)\}/2$. Thus the order of L and hence $C\ell(F) = p^s$.

A similar argument yields the following.

PROPOSITION 10.7 If $G \in k[x,y]$ is as in (10.6) and is such that $p \le \deg G < 2p$, then $C\ell(F\colon\ z^p = G)$ has order p^s, where $s \le g(g - 1)/2 - (p - 2)$ and where $g = \deg G$.

Next we will show that some of the results that we have obtained thus far extend to the several variable case. The following results are to be found in Fossum (1973).

10.8 Let $A = A_0 + A_1 + \cdots$ be a graded noetherian Krull domain such that A_0 is a field. Let $m = A_1 + \cdots$. Then $C\ell(A) \to C\ell(A_m)$ is a bijection.

10.9 Let $A = A_0 + A_1 + \cdots$ be a graded Krull domain such that A_0 is a field, k. Let k' be an extension field of k. Suppose that $A \otimes_k k' = A'$ is a Krull domain. Then A' is a faithfully flat A-module and the induced homomorphism

$$C\ell(A) \to C\ell(A')$$

is an injection.

10.10 Let G be a finite group of derivations of a Krull domain B of characteristic $p > 0$. Let A be the fixed subring of G. Let $D_1, \ldots, D_r$ be a basis for G over $\mathbf{Z}/p\mathbf{Z}$. Then the kernel of the homomorphism $C\ell(A) \xrightarrow{\phi} C(B)$ is isomorphic to a subgroup of V_0/V_0', where $V_0 = \{(t^{-1}D_1t, \ldots, t^{-1}D_rt)\colon\ t \in qt(B) \text{ and } t^{-1}D_it \in B \text{ for all } i = 1, \ldots, r\}$, and $V_0' = \{(u^{-1}D_1u, \ldots, u^{-1}D_ru)\colon\ u \in B^*\}$ with B^* the units of B.

10.11 Let us describe this injection in (10.10). Let I be a divisorial ideal of A whose class is in the kernel of ϕ: $C\ell\ A \to C\ell\ B$. Then B: (B:IB) is a principal ideal, say xB for some x in B. We then map I to $(x^{-1}D_1x, \ldots, x^{-1}D_rx)$ in V_0/V_0'.

10.12 Throughout the remainder of this section we assume that k is an algebraically closed field of characteristic $p > 0$, $G(x_1, \ldots, x_n) \in k[x_1, \ldots, x_n]$ is a polynomial in n variables, and $F \subseteq \mathbf{A}^{n+1}$ is the surface defined by the equation $z^p = G(x_1, \ldots, x_n)$. Also, we assume that the polynomial $\partial G/\partial x_1 \neq 0$ and that the greatest common divisor of the n-tuple of polynomials $(\partial G/\partial x_1, \ldots, \partial G/\partial x_n)$ in $k[x_1, \ldots, x_n]$ is 1. [This condition on G is equivalent to saying that the surface F is normal (see Matsumura, 1970, p. 125).]

LEMMA 10.13 The coordinate ring of F is isomorphic to $A = k[x_1^p, \ldots, x_n^p, G]$.

Proof: Completely analogous to lemma 2.1.

10.14 For each $i = 1, 2, \ldots, n-1$, let $D_i\colon k(x_1,\ldots,x_n) \to k(x_1,\ldots,x_n)$ be the k-derivation defined by

$$D_i = \frac{\partial G}{\partial x_i}\frac{\partial}{\partial x_1} - \frac{\partial G}{\partial x_1}\frac{\partial}{\partial x_{i+1}}$$

LEMMA 10.15 $\bigcap_{i=1}^{n-1} D_i^{-1}(0) \cap k[x_1,\ldots,x_n] = A$.

Proof: We have that $k(x_1,\ldots,x_n) \supsetneqq D_1^{-1}(0) \supsetneqq D_1^{-1}(0) \cap D_2^{-1}(0) \supsetneqq \cdots \supsetneqq D_1^{-1}(0) \cap \cdots \cap D_{n-1}^{-1}(0) \supset qt(A)$, because for each $j = 1, \ldots, n-1$ we have that $x_{j+1} \in (D_1^{-1}(0) \cap \cdots \cap D_{j-1}^{-1}(0)) \setminus D_j^{-1}(0)$. Since $[k(x_1,\ldots,x_n)\colon qt(A)] = p^{n-1}$, it follows that $qt(A) = \bigcap_{i=1}^{n-1} D_i^{-1}(0)$. Since A is integrally closed, the result follows.

LEMMA 10.16 Let $V = \{(t^{-1}D_1t,\ldots,t^{-1}D_{n-1}t)\colon t \in k(x_1,\ldots,x_n)$ and $t^{-1}D_it \in k[x_1,\ldots,x_n]\}$. Then C (F) injects into V.

Proof: Use (10.10).

PROPOSITION 10.17 $C\ell(F)$ is a p-group of type $(p,\ldots,p)$ and of order p^f, where $f \le (n-1)g(g-1)/2$, where $g = \deg G$.

Proof: Let $(t_1,\ldots,t_{n-1}) \in V$. Then the degree of each $t_i \le g-2$. We will now show that there are at most $p^{g(g-1)/2}$ possible such t_i for each $i = 1, 2, \ldots, n-1$. We begin with t_1. Recall that

$$D_1 = \frac{\partial G}{\partial x_2}\frac{\partial}{\partial x_1} - \frac{\partial G}{\partial x_1}\frac{\partial}{\partial x_2}$$

We have that $k(x_1,\ldots,x_n)$ is a purely inseparable extension of $D_1^{-1}(0)$ of degree p. By (1.9) there exists an $a \in k[x_1,\ldots,x_n] \cap D_1^{-1}(0)$ such that $D_1^p = aD_1$ and t_1 is a logarithmic derivative of D_1 if and only if

$$D_1^{p-1}(t_1) - at_1 = -t_1^p \tag{10.17.1}$$

We write $t_1 = \Sigma_{0\le r+s\le g-2}\, \alpha_{rs}x_1^rx_2^s$, where $\alpha_{rs} \in k[x_3,\ldots,x_n]$. Substituting this expression for t_1 into (10.17.1) and comparing coefficients, we obtain for each pair (r,s) with $0 \le r+s \le g-2$

an equation of the form

$$\ell_{rs} = \alpha_{rs}^p \quad \text{where } \ell_{rs} \text{ is a linear equation} \tag{10.17.2}$$

in the α_{rs} with coefficients in $k[x_3,\ldots,x_n]$. We may also obtain linear equations in addition to the equations in (10.17.2).

Let L be an algebraic closure of $k(x_3,\ldots,x_n)$. Then using the usual argument (see lemma 2.8), we see that the equations of (10.17.2) have at most $p^{g(g-1)/2}$ solutions in L, hence in $k[x_3,\ldots,x_n]$. Thus there are at most $p^{g(g-1)/2}$ possible t_1's. Similarly, there are at most $p^{g(g-1)/2}$ possibilities for each t_i, i = 2, ..., n - 1. Thus V has order p^f where $f \le (n-1)g(g-1)/2$. Since each element of V has p-torsion and $C\ell(F) \subseteq V$, the result follows.

The next proposition generalizes proposition (10.6) to the n-variable case.

PROPOSITION 10.18 For each i = 2, 3, ..., n, let m_i be a positive integer such that $2pm_i \le \deg_{x_1,x_i}(G) \le 2(p+1)m_i$. Assume that the gcd $(\partial G/\partial x_1, \partial G/\partial x_i) = 1$. Then the order of $C\ell(F)$ is p^s, where $s \le \{g(g-1)(n-1) - (p-1)\,\Sigma_{i=2}^n m_i(m_i+1)\}/2$.

Proof: Let $(t_1,\ldots,t_{n-1}) \in V$. For each j = 1, ..., n - 1, we have that t_j is a logarithmic derivative of D_j if and only if

$$\nabla_j(G^q t_j) = \begin{cases} t_j^p & \text{for } q = p-1 \\ 0 & \text{for } 0 \le q \le p-2 \end{cases} \tag{10.18.1}$$

where $\nabla_j = \partial^{2(p-1)}/\partial x_1^{p-1}\partial x_{j+1}^{p-1}$. This is by (1.9) and Ganong's formula.

If we just repeat the argument used in (10.6), we have that there are p^{s_j} possibilities for t_j, where $s_j \le [g(g-1) - (p-1)m_{j+1}(m_{j+1}+1)]$, from which the result follows.

Note that if $\partial G/\partial x_j = 0$ for some j = 2, ..., n - 1, we can replace D_j with $\partial/\partial x_j$ and still have that

$$\bigcap_{i\neq j} D_i^{-1}(0) \cap \frac{\partial}{\partial x_j}^{-1}(0) = A$$

Since $\partial/\partial x_j$ has no logarithmic derivatives except 0, we have the following results

PROPOSITION 10.19 If G is such that $\partial G/\partial x_j = 0$ for $j = r, \ldots, n$ for some $r > 2$, then $C\ell(F)$ injects into $V' = \{(t_1^{-1}D_1(t_1), \ldots, t_{r-1}^{-1}D_{r-1}(t_{r-1})$: $t_i \in k(x_1,\ldots,x_n)$ and $t_i^{-1}D(t_i) \in k[x_1,\ldots,x_n]\}$.

Proof: Use (10.10) and the observation above.

COROLLARY 10.20 With G as in (10.19), we have that the order of $C\ell(F) = p^s$, where $s \leq (r - 1)g(g - 1)/2$.

Proof: Use the same argument as in (10.17).

Below we calculate $C\ell(F)$ for the case when G is homogeneous of degree *not* divisible by p, thus generalizing proposition (7.1).

PROPOSITION 10.21 Let $h_1, h_2, \ldots, h_r$ be distinct homogeneous irreducible polynomials in $k[x_1,\ldots,x_n]$ the sum of whose degrees is g with g *not* divisible by p. Let $G = h_1\cdots h_r$ and F be the hypersurface defined by the equation $z^p = G$. Then F is normal and $C\ell(F)$ has order p^{n-1} generated by the height 1 primes $GA + h_i^pA$ in $A = k[x_1^p,\ldots,x_n^p,G]$.

Proof: By Euler's formula $\Sigma_{i=1}^n x_i(\partial G/\partial x_i) = gG$. If h divides each $\partial G/\partial x_i$, then h divides G and must be a multiple factor of G. Therefore, $\gcd(\partial G/\partial x_1,\ldots,\partial G/\partial x_n) = 1$ and F is normal.

Without loss of generality we can assume that the elements $\partial G/\partial x_1$, $\partial G/\partial x_2$, ..., $\partial G/\partial x_m$ form a basis for the $\mathbb{Z}/p\mathbb{Z}$-vector space spanned by $\partial G/\partial x_i$, $i = 1, \ldots, n$.

For each pair (j,e) with $1 \leq j \leq e \leq n$ and $j \leq m$, let D_{je} be the $k[x_1,\ldots,x_n]$-derivation defined by

$$D_{je} = \frac{\partial G}{\partial x_e}\frac{\partial}{\partial x_j} - \frac{\partial G}{\partial x_j}\frac{\partial}{\partial x_e}$$

Clearly, $\bigcap D_{je}^{-1}(0) \supseteq A$.

By (10.15), $\cap D_{1e}^{-1}(0) = A$. Thus A is the fixed subring of the D_{je}. Also we have that the D_{je} are $\mathbb{Z}/p\mathbb{Z}$-independent. For if $\Sigma\, c_{ue}D_{ue} = 0$, then $\Sigma\, c_{je}D_{je}(x_n) = 0$, which implies that $\Sigma\, c_{jn} \times \partial G/\partial x_j = 0$. Thus each $c_{jn} = 0$. If we evaluate $\Sigma\, c_{je}D_{je}(x_{n-1})$, we conclude that each $c_{j(n-1)} = 0$. Continuing in this way, we see that each $c_{je} = 0$.

Let V_0 be the additive group that the derivations D_{je} define as in (10.10). By (10.10), $C\ell(F)$ injects into V_0.

10.21.1 Let $(v_{je}) \in V_0$. We show that there exists a homogeneous polynomial $\tilde{f}$ such that $v_{je} = D_{je}(\tilde{f})/\tilde{f}$ for each pair (j,e).

Temporarily fix a pair (j,e). Suppose that $D_{je}(f)/f = v \in k[x_1,\ldots,x_n]$. Multiplying by an element in $k[x_1^p,\ldots,x_n^p]$, we can assume that f belongs to $k[x_1,\ldots,x_n]$.

Let v_1 and f_1 be the lowest-degree forms of v and f, respectively. Let v_2 and f_2 be the highest-degree forms of v and f, respectively.

We have that $\deg(f_1) + g - 2 \le \deg(D_{je}(f)) \le \deg(f_2) + g - 2$. If we compare the degrees of both sides of the equality $D_{je}(f) = fv$, we have that $\deg(v_2) + \deg(f_2) \le \deg(f_2) + g - 2$ and $\deg(v_1) + \deg(f_1) \le \deg(f_1) + g - 2$. This implies that $g - 2 \le \deg(v_1) \le \deg(v_2) \le g - 2$ and that v is homogeneous. Thus it must be that $\deg(D_{je}(f)) = \deg(f_2) + g - 2$ and hence $D_{je}(f_2) = f_2 v$.

It follows that if $(v_{je}) \in V_0$ and $D_{je}(f)/f = v_{je}$ for each (j,e), we can first assume that f is a polynomial and the highest-degree form of f, say $\tilde{f}$, yields (v_{je}).

Note that if $f_1^{e_1} f_2^{e_2} \cdots f_s^{e_s} = \tilde{f}$ is a prime factorization of $\tilde{f}$, then $e_1 D_{je}(f_1)/f_1 + \cdots + e_s D_{je}(f_s)/f_s = D\tilde{f}/\tilde{f}$, where $D_{je}(f_1)f_1, \ldots, D_{je}(f_s)/f_s \in k[x_1,\ldots,x_n]$. We conclude that

10.21.2 V_0 is generated by elements of the form $(D_{je}(\tilde{f})/\tilde{f})$ with $\tilde{f}$ homogeneous and irreducible.

Let t be a homogeneous irreducible polynomial such that t divides $D_{je}(t)$ for each (j,e). Then t divides

$$x_2D_{12}(t) + \cdots + x_nD_{1n}(t) = x_2\left(\frac{\partial G}{\partial x_2}\frac{\partial t}{\partial x_1} - \frac{\partial G}{\partial x_1}\frac{\partial t}{\partial x_2}\right) + \cdots$$
$$+ x_n\left(\frac{\partial G}{\partial x_n}\frac{\partial t}{\partial x_1} - \frac{\partial G}{\partial x_1}\frac{\partial t}{\partial x_n}\right)$$
$$= gG\frac{\partial t}{\partial x_1} - (\deg t)\frac{\partial G}{\partial x_1}t$$

by Euler's formula. Thus t divides $G(\partial t/\partial x_1)$.

Note that t also divides

$$-x_1D_{12}(t) + x_3D_{23}(t) + \cdots + x_nD_{2n}(t) = gG\frac{\partial t}{\partial x_2} - (\deg t)\frac{\partial G}{\partial x_2}t$$

by Euler's formula. Thus t divides $G(\partial t/\partial x_2)$.

10.21.3 Continuing in this way, we see that for each j such that $1 \le j \le m$ and $j < n$, t divides $G(\partial t/\partial x_j)$.

We claim that (10.21.3) implies that t divides G. If t does *not* divide G, then from (10.21.3) it follows that $\partial t/\partial x_j = 0$ and t divides $\partial G/\partial x_j$ for each $i \le j \le m$ with $j < n$.

We consider two cases.

Case 1: $\partial G/\partial x_1, \ldots, \partial G/\partial x_n$ are independent over $\mathbb{Z}/p\mathbb{Z}$ (i.e., $m = n$). Then t divides $\partial G/\partial x_1, \ldots, \partial G/\partial x_{n-1}$ and $\partial t/\partial x_1 = \cdots = \partial t/\partial x_{n-1} = 0$. By Euler's formula $(\deg t)t = x_n(\partial t/\partial x_n)$. Thus either t divides x_n, in which case t divides $x_1(\partial G/\partial x_1) + \cdots + x_n(\partial G/\partial x_n) = gG$, or $\partial t/\partial x_n = 0$, in which case t is not prime.

Case 2: $\partial G/\partial x_1, \ldots, \partial G/\partial x_n$ are $\mathbb{Z}/p\mathbb{Z}$-dependent (i.e., $m < n$). Since t divides $\partial G/\partial x_1, \ldots, \partial G/\partial x_m$ it must be that t divides $\partial G/\partial x_i$, $i = 1, \ldots, n$. Thus $t \in k$, contradicting the fact that t is prime.

Thus t divides G. Combining this fact with (10.21.2), we have that

10.21.4 V_0 is generated by the elements $u_i = (D_{je}(h_i)/h_i)$, $i = 1, 2, \ldots, r$.

We will demonstrate that $\{u_1, u_2, \ldots, u_{r-1}\}$ form a basis for V_0 over $\mathbb{Z}/p\mathbb{Z}$. Note that $u_1 + \cdots + u_r = (D_{je}(G)) = (0,0,\ldots,0)$. Thus $\{u_1, \ldots, u_{r-1}\}$ generate V_0.

If $d_1, \ldots, d_{r-1} \in \mathbf{Z}^+$ are such that $d_1u_1 + \cdots + d_{r-1}u_{r-1} = 0$, then $D_{je}\left(h_1^{d_1}\cdots h_{r-1}^{d_{r-1}}\right) = 0$ for each pair (j,e) and thus $h = h_1^{d_1}\cdots h_{r-1}^{d_{r-1}}$ belongs to A. We then have that $k(x_1^p,\ldots,x_n^p) \subset k(x_1^p,\ldots,x_n^p,h) \subset qt(A)$. If $k(x_1^p,\ldots,x_n^p,h) = qt(A)$, then $\exists\ \alpha_0, \ldots, \alpha_p \in k[x_1^p,\ldots,x_n^p]$ such that $\alpha_0 G = \alpha_1 + \alpha_2 h + \cdots + \alpha_p h^{p-1}$. Since G and h are homogeneous, we can assume that $\alpha_0, \ldots, \alpha_p$ are also. Since the $\deg(\alpha_i h^{i-1})$ is congruent to $(i - 1)$ deg h (mod p) and $\deg(\alpha_0 G) = g$ (mod p), we have that only one of $\alpha_1, \ldots, \alpha_p \neq 0$ and thus $\alpha_0 G = \alpha_i h^{i-1}$ for some $i = 1, \ldots, p$, which is clearly impossible.

Thus $k(x_1^p,\ldots,x_n^p,h) = k(x_1^p,\ldots,x_n^p)$ and $d_1 = \cdots = d_{r-1} = 0$ (mod p).

Finally, we note that the nonprincipal height 1 primes $Q_i = GA + h_i^pA$ in A map to the elements u_i for $i = 1, \ldots, r$, under the homomorphism described in (10.11). It follows that this mapping (10.11) is an isomorphism and that $C\ell(F)$ has order p^{r-1} generated by $Q_1, \ldots, Q_{r-1}$.

REMARK 10.22 Proposition (10.21) is not valid when p divides g. For we will see in a moment that the surface $z^p = x_1x_2\cdots x_p + x_0^p$ has nontrivial class group, although $x_1x_2\cdots x_p + x_0^p$ is irreducible in $k[x_0,x_1,\ldots,x_n]$.

We can use (10.21) to attack some special cases when p divides g.

COROLLARY 10.23 Let $h_1, h_2, \ldots, h_r$ be distinct homogeneous irreducible polynomials in $k[x_1,\ldots,x_n]$. Let $G = x_0h_1h_2\cdots h_r$. Then $F \subseteq \mathbf{A}^{n+2}$ is normal and the order of $C\ell(F) = p^r$.

Proof: Let g = deg G. If p does not divide g, the result follows by (10.21).

Suppose than that g = pm. Then $\gcd(\partial G/\partial x_1,\ldots,\partial G/\partial x_n) = x_0$. Thus $\gcd(\partial G/\partial x_0,\ldots,\partial G/\partial x_n) = 1$ and F is normal.

Let $R = k[x_0,\ldots,x_n,z]$, $z^p = G$. Then R is the coordinate ring of F.

We have that $R[1/x_0] \cong R_1$, where $R_1 = k[x_0',x_1',\ldots,x_n',z',1/x_0']$, $(z')^p = h_1'\cdots h_r'$, where $h_j' = h_j(x_j',\ldots,x_n')$ for $j = 1, \ldots, r$. The map $R_1 \to R$ is given by $x_0' \to x_0$, $z' \to z/x_0^m$, $x_i' \to x_i/x_0$ for $i = 1, \ldots, n$.

By (1.4), $C\ell(R_1) \approx C\ell(R_2)$, where $R_2 = k[x_0',x_1',\ldots,x_n',z']$, $(z')^p = h_1'\cdots h_r'$. By (1.6), $C\ell(R_2) \simeq C\ell(R_3)$, where $R_3 = k[x_1',x_2',\ldots,x_n',z']$, $(z')^p = h_1'\cdots h_r'$. By (10.21), the order of $C\ell(R_3)$ and hence $C\ell(R[1/x_0])$ is p^{r-1}.

Again by (1.4) we have an exact sequence

$$0 \to \ker \phi \to C\ell(R) \to C\ell\ R\left[\frac{1}{x_0}\right] \to 0 \qquad (10.23.1)$$

where $\ker \phi$ is generated by those height 1 primes in R that contain x_0. Such a prime would have to contain z also and hence the ideal $x_0R + zR$, which is clearly a nonprincipal height 1 prime. Thus $\ker \phi \simeq \mathbb{Z}/p\mathbb{Z}$, from which it follows that $C\ell(R)$ has order p^r.

COROLLARY 10.24 The divisor class group of the surface $F \subseteq \mathbf{A}^{np+1}$ defined by the equation $z^p = x_1x_2\cdots x_{np}$ is a direct sum of $np - 1$ copies of $\mathbb{Z}/p\mathbb{Z}$.

Proof: Use (10.23).

REMARK 10.25 The surface in (10.22) is isomorphic to the surface $z^p = x_1\cdots x_p$, which has nontrivial class group by (10.24).

In Section 9 we studied the divisor class group of surfaces defined by equations of the form $z^{p^n} = G(x,y)$, where $G(x,y)$ represents a polynomial in two variables. Below we look at the several variable case.

10.26 Let k and $G(x_1,\ldots,x_n)$ be as in (10.12). For each positive integer m we let $F_m \subseteq \mathbf{A}_k^{n+1}$ be the surface (necessarily normal) defined by the equation $z^{p^m} = G(x_1,\ldots,x_n)$.

LEMMA 10.27 For each m, the coordinate ring of F_m is isomorphic to $A_m = k[x_1^{p^m},\ldots,x_n^{p^m},G]$.

Proof: Similar to (2.1).

Just as in Section 9, we obtain an inductive method for studying $C\ell(F_m)$. For each m = 1, 2, ..., let $B_m = k[x_1^{p^{m+1}}, x_2^{p^{m+1}}, \ldots, x_n^{p^{m+1}}, G^p]$. Then B_m and A_m are isomorphic and we have that $B_m \subset A_{m+1} \subset A_m$ with A_m integral over B_m and $[qt(A_m) : qt(A_{m+1})] = p^{n-1}$ and $[qt(A_{m+1}) : qt(B_m)] = p$.

By (1.3) there exist group homomorphisms θ_m: $C\ell(B_m) \to C\ell(A_{m+1})$ and ϕ_m: $C\ell(A_{m+1}) \to C\ell(A_m)$. Let E_m be the restriction of the $k(x_1, \ldots, x_n)$ derivation to A_{m+1}.

LEMMA 10.28 E_m maps A_{m+1} into A_{m+1} and has kernel B_m.
Proof: Identical to the proof of (9.2).

PROPOSITION 10.29 For each positive integer m, $C\ell(F_m)$ injects into $C\ell(F_{m+1})$.
Proof: Identical to the proof of (9.3).

To study ϕ_m: $C\ell(A_{m+1}) \to C\ell(A_m)$ we define derivations D_{mj}: $A_m \to A_m$. Given $\alpha \in A_m$, there exist unique $\alpha_i \in k[x_1, \ldots, x_n]$ such that $\alpha = \Sigma_{i=0}^{p^m-1} \alpha_i^{p^m} G^i$.

10.30 Define $D_{mj}\alpha = \Sigma_{i=0}^{p^m-1} (D_j\alpha_i)^{p^m} G^i$, where

$$D_j = \frac{\partial G}{\partial x_{j+1}} \frac{\partial}{\partial x_1} - \frac{\partial G}{\partial x_1} \frac{\partial}{x_{j+1}} \qquad \text{for } j = 1, \ldots, n-1$$

We have the following analogs to (9.5)-(9.10). The proofs generalize nicely.

LEMMA 10.31 The mappings D_{mj} defined above are derivations.

LEMMA 10.32 Let D_{mj}: $A_m \to A_m$ be as in (10.30).

(i) $A_{m+1} = \ker D_{m1} \cap \cdots \cap \ker D_{m(n-1)}$.

(ii) Let $V_m = \{(t^{-1}D_{m1}t, \ldots, t^{-1}D_{m(n-1)}t)$: $t \in qt(A_m)$ and $t^{-1}D_{mj}t \in A_m\}$. Then $\ker \phi_m$ injects into V_m.

(iii) Let $a_j \in D_j^{-1}(0) \cap k[x_1,\ldots,x_n]$ be such that $D_j^p = a_j D_j$. Then

$$D_{mj}^p = a_j^{p^m} D_{mj},\ j = 1,\ \ldots,\ n - 1.$$

PROPOSITION 10.33 For each $j = 1, \ldots, n - 1$, let $t_j = \alpha_{j0}^{p^m} + \alpha_{j1}^{p^m} G + \cdots + \alpha_{j(p^m-1)}^{p^m} G^{p^m-1} \in A_m$. Then $(t_1,\ldots,t_{n-1}) \in V_m$ if and only if for each j:

(1) $\nabla_j(G^q \alpha_{ji}) = 0$ for $0 \le q \le p - 1$, $0 \le i \le p^m - 1$ and $i \not\equiv 0 \pmod p$.

(2) $\nabla_j(G^q \alpha_{j(ep)}) = \alpha^p_{j(1+(p-(q+1))p^{m-1})}$ for $e = 0, 1, \ldots, p^{m-1} - 1$,

where $\nabla_j = \partial^{2(p-1)}/\partial x_1^{p-1} \partial x_{j+1}^{p-1}$.

LEMMA 10.34 With $(t_1,\ldots,t_{n-1})$ as in (10.33), we have that each α_{ji} is determined by α_{j0} for $i = 0, \ldots, p^m - 1$, $i = 1, \ldots, n - 1$.

THEOREM 10.35 For each m, ker ϕ_m is a p-group of type $(p,\ldots,p)$ of order p^f, where $f \le (n - 1)g(g - 1)/2$ with $g = \deg G$.

THEOREM 10.36 For each m, $C\ell(F_m)$ is a finite p-group of type $(p^{i_1}, \ldots, p^{i_r})$ where each $i_j \le m$. The order of $C\ell(F_m) \le p^{m(n-1)g(g-1)/2}$, where $g = \deg G$.

REMARK 10.37 Just as we saw in Section 9 [see (9.11)], we now have an inductive method for studying $C\ell(F_m)$. We remind ourselves of the following diagram for each m.

$$C\ell(B_m) \xrightarrow{\theta_m} C\ (A_{m+1}) \xrightarrow{\phi_m} C\ell(A_m)$$

$$D_{mj}:\ A_m \to A_m \tag{10.38}$$

$$\ker \phi_m \hookrightarrow V_m = \{(t^{-1}D_{m1}t,\ldots,t^{-1}D_{m(n-1)}t):\ t \in qt(A_m)$$
$$\text{and } t^{-1}D_{mj}t \in A_m,\ j = 1, 2, \ldots, n - 1\}$$

PROPOSITION 10.39 Let $h_1, \ldots, h_r$ be distinct homogeneous irreducible polynomials in $k[x_1,\ldots,x_n]$, the sum of whose degree is g *not* divisible by p. For each m, let $F_m \subsetneq \mathbf{A}^{n+1}$ be the hypersurface defined by $z^{p^m} = G(x_1,\ldots,x_n)$, where $G = h_1 \cdots h_r$. Then F_m is normal and $C\ell(F_m)$ is a direct sum of r - 1 copies of $\mathbb{Z}/p^m\mathbb{Z}$, generated by the height 1 primes $GA_m + h_i^{p^m}A_m$ in A_m, i = 1, ..., r - 1.

Proof: As in (10.21) we have that gcd $(\partial G/\partial x_1,\ldots,\partial G/\partial x_n) = 1$. We can assume $\{\partial G/\partial x_1,\ldots,\partial G/\partial x_s\}$ form a basis for the $\mathbb{Z}/p\mathbb{Z}$-vector space spanned by $\partial G/\partial x_1, \ldots, \partial G/\partial x_n$, where $s \le n$.

For each pair (j,e) with $1 \le j \quad e \le n$ and $j \le s$, let

$$D_{je} = \frac{\partial G}{\partial x_e}\frac{\partial}{\partial x_j} - \frac{\partial G}{\partial x_j}\frac{\partial}{\partial x_e}$$

Define a derivation $D_{je}^{(m)}\colon A_m \to A_m$ by

$$D_{je}^{(m)}(\alpha_0^{p^m} + \alpha_1^{p^m}G + \cdots + \alpha_{p^m-1}^{p^m}G^{p^m-1})$$
$$= (D_{je}\alpha_0)^{p^m} + (D_{je}\alpha_1)^{p^m}G + \cdots + (D_{je}\alpha_{p^m-1})^{p^m}G^{p^m-1} \qquad (10.39.1)$$

Just as in (10.21), the $D_{je}^{(m)}$ have fixed subring A_{m+1} and are $\mathbb{Z}/p\mathbb{Z}$-independent, since the D_{je} are.

Let V_m be the additive group that the derivations $D_{je}^{(m)}$ define (10.10). Fix a pair (j,e). Suppose that $D_{je}^{(m)}(f)/f = v \in A_m$, where $f \in qt(A_m)$. Clearly, we can assume that $f \in A_m$. Let f_1 and v_1 be the lowest-degree forms and f_2 and v_2 be the highest-degree forms of f and v, respectively. Then f_1, v_1, f_2, and v_2 all belong to A_m and either $D_{je}^{(m)}(f_2) = 0$ or $\deg_{je}^{(m)}(f_2) = (\deg f_2) + (g - 2)p^m$, similarly for f_1. Thus

$$(\deg v_2) + (\deg f_2) \le (\deg f_2) + (g - 2)p^m$$
$$(\deg v_1) + (\deg f_1) \ge (\deg f_1) + (g - 2)p^m$$

from which we have that

$$(g - 2)p^m \le \deg v_1 \le \deg v_2 \le (g - 2)p^m$$

Therefore, v is homogeneous of degree $(g - 2)p^m$. Thus v can only be of the form $v = u^{p^m}$ for some $u \in k[x_1,\dots,x_n]$ of degree g - 2. By (1.9b), v is a logarithmic derivative of $D_{je}^{(m)}$ if and only if $(D_{je}^{(m)})^{p-1}(v) - a_{je}^{(m)}v = -v^p$, where $a_{je}^{(m)}$ is the element of A_{m+1} such that $(D_{je}^{m})^p = a_{je}^{(m)}D_{je}$.

From (10.32) we have that $a_{je}^{(m)} = a_{je}^{p^m}$, where $a_{je} \in k[x_1,\dots,x_n]$ is such that $(D_{je})^p = a_{je}D_{je}$. Using (10.39.1) we have that v is a logarithmic derivative of $D_{je}^{(m)}$ if and only if $D_{je}(u) - a_{je}u = -u^p$, that is, if and only if u is a logarithmic derivative of D_{je}.

Thus we have that

10.39.2 $(v_{je}) \in V_m$ if and only if there exist $u_{je} \in k[x_1,\dots,x_n]$ such that $(u_{je}) \in V_0$, where V_0 is as described in the proof of (10.21), with $v_{je} = u_{je}^{p^m}$.

From this fact it follows that the mapping $(w_{je}) \to (w_{je}^{p^m})$ from $V_0 \to V_m$ is an isomorphism.

In (10.21) we showed that V_0 has order p^{r-1}; hence the ker ϕ_m has order p^{r-1}. We have that the height 1 primes $p_i = GA_m + h_1^{p^m}A_m$ have ramification index 1 over their contractions $Q_1 = GA_{m+1} + h_i^{p^{m+1}}$ in A_{m+1} and Q_i has ramification index p over their contractions $P_i' = G^pB_m + h^{p^{m+1}}B_m$ in B_m, i = 1, ..., r - 1.

Using induction we have that the primes P_i generate $C\ell(A_m)$ and are each of order p^m; hence the same holds true of the primes P_i' in B_m.

Since θ_m: $C\ell(B_m) \to C\ell(A_{m+1})$ is injective by (10.29), we see that the elements p^mQ_i are a $\mathbb{Z}/p\mathbb{Z}$-basis for the ker ϕ_m. Since the ramification indexes $e(P_i:Q_i) = 1$ we have that ϕ_m is surjective, from which the theorem follows.

COROLLARY 10.40 Let $h_1, \dots, h_r$ be distinct irreducible homogeneous polynomials in $k[x_1,\dots,x_n]$. Let $G = x_0h_1\cdots h_r$ and $F_m \subsetneq \mathbf{A}^{n+2}$ be the hypersurface defined by $z^{p^m} = G$. Then $C\ell(F_m)$ is a direct sum of r copies of $\mathbb{Z}/p^m\mathbb{Z}$, generated by the nonprincipal height 1 primes $Q_i = h_i^{p^m} A_m + GA_m$, $i = 1, \ldots, r$.

Proof: Let $h = h_1 \cdots h_r$ and $R = k[x_0, x_1, \ldots, x_n, z]$, $z^{p^m} = G$, which is the coordinate ring of F_m. By (1.4) we have an exact sequence

$$0 \to H \to C\ell(R) \to C\ell\left(R\left[\frac{1}{h}\right]\right) \to 0 \qquad (10.40.1)$$

where H is the subgroup of $C\ell(R)$ generated by those nonprincipal height 1 primes in R that contain some power of h.

We have that $R[1/h] \cong k[z^{p^m}/h, x_1, \ldots, x_n, z, 1/h] = k[x_1, \ldots, x_n, z, 1/h]$. Since $k[x_1, \ldots, x_n, z]$ is factorial, $k[x_1, \ldots, x_n, 1/h]$ is also, again by (1.4). Therefore, $C\ell(R[1/h]) = 0$. From (10.40.1) we have that $C\ell(R)$ is isomorphic to H. It follows that $C\ell(A_m)$ is generated by those height 1 primes in A_m containing some power of h. Let $Q \subset A_m$ be one such prime. Then there is a unique principal height 1 prime, $fk[x_1, \ldots, x_n]$ say, in $k[x_1, \ldots, x_n]$ that lies over Q. f must divide h, thus f must be a k-multiple of h_i for some $i = 1, \ldots, r$. Thus $Q = Q_i$ for some $i = 1, \ldots, r$, and $C\ell(A_m)$ is generated by these Q_i.

By (10.36), $p^m Q_i = 0$ in $C\ell(A_m)$ for $i = 1, \ldots, r$. We now show that the Q_i are $\mathbb{Z}/p^m\mathbb{Z}$-independent. The $m = 1$ case is covered by (10.21). We proceed by induction on m. We will be done if we show that the elements $p^{m-1}Q_i$ are independent over $\mathbb{Z}/p\mathbb{Z}$.

Let $P_i' = Q_i \cap B_m = h_i^{p^m} B_m + G^p B_m$. The ramification index of R_i over P_i' is p for $i = 1, \ldots, r$. By induction $\{p^{m-2}P_1, \ldots, p^{m-2}P_r\}$ are $\mathbb{Z}/p\mathbb{Z}$-independent in $C\ell(B_m)$. Since $\theta_{m-1}(p^{m-2}P_i) = p^{m-1}Q_i$ for $i = 1, \ldots, r$ and θ_{m-1} is an injection, the elements $p^{m-1}Q_i$ are independent over $\mathbb{Z}/p\mathbb{Z}$. The result follows.

COROLLARY 10.41 The divisor class group of the hypersurface $F_m \subseteq \mathbb{A}^{np+1}$ defined by the equation $z^{p^m} = x_1 x_2 \cdots x_{np}$ is a direct sum of $np - 1$ copies of $\mathbb{Z}/p^m\mathbb{Z}$.

Proof: Use (10.40).

REMARK 10.42 Note that if we let Q be the origin of the surface F_m in (10.39), then the divisor class group of the local ring of F_m at Q, $C\ell((F_m)_Q)$, and $C\ell(F_m)$ are isomorphic by (10.8). Of course, the same holds true of the surfaces in (10.40) and (10.41).

We come to the last topic to be discussed in this chapter. We show that if A is an integrally closed domain such that $k[x_1^{p^m},\ldots,x_n^{p^m}] \subsetneq A \subseteq k[x_1,\ldots,x_n]$, where k is an algebraically closed field of characteristic $p > 0$, then $C\ell(A)$ is a finite p-group of type $(p^{i_1},p^{i_2},\ldots,p^{i_r})$ with each $i_j \le m$. We will use the following fact, to be found in Jacobson (1964, p. 185, exercise 3).

LEMMA 10.43 Let P and L be fields such that P is purely inseparable of exponent one over L and $[P:L] = p^m < \infty$. Then there exists a derivation D of P/L such that $D^{-1}(0) = L$.

LEMMA 10.44 Let k be an algebraically closed field of characteristic $p > 0$. Let B be an integrally closed finitely generated k-subalgebra of $k[x_1,\ldots,x_n]$ and D a qt(B)/k derivation such that $[qt(B): D^{-1}(0)] = p$. Let $A = D^{-1}(0) \cap B$. Then the homomorphism ϕ: $C\ell(A) \to C\ell(B)$ of (1.3) has kernel of finite order and type $(p,\ldots,p)$.

Proof: There exist $f_1, f_2, \ldots, f_r \in k[x_1,\ldots,x_n]$ such that $B = k[f_1,\ldots,f_r]$. We can ensure by multiplying D by an appropriate element of B that $D(B) \subset B$. Let $L \subsetneq B$ be the group of logarithmic derivatives of D. That is, $L = \{t^{-1}Dt \mid t \in qt(B) \text{ and } t^{-1}Dt \in B\}$. By (1.8a) there exists an injection $\ker \phi \to L$. Let $d = \max\{\deg(Df_i) - \deg(f_i)\}$. If $h \in L$, then $\exists\ t \in qt(B)$ such that $t^{-1}Dt = h$. We can assume $t \in B$, for we can multiply t by a pth power of an element in B to arrange this.

We have that $\deg(Dt) \le \deg t + d$, which implies that $\deg h \le d$.

Thus L is contained in the k-vector space of polynomials of degree $\le d$, which has dimension $< \infty$.

10.44.1 If $h_1, \ldots, h_s$ are in L and are independent over $\mathbb{Z}/p\mathbb{Z}$, then $h_1, \ldots, h_s$ are independent over k.

We prove (10.44.1) by induction on s. The case s = 1 is obvious. Suppose that $\alpha_1 h_1 + \alpha_2 h_2 + \cdots + \alpha_s h_s = 0$ with $\alpha_i \in k$ for each i and $\{h_1, \ldots, h_s\}$ independent over $\mathbf{Z}/p\mathbf{Z}$.

By (1.9), $\exists$ $a \in A$ such that $D^p = aD$. We also have by (1.9) that

$$\sum_{i=1}^{s} \alpha_i h_i^p = -\sum_{i=1}^{s} (D^{p-1} - aI)\alpha_i h = -(D^{p-1} - aI) \sum_{i=1}^{s} \alpha_i h = 0 \tag{10.44.2}$$

where I is the identity map. Thus $\Sigma_{i=1}^{s} (\alpha_i)^{1/p} h_i = 0$.

Suppose that $\alpha_s \neq 0$. Then

$$(\alpha_s)^{1/p} \sum_{i=1}^{s} \alpha_i h_i - \alpha_s \sum_{i=1}^{s} (\alpha_i)^{1/p} h_i$$

$$= \sum_{i=1}^{s-1} [(\alpha_s)^{1/p}\alpha_i - \alpha_s(\alpha_i)^{1/p}]h_i = 0$$

By induction, $(\alpha_s)^{1/p}\alpha_i - \alpha_s(\alpha_i)^{1/p} = 0$ for $1 \leq i \leq s - 1$. This implies that $(\alpha_i/\alpha_s)^p = \alpha_i/\alpha_s$ and $\alpha_i/\alpha_s \in \mathbf{Z}/p\mathbf{Z}$ for each i.

Hence $\Sigma\, \alpha_i h_i = 0$ implies that $\Sigma_{i=1}^{s} (\alpha_i/\alpha_s)h_i = 0$, which contradicts the fact that the h_i are $\mathbf{Z}/p\mathbf{Z}$-independent. We conclude that α_s and hence all $\alpha_i = 0$.

Since $L \subset B$ each element of L has p-torsion. By (10.44.1), L has no more than a finite number of independent elements. The desired conclusion follows.

PROPOSITION 10.45 Let k be an algebraically closed field of characteristic p > 0 and A be an integrally closed domain such that $k[x_1^{p^m}, \ldots, x_n^{p^m}] \subsetneq A \subseteq k[x_1, \ldots, x_n]$. Then $C\ell(A)$ is a finite p-group of type $(p^{i_1}, \ldots, p^{i_r})$ with each $i_j \leq mn - 1$.

Proof: $k(x_1, \ldots, x_n)$ is a purely inseparable extension of qt(A) of finite degree p^s where $s \leq nm - 1$. There exists fields $k(x_1, \ldots, x_n) = L_0 \supset L_1 \supset \cdots \supset L_s = qt(A)$, with L_1/L_{i+1} a purely inseparable extension of degree p.

For each $i = 0, 1, \ldots, s$, let $A_i = k[x_1,\ldots,x_n] \cap L_i$. Then $A_s = A$ and each A_i is a finite $k[x_1^{p^m},\ldots,x_n^{p^m}]$-module. Thus each A_i is noetherian and a Krull domain (A_i is an intersection of Krull domains). Hence each A_i is integrally closed (see Samuel, 1964a, p. 5).

By (10.43), there exists derivations D_i: $L_i \to L_i$ such that $D_i^{-1}(0) = L_{i+1}$. By (10.44) the homomorphisms ϕ_i: $C\ell(A_{i+1}) \to C\ell(A_i)$ has kernel of finite order and of type $(p,p,\ldots,p)$. Inductively, we see that each A_i has class group of finite order and of type $(p^{r_1},\ldots,p^{r_q})$, where each $r_i \le i$.

REFERENCES

Atiyah, M., and I. Macdonald, *Introduction to Commutative Algebra*, Addison-Wesley, Reading, Mass., 1969.

Baba, K., On p-radical descent of higher exponent, to appear in *Osaka J. Math.*, 18 (1981).

Blass, P., Zariski surfaces, to appear in *Diss. Math.* Vol. 200 (1984).

Blass, P., Some geometric applications of a differential equation in characteristic p > 0 to the theory of algebraic surfaces, *Contemporary Mathematics*, AMS 13 (1982) (with the cooperation of James Sturnfield and Jeffrey Lang).

Fossum, T., *The Divisor Class Group of a Krull Domain*, Springer-Verlag, New York, 1973.

Fujita, T., On the Zariski problem, *Proc. Japan Acad.*, 55A, No. 3 (1979).

Ganong, R., Plane Frobenius sandwiches, *Proc. Amer. Math. Soc.*, to appear.

Ganong, R., On plane curves with one place at infinity, *J. Reine Angew. Math.*, 307 (1979), 173-193.

Gould, H. W., *Combinatorial Identities*, Morgantown, W.Va., 1959-1960.

Griffiths, P., and J. Harris, *Principles of Algebraic Geometry*, Wiley, New York, 1978.

Grothendieck, A., *E. G. A., II*, IHES Publ. Math. No. 8, 1961.

Hallier, N., Quelques propriétés arithmétiques des dérivations, *C. R. Acad. Sci.*, 258 (1964), 6041-6044.

Hallier, N., Étude des dérivations de certains corps, *C. R. Acad. Sci.*, 261 (1965b), 3716-3718.

Hallier, N., Utilisation des groupes de cohomologie, *C. R. Acad. Sci.*, 261 (1965b), 3922-3924.

Hallier, N., Quelques propriétés d'une dérivation particulière, *C. R. Acad. Sci.*, 262 (1966), 553-556.

Jacobson, N., *Lectures in Abstract Algebra*, Vol. III, *Theory of Fields and Galois Theory*, Van Nostrand, New York, 1964.

Lang, J. An example related to the affine theorem of Castelnuovo, *Mich. Math. J.*, 28 (1981).

Lang, J., The divisor class group of the surface $z^{p^n} = G(x,y)$ over fields of characteristic $p > 0$, *J. Algebra*, 84, No. 2 (1983).

Lipman, J., Desingularization of two-dimensional schemes, *Ann. Math.*, 107 (1978), 151-207.

Lipman, J., *Rational Singularities*, IHES Publications, No. 36, 1969, 195-279.

Matsumura, H., *Commutative Algebra*, New York, W. A. Benjamin, Menlo Park, Calif., 1970.

Miyanishi, M., Regular subrings of a polynomial ring, *Osaka J. Math.*, 17 (1980).

Miyanishi, M., and P. Russell, Purely inseparable coverings of the affine plane of exponent one, to appear in *J. Pure Appl. Algebra*, 28 (1983).

Miyanishi, M., and T. Sugie, Affine surfaces containing cylinderlike open sets, *J. Math. Kyoto Univ.*, 20 (1980).

Russell, P., Hamburger-Noether expansion and approximate roots, to appear in *Manuscripta Math.*, 31 (1980).

Russell, P., Affine ruled surfaces, to appear in *Math. Ann.*, 255 (1981)

Samuel, P., Lectures on Unique Factorization Domains, Tata Lecture Notes, 1964a.

Samuel, P., Classes de diviseurs et derivées logarithmiques, *Topology*, 3 (1964b), 81-96.

Singh, B., On a conjecture of Samuel, *Math. Z.*, 105 (1968), 157-158.

Yuan, S., On logarithmic derivatives, *Bull. Soc. Math. France*, 96 (1968), 41-52.

Zariski, O., On Castelnuovo's criterion of rationality $p_a = p_g = 0$ of an algebraic surface, *Ill. J. Math.*, 2, No. 3 (1958), 303.

APPENDIX: LOCALLY FACTORIAL GENERIC ZARISKI SURFACES ARE FACTORIAL†

In this appendix we study normal affine surfaces in $\mathbf{A}_k^3$ defined by equations of the form $z^p = G(x,y)$, where the ground field k is

---

†By Jeffrey Lang and Piotr Blass; reprinted from Journal of Algebra, Vol. 106 (1987).

assumed to be algebraically closed of characteristics $p \neq 0$. If G is such that G_x, G_y, and $G_{xx}G_{yy} - G_{xy}^2$ have no points in common, such surfaces are called Zariski surfaces [a title given to them by P. Blass in Blass (1983)]. If we also assume that $z^p = G$ has the maximum possible number of singularities [$(\deg G - 1)^2$ if p does not divide deg G or $(\deg G)^2 - 3(\deg G) + 3$ otherwise], then we call such a surface a generic Zariski surface because $z^p = G$ is of this type for a generic choice of G. In this appendix we develop a technique for calculating the divisor class group of Zariski surfaces at a singular point. We then employ this technique to show that locally factorial generic Zariski surfaces are factorial.

0. Notation

k algebraically closed field of characteristic $p \neq 0$

$\mathbf{A}_k^3$ affine 3-space over k

Surface irreducible, reduced, two-dimensional, quasi-projective variety over k

The notation F: $f(x,y,z) = 0$ means

$$F = \operatorname{Spec} \frac{k[x,y,z]}{f(x,y,z)} : F \subsetneq \mathbf{A}_k^3$$

If A is a Krull ring, we denote by $C\ell(F)$ the divisor class group of A.

If F is a surface, we denote by $C\ell(F)$ the divisor class group of the coordinate ring of F.

For $f \in k[x,y,z]$:

deg f the total degree of f

$\deg_x f$ the degree of f in x

$\deg_y f$ the degree of f in y

1. Preliminaries

The following results, (1.1)-(1.6) can be found in P. Samuel's 1964 Tata notes (Samuel, 1964a). For the definition of a Krull ring and the divisor class group of a Krull ring, the reader is referred either to Samuel (1964a) or Bourbaki (1972, chap. VII, sec. I).

The rings that are studied in this appendix are coordinate rings or localizations of coordinate rings of normal affine surfaces (hence are noetherian integrally closed domains) and are thus Krull rings. Let $A \subset B$ be rings. Let $p \subset A$ and $q \subset B$ be prime ideals. We write q/p if $q \quad A = p$ and we say that q lies over p.

THEOREM 1.1 Let $A \subset B$ be Krull rings. Suppose that either B is integral over A or B is A flat. Then there is a well-defined group homomorphism φ: $C\ell(A) \to C\ell(B)$ (see Samuel, 1964a, pp. 19-20).

The homomorphism of theorem (1.1) is defined in the following manner. If q and p are height 1 primes of B and A with $q|p$, we let $e(q:p)$ be the ramification index of q over p. Then for each height 1 prime p of A we define $\varphi(p) = \Sigma_{q/p}\, e(q:p)q$, the sum taken over all height 1 primes in B lying over p. The sum is always finite since B is a Krull ring. We then extend φ by linearity.

Let B be a Krull ring of characteristic $p \neq 0$. Let Δ be a derivation of qt(B) such that $\Delta(B) \subset B$. Let $K = \ker \Delta$ and $A = B \cap K$. Then A is a Krull ring with B integral over A. Thus we have a map φ: $C\ell(A) \to C\ell(B)$. Set $L = \{t^{-1}\Delta t : t \in qt(B) \text{ and } t^{-1}\Delta t \in B\}$. Set $L^1 = \{u^{-1}\Delta u\text{: } u \text{ is a unit in } B\}$. Then L^1 is a subgroup of L.

THEOREM 1.2 (a) There exists a canonical monomorphism $\bar{\varphi}$: $\ker \varphi \to L/L^1$. (b) If $[qt(B):K] = p$ and $\Delta(B)$ is not contained in any height 1 prime of B, then $\bar{\varphi}$ is an isomorphism (Samuel, 1964a, p. 62).

THEOREM 1.3 (a) If $[qt(B):K] = p$, there exists an $a \in A$ such that $\Delta^p = a\Delta$. (b) an element $t \in B$ is in L if and only if $\Delta^{p-1}t - at = -t^p$ (Samuel, 1964a, pp. 63-64).

REMARK 1.4 $\bar{\varphi}$: $\ker \varphi \to L/L^1$ is described as follows. If $\beta \in \ker \varphi$, then $\varphi(\beta) = tB$ for some $t \in qt(B)$. $\bar{\varphi}$ takes β to $t^{-1}\Delta t$.

THEOREM 1.5 Let A be a Krull ring and S a multiplicatively closed subset in A. Then $S^{-1}A$ is an A-flat Krull ring and ϕ: $C\ell\ A \to C\ell\ S^{-1}A$ is surjective where the $\ker \phi$ is generated by the divisor classes of height 1 primes that intersect S (Samuel, 1964a, p. 21).

THEOREM 1.6 Let A be a noetherian ring and m an ideal contained in the Jacobson radical of A. Let A have the m-adic topology and $\hat{A}$ be the completion of A. Then if $\hat{A}$ is a Krull ring, so is A. Also ϕ: $C\ell(A) \to C\ell(A)$ is an injection (Samuel, 1964a, p. 23).

REMARK 1.7 By the jacobian criterion F: $z^p = G(x,y)$ is normal if and only if G satisfies the condition that G_x and G_y have no common factors. Hereafter we will always assume that G satisfies this condition.

DEFINITION 1.8 Let D: $k(x,y) \to k(x,y)$ be the k-derivation defined by $D = G_y(\partial/\partial x) - G_x(\partial/\partial y)$. D is called the *jacobian derivation* of G.

LEMMA 1.9 Let L be the group of logarithmic derivatives of D in $k[x,y]$. Then if $t \in L$, $\deg t \le \deg G - 2$ (Lang, 1983a, p. 394).

THEOREM 1.10 Let $A = [x^p, y^p, G]$ and be the coordinate ring of F: $z^p = G$. Then (a) $A = D^{-1}(0) \cap k[x,y]$; (b) $\mathcal{O} \simeq A$; and (c) $C\ell(A) \simeq L = \{f^{-1}Df\colon\ f \in k(x,y) \text{ and } f^{-1}Df \in k[x,y]\}$, the group of logarithmic derivatives of D in $k[x,y]$ (Lang, 1983a, pp. 393-394).

THEOREM 1.11 Let D be the jacobian derivation of G. Then $\forall\ \alpha \in k(x,y)$, $D^{p-1}\alpha - a\alpha = -\Sigma_{i=0}^{p-1} \nabla G^i\ (G^{p-i-1})$, where $\nabla = \partial^{2p-2}/\partial x^{p-1}\partial y^{p-1}$, and $D^p = aD$ (lang, 1983a, p. 395)

THEOREM 1.12 Let Q be a singularity of the surface F: $z^p = G$. Then $C\ell(\mathcal{O}_Q) \simeq L/L_Q$, where L is the group of logarithmic derivatives of D in $k[x,y]$ and $L_Q = \{f^{-1}Df\colon\ f^{-1}Df \in k[x,y],\ f \in k(x,y)$, f is defined at Q and $f(Q) \ne 0$.

Proof: By (1.5) the sequence $0 \to \ker\phi \to C\ell(A) \to C\ell(S^{-1}A) \to 0$ is exact. Let (p) be the divisor class of a height 1 prime $\underline{p}$ such that $\underline{p} \cap S \ne \emptyset$, where S is the complement of the maximal ideal M_Q of $A = k[x^p, y^p, G]$ corresponding to the singularity Q of F. Let $f \in k[x,y]$ be the unique height 1 prime of $k[x,y]$ lying over $\underline{p}$. $\underline{p} \cap S \ne \emptyset$ implies that $f(Q) \ne 0$. By (1.4), $f^{-1}Df \in L_Q$. Therefore, the restriction of the isomorphism $C\ell(A) \xrightarrow{\simeq} L$ of (1.10) to $\ker\phi$ is an injection of ker ϕ into L_Q. It is easy to see that the restriction is surjective as well. Thus we have a commutative diagram of exact sequences:

$$\begin{array}{ccccccccc} 0 & \longrightarrow & L_Q & \longrightarrow & L & & & & \\ & & \uparrow\cong & & \uparrow\cong & & & & \\ 0 & \longrightarrow & \ker\phi & \longrightarrow & Cl(A) & \longrightarrow & Cl(S^{-1}A) & \longrightarrow & 0 \end{array}$$

We conclude that $C\ell(\mathcal{O}_Q) \simeq C\ell(S^{-1}A) \simeq L/L_Q$.

REMARK 1.13 P. Blass (1983) showed that if G_x, G_y, and $W = G_{xx}G_{yy} - G_{xy}^2$ have no points in common, then F: $z^p = G$ has only isolated double-point singularities with local equation of the form $z^p = xy +$ (higher-degree terms). Hereafter we will refer to this condition on G as *condition* (B). It can also be shown that for a generic G, the surface G: $z^p = G$ has $(g - 1)^2$ distinct singularities if p does not divide G and $g^2 - 3g + 3$ distinct singularities if p divides G. When we assume that both of these conditions are satisfied [i.e., (B) and F have the maximum possible number of singularities], we will call F a *generic Zariski surface*.

THEOREM 1.14 $C\ell(\mathcal{O}_Q) \simeq 0$ or $\mathbb{Z}/p\mathbb{Z}$ if G satisfies condition (B) (see Lang, 1983b, p. 632).

REMARK 1.15 Lang (1983a) described a technique for computing the divisor class group of $\mathcal{O}$. We now extend this method to provide a computational technique for determining $C\ell(\mathcal{O}_Q)$. First, we determine L, the group of logarithmic derivatives of D in k[x,y]. Then we evaluate each of these at Q; if we obtain only zeros, then $C\ell(\mathcal{O}_Q) = 0$; otherwise, $C\ell(\mathcal{O}_Q) \simeq \mathbb{Z}/p\mathbb{Z}$. We prove this result in several steps.

STEP 1: LEMMA 1.16 Let G satisfy condition (B), D be the jacobian derivation of G, and $a = -\Sigma_{i=0}^{p-1} G^i\nabla(G^{p-i-1})$, where $\nabla = \partial^{2p-2}/\partial x^{p-1} \times \partial y^{p-1}$. Then (a) $D^p = aD$ and (b) $a(Q) \neq 0$ at each singularity Q of F.
Proof: In (1.11) if we let $\alpha = 1$ we obtain the desired formula for (a). To prove (b) we can assume after a linear change of coordinates that Q = (0,0) and F has the form $z^p = xy +$ (higher-degree terms),

by (1.11). Then $a(Q) = a(0,0) = \nabla(G^{p-1})(0,0) = -[1 + (\text{higher-degree terms})](0,0) = -1$.

DEFINITION 1.17 For each singularity Q of F, let H_Q be the additive subgroup of k consisting of the roots of the polynomial $h(t) = t^p - a(Q)t$. Note that $H_Q \approx \mathbb{Z}/p\mathbb{Z}$ by (1.16).

STEP 2: LEMMA 1.18 For each singularity Q of F, $t \in L$ implies that $t(Q) \in H_Q$.

Proof: If Q is a singularity of F and $h \in k[x,y]$, then $(Dh)(Q) = h_x(Q)G_y(Q) - h_y(Q)G_x(Q) = 0$. If $t \in L$, $D^{p-1}t - at = -t^p$ by (1.4). If we evaluate this expression at Q, we have $a(Q)t(Q) = (t(Q))^p$. Therefore, $t(Q) \in H_Q$.

THEOREM 1.18 Let G satisfy condition (B). Then $C\ell(\mathcal{O}_Q) \cong \mathbb{Z}/p\mathbb{Z}$ if and only if the map L to H_Q of (1.18) is not the zero mapping.

Proof: Again we may assume that Q = (0,0) and $z^p = xy + $ (higher-degree terms). We have the following commutative diagram by (1.5), (1.6), (1.10), and (1.12), where $\hat{L} = \{Df/f \in k[x,y]\colon\ f \subsetneq 1)x,u)\}$ and $\hat{L}^1 = \{Df/f\colon\ f \text{ is a unit in } k[x,y]\}$:

$$\begin{array}{ccccc}
\mathrm{Cl}(\mathcal{O}) & \xrightarrow{\text{surjection}} & \mathrm{Cl}(\mathcal{O}_Q) & \xrightarrow{\text{injection}} & \mathrm{Cl}(\mathcal{O}_Q) \simeq \mathbb{Z}/p\mathbb{Z} \\
\downarrow \cong & & \downarrow \cong & & \downarrow \cong \\
L & \xrightarrow{\text{surjection}} & L/L_Q & \xrightarrow{\text{injection}} & \widetilde{L}/\widetilde{L}^1 \simeq \mathbb{Z}/p\mathbb{Z}
\end{array}$$

We showed (Lang, 1983b, p. 632) that G can be factored in k[x,y] into factors u and v, where u = x + (higher-degree terms) and v = y + (higher-degree terms). We also showed that Du/u generates

$$\frac{L}{L^1}\ \frac{Du}{u} = \frac{U_xG_y - U_yG_x}{u} = \frac{x + (\text{higher-degree terms})}{x + (\text{higher-degree terms})}\ \frac{1}{2}$$

This implies that Du/u = 1 + (higher-degree terms) in k[x,y].

(⇒) If $C\ell(\mathcal{O}_Q) \cong \mathbb{Z}/p\mathbb{Z}$, then from the commutative diagram we have that $L/L_Q \approx \tilde{L}/\tilde{L}^1$. Therefore, there is a $t \in L$ such that $Du/u - t \in \tilde{L}^1$. If h is a unit in k[x,y], then $Du(0,0)/h = 0$. Therefore,

$Du(0,0)/u - t(0,0) = 0$. This implies that $t(0,0) = 1$ and the map from L to H_Q is not trivial.

($\Leftarrow$) If $t \in L$ and $t(0,0) \neq 0$, then $t \in L_Q$. Therefore, $C\ell(\mathcal{O}_Q) = \mathbb{Z}/p\mathbb{Z}$. This proves the theorem.

THEOREM 1.19 Let G satisfy condition (B). If F is locally factorial, then F is factorial.

Proof (by contradiction): Suppose that F is not factorial. Then by (1.10) there is a nonzero $t \in L$. Let $t = S_1S_2 \cdots S_n$ be a factorization of t into prime factors. Each S_i is relatively prime to either G_x or G_y since G_x and G_y have no common factors. We may assume that $S_1, S_2, \ldots, S_r$ are prime to G_x and that $S_{r+1}, S_{r+2}, \ldots, S_n$ are prime to G_y. Then for each $i = 1, 2, \ldots, r$, S_i and G_x intersect in at most $(\deg S_i)r(g - 1)$ distinct points, where $g = \deg G$, by Bezout's theorem. For $i = r + 1, \ldots, n$, S_i and G_y have at most $(\deg S_i)(g - 1)$ points of intersection. The total numbers of these intersection points are at most $\Sigma_{i=1}^n (g - 1) \deg S_i = (g - 1) \times \Sigma_{i=1}^n \deg S_i = (g - 1) \deg t \leq (g - 1)(g - 2)$ by (1.9). Since G_x and G_y intersect in at least $g^2 - 3g + 3$ distinct points [see (1.13)], we can choose Q, a singularity of F that does not satisfy any of the S_i's. Then $t(Q) \neq 0$. By (1.18), $C\ell(\mathcal{O}_Q) \neq 0$ and F is not locally factorial.

In the proof of (1.19) we proved the following facts.

LEMMA 1.20 If $t \in L$, then t is not zero at at least $(g - 1)^2 - (\deg t)(g - 1)$ singularities if g is not divisible by p, $(g^2 - 3g + 3) - (\deg t)(g - 1)$ singularities if g is divisible by p.

THEOREM 1.21 Let n(F) denote the number of singularities of F. The map $L \to \oplus_{Q \in \mathrm{Sing}(F)} \mathcal{O}_Q$ defined by $t \to (t(Q_1), t(Q_2), \ldots, t(Q_{n(F)}))$ is an injection.

EXAMPLE 1.22 The surface $z^p = xy + x^{p+1} + y^{p+1}$ was studied in Lang (1983a) and shown to have class group of order p^{2p-1}. Since $t = 1 \in L$, we have that $C\ell(\mathcal{O}_Q) \cong \mathbb{Z}/p\mathbb{Z}$ at each singularity (Lang, 1983a, p. 401).

EXAMPLE 1.23 The surface $z^5 = xy + x^3 + y^3 + x^2y^2$ has class group of order p^3. L is generated by $1 + xy$, $x + y$, and $x + \ y$, where is a primitive ninth root of unity. Since these three logarithmic derivatives have no points in common, $C\ell(\mathcal{O}_Q) = \mathbb{Z}/p\mathbb{Z}$ at each singularity.

REMARK 1.24 We have not as yet been able to find an example of a generic Zariski surface F: $z^p = G$ that has nontrivial class group but is factorial at a singularity.

REFERENCES

Blass, P., Some geometric applications of a differential equation in characteristics $p > 0$ to the theory of algebraic surfaces, *Contemporary Mathematics*, AMS 13 (1982) (with the cooperation of James Sturnfield and Jeffrey Lang).

Blass, P., Zariski surfaces, *Diss. Math.*, 200 (1983).

Blass, P., Picard groups of Zariski surfaces I, *Compositio Math.*, 54 (1985), 3-36.

Bourbaki, N., *Commutative Algebra*, Addison-Wesley, Reading, Mass., 1972.

Fossum, R., *The Divisor Class Group of a Krull Domain*, Springer-Verlag, New York, 1973.

Lang, J., An example related to the affine theorem of Castelnuovo, *Mich. Math. J.*, 28 (1981).

Lang, J., The divisor classes of the hypersurface $z^{p^n} = G(x_1,\ldots,x_n)$ in characteristics $p > 0$, *Trans. Am. Math. Soc.*, 279, No. 2 (1983a).

Lang, J., The divisor class group of the surface $z^{p^n} = G(x,y)$ over Fields of characteristics $p > 0$, *J. Algebra*, 84, No. 2 (1983b).

Lang, J., The divisor classes of the $z^p = G(x,y)$ a programmable problem, to appear in *J. Algebra*, *100*, No. 2 (1986).

Samuel, P., Lectures on Unique Factorization Domains, Tata Lecture Notes, 1964a.

Samuel, P., Classes de diviseurs et derivees logarithmiques, *Topology*, 3 (1964b), 81-96.

4

Picard Groups of Generic Zariski Surfaces†

INTRODUCTION

This chapter originated in the following philosophical framework. Inspired by M. Noether's theorem (Hartshorne, 1975) and facts about generic double planes over C (Steenbrink, 1982) as well as by Šafarevič's computations in characteristic 2 (Rudakov and Šafarevič, 1979), we were led to suspect that the ring $k[x,y,z]/(z^p - f(x,y))$ is a UFD provided that f is a sufficiently general polynomial of degree m. Here $k = \bar{k}$, characteristic $k = p \geq 5$, $m \geq 5$, say. The conjecture above still remains open, but here we offer a result that in our opinion renders it extremely probable, to say the least, in the case when p divides m.

We consider in the paper a generic polynomial with indeterminate coefficients.

$F = \Sigma\, T_{ij}X^iY^j$, where $0 \leq i + j \leq p$, and the ring $R = L[X,Y,Z]/(Z^p - F)$, $L = \overline{k(T_{ij})}$, where T_{ij} are algebraically independent transcendentals over k. deg F = p. We show that R is a UFD. This is our main theorem (MTH) in its algebraic form. In a subsequent paper we hope to prove our original conjecture and also to generalize the result of this paper to degrees of $f \neq p$.

The proof given here follows very closely an outline shown to the author by Deligne. It is conceptually akin to SGA VII, exp X|X.

---

†By Piotr Blass; most of this chapter is reprinted from *Compositio Mathematica*, 54 (1985), 3-36. 1985 Martinus Hijhoff Publishers, Dordrecht, The Netherlands.

(Deligne and Katz, 1973). A new feature is, for example, DRL + MTH (see Section 5). Let us outline the main idea of the proof, which is simple, if somewhat lengthy in detail.

Let

$$\bar{S} = \text{Proj } L[X,Y,Z,Z_0]/(Z^p - F(X,Y,Z_0))$$

$\bar{S}$ has $N = p^2 - 3p + 3$ rational double points of type A_{p-1}, (Sing $\bar{S}$). The Galois group $G = \text{Gal}(\overline{k(T_{ij})}: k(T_{ij}))$ acts on $\bar{S}$ and it permutes the N elements of Sing $\bar{S}$.

Step 1: Show that G induces the full symmetric group on Sing $\bar{S}$.

Step 2: Refine the above and analyze more closely the action of G on Pic $\tilde{S}$ ($\tilde{S}$ is the minimal desingularization of $\bar{S}$). It turns out that there are elements of G which induce the identity on Sing $\bar{S}$ but nevertheless act nontrivially on certain exceptional curves contained in $\tilde{S}$. See the double reversal lemma (DRL) and Theorem G2' below.

Step 3: Deduce from DRL the fact that Pic $\tilde{S}$ is generated by the obvious curves, namely, the exceptional curves for $\tilde{S} \to S$ and a very ample curve on $\bar{S}$ pulled back to $\tilde{S}$.

Step 3 provides us with the proof of our main theorem in its geometric form (MTHG). The equivalence of the geometric form with the algebraic form (R is a UFD) follows from a simple exact sequence which we had introduced in a previous paper (Blass, 1982); see (0.5.1) below.

Finally, let us remark that J. Lang has proven a number of related results in his Ph.D. thesis (Lang, 1981). His method was to use differential equations in characteristic $p > 0$.

At a crucial point in our proof we use the fact that Pic($\tilde{S}$) has no p-torsion. We had conjectured it, but the first proof is due to W. E. Lang, whose result we quote from Lang (1984).

It is not hard to extend our main theorem to the case where F is replaced by the polynomial not of degree p but rather of degree pe, e = 1, 2, ... We leave this extension as an exercise for the reader. In a subsequent paper we hope, among other things, to examine the transition from "generic" to "general" and to prove our original conjecture at least in the case when $p|m$.

0. NOTATION

0.1 $k = \bar{k}$ is an algebraically closed field of characteristic $p \geq 5$. T_{ij} are indeterminates algebraically independent over k, $0 \leq i + j \leq p$.

$$F(X,Y) = \sum_{0\leq i+j\leq p} T_{ij}X^iY^j$$

Σ stands for $\Sigma_{0\leq i+j\leq p}$ unless stated to the contrary. F_x, F_y means $\partial F/\partial x$, $\partial F/\partial y$, etc.

$$H(F) = \text{hessian of } F = \det\begin{vmatrix} F_{xx} & F_{xy} \\ F_{yx} & F_{yy} \end{vmatrix}$$

$L = \overline{k(T_{ij})}$, the algebraic closure of $k(T_{ij})$
$G = \mathrm{Gal}(\overline{k(T_{ij})}: k(T_{ij}))$
$A = \mathrm{Spec}\ k[T_{ij}]$
$E = \mathrm{Spec}(k[T_{ij}][X,Y]/(F_x,F_y))$
$D = \mathrm{Spec}(k[T_{ij}][X,Y,W]/(F_x,F_y, W^2 - H(F))$

We have the natural morphisms

$$D \xrightarrow{\phi} E \xrightarrow{\pi} A$$

If $X \to A$ is a morphism we denote by E_X the scheme $E \times_A X$ and by π_X the projection π_X: $E_X \to X$.

If $U \subset A$ is open or closed, π_U: $E_U \to U$ has the foregoing meaning with respect to the inclusion map $U \to A$. We apply the same convention to maps $D \to E$, $D \to A$, $X \to E$, $X \to A$.

0.2 We will identify closed points of A with polynomials of degree p in $k[X,Y]$. The following subset of A will be important to us: $V \subset A$ corresponding to polynomials $g \in k[X,Y]$ of degree p such that the surface $Z^p = g$ has no singularities at infinity. It is not hard to prove that V is open and dense in A. This follows from the following fact.

Let us write out $g(X,Y)$ in terms of homogeneous parts:

$$g = g_p + g_{p-1} + \cdots + g_1 + g_0$$

Then $g \in V$ if and only if the system of equations $g_{p-1} = 0$, $\partial g_p/\partial x = 0$, $\partial g_p/\partial y = 0$ has only the solution $x = 0$, $y = 0$. Further, we define a subset $U \subset V$ as follows: $g \in U$ if and only if $g \in V$ and g has only nondegenerate singularities (i.e., $g_x = g_y = 0$ implies hessian of $g \neq 0$). We will show below that U is open and dense in V [see (3.1.3)].

0.3 In this chapter we will constantly discuss an algebraic surface over L. Let us fix once and for all some notation relative to that surface. $R = L[X,Y,Z]/(Z^p - F(X,Y))$; $S^{affine} = S^{aff} = \mathrm{Spec}(L[X,Y,Z]/(Z^p - F(X,Y)))$, where F is as above ($S^{aff} = \mathrm{Spec}\ R$); $\bar{S} = \mathrm{Proj}(L[X,Y,Z,Z_0]/(Z^p - F(X,Y,Z_0)))$, where $F(X,Y,Z_0)$ is F homogenized [sometimes we will write $F(X,Y,X_0)$]; η: $\tilde{S} \to \bar{S}$ a minimal desingularization of $\bar{S}$ over L; $\mathbf{P}^n_L$ denotes the projective n-space over L.

The following facts can be shown about S^{aff} and $\bar{S}$. There is a natural projection, a finite map:

$$\mu:\ \bar{S} \to \mathbf{P}^2_L \qquad \text{(see Blass, 1982, sec. 1)}$$

We will denote by ℓ the element of Pic $\tilde{S}$ given by

$$\eta^*\mu^*(O_{\mathbf{P}^2_L}(1))$$

$\bar{S}$ has a set of isolated singularities denoted $\mathrm{Sing}(\bar{S})$. One can show that $\mathrm{Sing}(\bar{S}) \subset S^{aff}$ [see lemma (2.4.1)]. Thus $\mathrm{Sing}(S) = \mathrm{Sing}(\bar{S})$. Also, all the singularities are rational double points of type A_{p-1} [see (3.1.14)]. Further, the number of singularities is $p^2 - 3p + 3$ [see (3.1.12)].

The elements of Sing S will be denoted by Greek letters α, β, γ, etc. If we need to write out their coordinates in $\mathbf{A}^3_L$, we will write $\alpha = (\alpha_1,\alpha_2,\alpha_3)$, etc. In particular, we define

$$H_\alpha = (F_{xx}F_{yy} - F^2_{xy})(\alpha_1,\alpha_2)$$

0.4 The following facts about the minimal resolution map $\eta\colon\ \tilde{S} \to \bar{S}$ are well known (see Blass, 1982, sec. 1).

Let $\alpha \in \operatorname{Sing} \bar{S}$, then $\eta^{-1}(\alpha)$ consists of a tree of $p - 1$ rational curves over L with intersection matrix (in Pic $\tilde{S}$)

$$\begin{bmatrix} -2 & 1 & & & \\ 1 & -2 & \ddots & & 0 \\ & 1 & \ddots & \ddots & \\ & & \ddots & \ddots & 1 \\ 0 & & & 1 & -2 \end{bmatrix} \tag{0.4.1}$$

The tree can be pictured as follows:

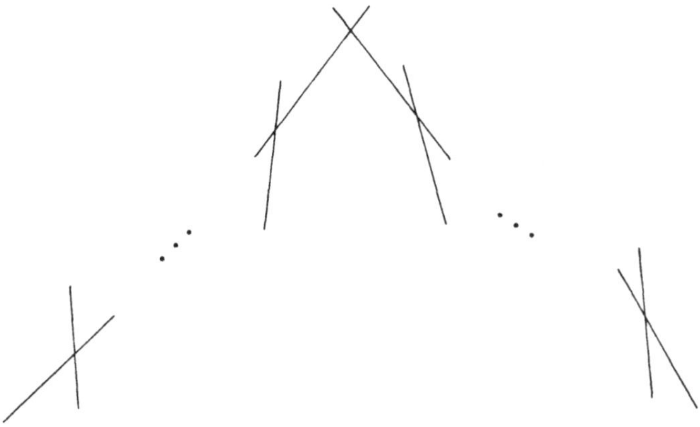

We wish to emphasize the following subtle point, which is of key importance in this paper. There are two possible natural choices of ordering the curves of the tree (from left to right or from right to left). We choose from the outset one orientation for each singularity $\alpha \in \operatorname{Sing} \bar{S}$ and we number the curves accordingly. Thus the tree will now be labeled:

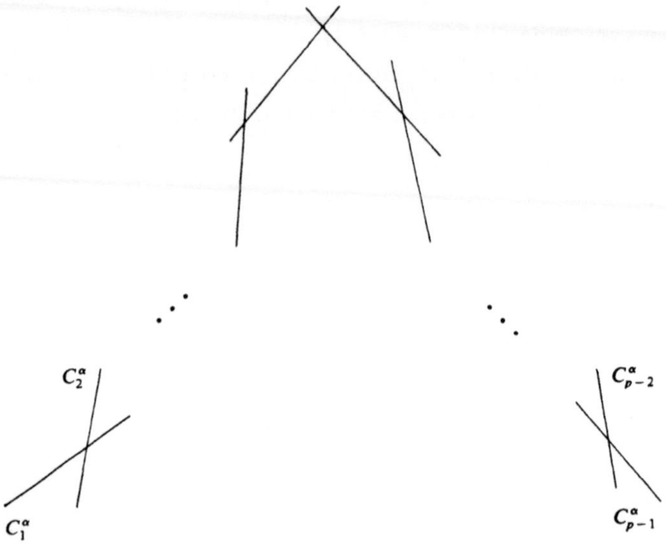

(0.4.2)

T_α stands for the ordered (tree) sequence $C_1^\alpha, C_2^\alpha, \ldots, C_{p-1}^\alpha$. T_α^{opp} stands for $C_{p-1}^\alpha, C_{p-2}^\alpha, \ldots, C_1^\alpha$.

For each $\alpha \in \operatorname{Sing} \bar{S}$ we now define a special element of Pic $\tilde{S}$,

$$D_\alpha = C_1^\alpha + 2C_2^\alpha + \cdots + (p-1)C_{p-1}^\alpha \tag{0.4.3}$$

and

$$D_\alpha^{opp} = C_{p-1}^\alpha + 2C_{p-2}^\alpha + 3C_{p-3}^\alpha + \cdots + (p-1)C_1^\alpha \tag{0.4.4}$$

D_α has the following properties. First, $D_\alpha \cdot C_j^\alpha \equiv 0(p)$ for all j, $1 \le j \le p - 1$. Also

$$D_\alpha^2 = -p(p-1) \tag{0.4.5}$$

$$\begin{aligned} D_\alpha + D_\alpha^{opp} &= pC_\alpha = p(C_1^\alpha + C_2^\alpha + \cdots + C_{p-1}^\alpha) \\ D_\alpha - D_\alpha^{opp} &= 2D_\alpha - pC_\alpha \end{aligned} \tag{0.4.6}$$

0.5 We will denote by

$$\operatorname{Pic}^{\text{obvious}}(\tilde{S}) \qquad \text{or} \qquad \operatorname{Pic}^{\text{ob}}(\tilde{S})$$

the subgroup generated by ℓ and the C_j^{α}, $\alpha \in \mathrm{Sing}(S)$, $1 \le j \le p - 1$. There is an intersection form on Pic $\tilde{S}$; we denote it by $A \cdot B$, $A,B \in$ Pic $\tilde{S}$. It can be shown that $\ell \cdot \ell = \ell^2 = p$. Also, we have the exact sequence

$$0 \to \mathrm{Pic}^{ob}\tilde{S} \to \mathrm{Pic}\ \tilde{S} \to C\ell\ R \to 0 \tag{0.5.1}$$

$C\ell$ R is a finite elementary p group (see Lang, 1981, or Samuel, 1964). It was shown by W. E. Lang that Pic($\tilde{S}$) has no torsion (Lang, 19).

0.6 Let X be a scheme (noetherian), Et(X) the category of finite étale coverings of X.

Let Ω be an algebraically closed field b: Spec $\Omega \to X$, a geometric point of X. Let $Y \in \mathrm{Et}(X)$.

$F_b^X(Y)$ is the set of liftings

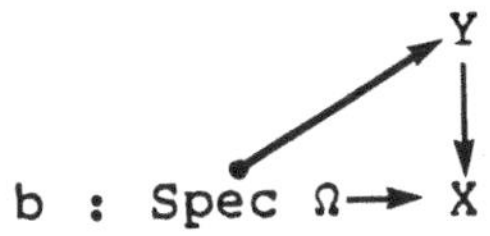

If $W \to X$ is a morphism, we get the base change functor Et X $\to$ Et W denoted R_W or just R.

If X, Y are schemes, we denote by $X \coprod Y$ the disjoint union of X and Y.

0.7 In the following definition the ground field is assumed algebraically closed of characteristic $\neq 2$. A, B are smooth curves, B irreducible, π finite.

DEFINITION 0.8 $A \xrightarrow{\pi} B$ is called *very simple* if there exists a point $\bar{q} \in A$ with $\pi(\bar{q}) = q$ and such that $e(\bar{q}) = 2$ (ramification is 2); further, if $B° = B - \{q\}$, $A° = \pi^{-1}(B°)$, then $A° \to B°$ is finite and étale; also π is étale on $A - \{q\}$.

DEFINITION 0.9 S, T are two finite sets, m: $S \to T$ a two-to-one and onto mapping. We call the following commutative diagram a *double flip*:

$$\begin{array}{ccc} S & \overset{m}{\dashrightarrow} & T \\ {\scriptstyle \sigma_S}\downarrow & & \downarrow{\scriptstyle id_T} \\ S & \overset{m}{\dashrightarrow} & T \end{array}$$

where σ_S is an automorphism of S such that there are two elements $A, B \in T$ with preimages $\{A_1, A_2\}$, $\{B_1, B_2\}$ in S and such that:

$$\sigma_S(A_1) = A_2 \qquad \sigma_S(A_2) = A_1$$

$$\sigma_S(B_1) = B_2 \qquad \sigma_S(B_2) = B_1$$

Also $\sigma_S | S - \{A_1, A_2, B_1, B_2\}$ is the identity. (Intuitively, we think of T as the set of positions of a certain number of coins; each coin has "heads" and "tails," hence the two-to-one map $S \xrightarrow{m} T$. A "double flip" corresponds to turning over exactly two coins without altering the order of the coins.)

1. STATEMENT OF THE MAIN THEOREM AND THE PRINCIPAL THEOREMS AND LEMMAS USED IN ITS PROOF

Our main theorem is this:

MAIN THEOREM (MTH) 1.1 R is a unique factorization domain.

We will restate the theorem and in fact prove it in its geometric form.

MAIN THEOREM (GEOMETRIC FORM) (MTHG) 1.2 $\mathrm{Pic}(\tilde{S}) = \mathrm{Pic}^{ob}(\tilde{S})$.

REMARK Since there is no torsion we will, in fact, prove that $p\,\mathrm{Pic}(\tilde{S}) = p\,\mathrm{Pic}^{ob}(\tilde{S})$.

The equivalence of MTH and MTHG is an immediate consequence of the exact sequence (0.5.1). We will deduce MTHG from the following theorems, to which we give descriptive names G1, G2, G2', and DRL (double reversal lemma).

Here are the statements:

THEOREM G1 1.3 $G = \mathrm{Gal}(k(\overline{T_{ij}}) : k(T_{ij}))$ acts on $\mathrm{Sing}(\bar{S})$ as the full symmetric group.

THEOREM G2 1.4 There exists a $\sigma \in G$ such that σ induces the identity on Sing S but for some pair of singularities $\alpha, \beta \in$ Sing S, $\sigma(\sqrt{H_\alpha}) = -\sqrt{H_\alpha}, \sigma(\sqrt{H_\beta}) = -\sqrt{H_\beta}$ and for all $\beta \neq \gamma \neq \alpha$, $\gamma \in$ Sing S, $\sigma(\sqrt{H_\gamma}) = \sqrt{H_\gamma}$.

THEOREM G2' 1.5 For every pair of singularities $\alpha, \beta \in$ Sing S there exists a $\sigma \in G$ such that σ induces the identity on Sing S but $\sigma(\sqrt{H_\alpha}) = -\sqrt{H_\alpha}, \sigma(\sqrt{H_\beta}) = -\sqrt{H_\beta}$ and $\sigma(\sqrt{H_\gamma}) = \sqrt{H_\gamma}$ for $\beta \neq \gamma \neq \alpha$, $\gamma \in$ Sing S.

DOUBLE REVERSAL LEMMA (DRL) 1.6 For every pair of singularities α, β, $\alpha \neq \beta$ in Sing $\bar{S}$ there is an element $\sigma \in G$ such that the induced isometry $i(\sigma)$ of $\mathrm{Pic}(\tilde{S})$ preserves $\mathrm{Pic}^{ob}(\tilde{S})$ and satisfies: $i(\sigma)(D_\alpha) = D_\alpha^{op}$, $i(\sigma)(D_\beta) = D_\beta^{op}$, $i(\sigma)(\ell) = \ell$, and for all $\gamma \in$ Sing $\bar{S}$ with $\alpha \neq \gamma \neq \beta$ we have $i(\sigma)(D_\gamma) = D_\gamma$.

The logical structure of the proof will be as follows:

```
easy exercise
      |
      v
G1,G2 => G2' => DRL => MTHG => MTH
               ↓       ↓
              show    show
```

The implication G1,G2 => G2' is a simple exercise and will not be described. The implication MTHG => MTH has been indicated above. Thus, our chapter focuses on proving G1, G2 (see Section 4). Further, in showing G2' => DRL, this is sketched at the end of Section 4, and finally we prove the implication DRL => MTHG in Section 5.

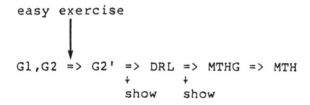

2. PRELIMINARIES

1. Fundamental Group Facts (see SGA I, Grothendieck, 1971)

Let $i\colon\ Y \to X$ be a morphism of locally noetherian connected (regular) schemes. Let $b\colon\ \operatorname{Spec} \Omega \to Y$ be a geometric point of Y. Let us abuse notation and denote by b also the corresponding geometric point of X.

2.1.0 We recall the definition of the induced homomorphism:

$$i_*\colon\ \pi_1(Y,b) \to \pi_1(X,b)$$

$$\begin{array}{ccc} Et(Y) & \overset{R}{\dashleftarrow} & Et(X) \\ \Big\downarrow F_b^Y & & \Big\downarrow F_b^Y \\ ENS & & ENS \end{array}$$

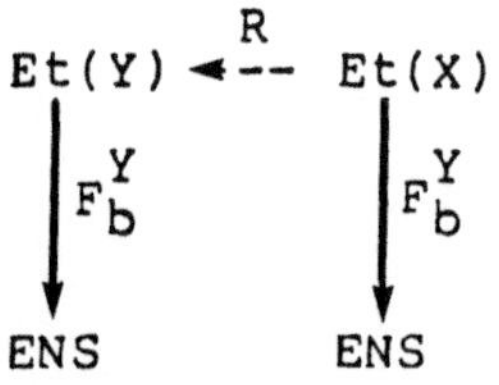

Now

$$\pi_1(Y,b) = \operatorname{Aut}(F_b^Y)$$

$$\pi_1(X,b) = \operatorname{Aut}(F_b^Y)$$

By SGA I (Grothendieck, 1971, p. 142), we have the isomorphism of functors:

$$F_b^Y \circ R \underset{\tau}{\overset{\mu}{\approx}} F_b^X \qquad \text{where } \mu\tau = \mathrm{id}F_b^X, \quad \tau\mu = \mathrm{id}(F_b^Y \circ R)$$

If $\sigma \in \pi_1(Y,b) = \operatorname{Aut}(F_b^Y)$, we define $\bar{\sigma} = i_*(\sigma)$ by the diagram

$$\begin{array}{ccc} F_b^X(W) & \overset{\tau}{\dashrightarrow} & F_b^Y(W_Y) \\ \Big\downarrow \bar{\sigma}_W & & \Big\downarrow \sigma_{W_Y} \\ F_b^X(W) & \overset{\mu}{\dashleftarrow} & F_b^Y(W_Y) \end{array}$$

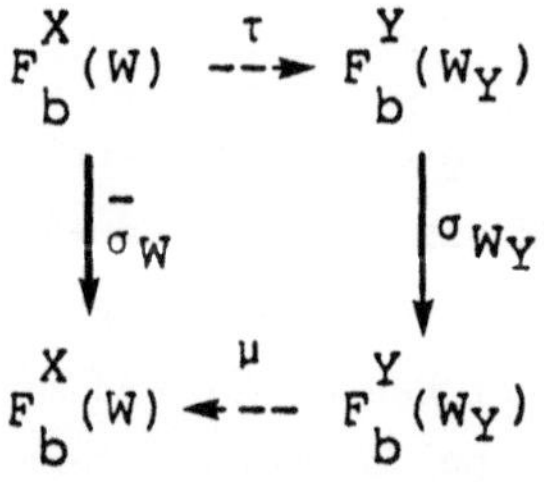

PROPOSITION 2.1.1 If $W \in Et(X)$ is irreducible, then $\pi_1(X,b)$ acts transitively on $F_b^X(W)$ for any base point b in X.

Proof: W is connected, so proposition 2.1.1 follows from SGA I (Grothendieck, 1971, p. 140).

PROPOSITION 2.1.2 Assume $W \in Et(X)$ irreducible and that $R(W) = W_Y$ decomposes $W_Y = s(Y) \sqcup T$, where $s\colon Y \to W_Y$ is a section and T irreducible. Then for any base point $b \in X$ the action of $\pi_1(X,b)$ on $F_b(W)$ is transitive and twice transitive.

Proof: Transitivity was shown above. Also, the statement is independent of base point (by SGA I). So choose the base point to be in Y.

$$\begin{array}{ccc} Et(Y) & \xleftarrow{R} & Et(X) \\ \downarrow{\scriptstyle F_b^X} & & \downarrow{\scriptstyle F_b^Y} \\ ENS & & ENS \end{array}$$

We have $F_b^Y \circ R \approx F_b^Y$ (see Grothendieck, 1971, p. 142).

$$\begin{aligned} (F_b^Y \circ R)(W) &= F_b^Y(W_Y) \\ &= F_b^Y(s(Y)) \sqcup F_b^Y(T) \\ &= \{A\} \sqcup F_b^Y(T) \end{aligned}$$

where $F_b^Y(s(Y))$ is the one-element set $\{A\}$.

Now $\pi_1(Y,b)$ acts on $F_b^Y(W_Y)$, fixes A, and acts transitively on $F_b^Y(T)$ since T is irreducible (see Grothendieck, 1971, p. 140). Now the identification $F_b^Y(W) = F_b^Y(W_Y)$ shows that for some element A_1 of $F_b^Y(X)$, $\pi_1(X,b)$ acts on $F_b^Y(W)$, so that the stabilizer of A_1 acts transitively on the complement of A_1 in $F_b^Y(W)$. This, together with transitivity, established twice transitivity.

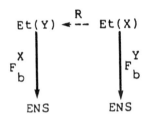

PROPOSITION 2.1.3 Let $W \in Et(X)$, $R(W) = W_Y \in Et(Y)$, and let b be a base point in Y. Suppose that the action of $\pi_1(Y,b)$ on $F_b^Y(W_Y)$ includes a transposition; then so does the action of $\pi_1(X,b)$ on $F_b^Y(W)$. Also, if $\bar{b}$ is any other base point in X not necessarily in Y, the action of $\pi_1(X,\bar{b})$ on $F_{\bar{b}}^X(W)$ also includes a transposition.

Proof: Follows immediately from the definition of the induced homomorphism $\pi_1(Y,b) \to \pi_1(X,b)$ and the fact that the functors F_b^Y, $F_{\bar{b}}^X$ are isomorphic.

PROPOSITION 2.1.4 Let $V,W \in Et(X)$, $m\colon\ V \to W$ an étale covering of degree 2; set

$$V_Y = R(V) \qquad W_Y = R(W)$$

$$m_Y\colon\ V_Y \to W_Y$$

the induced covering. Suppose that there is a $\sigma \in \pi_1(Y,b)$ such that the diagram

$$\begin{array}{ccc} F_b(V_Y) & \dashrightarrow & F_b(W_Y) \\ \downarrow{\scriptstyle \sigma_{V_Y}} & & \downarrow{\scriptstyle \sigma_{W_Y}} \\ F_b(V_Y) & \dashrightarrow & F_b(W_Y) \end{array}$$

is a double flip. Then the diagram

$$\begin{array}{ccc} F_b(V) & \dashrightarrow & F_b(W) \\ \downarrow{\scriptstyle \bar{\sigma}_{V}} & & \downarrow{\scriptstyle \bar{\sigma}_{W}} \\ F_b(V) & \dashrightarrow & F_b(W) \end{array}$$

is also a double flip.

Proof: Follows from the definition of $\bar{\sigma}$ and a diagram chase.

REMARK 2.1.4.1 Under the foregoing assumptions, for any base point $\bar{b}$ in X, the action of $\pi_1(X,\bar{b})$ on $F_{\bar{b}}(V) \to F_{\bar{b}}(W)$ includes a double flip.

Proof: Diagram chase using the fact that F_b and $F_{\bar{b}}$ are isomorphic functors.

2. Preliminaries on Curves

THEOREM C1 2.2.1 Let $A \to B$ be a very simple covering; then the action of $\pi_1(B^\circ, b)$ on $F_b(A^\circ)$ contains a transposition.

THEOREM C2 2.2.2 Let $D \xrightarrow{\phi} A \xrightarrow{\pi} B$ be two very simple coverings. Assume that $\bar{\bar{q}} \in D$, $\bar{q} \in A$, $q \in B$ are such that $e_\phi(\bar{\bar{q}}) = 2$, $e_\pi(\bar{q}) = 2$, $\phi(\bar{\bar{q}}) = \bar{q}$, $\pi(\bar{q}) = q$. Let $D^\circ \xrightarrow{\phi^\circ} A^\circ \xrightarrow{\pi^\circ} B^\circ$ be the induced étale coverings where $B^\circ = B - \{q\}$, $A^\circ = \pi^{-1}(B^\circ)$, $D^\circ = \phi^{-1}(B^\circ)$. Then for any base point b in B° the action of $\pi_1(B^\circ, b)$ on the diagram $F_b(D^\circ) \to F_b(A^\circ)$ includes a double flip.

Proof of theorem C1: We let $f\colon A \to B$, $\bar{q} \in A$, $f(\bar{q}) = q$, $e_f(\bar{q}) = 2$

$B^\circ = B - \{q\}$ $\quad A^\circ = f^{-1}(B^\circ)$

$f^\circ\colon A^\circ \to B^\circ$ the induced étale cover

Let $\mathcal{O}_q$ be the local ring of q in B. Let $\tilde{\mathcal{O}}_q$ be the henselization of $\mathcal{O}_q$. We may take $\tilde{\mathcal{O}}_q \subset \overline{k(B^\circ)}^{sep}$. Consider the cartesian diagram

$$\begin{array}{ccc} \operatorname{Spec} \tilde{\mathcal{O}}_q \times_B A & \dashrightarrow & A \\ \downarrow & & \downarrow \\ \operatorname{Spec} \tilde{\mathcal{O}}_q & \dashrightarrow & B \end{array}$$

If $\tilde{K}$ is the field of fractions of $\tilde{\mathcal{O}}_q$, we get the induced diagram

$$\begin{array}{ccc} \operatorname{Spec} \tilde{K} \times_B A & \dashrightarrow & A^\circ \\ \downarrow & & \downarrow \\ \operatorname{Spec} \tilde{K} & \dashrightarrow & B^\circ \end{array}$$

Now $\operatorname{Spec} \tilde{\mathcal{O}}_q \times B \underset{\text{over } q}{\approx} S \sqcup \operatorname{Spec} \tilde{\mathcal{O}}_q(\sqrt{t})$, where S is a union of sections over $\operatorname{Spec} \tilde{\mathcal{O}}_q$, thus a trivial covering, and t is a uniformizing parameter in $\tilde{\mathcal{O}}_q$. This follows from the elementary theory of henselian rings. Correspondingly, we can write the second diagram as

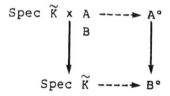

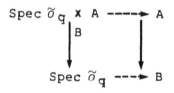

$$\begin{array}{ccc} \tilde{S} \sqcup \operatorname{Spec} \tilde{K}(\sqrt{t}) & \dashrightarrow & A^\circ \\ \downarrow & & \downarrow \\ \operatorname{Spec} \tilde{K} & \dashrightarrow & B^\circ \end{array}$$

where again $\tilde{S}$ is a union of several copies of Spec $\tilde{K}$.

Now we have Spec $\overline{k(B^\circ)} \to \operatorname{Spec} \tilde{K} \to \operatorname{Spec} k(B^\circ) \to B^\circ$. Thus we get a geometric point b of Spec $\tilde{K}$ and the corresponding geometric point b_1 of B°. $\pi_1(\operatorname{Spec} \tilde{K}, b) = \operatorname{Gal}(\overline{k(A^\circ)}:\tilde{K})$ acts on $F_b^{\operatorname{Spec} \tilde{K}}(\tilde{S} \sqcup \operatorname{Spec} \tilde{K}(\sqrt{t}))$ and it is clear that it induces a transposition. Consequently, $\pi_1(B^\circ, b_1)$ induces a transposition in $F_{b_1}^{B^\circ}(A^\circ)$ by our discussion of the fundamental group (2.1.3).

Proof of theorem C2: Choose a henselization

$$\mathcal{O}_q \subset \tilde{\mathcal{O}}_q \subset \overline{k(B^\circ)}$$

and consider the base change:

$$\begin{array}{ccccc} S_1 \sqcup S_2 \sqcup \operatorname{Spec} \tilde{\mathcal{O}}_q(\sqrt[4]{t}) & \approx & \operatorname{Spec} \tilde{\mathcal{O}}_q \times D & \to & D \\ \downarrow & & \downarrow & & \downarrow \\ S \sqcup \operatorname{Spec} \tilde{\mathcal{O}}_q(\sqrt{t}) & \approx & \operatorname{Spec} \tilde{\mathcal{O}}_q \times_B A & \to & A \\ \downarrow & & \downarrow & & \downarrow \\ \operatorname{Spec} \tilde{\mathcal{O}}_q & \approx & \operatorname{Spec} \tilde{\mathcal{O}}_q & \dashrightarrow & B \end{array}$$

We get the leftmost column from elementary properties of henselian rings. S is a union of several copies of Spec $\tilde{\mathcal{O}}_q$, thus a trivial covering. Similarly, S_1 and S_2 are trivial coverings of S.

We get the induced diagram, where K is the field of fractions of $\tilde{\mathcal{O}}_q$.

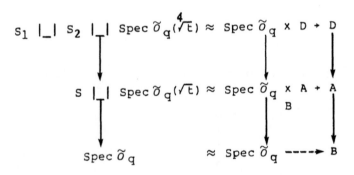

$$\begin{array}{ccc} \tilde{S}_1 \sqcup \tilde{S}_2 \sqcup \operatorname{Spec} \tilde{K}(\sqrt[4]{t}) & \longrightarrow & D^\circ \\ \downarrow & & \downarrow \\ S \sqcup \operatorname{Spec} \tilde{K}(\sqrt{t}) & \dashrightarrow & A^\circ \\ \downarrow & & \downarrow \\ \operatorname{Spec} \tilde{K} & \dashrightarrow & B^\circ \end{array}$$

Choose an embedding $\tilde{K} \subset \overline{K(B^\circ)}$. This defines a geometric point b of Spec $\tilde{K}$ (and b_1 of B°). By Grothendieck (1971, p. 143, proposition 8.1) we have $\pi_1(\operatorname{Spec} \tilde{K}, b) = \operatorname{Gal}(\overline{k(B^\circ)}:\tilde{K})$ and this group acts on

$$\begin{array}{c} F_b(\tilde{S}_1 \sqcup \tilde{S}_2 \sqcup \operatorname{Spec} \tilde{K}(\sqrt[4]{t})) \\ \downarrow \\ F_b(\tilde{S} \sqcup \operatorname{Spec} \tilde{K}(\sqrt{t})) \end{array}$$

and we get a double flip by simple Galois theory (just send $\sqrt[4]{t}$ to $-\sqrt[4]{t}$). If b_1 is the corresponding base point of B°, we conclude that the $\pi_1(B^\circ, b_1)$ action on $F_{b_1}(D^\circ) \to F_{b_1}(A^\circ)$ includes a double flip [from our discussion of the fundamental group (2.1.4)].

3. Irreducibility of the Hessian

We will need the following proposition in order to prove that D is irreducible and normal and ϕ: $D \to E$ is finite. A related question is discussed in Coolidge's book about curves (Coolidge, 1931, Chapter IX, section 3).

PROPOSITION 2.3.1 $F_{xx}F_{yy} - (F_{xy})^2 = H(F)$ is irreducible in $R_1 = k[T_{ij}, X, Y]$.

LEMMA 2.3.2 F_{xx} is irreducible and so is F_{yy}.

Proof: $F_{xx} = 2T_{20} + S_{xx}$, where S_{xx} does not involve T_{20}. $R_1/(F_{xx}) = R_2$; $R_2 = R_1/(T_{20})$, a domain.

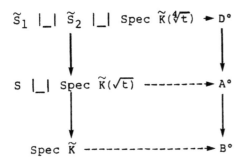

LEMMA 2.3.3 F_{yy} does not divide $H(F)$.

Proof: F_{yy} contains the term $2T_{02}$ and $(F_{xy})^2$ does not contain T_{02}.

COROLLARY 2.3.4 Let V: $H(F) = 0$ so that $k[V] = R_1/H(F)$; then F_{yy} is not a zero divisor in $k[V]$.

Proof: Obvious.

COROLLARY 2.3.5 $k[V]$ injects into the localization $k[V][1/F_{yy}]$.

LEMMA 2.3.6 $k[V][1/F_{yy}]$ is a domain.

Proof: Define a homomorphism h of R_1 into R_1 by $h(X) = X$, $h(Y) = Y$, $h(T_{ij}) = T_{ij}$ for $(i,j) \neq (2,0)$ and $h(T_{20}) = H(F) = T_{20}F_{yy} + S_{xx}F_{yy} - (F_{xy})^2$. Note that F_{yy}, S_{xx}, F_{xy} do not involve T_{20}. Also, $h(F_{yy}) = F_{yy}$. h induces an injective map $\bar{h}$: $R_1/(T_{20}) \to R_1/(H(F))$, which takes the class of F_{yy} to the class of F_{yy}. We therefore get the induced map:

$$h_1: \quad R_1(T_{20})\left[\frac{1}{[F_{yy}]}\right] \qquad k[V]\left[\frac{1}{[F_{yy}]}\right]$$

which is still injective. But now the image of h_1 includes not only T_{ij}, X, Y for $(i,j) \neq (2,0)$ and $S_{xx}F_{yy} - (F_{xy})^2$, hence also $T_{20}F_{yy}$, but also T_{20} since we have inverted F_{yy} in the image. Thus h_1 is an isomorphism.

This proves proposition 2.3.1.

4. Nonsingularity of $\bar{S}$ at Infinity

For the sake of completeness we include a simple lemma about the singularities of $\bar{S}$.

LEMMA 2.4.1 There are no singularities of $\bar{S}$ in $\bar{S} - S$.

Proof: At such a singular point we would have

$$\sum_{i+j=p-1} T_{ij}X^iY^j = 0$$

$$\sum_{\substack{i+j=p\\ i,j>0}} iT_{ij}X^{i-1}Y^j = 0 \qquad\qquad \sum_{\substack{i+j=p\\ i,j>0}} jT_{ij}X^iY^{j-1} = 0$$

But it is clear that the only common solution in L is X = Y = 0; but then also Z = 0 and that does not define a point of the projective space over L.

3. GEOMETRY OF THE MAPPINGS $E \to A$, $E_V \to V$, $E_U \to U$, $D \to E$

Generalities

The main results of this section are theorem (3.2.12) (existence of a special pencil) and the twice-transitivity result (3.3.1). To prove these results, we have to establish certain basic facts about the maps π: $E \to A$ and ϕ: $D \to E$. These facts should also be of some independent interest in the theory of Zariski surfaces. Theorem (3.2.12) will be combined with basic facts about the fundamental group to prove the crucial theorem G2 (1.4). The twice-transitivity result will be essential in proving the two-transitivity in theorem G1 (1.3).

PROPOSITION 3.1.1 E is smooth, irreducible; in fact, E is isomorphic to an affine space over k of dimension equal to the dimension of A.

Proof: If we write out F_x and F_y, we easily see that

$$E = \operatorname{Spec} \frac{k[T_{00},T_{10},T_{01},\ldots][X,Y]}{(F_X,F_Y)}$$

is isomorphic to

$$E = \operatorname{Spec} k[T_{00},T_{20},T_{11},T_{02},\ldots][X,Y]$$

REMARK 3.1.1.1 The observation above is due to S. Mori. We will use his idea again in (3.2.10).

REMARK 3.1.1.2 π has at least one finite fiber.

Proof: This follows from example (3.1.6) below.

COROLLARY 3.1.1.3 π is dominating and generically finite. The field extension $k(A) \subset k(E)$ is a finite algebraic extension.

Proof: This follows from the remark above and semicontinuity of dimension.

PROPOSITION 3.1.2 π_V: $E_V \to V$ is a finite map.

Proof: It is enough to show that π_V is quasi-finite and projective (Hartshorne, 1977, exercise 11.2, p. 280).

(i) π_V is quasi-finite. If $g \in V$, then the surface $Z^p = g$ can only have finitely many singularities. Otherwise, there would be a singularity at infinity which contradicts the definition of V.

(ii) π_V is projective. We have the commutative diagram

$$\begin{array}{ccccc} \mathrm{Sing}(\bar{S}_V)_{\mathrm{red}} & \dashrightarrow & \bar{S}_V & \subset & P^3_V \\ \downarrow & & \downarrow & & \downarrow \\ E_V & \dashrightarrow & P^2_V & \dashrightarrow & V \end{array}$$

where $\bar{S}_V$ is defined in P^3_V by $Z^p - F(X,Y,Z_0)$ and $\mathrm{Sing}(\bar{S}_V)$ is defined by the homogeneous polynomial above and its partials with respect to X, Y, Z, Z_0. The middle vertical arrow corresponds to projecting from the point at infinity on the Z axis in each fiber. The map $E_V \to P^2_V$ is projective because $\mathrm{Sing}(\bar{S}_V) \subset \bar{S}_V$ is closed and E_V is the reduced *image* of $\mathrm{Sing}(\bar{S}_V)$, thus closed in P^2_V. Therefore, $E_V \to V$ is projective.

COROLLARY 3.1.3 $U \subset V$ is open and dense. π_U: $E_U \to U$ is étale.

Proof: E_V is open and dense in E. The polynomial H(F) is not identically zero on E. Thus it defines a proper closed subscheme $\sigma_V \subset E_V$. Since π_V is finite, $\pi_V(\sigma_V)$ is a proper and closed subset of V. Now $U = V - \pi_V(\sigma_V)$. This shows that U is open and dense in V. The fact that π_U is étale follows from the jacobian criterion.

COROLLARY 3.1.4 For any base point b in U, the action of $\pi_1(U,b)$ on $F_b(E_U)$ is transitive.

Proof: E_U is open and dense in E and thus it is irreducible and therefore connected. The corollary follows now from SGA I (Grothendieck, 1971, p. 140).

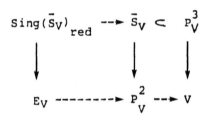

PROPOSITION 3.1.5 D is irreducible and normal. ϕ: $D \to E$ is finite of degree 2.

Proof: This follows from the irreducibility of the polynomial H(F) in the ring $k[E] = k[T_{00}, T_{20}, T_{11}, T_{01}, \ldots, X, Y]$ [see (2.3.1) and Hartshorne, 1977, p. 147, exercise 6.4].

Next we give an example of a point $\tau \in U$. By counting the points in $\pi^{-1}(\tau)$ we will find the degree of π to be $p^2 - 3p + 3$. This result was originally proven by J. Sturnfield (see Blass, 1982).

EXAMPLE 3.1.6 For almost every choice of $A \in k$, the polynomial $h(x,y) = xy + Ax^{p-1} + y^{p-1} + xy^{p-1} + yx^{p-1}$ defines a point $\tau \in U$. Further, we have that the cardinality of $\pi^{-1}(\tau)$ is $p^2 - 3p + 3$. The proof will follow from the following three computational lemmas:

LEMMA 3.1.7 Let $g = xy + Ax^{p-1} + y^{p-1} + xy^{p-1} + yx^{p-1}$. For all but finitely many choices of $A \in k$, the system of equations

$$g_x = 0 \qquad g_y = 0$$

$$\det \begin{vmatrix} g_{xx} & g_{xy} \\ g_{yx} & g_{yy} \end{vmatrix} = 0$$

has no solution.

LEMMA 3.1.8 For all but finitely many $A \in k$, the intersection number of $g_x = 0$ and $g_y = 0$ at infinity is $p - 2$.

Proof of lemma 3.1.7: Consider the system

$$y - Ax^{p-2} + y^{p-1} - yx^{p-2} = 0 \qquad (*)$$

$$x - y^{p-2} - xy^{p-2} + x^{p-1} = 0 \qquad (**)$$

$$(2Ax^{p-3} + 2yx^{p-3})(2y^{p-3} + 2xy^{p-3}) - (1 - y^{p-2} - x^{p-2})^2 = 0 \qquad (***)$$

$x = 0$ implies that $y = 0$, but $(x,y) = (0,0)$ is not a solution. Assume that $x \neq 0$, and rewrite equation (***):

$$4(Ax^{p-2} + yx^{p-2})(y^{p-3} + xy^{p-3})$$
$$- (1 - y^{p-2} - x^{p-2})^2 x = 0$$

Eliminate Ax^{p-2}:

$$4(y + y^{p-1} - yx^{p-2} + yx^{p-2})(y^{p-3} + 2x^{p-3})$$
$$- (1 - y^{p-2} - x^{p-2})^2 x = 0$$

It is enough to show that

$$4(y + y^{p-1})(y^{p-3} + x^{p-3}) - (1 - y^{p-2} - x^{p-2})^2 = 0$$

and

$$x - y^{p-2} - xy^{p-2} + x^{p-1} = 0$$

have no common component. ($x = 0$ is excluded again.) We get finitely many values of A from equation (*) once this is established.

At common points at infinity of the two curves,

$$-xy^{p-2} + x^{p-1} = 0 \qquad \text{(i)}$$
$$4y^{p-1}(y^{p-3} + x^{p-3}) - (y^{p-2} + x^{p-2})^2 = 0 \qquad \text{(ii)}$$

If $x = 0$, then $4y^{p-1}(y^{p-3}) - (y^{p-2})^2 = 0$, or $4y^{2p-4} - y^{2p-4} = 0$. Since $p \neq 3$, this gives $y^{2p-4} = 0$, thus no point with $x = 0$. If $x \neq 0$, then $y^{p-2}/x^{p-2} = 1$. Set $T = y/x$ so that $T^{p-2} = 1$. Equation (i) gives $4T^{p-1}(T^{p-2} + 1) - (T^{p-2} + 1)^2 = 0$, so that $8T - 4 = 0$ or $T = 1/2$, so that $(2)^{p-2} = 1$ or $2 \equiv 1$, so that $2 \equiv 1$. Contradiction. Therefore, the two curves have no common point at infinity and they cannot have a common component.

Proof of lemma 3.1.8: This proof is based on an idea of J. Sturnfield. At a common point at infinity we must have

$$\begin{aligned} y(y^{p-2} - x^{p-2}) &= 0 \\ x(-y^{p-2} + x^{p-2}) &= 0 \end{aligned} \qquad (3.1.8.1)$$

Thus there are $p - 2$ common points: $y/x = \lambda$ where $\lambda^{p-2} = 1$. Homogenize g_x, $yW^{p-2} - Ax^{p-2}W + y^{p-1} - yx^{p-2}$. Set $y = 1$ and $\tau = 1/\lambda$.

Rewrite $W^{p-2} - Ax^{p-2}W + 1 - x^{p-2} = 0$ as

$$W^{p-2} - A[(x - \tau) + \tau]^{p-2}W + 1 - [(x - \tau) + \tau]^{p-2} = 0$$

The tangent line at $W = 0$, $x = \tau$ is

$$-A\tau^{p-2}W - (p - 2)\tau^{p-3}(x - \tau)$$

or

$$-A\tau^{p-2}W + 2\tau^{p-3}(x - \tau)$$

or

$$-A\tau W + 2(x - \tau) = 0$$

Similarly, for the second curve $g_y = 0$,

$$xW^{p-2} - W - x + x^{p-1} = 0$$

and we consider $W = 0$. We obtain

$$x^{p-1} - x = 0$$

$$x^{p-2} = 1$$

$$x = \tau$$

and we have

$$\tau^{p-1} - \tau = 0$$

We compute the tangent line at $W = 0$, $x = \tau$. From

$$-W - [(x - \tau) + \tau] + [(x - \tau) + \tau]^{p-1} + \cdots$$

we obtain

$$-W - (x - \tau) + (p - 1)\tau^{p-2}(x - \tau) = 0$$

or

$$-W + (x - \tau)(-2) = 0$$

Finally,

$$-W + (x - \tau)(-2) = 0$$

defines the tangent line. We compare the two tangent lines.

Suppose that

$$\det\begin{bmatrix} -1 & (-2) \\ -A\tau & 2 \end{bmatrix} = 0$$

so that $2 + 2\tau A = 0$. But choosing A so that $A \neq -1/\tau$ for every τ such that $\tau^{p-2} = 1$, we get a contradiction. Thus for all but finitely many $A \in k$ we get no common tangent.

LEMMA 3.1.9 For all but finitely many choices of $A \in k$, we have $g \in V$.

Proof: By computation. By (0.2) we must consider the system

$$\begin{aligned} Ax^{p-1} + y^{p-1} &= 0 \\ y^{p-1} - yx^{p-2} &= 0 \\ x^{p-1} - xy^{p-2} &= 0 \end{aligned} \tag{3.1.9.1}$$

Suppose that there is a solution with $x \neq 0$; then

$$Ax^{p-1} + y^{p-1} = 0 \quad \text{and} \quad y^{p-2} - x^{p-2} = 0$$

Set $T = y/x$ so that $T^{p-2} = 1$. But $AT^{p-1} = -1$ and we obtain

$$AT = -1$$

$$-\frac{1}{A} = T$$

$$\left(-\frac{1}{A}\right)^{p-2} = 1$$

Contradiction for almost every choice of A.

COROLLARY 3.1.10 There exists a point $g \in V$ such that $\pi^{-1}(g)$ consists of $p^2 - 3p + 3$ unramified points (at which π is étale). $\deg \pi = p^2 - 3p + 3$.

COROLLARY 3.1.10.1 π is generically étale.

COROLLARY 3.1.10.2 U is open and dense in A (and in V).

COROLLARY 3.1.11 The degree of the field extension $k(A) \subset k(E)$ is $p^2 - 3p + 3$.

COROLLARY 3.1.12 The surface S^{aff} has $p^2 - 3p + 3$ singularities at finite distance.

Proof: Sing S $\approx F_b^U(E_U)$ where b: $\overline{k(U)} \to$ Spec k(U) $\to$ Spec U. Now the cardinality of $F_b^U(E_U)$ is equal to the degree of π_U: $E_U \to U$ and that is $p^2 - 3p + 3$.

COROLLARY 3.1.13 The map $D_U \to E_U$ is finite and étale of degree 2. The degree of the field extension $k(A) \subset k(D)$ is $2p^2 - 6p + 6$.

Proof: Follows from definition of D and (3.1.11) above.

COROLLARY 3.1.14 All the singularities of S are nondegenerate.

Proof: The maps $D_U \to E_U \to U$ are étale coverings. Let b: Spec $\overline{k(U)} \to$ U be as above (3.1.12). Now we have m: $F_b(D_U) \to F_b(E_U)$ and this map is two-to-one and onto. But $F_b(D_U) \approx$ set of pairs

$$\langle\alpha, t\rangle \qquad \alpha \in \text{Sing } S,\ t \in L,\ t^2 = H_\alpha$$

Furthermore, $F_b(E_U) \approx$ Sing S, and m corresponds to projecting $\langle\alpha,t\rangle$ to α. By a counting argument we must have all $t \neq 0$, so all $H_\alpha \neq 0$.

3.2 Existence of a Special Pencil

To prove the main result of this section, theorem (3.2.12) below, our next objective is to find a point $q \in V - U$ such that the cardinality of $\pi^{-1}(q) = (\deg \pi) - 1$. We will do this by producing an explicit example of such a q.

EXAMPLE 3.2.1 [Construction of the Special Polynomial (Point of V) q] The point q will be constructed as follows:

$$q \leftrightarrow y^2 + x^3 + Ax^{p-1} + By^{p-1} + xy^{p-1} + yx^{p-1} = g(x,y)$$

where B = +1 if $p \neq 13$ and B = -1 if p = 13. $A \in k$ can be chosen so that the cardinality of $\pi^{-1}(q) = \deg \pi - 1$, $q \in V$. Also, the surface $Z^p = g$ has $(\deg \pi) - 1$ nondegenerate singularities and one degenerate singularity [in other words, H(F) is equal to zero at exactly one point of $\pi^{-1}(q)$].

LEMMA 3.2.2 $q \in V$ for almost every choice of $A \in k$.

Proof: Identical to the proof of 3.19.

LEMMA 3.2.3 For almost every $A \in k$ the system of equations (**) below has only the solutions $(0,0,t)$, $t \in k$.

$$(**) \quad \begin{aligned} g_{xx}g_{yy} - (g_{xy})^2 &= 0 \\ g_x &= 0 \\ g_y &= 0 \end{aligned} \tag{3.2.3.1}$$

Proof: Consists of the following computation:

$$g = y^2 + x^3 + Ax^{p-1} + By^{p-1} + xy^{p-1} + yx^{p-1}$$

$$g_x = 3x^2 - Ax^{p-2} + y^{p-1} - yx^{p-2}$$

$$g_y = 2y - By^{p-2} - xy^{p-2} + x^{p-1}$$

$$g_{xx} = 6x + 2Ax^{p-3} + 2yx^{p-3}$$

$$g_{yy} = 2 + 2By^{p-3} + 2xy^{p-3}$$

$$g_{xy} = -y^{p-2} - x^{p-2}$$

$$(**) \quad \begin{aligned} g_{xx}g_{yy} - (g_{xy})^2 &= 0 \\ g_x &= 0 \\ g_y &= 0 \end{aligned}$$

$(0,0,t)$ is always a root. Also, $x = 0 \Rightarrow y = 0$.

Let us find all solutions with $x \neq 0$. We eliminate A from the first two equations as follows:

$$Ax^{p-2} = 3x^2 + y^{p-1} - yx^{p-2}$$

$$xg_{xx}g_{yy} - x(g_{xy})^2 = 0$$

or

$$(6x^2 + 2Ax^{p-2} + 2yx^{p-2})g_{yy} - x(g_{xy})^2 = 0$$

and we obtain

$$2(3x^2 + 3x^2 + y^{p-1} - yx^{p-2} + yx^{p-2})g_{yy} - x(g_{xy}) = 0$$
$$g_y = 0$$

We wish to show that the curves

$$(***) \quad 2(6x^2 + y^{p-1})(2 + 2By^{p-3} + 2xy^{p-3}) - x(y^{p-2} + x^{p-2})^2 = 0 \tag{3.2.3.2}$$

and

$$g_y = 2y - By^{p-2} - xy^{p-2} + x^{p-1} = 0$$

have only finitely many points in common. Again $x = 0$ implies $y = 0$, so it is enough to prove this for $x \neq 0$.

LEMMA 3.2.4 The curve $g_y = 0$ is smooth and irreducible.

Proof: Consider the projective curve (closure of $g_y = 0$).

$$2yW^{p-2} - By^{p-2}W - xy^{p-2} + x^{p-1} = 0$$

At points at infinity, i.e., where $W = 0$, we have:

$$x(x^{p-2} - y^{p-2}) = 0$$
$$W = 0 \quad x = 0 \quad y = 1$$

or if

$$x \neq 0 \quad \left(\frac{y}{x}\right)^{p-2} = 1$$

$$\frac{\partial}{\partial W} = 0 \text{ at infinity implies } y = 0$$

Thus, there are no singularities at infinity. We dehomogenize:

$$2y - By^{p-2} - xy^{p-2} + x^{p-1} = 0$$

and check for singularities at finite distance:

$$2y - By^{p-2} - xy^{p-2} + x^{p-1} = 0$$
$$2 + 2By^{p-3} + 2xy^{p-3} = 0 \tag{3.2.4.1}$$
$$-y^{p-2} - x^{p-2} = 0$$

First of all $x = 0$ is impossible if $y = 0$ then $x = 0$, contradiction. Assume $y \neq 0$ and transform

$$\begin{aligned} &2y - By^{p-2} - xy^{p-2} + x^{p-1} = 0 \\ &2y + 2By^{p-2} + 2xy^{p-2} = 0 \\ &y^{p-2} = -x^{p-2} \end{aligned} \qquad (3.2.4.2)$$

$$\begin{aligned} &2y + Bx^{p-2} + x^{p-1} + x^{p-1} = 0 \\ &2y - 2Bx^{p-2} - 2x^{p-1} = 0 \\ &y^{p-2} = -x^{p-2} \end{aligned} \qquad (3.2.4.3)$$

$$\begin{aligned} &Bx^{p-2} + 2x^{p-1} = -2Bx^{p-2} - 2x^{p-1} \\ &y^{p-2} = -x^{p-2} \\ &2y = 2Bx^{p-2} + 2x^{p-1} \end{aligned} \qquad (3.2.4.4)$$

$$\begin{aligned} &3Bx^{p-2} + 4x^{p-1} = 0 \\ &y^{p-2} = -x^{p-2} \\ &y = Bx^{p-2} + x^{p-1} \end{aligned} \qquad (3.2.4.5)$$

$x = 0$ is excluded. We obtain

$$x = \frac{-3B}{4} \qquad x^{p-2} = \frac{-4}{3B} \qquad x^{p-1} = 1$$

$$y = \frac{-4}{3} + 1 = -\frac{1}{3} \qquad y^{p-2} = -3$$

Now we must have

$$-3 = +\frac{4}{3B} \qquad \text{or} \qquad 9B = -4$$

If $p \neq 13$, then $B = 1$ by our assumption and $13 \equiv 0$, a contradiction. If $p = 13$, then $B = -1$, $-9 \equiv -4$ or $5 \equiv 0$, again a contradiction. Q.E.D. for lemma (3.2.4).

Proof of (3.2.3) Conclusions: To complete the proof of (3.2.3) it is enough to find one point on the curve $g_y = 0$ which is not on the first one in the system (***) (3.2.3.2).

Take a point with $x = 0$, $2y = By^{p-2}$ and with $y \neq 0$, so that $2 = By^{p-3}$. The first curve gives

$$(2y^{p-1})(2 + 2By^{p-3}) = 0$$

or

$$4y^{p-1} + 4By^{2p-4} = 0$$

Since $y \neq 0$, $1 + By^{p-3} = 0$ and $By^{p-3} = -1$. Thus $2 \equiv -1$ or $3 \equiv 0$, a contradiction since $p \geq 5$.

COROLLARY 3.2.5 For almost every $A \in k$ (**) (3.2.3.1) has only the (0,0) solution in (x,y).

LEMMA 3.2.6 For almost every choice of $A \in k$ the intersection number of $g_x = 0$ and $g_y = 0$ at the points at infinity is $p - 2$.
Proof: Analogous to (3.1.8). We omit it.

LEMMA 3.2.7 Assumptions as in (3.2.6). The intersection number at the origin is 2.
Proof: Simple computation using $p \neq 2$ and $p \neq 3$ (omitted).

COROLLARY 3.2.8 The cardinality of $\pi^{-1}(q)$ is $p^2 - 3p + 2 = \deg \pi - 1$. The hessian is nonzero at all but one point of $\pi^{-1}(q)$. It is zero at one point which we call $\bar{q}$, $\bar{q} \in \pi^{-1}(q)$.
Proof: The total intersection number of the curves $g_x = 0$, $g_y = 0$ is $(p - 1)^2$ by Bézout's theorem. The number of intersections at finite distance is $(p - 1)^2 - (p - 2) - 1$ by the above. Thus it is $p^2 - 3p + 2$. Also, the local intersection number is equal to one at every point where the hessian is nonzero.

PROPOSITION 3.2.9 Let A be as in example (3.2.1). Let L be the line in $A = \operatorname{Spec} k[T_{ij}]$ corresponding to polynomials of the form

$$y^2 + x^3 + \lambda x + Ax^{p-1} + By^{p-1} + yx^{p-1} + xy^{p-1}$$

Then $\pi_L\colon E_L \to L$, and E_L is an irreducible and smooth curve. There is exactly one ramified point lying over q; let us call it $\bar{q}$ with ramification index $e(\bar{q}) = 2$.

LEMMA 3.2.10 E_L is smooth and irreducible.

Proof:

$$E_L \simeq \operatorname{Spec} \frac{k[x,y,\lambda]}{(\lambda + 3x^2 - Ax^{p-2} - yx^{p-2} + y^{p-1},\ 2y - By^{p-2} + x^{p-1} - y^{p-2}x)}$$

$$\simeq \operatorname{Spec} \frac{k[x,y]}{(2y - By^{p-2} + x^{p-1} - y^{p-2}x)}$$

Now we have shown above that the last curve is smooth and irreducible; see (3.2.4). $E_L \to L$ is isomorphic to the projection to the Spec $k[\lambda]$ axis of the space curve defined in Spec $k[x,y,\lambda]$ by the two equations

$$\lambda + 3x^2 - Ax^{p-2} + y^{p-1} - yx^{p-2} = 0$$
$$2y - By^{p-2} - xy^{p-2} + x^{p-1} = 0$$

q corresponds to the point $\lambda = 0$. The matrix of partials with respect to x, y, λ is

$$\begin{bmatrix} 6x + 2Ax^{p-3} + 2yx^{p-2} & -y^{p-2} - y^{p-2} & 1 \\ -y^{p-2} - x^{p-2} & 2 + 2By^{p-3} - 2xy^{p-3} & 0 \end{bmatrix}$$

or

$$\begin{bmatrix} g_{xx} & g_{xy} & 1 \\ g_{yx} & g_{yy} & 0 \end{bmatrix}$$

We have shown above that if $\lambda = 0$, then

$$\det \begin{bmatrix} g_{xx} & g_{xy} \\ g_{yx} & g_{yy} \end{bmatrix} \neq 0$$

for every point of the space curve except $\lambda = x = y = 0$. Thus all of those points are smooth. Also, $\lambda = x = y = 0$, the point $\bar{q}$ is smooth, and x is a uniformizing parameter at $\bar{q}$. Finally, we see that the hessian has a simple zero at $\bar{q}$ and $e(\bar{q}) = 2$. All this follows from the matrix.

Similarly, we consider the map

$$D_L \to E_L \to L$$

D_L is isomorphic to the curve in 4-space Spec $k[x,y,\lambda,W]$:

$$\lambda + 3x^2 - Ax^{p-2} + y^{p-1} + yx^{p-2} = 0$$

$$2y - By^{p-2} - xy^{p-2} + x^{p-1} = 0$$

$$W^2 = g_{xx}g_{yy} - (g_{xy})^2$$

The jacobian matrix is

$$\begin{bmatrix} g_{xx} & g_{xy} & 1 & 0 \\ g_{yx} & g_{yy} & 0 & 0 \\ \alpha & \beta & 0 & 2W \end{bmatrix} \qquad (3.2.10.1)$$

$$\alpha = g_{xxx}g_{yy} + g_{xx}g_{yyx} - 2g_{xy}g_{xxy}$$
$$\beta = g_{xxy}g_{yy} + g_{xx}g_{yyy} - 2g_{xy}g_{xyy}$$

All points with $W \neq 0$ are clearly smooth. If $\lambda = 0$ and $W = 0$, then we must be at the point $\bar{\bar{q}}$ given by $x = y = \lambda = W = 0$ by (3.2.9). We compute the matrix there:

$$\begin{bmatrix} 0 & 0 & 1 & 0 \\ 0 & 2 & 0 & 0 \\ 12 & 0 & 0 & 0 \end{bmatrix}$$

It has rank 3; thus the point is nonsingular.

W is a uniformizing parameter at $\bar{\bar{q}}$ and $g_{xx}g_{yy} - (g_{xy})^2 = W^2$ has a zero of order 2 at $\bar{\bar{q}}$. From this we see that the projection $\pi_L \circ \phi_L$: $D_L \to L$ (L corresponds to the Spec $k[\lambda]$ axis) has ramification equal to 4 at $\bar{\bar{q}}$. Also, $\phi_L(\bar{\bar{q}}) = \bar{q}$ and $e_{\pi_L}(\bar{q}) = 2$; therefore, $e_{\phi_L}(\bar{q}) = 2$.

LEMMA 3.2.11 D_L is irreducible.

Proof: ϕ_L: $D_L \to E_L$ is finite and generically 2 to 1. Thus D_L may have at most two components. Let C_L be the component that contains $\bar{\bar{q}}$. Now $\bar{\bar{q}}$ is a smooth point of C_L and $e(\bar{\bar{q}}) = 2$. Also, $C_L \to E_L$ is a finite map (proper and quasi-finite); thus $C_L \to E_L$ must have degree 2. Therefore, $C_L = D_L$ [inclusion of a closed subscheme is proper (see Hartshorne, 1977, p. 102, 48(a)].

The following theorem summarizes what we have shown.

THEOREM 3.2.12 (Existence of a Special Pencil) There exists a point in $q \in V - U$ and a smooth rational curve L closed in V such that $q \in L$ and $L_U = L \cap U$ is open and dense in L and closed in U. Further, let L_1 be the open subset of L defined by $L_1 = L_U \cup \{q\}$. We have the induced coverings

$$\begin{array}{ccccc} D_L & \to & E_L & \to & L \\ \cup & & \cup & & \cup \\ D_{L_1} & \to & E_{L_1} & \to & L_1 \\ \cup & & \cup & & \cup \\ D_{L_U} & \to & E_{L_U} & \to & L_U \end{array}$$

$$D_{L_1} \to E_{L_1} \quad \text{and} \quad E_{L_1} \to L_1$$

are very simple coverings

$$\bar{\bar{q}} \in D_{L_1} \qquad \bar{\bar{q}} \to \bar{q} \qquad \bar{q} \to q$$

$$e(\bar{\bar{q}}) = 2 \qquad e(\bar{q}) = 2$$

REMARK 3.2.12.1 $L_U \subset U$ is closed and the bottom line may be induced directly by the base change $L_U \to U$ from E_U. $D_{L_U} \to E_{L_U} \to L_U$ are étale maps.

REMARK 3.2.12.2 E_L, E_{L_1}, D_{L_1} are smooth irreducible curves.

Proof: To obtain $L_U = L \cap U$, remove from L the finitely many points over which π: $E_L \to L$ is ramified [π is not everywhere ramified; for example, it is not ramified at N - 2 points of $\pi^{-1}(q)$]. Now q corresponds to $\lambda = 0$. The smoothness of D_{L_U} can be seen from the matrix (3.2.10.1). All the other assertions have been shown before.

COROLLARY 3.2.13 Let b: Spec $\Omega \to L_U$ be any geometric base point; then

(a) The action of $\pi_1(L_U,b)$ on $F_b^{D_U}(E_{L_U})$ includes a transposition.

(b) The action of $\pi_1(L_U,b)$ on $F_b^{L_U}(D_{L_U}) \to F_b^{L_U}(E_{L_U})$ includes a double flip.

Proof: (a) See theorem C1 (2.2.1); (b) see theorem C2 (2.2.2).

COROLLARY 3.2.14 For any geometric base point b in U, we have the following:

(a) The action of $\pi_1(U,b)$ on $F_b(E_U)$ includes a transposition.

(b) The action of $\pi_1(U,b)$ on $F_b(D_U) \to F_b(E_U)$ includes a double flip.

Proof: See propositions (2.1.3), (2.1.4), and (2.1.4.1).

3.3 Twice Transitivity of the Action of $\pi_1(U,b)$ and $F_b(E_U)$

THEOREM 3.3.1 For any base point b in Z_U, $\pi_1(Z_U,b)$ acts on $F_b^{Z_U}(E_{Z_U})$ in such a way that there is an element $A \in F_b^{Z_U}(E_{Z_U})$ whose stabilizer in $\pi_1(Z_U,b)$ acts transitively on $F_b(E_{Z_U}) - \{A\}$.

COROLLARY 3.3.2 $\pi_1(U,b)$ acts on $F_b(E_U)$ transitively and twice transitively for any base point b in U.

Proof: Transitivity has been shown before. If b is in Z_U, we have $F_b^{Z_U}(E_{Z_U}) \simeq F_b^U(E_U)$ and we recall the induced homomorphism $\pi_1(Z_U,b) \to \pi_1(U,b)$, which is such that the actions of $\pi_1(U,b)$ on $F_b^U(E_U)$ and of $\pi_1(Z_U,b)$ on $F_b^{Z_U}(E_{Z_U})$ are compatible, so that by theorem (3.3.1) there is an element $\bar{A} \in F_b^U(E_U)$ whose stabilizer in $\pi_1(U,b)$ acts transitively on $F_b^U(E_U) - \{\bar{A}\}$. This shows transitivity and twice transitivity. If b is not in Z_U, then the conclusion still holds since all the functors F_b^U are isomorphic by Grothendieck (1971).

Proof of theorem 3.3.1: $Z = \text{Spec } k[T_{00},T_{20},T_{11},T_{02},\ldots][X,Y]$.

$$E_Z = \operatorname{Spec} \frac{k[T_{00},T_{20},T_{11},T_{02},\ldots][X,Y]}{(2T_{20}X + T_{11}Y + \cdots,\ 2T_{02}Y + T_{11}X + \cdots)}$$

We get a section from sending $X \to 0$, $Y \to 0$. Let $S\colon\ Z_U \to E_{Z_U}$ be the section. By SGA I (Grothendieck, 1971, p. 7, corollary 5.3) we have

$$E_{Z_U} = S(Z_U) \sqcup T$$

T is reduced, smooth by SGA I (Grothendieck, 1971, p. 16, 9.2). We need only to show that T is topologically connected. Now T is covered by two opens: T_x where $x \neq 0$ and T_y where $y \neq 0$. T_x is homeomorphic to Spec $k[T_{00},T_{02},\ldots][X,Y][1/X]$, thus connected. T_y is homeomorphic to Spec $k[T_{00},T_{20},\ldots][X,Y][1/Y]$, again connected.

To see that $T_x \cap T_y \neq 0$, note that for the point τ from example (3.1.6), at least one point in $\pi^{-1}(\tau)$ has $X \neq 0$, $Y \neq 0$. Thus T is connected and therefore irreducible by SGA I (Grothendieck, 1971, p. 21, proposition 10.1). Now

$$F_b^{Z_U}(E_{Z_U}) = F_b^{Z_U}(S(Z_U)) \sqcup F_b^{Z_U}(T)$$
$$= \{A\} \sqcup F_b^{Z_U}(T)$$

Now $\pi_1(Z_U,b)$ stabilizes A and acts transitively on $F_b^{Z_U}(T)$ by SGA I (Grothendieck, 1971, p. 140).

REMARK 3.3.3 Theorem (3.3.1) was inspired by reading J. Harris's paper Galois Groups of Enumerative Problems (Harris, 1980).

4. PROOF OF THEOREMS G1, G2, AND G2', AND OF DRL

THEOREM 4.1 For any geometric base point b: Spec $\to$ U the action of $\pi_1(U,b)$ on $F_b(E_U)$ is the full symmetric group.

Proof: It is transitive by (3.1.4). It is twice transitive by (3.3.2). It contains a transposition by (3.2.14). All the properties above are independent of the choice of the base point.

THEOREM 4.2 In the notation of Theorem 4.1, the action of $\pi_1(U,b)$ on $F_b^U(D_U) \to F_b^U(E_U)$ contains a double flip.

Proof: This is a restatement of (3.2.14(b)).

Proof of theorem G1 (1.3): Let b: Spec $\overline{k(T_{ij})} \to k(T_{ij})$ be the base point of U. We have a surjective map $G \to \pi_1(U,b) \to (e)$ by SGA I (Grothendieck, 1971, p. 143). Also, Sing S - $F_b(E_U)$ and this identification is G-equivariant, where G acts on $F_b(E_U)$ via $G \to \pi_1(U,b)$. This follows from the proof on page 143 of SGA I (Grothendieck, 1971). This proves G1.

Proof of theorem G2 (1.4): Choose b as in the proof of G1. Now $F_b(D_U)$ - set of pairs $\langle\alpha,t\rangle$, where $\alpha \in$ Sing S, $t^2 = H_\alpha$, $t \in L = \overline{k(T_{ij})}$ (see 0.3 for the definition of H_α). Let us call this set of pairs P. The map

$$\begin{array}{c} F_b(D_U) \\ \downarrow \\ F_u(E_U) \end{array}$$

corresponds to projection $\langle\alpha,t\rangle$ to α. The diagram

$$\begin{array}{ccc} P & \approx & F_b(D_U) \\ \downarrow & & \downarrow \\ \text{Sing } S & \approx & F_b(E_U) \end{array}$$

where G acts on $F_b(D_U)$, $F_b(E_U)$ via the map $G \to \pi_1(U,b) \to (e)$. Theorem G2 follows immediately from theorem (4.2).

Proof of theorem G2' (1.5): Follows immediately from G1, G2.

Proof of DRL (1.6): Since we have already proven G2' it is enough to prove the implication G2' → DRL. The proof is somewhat tedious; thus we only sketch it here.

Sketch of proof that G2' → DRL: Let σ be as in G2'; then σ induces an automorphism of $\bar{S}$ and it can be shown that there exists a commutative diagram

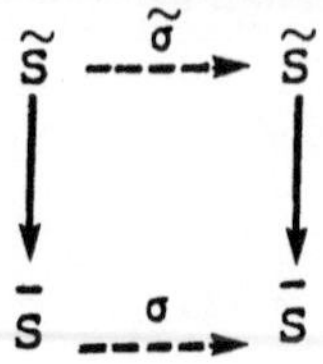

with $\tilde{\sigma}$ an automorphism. It can be shown that $\tilde{\sigma}$ induces the desired isometry as in DRL. We omit the details of the proof except to point out that $\tilde{S}$ is the result of $(p - 1)/2$ blowups.

$$S^{\left(\frac{p-1}{2}\right)} = \tilde{S} \xrightarrow{\ \tilde{\sigma}\ =\ \sigma^{\left(\frac{p-1}{2}\right)}\ } \tilde{S} = S^{\left(\frac{p-1}{2}\right)}$$

$$\downarrow \qquad\qquad\qquad\qquad\qquad\qquad \downarrow$$

$$\vdots \qquad\qquad\qquad\qquad\qquad\qquad \vdots$$

$$S^{(2)} \xrightarrow{\ \sigma^{(2)}\ } S^{(2)}$$

$$\downarrow \qquad\qquad\qquad\qquad\qquad\qquad \downarrow$$

$$S^{(1)} \xrightarrow{\ \sigma^{(1)}\ } S^{(1)}$$

$$\downarrow \qquad\qquad\qquad\qquad\qquad\qquad \downarrow$$

$$\bar{S} \xrightarrow{\ \bar{\sigma}\ } \bar{S}$$

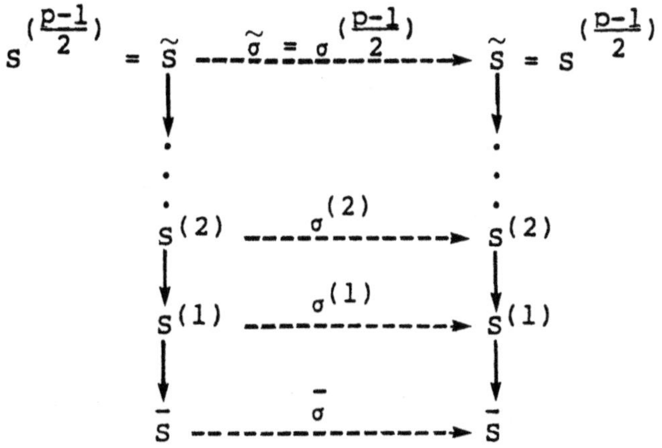

In fact we can construct $\sigma((p - 1)/2) = \tilde{\sigma}$ by induction. The influence of $\tilde{\sigma}$ on T_α, T_β, T_γ can then be seen if we write out the local equations for the resolution.

5. PROOF OF THE MAIN THEOREM

We need only to show the implication:

    DRL => MTHG

We begin with an easy lemma. Let $\alpha \in$ Sing S. Let P_α be the subgroup of Pic($\tilde{S}$) generated by the curves C_i^α, $i = 1, 2, \ldots, p - 1$. P_α inherits the intersection form from Pic($\tilde{S}$).

LEMMA 5.1 Let $x_\alpha \in P_\alpha$ have the property that $x_\alpha \cdot w \equiv o(p)$ for every $w \in P_\alpha$. Then $x_\alpha = n_\alpha D_\alpha + pz$, where $0 \leq n_\alpha \leq p - 1$, $z \in P_\alpha$.

Proof: Let $\bar{P}_\alpha = P_\alpha/pP_\alpha$. This is a $\mathbf{Z}/p\mathbf{Z}$ vector space. Let u: $P_\alpha \to \bar{P}_\alpha = P_\alpha/pP_\alpha$ be the natural map. We have a commutative diagram

$$\begin{array}{ccc} P_\alpha \times P_\alpha & \overset{i}{\dashrightarrow} & \mathbf{Z} \\ {\scriptstyle u \times u}\downarrow & & \downarrow{\scriptstyle v} \\ \bar{P}_\alpha \times \bar{P}_\alpha & \overset{i_1}{\dashrightarrow} & \mathbf{Z}/p\mathbf{Z} \end{array}$$

where i is the intersection form and i_1 is the intersection form modulo p.

Now i_1 has a one-dimensional kernel over $\mathbf{Z}/p\mathbf{Z}$ because the intersection matrix of i_1 with respect to the basis $u(C_1^\alpha)$, $u(C_2^\alpha)$, ..., $u(C_{p-1}^\alpha)$, which is

$$\begin{bmatrix} -2 & 1 & & & 0 \\ 1 & -2 & \ddots & & \\ & \ddots & \ddots & \ddots & \\ & & \ddots & \ddots & 1 \\ 0 & & & 1 & -2 \end{bmatrix}$$

of size $(p - 1) \times (p - 1)$, has rank $p - 2$ over $\mathbf{Z}/p\mathbf{Z}$. $u(D_\alpha)$ is in the kernel and is $\neq 0$ in $\bar{P}_\alpha$. Thus we have $u(x_\alpha) = n_\alpha u(D_\alpha)$ for some $0 \le n_\alpha \le p - 1$. Consequently, $x_\alpha - n_\alpha D_\alpha = pz$ for some $z \in P_\alpha$, as asserted.

COROLLARY 5.2 Any element $x \in \mathrm{Pic}^{ob}(\tilde{S})$ which satisfies the condition that $x \cdot C_j^\alpha \equiv o(p)$ for all $\alpha \in$ Sing S and $1 \le j \le p - 1$ can be written in the form

$$x = n_\ell \ell + \Sigma\, m_\gamma D_\gamma + py$$

where $y \in \mathrm{Pic}^{ob}(\tilde{S})$ and $0 \le n_\ell, m_\gamma \le p - 1$, and Σ is taken over all the singularities of S.

Proof of the main theorem (1.1), (1.2): We need only to show that $p\,\mathrm{Pic}(\tilde{S}) \subset p\,\mathrm{Pic}^{ob}(\tilde{S})$ since there is no torsion.

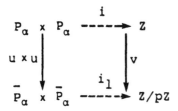

Let $x \in p\ \mathrm{Pic}(\tilde{S})$; then $x \in \mathrm{Pic}^{ob}(\tilde{S})$ because of (0.5.1). Write

$$x = n_0 \ell + \Sigma\, n_\alpha D_\alpha + py \qquad \alpha \in \mathrm{Sing}\ S$$

where $0 \le n_0, n_\alpha \le p - 1$, $y \in \mathrm{Pic}^{ob}(\tilde{S})$. This representation is possible because of (5.2). We need only to show that $n_\alpha = 0$ for all $\alpha \in \mathrm{Sing}\ S$ and that $n_0 = 0$.

Pick any pair of singularities $\alpha \neq \beta$ and apply the DRL; thus we get a $\sigma \in G$ and if $i(\sigma)$ is the induced isometry of $\mathrm{Pic}(\tilde{S})$, we have

$$\begin{aligned} x - i(\sigma)(x) &= n_\alpha(D_\alpha - D_\alpha^{op}) + n_\beta(D_\beta - D_\beta^{op}) + p(y - i(\sigma)(y)) \\ &= 2n_\alpha D_\alpha + 2n_\beta D_\beta - pn_\alpha\, \Sigma\, C_i^\alpha - pn_\beta\, \Sigma\, C_j^\beta + (y - i(\sigma)(y)) \end{aligned}$$

In any case, since $p \neq 2$ we conclude that

$$n_\alpha D_\alpha + n_\beta D_\beta \in p\ \mathrm{Pic}(\tilde{S})$$

and then p^2 must divide

$$(n_\alpha D_\alpha + n_\beta D_\beta)^2 = (n_\alpha^2 + n_\beta^2)p(1 - p)$$

by (0.4.5), so that p must divide $n_\alpha^2 + n_\beta^2$; thus for any pair of singularities α, β, $\alpha \neq \beta$, we have shown that $n_\alpha^2 \equiv -n_\beta^2$ modulo p. But there are at least three singularities and we conclude that in fact this is possible only if $n_\alpha = 0$ for all $\alpha \in \mathrm{Sing}\ S$.

Thus $x - py = n_0\ell$ belongs to $p\ \mathrm{Pic}(\tilde{S})$. Squaring again we get that p^2 divides $(n_0\ell)^2 = pn_0^2$, so that $n_0 = 0$.

Acknowledgments In addition to P. Deligne, to whom the main idea of this chapter is due, I wish to thank Bill Fulton for his patient illumination of Deligne's outline and for listening to my several early versions. With two such guides I could hardly go astray.

In addition, I would like to acknowledge with pleasure conversations with J. S. Milne, A. Fauntleroy, R. Donagi, R. Lazarsfeld, L. Tu, L. Ein, L. Illusie, M. Raynaud, J. Denef, J. Milnor, S. Mori, and S. Yau.

I was also greatly encouraged by frequent conversations with my former student and present collaborator, Jeffrey Lang. He had solved several questions related to this paper.

I am deeply grateful to the Institute for Advanced Study in Princeton for providing the perfect environment for this effort during the Algebraic Geometry Year 1981-1982.

REFERENCES

Blass, P., Some geometric applications of a differential equation in characteristics p > 0 to the theory of algebraic surfaces, *Contemporary Mathematics*, AMS 13 (1982) (with the cooperation of James Sturnfield and Jeffrey Lang).

Coolidge, J., *Treatise on Algebraic Plane Curves*, Oxford, The Clarendon Press, 1931.

Deligne, P., and N. Katz, *SGA VII, Part II*, Lecture Notes in Mathematics, No. 340, Springer-Verlag, New York, 1973.

Grothendieck, A., *SGA I*, Lecture Notes in Mathematics, No. 224, Springer-Verlag, New York, 1971.

Harris, J., Galois groups of enumerative problems, *Duke J. Math.* (1980).

Hartshorne, R., Equivalence relations on algebraic cycles and subvarieties of small codimension, *Proc. Symp. Pure Math.*, 29 (Arcata), American Mathematical Society, Providence, R.I., 1975, corollary 3.5.

Hartshorne, R., *Algebraic Geometry*, Graduate Texts in Mathematics, Springer-Verlag, New York, 1977.

Lang, J., Ph.D. thesis, Purdue University, 1981.

Lang, W. E., Remarks on p-torsion of algebraic surfaces, *Compositio Math.*, 52(2), (1984), 197-202.

Rudakov, A. N., and I. R. Safarevic, Supersingular K3 surfaces over fields of characteristic two, *Math. USSR Izv.* 13, No. 1 (1979), 147-165.

Samuel, P., Lectures on Unique Factorization Domains, Tata Lecture Notes, 1964.

Steenbrink, J., On the Picard group of certain smooth surfaces in weighted projective spaces, in *Algebraic Geometry*, Proceedings La Rabida 1981. Lecture Notes in Mathematics 961, Springer-Verlag, Berlin etc., 1982, pp. 302-313.

Zariski, O., On Castelnuovo's criterion of rationality $p_a = p_g = 0$ of an algebraic surface, *Ill. J. Math.*, 2, No. 3 (1958), 303.

APPENDIX 1: FROM GENERIC TO GENERAL[†]

In this appendix we show how to pass from *generic* to *general* and how to prove our conjecture stated in the introduction to Blass (1983, 1985) [see theorem 6(2) below].[††]

We use techniques of P. Samuel and J. Lang and of course the main result of Blass (1985) which was proven with the help of Deligne.

We begin by introducing some notation which is analogous to Lang (1983). If R is a normal noetherian domain we denote by $C\ell\ R$ its divisor class group. Let k denote an algebraically closed field of characteristic $p \geq 5$, and let T_{ij} be indeterminates. We consider the polynomial ring $k[T_{ij}]$, $0 \leq i + j \leq p$, and two polynomials

$$F(x,y) = \Sigma\ T_{ij}X^iY^j \qquad 0 \leq i + j \leq p$$

and

$$T(x,y) = \Sigma\ 1_{\alpha,\beta}X^{\alpha}Y^{\beta} \qquad 0 \leq \alpha + \beta \leq p - 2$$

with $1_{\alpha,\beta}$ also indeterminates over k. We denote $\nabla = \partial^{2p-2}/\partial x^{p-1} \times \partial y^{p-1}$. We consider the system of equations:

$$\text{(LS)} \quad \nabla(F^jT) = 0 \qquad j = 0, 1, 2, \ldots, p - 2$$

$$\text{(PLS)} \quad \nabla(F^{p-1}T) = T^p$$

We consider the above equations as equalities of polynomials in X, Y. By comparing coefficients of the various monomials in X and Y we get an equivalent system

$$\text{(LS1)} \quad \Sigma\ \underline{P}^{\gamma,\delta}_{\alpha,\beta}1_{\alpha,\beta} = 0$$

$$\text{(PLS1)} \quad \Sigma\ Q^{\gamma,\delta}_{\mu,\nu}1_{\gamma,\delta} = 1^p_{\mu,\nu}$$

$$\underline{P},Q \in k[T_{ij}]$$

[†]By Piotr Blass. Reprinted with slight modifications from *Compositio Mathematica*, 54 (1985), 37-40.

[††]We have recently developed a different proof for passing from generic to general, based on linear algebra. Details will appear elsewhere.

If we specialize the indeterminates T_{ij} to have values $(c_{ij}) \in$ Spec $k[T_{ij}]$, a closed point, then we denote the corresponding system $LS1(c_{ij}) + PLS1(c_{ij})$.

We use the following facts.

THEOREM 1 (J Lang) Let A be any algebraically closed field of characteristic $p > 0$. Let $G(x,y) \in A[x,y]$ be a polynomial such that $\partial G/\partial x$ and $\partial G/\partial y$ are relatively prime polynomials in $A[x,y]$. Then the surface S: $z^p = G(x,y)$ is normal and its divisor class group $C\ell$ S is isomorphic to the set of polynomial solutions $t(x,y) \in A[x,y]$ of degree deg $t \leq$ deg $G - 2$ of the following system of equations:

$$\nabla(G^j t) = 0 \qquad j = 0, 1, 2, \ldots, p - 2$$

$$\nabla(G^{p-1}t) = t^{\underline{p}}$$

Proof: See Lang (1983, 2.1, 2.3, 2.9.1).

THEOREM 2 (Blass-Deligne) The only solution of (LS1) + (PLS1) in $L = k(T_{ij})$, i.e., with $t_{\alpha,\beta} \in L$, is the identically zero solution $t_{\alpha,\beta} = 0$.

Proof: The set of solutions is isomorphic to

$$C\ell \frac{L[X,Y,Z]}{(Z^p - \Sigma\, T_{ij}X^iY^j)}$$

by Theorem 1, but the latter group is shown to be zero in Blass (1985). From now on Σ means $\Sigma_{0 \leq i+j=p}$.

The following is simple.

LEMMA 3 If $q = (c_{ij}) \in$ Spec $k[T_{ij}]$, then the set of solutions $(t_{\alpha,\beta})$ of $LS(c_{ij}) + PLS(c_{ij})$ is finite.

Proof: Lang (1983, proof of Lemma 2.8).

In what follows, let H be the subscheme of Spec $k[T_{ij}] \times$ Spec $k[t_{\alpha,\beta}]$ defined by (LS + PLS) or equivalently by (LS1 and PLS1).

Consider the projection $H_{red} \to H \to$ Spec $k[T_{ij}]$. We denote by $\kappa^{-1}(q)$ the set (group) of closed points of H_{red} that map to q.

REMARK 4 We point out that if $q = (c_{ij}) \in \operatorname{Spec} k[T_{ij}]$ is a closed point, then $\kappa^{-1}(q)$ is in one to one correspondence with the solution set of equations $LS1(c_{ij}) + PLS1(c_{ij})$.

PROPOSITION 5 There exists an open and dense subset $\mathcal{O}_{\mathbb{P}}$ of Spec $k[T_{ij}]$ such that for $q \in \mathcal{O}_{\mathbb{P}}$. $\kappa^{-1}(q)$ consists of a single point with coordinates $t_{\alpha,\beta} = 0$ for all α, β.

LEMMA 6 Let $Z \subset H_{red}$ be the subset of $\operatorname{Spec} k[T_{ij}] \times \operatorname{Spec} k[1_{\alpha,\beta}]$ defined by $t_{\alpha,\beta} = 0$ (all α, β). Let C be any irreducible component of H_{red} whose image $\kappa(C)$ is dense in Spec $k[T_{ij}]$. Then $C = Z$.

Proof: First of all, $\dim C = \dim Z = \dim k[T_{ij}]$ because of lemma 3. Consider the diagram

$$\begin{array}{ccc} \mathcal{O}(C) & \longleftarrow & k[T_{ij}] \\ & \searrow^{m} & \downarrow \\ & & \overline{k(T_{ij})} = L \end{array}$$

Let $[t_{\alpha,\beta}]$ be the class of $t_{\alpha,\beta}$ in $\mathcal{O}(C)$. $\mathcal{O}(C)$ has fraction field which is finite algebraic over $k[T_{ij}]$. Hence we get an injective map m.

Suppose that for some α, β, $[t_{\alpha,\beta}] \neq 0$ in $\mathcal{O}(C)$ then $m([t_{\alpha,\beta}]) \neq 0$ and we would get a nontrivial solution of (LS) + (PLS) in L which contradicts the Blass-Deligne theorem. Thus $[t_{\alpha,\beta}] = 0$ in $\mathcal{O}(C)$ for all α, β; i.e., $C \subsetneq Z$ and consequently $C = Z$ since Z is irreducible.

Proof of proposition 5: Let $H_{red} = Z \cup C_1 \cup \cdots \cup C_3$ be a decomposition of H_{red} into irreducible components.

We have $\overline{\kappa(C_j)} \subset \operatorname{Spec} k[T_{ij}]$ by lemma 6. Thus set $\mathcal{O}_{\mathbb{P}} = \operatorname{Spec} k[T_{ij}] - \cup_{j=1}^{3} \overline{(c_j)}$. For every $q \in \mathcal{O}_{(\mathbb{P})}$, $\kappa^{-1}(q)$ is a single point of Z. Q.E.D.

REMARK 7 There exists an open and dense subset of Spec $k[T_{ij}]$, for example, the subset 1' defined in Blass [1985, (0.2)] such that if $q = (c_{ij})$ belongs to it, then

$$\kappa^{-1}(q) \leftrightarrow C\ell\ \frac{k[x,y,z]}{(z^p - \Sigma\, c_{ij}x^iy^j)}$$

Proof: For $q \in V$, $q = (c_{ij})$, the polynomials $\partial(\Sigma\, c_{ij}x^iy^j)/\partial x$ and $\partial(\Sigma\, c_{ij}x^iy^j)/\partial y$ are relatively prime. Thus Remark 7 follows from Remark 4 and Theorem 1. Q.E.D.

THEOREM 8 There exists an open and dense subset $D \subset \operatorname{Spec} k[T_{ij}]$ such that for every closed point $q = (c_{ij}) \in D$.

(1) $\kappa^{-1}(a)$ consists of the single point.

(2) $\operatorname{Spec} k[x,y,z]/(z^p - \Sigma\, c_{ij}x^iy^j)$ is a UFD.

(3) $C\ell$ of the above ring in (2) is the zero group.

(4) The system $SL1(c_{ij}) + PSL1(c_{ij})$ has only the zero solution.

Proof: Set $D = V \cap \mathcal{O}_{\mathbb{P}}$. Then (1) follows from Proposition 5 and we deduce (3) and (2) from Remark 7. Finally (4) follows because the closed points of $\kappa^{-1}(q)$ are in one-to-one correspondence with the solutions of the system $SL1(c_{ij}) + PLS1(c_{ij})$. Q.E.D.

REFERENCES

Blass, P., Groupes de Picard des surfaces de Zariski, *C. R. Acad. Sci.*, 1-315 (1983).

Blass, P., Picard groups of Zariski surfaces I, *Compositio Math.*, 54 (1985), 3-36.

Lang, Jeffrey, The divisor class group of the surface $z^{p^n} = G(x,y)$ over fields of characteristic $p > 0$, *J. Algebra*, 84, No. 2 (1983).

APPENDIX 2: APPLICATIONS OF MONODROMY†

Introduction, Historical Comments††

This work is based on an outline by Pierre Deligne. My interest in Zariski surfaces was originally inspired by Hironaka. Let me recall the definition of a Zariski surface.

---

†By Piotr Blass, from a lecture given at the Algebraic Geometry Seminar, Harvard, November 9, 1982.

††This lecture is logically independent of the rest of the book. No references to it are given elsewhere in the text, although similar names of theorems are given in Chapter 4. Our purpose in including this lecture here is to give a quick if somewhat informal overview of the material in Chapter 4.

DEFINITION 1.1 Let $k = \bar{k}$, char $k > 0$. A smooth projective surface X over k is called a Zariski surface if X is birationally equivalent to

$$\frac{k[x,y,z]}{(z^p - f(x,y))} \tag{1.1.1}$$

Such surfaces were originally introduced by O. Zariski for purposes of birational geometry, as examples of unirational irrational surfaces. They were studied in my thesis completed in 1977, and later by Jeffrey Lang in his Purdue thesis, which deals with their Picard groups. Our main theorem in this lecture is motivated by the classical theorem of Max Noether: A generic surface of degree ≥ 4 in $\mathbb{P}^3$ has Pic = $\mathbb{Z}$.

We are especially motivated by the monodromy-type proofs given for this theorem first by S. Lefschetz in his classic book on topology of algebraic varieties and then in SGA VII by Deligne, who proved Noether's theorem in arbitrary characteristic.

Statement of the Main Theorem and Principal Lemmas.
Notation. Logical Skeleton of the Proof

Throughout the appendix $k = \bar{k}$, char $k = p \geq 5$, T_{ij} are indeterminates over k, $0 \leq i + j \leq p$, $L = \overline{k(T_{ij})}$, $R = L[X,Y,Z]/(Z^p - \Sigma\, T_{ij}X^iY^j)$.

MAIN THEOREM (ALGEBRAIC FORM) (MTH) 2.1

R is a UFD [or $C\ell(R) = 0$]

In order to prove MTH we need to restate it geometrically and use some Galois-theoretic and geometric lemmas. First, some geometric language for this:

$S = S^{\text{aff}} = \text{Spec } R$: $Z^p = \Sigma\, T_{ij}X^iY^j = F(X,Y)$ $\quad 0 \leq i + j \leq p$

$\bar{S}$ = closure of S in $\mathbf{P}^3_L$

$\tilde{S} \to \bar{S}$ is a minimal desingularization of $\bar{S}$. $\tilde{S}$ is a Zariski surface over L.

It can be shown that $\bar{S}$ has $p^2 - 3p + 3$ singular points (first computed by J. Sturnfield at Purdue), all of which are contained in S and are rational double points with resolution graph of type A_{p-1}. In other words, locally in the étale topology the singularities are given by $Z^p = UV$ (biplanar double points in Zariski's terminology). We omit the proof.

The resolution process is this: Let $\alpha \in \text{Sing } S$.

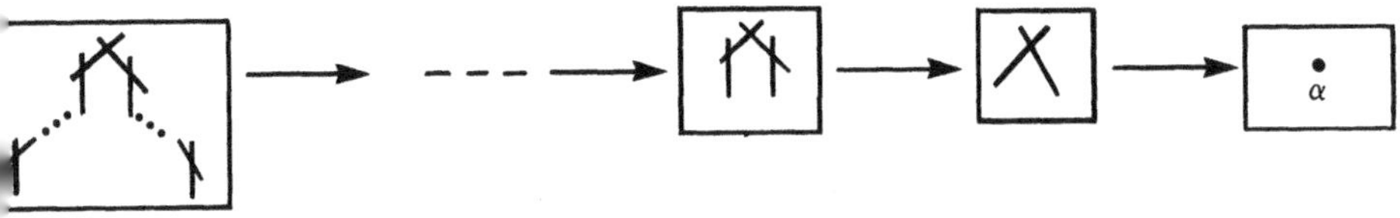

$(p - 1)/2$ blowups result in a tree of exceptional curves.

It is crucial for our proof to order the tree either "left to right" or "right to left." We fix an ordering and label the tree:

T_α is the ordered tree $C_1^\alpha, C_2^\alpha, \ldots, C_{p-1}^\alpha$. T_α^{op} is the ordered tree $C_{p-1}^\alpha, C_{p-2}^\alpha, \ldots, C_1^\alpha$. There is one more obvious curve on S, namely, $\pi^{-1}(\bar{S} - S) = \ell$. We have the exact sequence

$$0 \to \text{Pic}^{ob}(\tilde{S}) \to \text{Pic}(\tilde{S}) \to C\ell\ R \to 0 \tag{2.2}$$

where $\text{Pic}^{ob}(\tilde{S})$ is the group generated by ℓ and the exceptional curves C_i^α, $\alpha \in \text{Sing } S$, $1 \le i \le p - 1$. We omit the proof of the exactness of the sequence (2.2).

Thus we get an equivalent form of our MTH (2.1).

MAIN THEOREM (GEOMETRIC FORM) (MTHG) 2.3

$$\mathrm{Pic}(\tilde{S}) = \mathrm{Pic}^{ob}(\tilde{S})$$

We will need some Galois theory. (The motto of this paper from now on is: Galois Lives!) Let $G = \mathrm{Gal}(\overline{k(T_{ij})} : k(T_{ij}))$.

LEMMA GAL A 2.4 G acts on Sing S as the full symmetric group.

Even more is true!

LEMMA GAL B 2.5 For every pair of singularities $\alpha, \beta \in$ Sing S, with $\alpha \neq \beta$, there is a $\sigma \in G$ such that σ induces the identity on Sing S, but $\sigma(\sqrt{H_\alpha}) = -\sqrt{H_\alpha}$, $\sigma(\sqrt{H_\beta}) = -\sqrt{H_\beta}$, and $\sigma(\sqrt{H_\gamma}) = \sqrt{H_\gamma}$ for all $\gamma \in$ Sing S, $\beta \neq \gamma \neq \alpha$, where H_α, "the hessian at α," is defined as follows:

$$F = \Sigma\, T_{ij} X^i Y^j \qquad 0 \leq i + j \leq p$$

$$H(F) = F_{xx} F_{yy} - F_{xy}^2$$

If $\alpha \in$ Sing S, $\alpha = (\alpha_1, \alpha_2, \alpha_3)$, $H_\alpha = H(F)(\alpha_1, \alpha_2)$.

Finally, I need to restate Gal B (2.5) geometrically in terms of the action of G on Pic $\tilde{S}$.

DOUBLE REVERSAL LEMMA (DRL) 2.6 For every pair of singularities $\alpha, \beta \in$ Sing S, $\alpha \neq \beta$, there exists a $\sigma \in G$ and an induced isometry $i(\sigma)$ of Pic$(\tilde{S})$ such that

$$i(\sigma)(T_\alpha) = T_\alpha^{op} \qquad i(\sigma)(T_\beta) = T_\beta^{op}$$

$i(\sigma)(\ell) = \ell$ and $i(\sigma)(T_\gamma) = T_\gamma$ for all $\gamma \in$ Sing S, $\beta \neq \gamma \neq \alpha$. (An isometry is a group automorphism that preserves the intersection product in Pic $\tilde{S}$.)

The logical skeleton of the proof is this:

Gal A, Gal B $\rightarrow$ DRL $\rightarrow$ MTHG

3. Proof of the Implication DRL ⇒ MTHG

We need some background results:

3.1 $C\ell\ R$ is an elementary p-group (Samuel Tata Notes on UFDs).

3.2 $\operatorname{Pic} \tilde{S}$ has no torsion.

This was my conjecture proven recently by W. E. Lang (see Lang 1984). (An idea for a proof is given at the end of this paper.) Because of (3.2), all that we need to prove to establish MTHG (2.3) is

$$p \operatorname{Pic}(\tilde{S}) \subset p \operatorname{Pic}^{ob}(\tilde{S}) \tag{3.3}$$

Proof of (3.3): Let x be in $p \operatorname{Pic}(\tilde{S}) \subset \operatorname{Pic}^{ob}(\tilde{S})$ [last relation because of 3.1 ($C\ell\ R$ is an elementary p-group) and the exact sequence 2.2]. Write $x = n_\ell \ell + \Sigma_{\alpha \in \text{Sing } S}\, x_\alpha$, x_α supported on curves of the trees T_α. Now comes a great trick! Since $x \in p \operatorname{Pic} \tilde{S}$, we must have

$$x_\alpha \cdot C_j^\alpha \equiv 0(p) \qquad \text{for } 1 \le j \le p - 1$$

Now the quadratic form on the curves $C_1^\alpha, C_2^\alpha, \ldots, C_{p-1}^\alpha$ has the matrix

$$M_\alpha = \begin{bmatrix} -2 & 1 & & & 0 \\ 1 & \ddots & \ddots & & \\ & \ddots & \ddots & \ddots & \\ & & \ddots & \ddots & 1 \\ 0 & & & 1 & -2 \end{bmatrix} \qquad \text{size } (p - 1) \times (p - 1);\ \det M_\alpha = p$$

Over $\mathbb{Z}/p\mathbb{Z}$ that matrix has rank p - 2, so that the quadratic form over $\mathbb{Z}/p\mathbb{Z}$ has a one-dimensional kernel. A basis element is

$$\begin{aligned} D_\alpha &= C_1^\alpha + 2C_2^\alpha + 3C_3^\alpha + \cdots + (p - 1)C_{p-1}^\alpha \\ D_\alpha^{op} &= C_{p-1}^\alpha + 2C_{p-1}^\alpha + \cdots + (p - 1)C_1^\alpha \end{aligned} \tag{3.3.1}$$

We conclude that x can be written as follows:

$$x = n_0 \ell + \sum_{\text{Sing } S} n_\alpha D_\alpha + py$$

where $0 \leq n_0$, $n_\alpha \leq p - 1$, $y \in \text{Pic}^{ob}(\tilde{S})$. We need to show that n_0 and all $n_\alpha = 0$.

Let α, β be any pair of singularities, $\alpha \neq \beta$; apply DRL:

$$-py + pi(\sigma)(y) + x - i(\sigma)(x) = n_\alpha(D_\alpha - D_\alpha^{op}) + n_\beta(D_\beta - D_\beta^{op})$$

$$x - i(\sigma)(x) \in p \text{ Pic}(\tilde{S}) \tag{3.3.2}$$

so that

$$(n_\alpha^2 + n_\beta^2)(D_\alpha - D_\alpha^{op})^2 \equiv 0(p^2) \tag{3.3.3}$$

Now $D_\alpha - D_\alpha^{op} = 2D_\alpha - p \Sigma C_i^\alpha$, so that $(D_\alpha - D_\alpha^{op})^2 \equiv 4D_\alpha^2(p^2)$, but one easily computes

$$4D_\alpha^2 = 4p(1 - p) \equiv 4p(p^2)$$

also $p \neq 2$. Thus $(D_\alpha - D_\alpha^{op})^2$ is divisible by p but not p^2, and we conclude that $n_\alpha^2 \equiv -n_\beta^2(p)$. However, there are ≥ 3 singularities, and therefore $n_\alpha = 0$ for all $\alpha \in \text{Sing } S$.

To conclude, repeat this squaring for $x - py = n_0 \ell$, $x - py \in p \text{ Pic } S$, so that $(x - py)^2 = n_0^2 \ell^2 \equiv 0(p^2)$. But it can be shown that $\ell^2 = p$ (ℓ comes from a hyperplane section of $\tilde{S}$), so that $n_0 = 0$ and we are done.

4.   The Implication GB $\Rightarrow$ DRL

We just indicate the principle that makes this work. Let σ be as in GB. We get a commutative diagram,

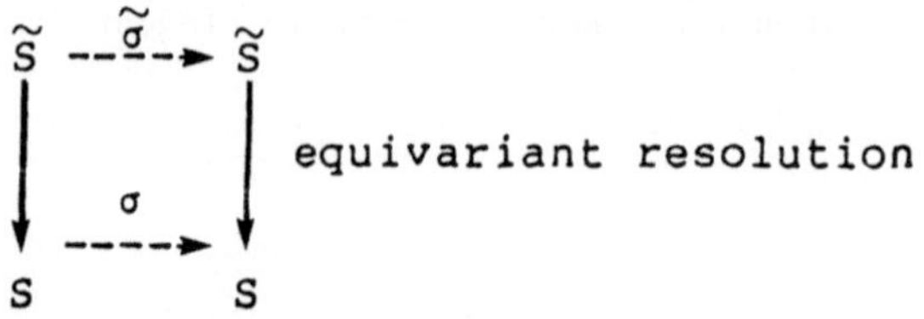

$$\begin{array}{ccc} \tilde{S} & \dashrightarrow{\tilde{\sigma}} & \tilde{S} \\ \downarrow & & \downarrow \\ S & \dashrightarrow{\sigma} & S \end{array} \qquad \text{equivariant resolution} \tag{4.1}$$

$\tilde{\sigma}$ induces the isometry $i(\sigma)$ of $\text{Pic } \tilde{S}$. This is how $\tilde{S}$ arises:

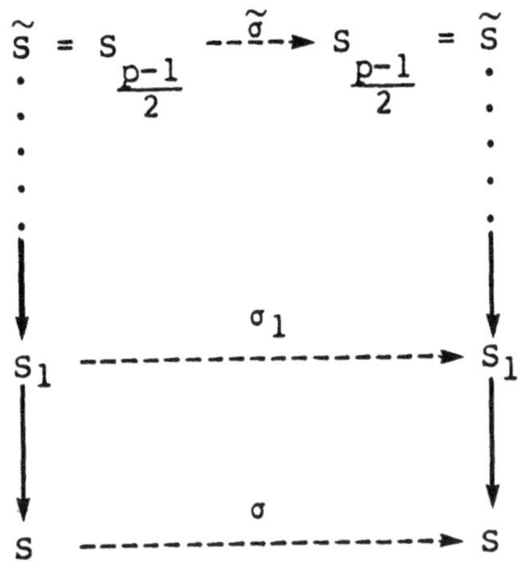

S is the p - 1/2 blowup

S_1 is the first blowup

To understand the action of $\tilde{\sigma}$, σ_1 on exceptional curves, let us write the local equation of S at a singular point $\gamma = (a,b,c)$:

$$(Z - c)^P = \frac{1}{2} F_{xx}(a,b)(x - a)^2 + F_{xy}(a,b)(x - a)(y - b)$$

$$= \frac{1}{2} F_{yy}(a,b)(y - b)^2 + \cdots + = uv + \cdots$$

[leading form factors over $k(T_{ij})(\sqrt{H_\gamma})$]. First blowup:

$$(Z - C)^{p-2} = \overset{uv}{\overset{\downarrow}{\text{same}}} + (Z - c)[\cdots] \tag{4.1.1}$$

and two more charts that I do not write down.

Pictorially, after one blowup:

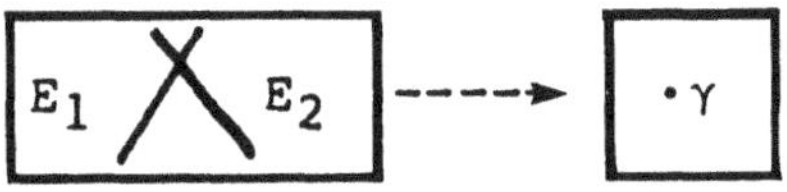

If $\sigma \in G$ and $\sigma(\sqrt{H_\gamma}) = \sqrt{H_\gamma}$, then $\sigma_1(E_1) = E_1$ and $\sigma_1(E_2) = E_2$ [since $\sigma(u) = u$ and $\sigma(v) = v$]; if $\sigma(\sqrt{H_\gamma}) = -\sqrt{H_\gamma}$, then $\sigma_1(E_1) = E_2$ and $\sigma(E_2) = E_1$]since $\sigma(v) = u$, $\sigma(u) = v$]. To prove GB ⇒ DRL, we apply this principle to the further blowups. Obviously, $i(\sigma)(\ell) = \ell$. ℓ is defined over $k(T_{ij})$.

I do not wish to go into all the details of that. In fact, I find this part very awkward to write down. So let me now pass to:

5. Proof of Gal A, Gal B

1. Auxiliary Schemes, Notations We need some notations and auxiliary schemes for this proof.

$$W = \text{Spec } k[T_{ij}]$$

(W is a parameter space for polynomials of degree p).

$$E = \frac{\text{Spec } k[T_{ij}][X,Y]}{(F_x,F_y)}$$

Recall that $F = \Sigma\, T_{ij}X^iY^j$, $0 \leq i + j \leq p$. We have a natural basic map π: $E \to W$ (generically finite and dominating). Closed points of W can be thought of as polynomials of degree p. We will say by abuse of notation $g(x,y) \in W$. Closed points of $E \approx$ pairs (g,ν), ν a critical point of g in the plane. One can show that E is isomorphic to an affine space over k, $E \approx \text{Spec } k[T_{00},T_{20},T_{02},\ldots,X,Y]$. We introduce two open and dense subsets of W, namely V, U:

$$\begin{array}{ccc} E & \xrightarrow{\pi} & W \\ \cup & & \cup \\ E_V & \xrightarrow{\pi_V} & V \\ \cup & & \cup \\ E_U & \xrightarrow{\pi_U} & U \end{array}$$

such that π_V is a finite map and π_U is an étale map. More specifically, $g \in V \overset{\text{def}}{\equiv}$ (the surface defined by $Z^p = g$) has no singularity at infinity in $\mathbb{P}^3_k$. $U \subset V$ is defined as follows: $g \in U$ iff $g \in V$ and in addition the surface defined by $Z^p = g$ has only biplanar double points [equivalently, $g_x = g_y = 0$ implies that $g_{xx}g_{yy} - (g_{xy})^2 \neq 0$]. One shows that U, V are both open and dense in W (we omit the proof).

2. Input from Fundamental Group, Monodromy Now we bring in monodromy in a way reminiscent of the Kazhdan-Margulis theorem in Deligne's Weil I paper (Deligne 1974).

We need some simple facts from the theory of the fundamental group. For any geometric point b: $\text{Spec } \Omega \to U$ of U we denote by

$F_b(E_u)$ the set of points of E_u lying over b (SGA I terminology of this type will be used in the sequel). Also, we will use some elementary theory of the fundamental group (from SGA I).

F_b is a functor from Et(U) to ENS (the category of finite sets).

$\text{Aut}(F_b) = \pi_1(U,b)$ (by SGA I)

For any base point b, $\pi_1(U,b)$ acts on $F_b(E_U)$ (the finite set).

In particular, if we choose the base point b_1 by taking

$\Omega = \overline{k(U)}$ then $F_{b_1}(E_U) = \text{Sing } S$

We have an exact sequence

$G \to \pi_1(U,b_1) \to (e)$ (SGA I)

$\pi_1(U,b_1)$ acts on $F_{b_1}(E_U) = \text{Sing } S$. [We identify $k(U) = k(T_{ij})$, $\overline{k(U)} = \overline{k(T_{ij})}$.]

ALTERNATIVE STATEMENT OF GAL A 5.1 For any base point b in U, $\pi_1(U,b)$ acts on $F_b(E_U)$ as the full symmetric group. We will show that the action is

(i) Transitive (5.1.1)

(ii) Twice transitive (5.1.2)

(iii) Contains a transposition (5.1.3)

All the statements above are independent of base point b. *Thus we will keep changing it in the proof!*

(i) Is true because E_U is irreducible (connected). It is dense in the affine space E.

(ii) Proof of (ii) uses an idea from J. Harris's paper "Galois Groups of Enumerative Problems" (J. Harris, 1979). We change the base to obtain a cartesian diagram:

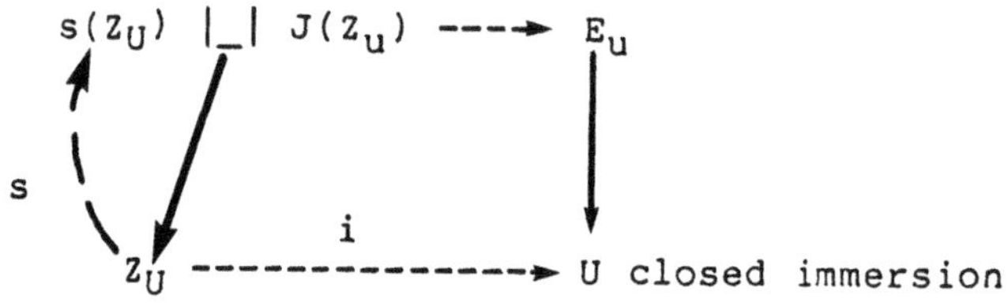

Z_u corresponds to polynomials g in U with singularity at the origin (nonvoid: e.g., I checked that $xy + Ax^{p-1} + y^{p-1} + xy^{p-1}yx^{p-1} \in Z_U$ for almost every $A \in k$). Also, s is a section, $J(Z_U)$ can be shown to be *irreducible* (omit the proof). Choose a base point b in Z_U: $\pi_1(Z_U,b)$ acts on

$$F_b^{Z_U}(s(Z_U) \sqcup J(Z_U)) = \underset{\substack{\uparrow \\ \text{singleton set}}}{\{B\}} \sqcup F_b^{Z_U}(J(Z_U))$$

It stabilizes B and acts transitively on $E_b^{Z_u}(J(Z_u))$. Now think of b as a geometric point of U; we already know that $\pi_1(U,b)$ acts transitively on $F_b(E_U)$. Also, $F_b(E_u) \approx \{B\} \sqcup F_b^{Z_U}(J(Z_U))$ and the identification is compatible with induced homomorphism $\pi_1(Z_U,b) \xrightarrow{i^*} \pi_1(U,b)$. We conclude that there is an element of $F_b(E_U)$ (corresponding to B) such that its stabilizer in $\pi_1(U,b)$ acts transitively on its complement, hence twice transitivity.

3. *Reduction to Curves (or One-Dimensional Henselian Rings)* We will deduce (iii) and Gal B from one common sublemma. We need one more auxiliary variety (I promise that this is the last one) to state the sublemma, namely:

$$D = E[\sqrt{H(F)}]$$

[recall that $H(F) = F_{xx}F_{yy} - F_{xy}^2$].
Fact: H(F) is irreducible in k[E] (easy proof is on page 279); also recall that E is an affine space. We conclude that $D \xrightarrow{\phi} E$ is a finite map, D normal. Also, one sees by the jacobian criterion that

$$D_U \xrightarrow{\phi_U} E_U \xrightarrow{\pi_U} U \qquad \phi_U,\ \pi_U \text{ étale, } \phi_U \text{ degree } 2$$

The idea of the sublemma is to change base from U to a curve or, more precisely, to a one-dimensional henselian ring. After the base change we easily get some information about monodromy.

SUBLEMMA 5.2 There exists a base change:

$$\begin{array}{ccc} \operatorname{Spec} \tilde{K}(\sqrt[4]{t}) \sqcup \tilde{\tilde{S}} & \dashrightarrow & D_U \\ \downarrow & & \downarrow \\ \operatorname{Spec} \tilde{K}(\sqrt{t}) \sqcup \tilde{S} & \dashrightarrow & E_U \\ \downarrow & & \downarrow \\ \operatorname{Spec} \tilde{K} & \dashrightarrow & U \end{array} \qquad (5.2.1)$$

where $\tilde{K}$ is the field of fractions of a one-dimensional henselian dvr $\tilde{\mathcal{O}}_q$ with uniformizing parameter t, and $\tilde{S}$, $\tilde{\tilde{S}}$ are disjoint unions of trivial coverings of Spec $\tilde{K}$ (disjoint union of sections over Spec $\tilde{K}$).

Let us first see how the sublemma implies both (iii) (5.1.3) and GB (2.5):

$$\text{Sublemma} \begin{array}{l} \nearrow \text{iii} \\ \searrow \text{GB} \end{array}$$

Choose a geometric point $\bar{b}$: Spec $\bar{K} \to$ Spec $\tilde{K}$ ($\bar{K}$ = algebraic closure of $\tilde{K}$). Taking $F_{\bar{b}}$ in diagram (5.2.1), we obtain

$$\begin{array}{ccc} F_{\bar{b}}(\operatorname{Spec} \tilde{K}(\sqrt[4]{t})) \sqcup F_{\bar{b}}(\tilde{\tilde{S}}) & = & X \\ \downarrow m & & \downarrow m \\ F_{\bar{b}}(\operatorname{Spec} \tilde{K}(\sqrt{t})) \sqcup F_{\bar{b}}(\tilde{S}) & = & Y \end{array} \qquad (5.2.2)$$

By SGA I, π_1(Spec $\tilde{K}$,b) = Gal($\bar{K}$:$\tilde{K}$) and the group acts on the diagram of finite sets (5.2.2). By Galois theory, using fourth roots of unity, we get an element μ which acts as follows:

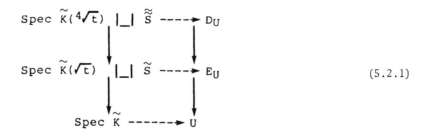

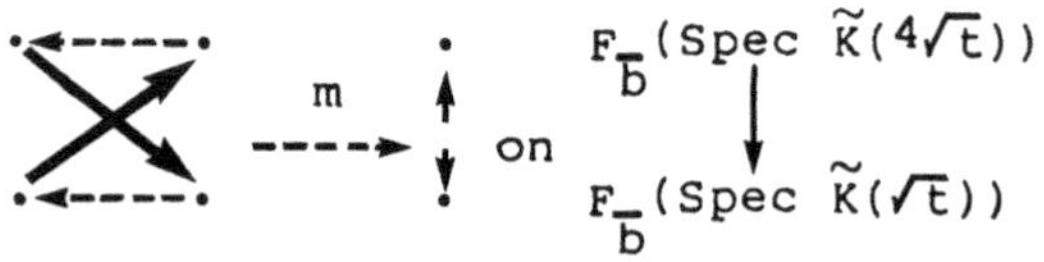

and induces the identity on $F_{\bar{b}}(\tilde{\tilde{S}}) \to F_{\bar{b}}(\tilde{S})$. In particular, σ induces a transposition on X. Let us also look at μ^2:

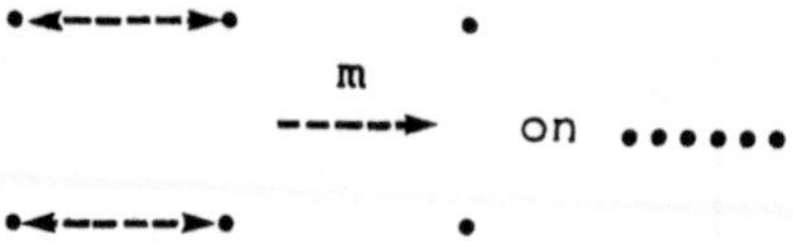

identity elsewhere in X and Y (an action like μ^2 will be called a *double flip*).

Proof of (iii) (5.1.3): There is an induced homomorphism π_1 (Spec $K,\bar{b}) \to \pi_1(U,\bar{b})$, and from μ we get an element $h(\mu)$ of $\pi_1(U,\bar{b})$ which induces a transposition in $F_{\bar{b}}(E_U)$ [since μ induces a transposition on X which is the pullback of $F_{\bar{b}}(E_U)$, and from the definition of the induced homomorphism SGA I].

Proof of GB (2.5): As above, using μ^2 we get an element $h(\mu^2)$ of $\pi_1(U,\bar{b})$ which acts on $F_{\bar{b}}(D_U) \xrightarrow{m_1} F_{\bar{b}}(E_U)$ as a double flip, i.e.,

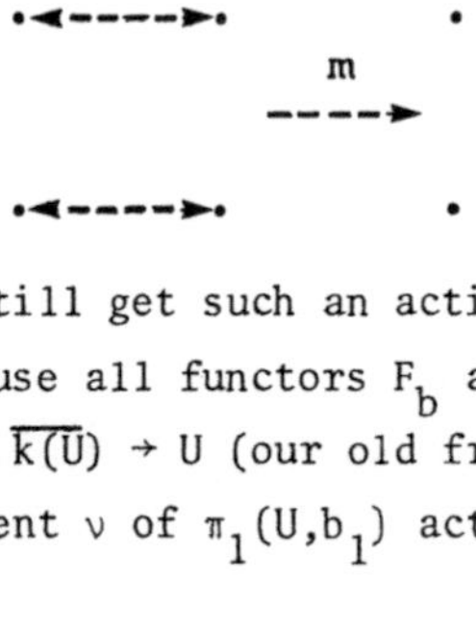

We still get such an action for any other geometric base point in U because all functors F_b are isomorphic. In particular, choose b_1: Spec $\overline{k(U)} \to U$ (our old friend). We get a double flip induced by an element ν of $\pi_1(U,b_1)$ acting on

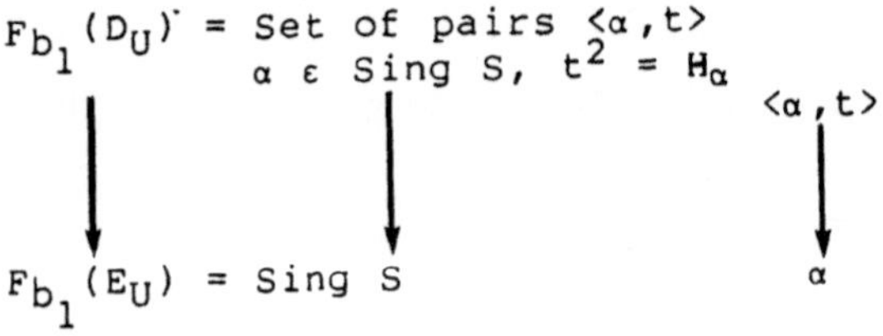

Again we use the surjection $G \to \pi_1(U,b_1) \to (e)$ and we will get the element σ of G as required in GB (σ maps to ν). [G acts on right-hand side sets, $\pi_1(U,b_1)$ on left-hand side sets compatibly with

the identification of left-hand- and right-hand side and with the map $G \to \pi_1 \to (e)$.]

Sketch of proof of sublemma 5.2: We pick a special point $q \in V - U$ [e.g., $q \leftrightarrow y^2 + x^3 + Ax^{p-1} + By^{p-1} + xy^{p-1} + yx^{p-1} = g(x,y)$ will do for almost every $A \in k$; we take $B = 1$ if $p \neq 13$ and $B = -1$ if $p = 13$] and a curve $L \subset V$, $q \in L$ [e.g., $g + \lambda x$, λ runs through k], so that $L_U = L \cap U$ is dense in L. If $L_1 = L_U \cup \{q\}$, the induced coverings are as follows:

$$\begin{array}{lll} D_{L_1} & \bar{\bar{q}} & e(\bar{\bar{q}}) = 4 \\ \Phi_{L_1}\big\downarrow & \big\downarrow & \\ E_{L_1} & \bar{q} & e(\bar{q}) = 2 \\ \pi_{L_1}\big\downarrow & \big\downarrow & \\ D_{L_1} & q & \end{array}$$

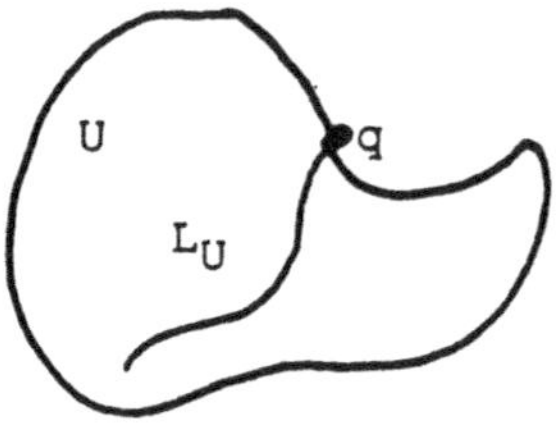

ϕ_{L_1}, π_{L_1} étale at all other points

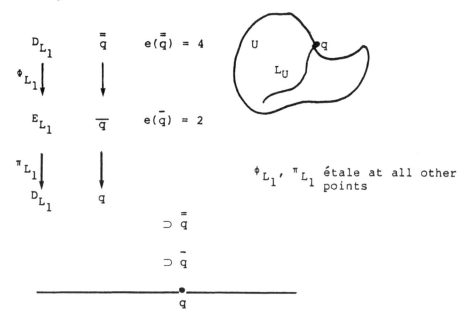

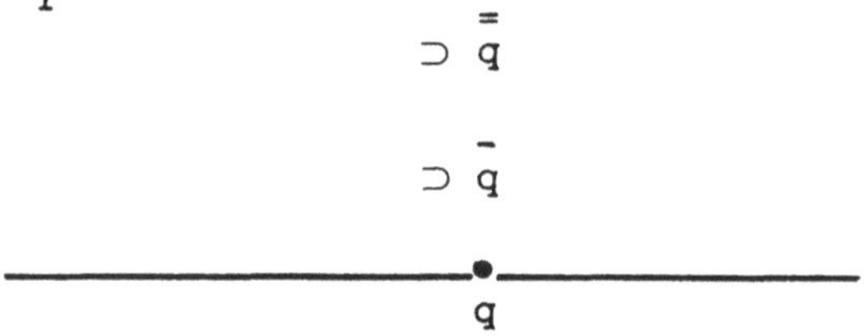

Also, D_{L_1}, E_{L_1} are irreducible, smooth curves. Henselize O_q, the local ring of q in L_1.

$$\begin{array}{l} O_q \subset \tilde{O}_q \subset \tilde{K} \supset k(L_u) = \text{fraction field of } O_q \\ \qquad\qquad\qquad\quad \big\downarrow \\ \tilde{K}\text{-field of fractions of } \tilde{O}_q \end{array}$$

Spec $\tilde{K} \to L_U \to U$ gives the desired base change by simple properties of henselian rings.

REMARK 1 Verifying that q and L mentioned above have all the desired properties is a lengthy computation which I shall not do here.

REMARK 2 As $\lambda \to 0$, $g + \lambda x$ behaves as follows: Two Morse singularities collapse to a cusp singularity (a catastrophe).

6. Conclusion

REMARK F1 (6.1) The proof extends to show that

$$R = \frac{L[x,y,z]}{(z^p - \sum_{0 \le i+j \le pe} T_{ij} x^i y^j)} \qquad p \ge 5,\ e = 1, 2, \ldots$$

is a UFD.

REMARK F2 (6.2) The MTH implies the following result:

$$R(c_{ij}) = \frac{k[x,y,z]}{(z^p - \sum_{0 \le i+j \le pe} c_{ij} x^i y^j)}$$

Then $R(c_{ij})$ is a UFD for a general choice of the constants c_{ij} in k, i.e., a Zariski open dense set. (See appendix 1 to Chapter 4.) Thus the situation is better than in the original theorem of Noether, where jump phenomena make passage from generic to general impossible. (We expect an Artin-type stratification on the coefficient space $U \subset W$.)

W. E. Lang Theorem: Pic $\tilde{S}$ has no Torsion

Idea of proof: $\tilde{S} \to \bar{S}$ minimal desingularization. Let $0 \neq L \in \text{Pic}\,\tilde{S}$ be torsion. Then by Artin's (1962) theorem,

$$L \equiv L_1 \subset \tilde{S} - \bigcup_{\text{all } \alpha, j} C_j^\alpha \qquad \text{because of rational singularities}$$

and we conclude that there must be a torsion element in Pic $\bar{S}$. But Pic$(\bar{S})$ has no torsion! For $\ell \neq p$ this is standard; for p-torsion prove $H^0(\Omega(1/\bar{S})) = 0$ and use the absence of logarithmic differentials.

REFERENCES

Artin, M., Some numerical criteria for contractability of curves on algebraic surfaces, *Amer. J. Math.*, 84 (1962), 485-496.

Deligne, P., La conjecture de Weil I, *Publ. Math., IHES*, 43 (1974), 273-307.

Deligne, P., and N. Katz, Groupes de Monodromie en Geometrie Algébrique, Lecture Notes in Mathematics 340, Springer-Verlag, 1973.

Harris, J., Galois groups of enumerative problems, *Duke Math. J.*, 46 (1979), 685-724.

Lang, W. E., Remarks on p-torsion of algebraic surfaces, *Comp. Math.*, 52(2) (1984), 197-202.

Lefschetz, S., *L'analysis situs et la geometrie algébrique*, Gauthier-Villars, 1924 (reprinted by Chelsea).

APPENDIX 3: CALCULATION OF THE PICARD GROUPS[†]

I. Énoncé du Théorème Principal

Notations algébriques: Soit k un corps algébriquement clos de caractéristique $p \geqq 5$. Soient X, Y, Z, T_{ij}, $0 \leqq i + j \leqq p$ des variables algébriquement indépendantes sur k; $L = \overline{k(T_{ij})}$ une clôture algébrique de $k(T_{ij})$:

$$R = L[X,Y,Z]/(Z^p - \sum_{0 \leqq i+j \leqq p} T_{ij}X^iY^j)$$

THÉORÈME PRINCIPAL 1.1 R est un anneau factoriel.

REMARQUE Nous laissons au lecteur l'exercice d'étendre la preuve ci-dessous en remplaçant p par n'importe quel multiple de p.

Notations Géométriques et Lemmes Galoisiens $S = \operatorname{Spec} R$, $S \subset \mathbf{A}^3_L$, $\bar{S}$ l'adhérence de S dans $\mathbb{P}^3_L$, π: $\tilde{S} \to \bar{S}$ une désingularisation minimale de $\bar{S}$ (la surface $\tilde{S}$ est une surface de Zariski sur L [1]), $G = \operatorname{Gal}(\overline{k(T_{ij})} : k(T_{ij}))$. On peut prouver que $\bar{S}$ est une surface normale sur L. De plus, $\operatorname{Sing} \bar{S} \subset S$, et le cardinal de Sing S est $p^2 - 3p + 3$ [1].

REMARQUE Nous pouvons définir $\bar{S}$ comme un sous-schema du fibré vectoriel associé à $\mathcal{O}_{\mathbb{P}^2_L}(1)$: $\bar{S}$ est défini par l'équation $z^p = \Sigma\, T_{ij}X^iY^jX_0^{p-i-j} = F_0(X,Y,X_0)$ où nous interprétons F_0 comme une

---

[†]By Piotr Blass. Reprinted from C. R. Acad. Sc. Paris, t. 297 (10 octobre 1983).

section de $\mathcal{O}_{\mathbb{P}^2_L}(p)$.

LEMME GAL A L'homomorphisme de G vers le groupe des permutations de l'ensemble fini Sing(S) est surjectif.

On démontre que localement pour la topologie étale chaque point singulier de S a la forme $w^p = uv$, donc le graphe de la résolution $\pi^{-1}(\alpha)$ pour chaque $\alpha \in$ Sing S est le suivant:

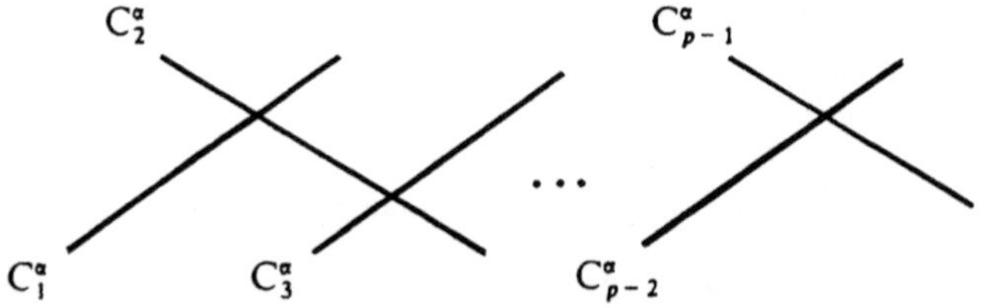

Soit T_α la suite $C_1^\alpha, C_2^\alpha, \ldots, C_{p-1}^\alpha$. La matrice d'intersection dans Pic $\tilde{S}$ est de type A_{p-1}, c'est-à-dire que:

$$(C_j^\alpha)^2 = -2 \qquad C_j^\alpha C_{j+1}^\alpha = 1 \qquad C_j^\alpha C_i^\alpha = 0 \quad \text{si } |i-j| \geq 2$$

Soit T_α^{opp} la suite opposée $C_{p-1}^\alpha, C_{p-2}^\alpha, \ldots, C_1^\alpha$. Si $\sigma \in G$ on designe par $i(\sigma)$ l'isométrie de Pic $\tilde{S}$, induite par σ.

LEMME GAL B Pour chaque paire de points singuliers α, β de Sing S, il existe $\sigma \in G$ tel que $i(\sigma)(T_\alpha) = T_\alpha^{opp}$, $i(\sigma)(T_\beta) = T_\beta^{opp}$ et $i(\sigma)(T_\gamma) = T_\gamma$ pour $\gamma \in$ Sing S distinct de α et β.

REMARQUE Soit 1 la classe du diviseur $\pi^{-1}(\bar{S} - S)$. Pour chaque $\sigma \in G$, $i(\sigma)(1) = 1$.

II. Preuve de L'implication

Gal A, Gal B ⇒ Théorème Principal. Soit $\text{Pic}^{ob}\tilde{S}$ le sous-groupe de Pic $\tilde{S}$ engendré par 1 et tous les diviseurs C_j^γ ($\gamma \in$ Sing S, $0 \leq j \leq p-1$). Nous avons une suite exacte:

$$0 \to \mathrm{Pic}^{ob}\tilde{S} \to \mathrm{Pic}\,\tilde{S} \to C\ell\, R \to 0$$

où $C\ell\, R$ est le groupe des classes de diviseurs de R. Le groupe $C\ell\, R$ est un p-groupe élémentaire (voir Samuel, 1964) et Pic $\tilde{S}$ n'a pas de p-torsion (voir Lang).

Pour prouver (1.1) il suffit donc de prouver:

THÉORÈME 2.1 $p\,\mathrm{Pic}\,\tilde{S} \subset p\,\mathrm{Pic}^{ob}\tilde{S}$.

Soit $x \in p\,\mathrm{Pic}\,\tilde{S}$. Alors $x \in \mathrm{Pic}^{ob}\tilde{S}$, donc $x = n_1 1 + \Sigma_{\alpha \in \mathrm{Sing}\, S}\, x_\alpha$ où $n_1 \in \mathbf{Z}$ et x_α est une combinaison linéaire des C_j^α. Nous devons avoir $(x_\alpha, C_j^\alpha) \equiv 0(p)$ pour chaque $\alpha \in \mathrm{Sing}\, S$ et $0 \leq j \leq p - 1$. Cela implique que x peut être récrit sous la forme $x = n_0 1 + \Sigma\, m_\alpha D_\alpha + py$ où $0 \leq n_0$, $m_\alpha \leq p - 1$, $y \in \mathrm{Pic}^{ob}\tilde{S}$ et $D_\alpha = C_1^\alpha + 2C_2^\alpha + \cdots + (p - 1) \times C_{p-1}^\alpha$. Notre but est de prouver que $n_0 = 0$ et $m_\alpha = 0$ pour chaque $\alpha \in \mathrm{Sing}\, S$. Soient α, β, γ, trois éléments distincts arbitraires de Sing S.

Nous appliquons les lemmes Gal B pour α, β.

$$x - i(\sigma)(x) = m_\alpha(D_\alpha - D_\alpha^{opp}) + m_\beta(D_\beta - D_\beta^{opp}) + p(y - i(\alpha)(y))$$

donc:

$$m_\alpha(D_\alpha - D_\alpha^{opp}) + m_\beta(D_\beta - D_\beta^{opp}) \in p\,\mathrm{Pic}\,\tilde{S}$$

ce qui, après calcul, implique que $m_\alpha D_\alpha + m_\beta D_\beta \in p\,\mathrm{Pic}\,\tilde{S}$. De même $m_\alpha D_\alpha + m_\gamma D_\gamma$ et $m_\beta D_\beta + m_\gamma D_\gamma$ sont dans p Pic $\tilde{S}$. Parce que $p \neq 2$ nous pouvons conclure que $m_\alpha D_\alpha \in p\,\mathrm{Pic}\,\tilde{S}$. Mais $(m_\alpha D_\alpha)^2 = m_\alpha^2(p)(p - 1) \equiv -pm_\alpha^2(p^2)$: nous devons alors avoir $m_\alpha = 0$. Puisque α est arbitraire dans Sing S, on a $x = n_0 1 + py$. Donc $n_0^2 1^2 = n_0^2 p \equiv 0(p^2)$ et par conséquent $n_0 = 0$.

III. Esquisse de la Preuve de Gal A, Gal B

Soient W l'espace affine des polynômes g(x,y) de degré $\leq p$ et $V \subset W$ l'ouvert de Zariski formé des g tels que la surface correspondante $z^p = g(x,y)$ n'ait pas de point singulier à l'infini. Soit $U \subset V \subset W$ défini par la condition que $g_x = g_y = 0$ implique $g_{xx}g_{yy} - g_{xy}^2 \neq 0$:

$g \in U$ si et seulement si tout point singulier de l'application g est quadratique ordinaire.

Alors $W = \operatorname{Spec} k[T_{ij}]$. Soit:

$$F(x,y) = \Sigma\, T_{ij} X^i Y^j \qquad E = \operatorname{Spec} \frac{k[T_{ij}][X,Y]}{(\partial F/\partial X, \partial F/\partial Y)}$$

$$H(F) = \left(\frac{\partial^2 F}{\partial X^2}\right)\left(\frac{\partial^2 F}{\partial Y^2}\right) - \left(\frac{\partial^2 F}{\partial X \partial Y}\right)^2 \qquad D = E[\sqrt{H(F)}]$$

Nous avons un diagramme cartésien:

$$\begin{array}{ccccc} D_u & \longrightarrow & D_v & \longrightarrow & D \\ \downarrow{\scriptstyle \psi_u} & & \downarrow{\scriptstyle \psi_v} & & \downarrow{\scriptstyle \psi} \\ E_u & \longrightarrow & E_v & \longrightarrow & E \\ \downarrow{\scriptstyle \varphi_u} & & \downarrow{\scriptstyle \varphi_v} & & \downarrow{\scriptstyle \varphi} \\ U & \hookrightarrow & V & \hookrightarrow & W \end{array}$$

Dans le diagramme ci-dessus, ψ_v, φ_v sont finis (projectifs) et ψ_u, φ_u sont étales. Dans ce qui suit nous employons des notations et des faits de Grothendieck (1971, p. 140). Soit Et(U) = catégorie des revêtements étales de U. Soit b un point géométrique de U. Pour un objet X de Et(U) soit $F_b(X)$ = ensemble des points géométriques de X au-dessus de b. Ainsi F_b devient un foncteur sur Et(U) à valeur dans la catégorie des ensembles finis. $\pi_1(U,b) = \operatorname{Aut}(F_b)$.

SOUS-LEMME 3.1 Pour chaque point géométrique b dans U l'action de $\pi_1(U,b)$ sur l'ensemble $F_b(E_u)$ est transitive.

Preuve: On démontre que E est isomorphe à l'espace affine sur k, conc E_u est connexe (Grothendieck, 1971, p. 140).

SOUS-LEMME 3.2 L'action de $\pi_1(U,b)$ sur $F_b(E_u)$ est doublement transitive.

Preuve: Soit Z l'ensemble fermé défini par $T_{10} = T_{01} = 0$. Nous considérons le diagramme cartésien:

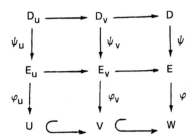

$$\begin{array}{ccc} E_Z & \longrightarrow & E_u \\ \varphi_Z \downarrow & & \downarrow \varphi_u \\ Z & \longrightarrow & U \end{array}$$

et démontrons que φ_Z a une section s: $Z \to E_Z$ et E_Z est la réunion disjointe de s(Z) et d'un schéma connexe J(Z). Pour conclure on choisit un point géométrique b_0 dans Z, on emploie l'homomorphisme induit $\pi_1(Z,b_0) \to \pi_1(U,b_0)$ et 3.1.

SOUS-LEMME 3.3 Il existe un anneau de valuation discrète hensélien d'uniformisante t et de corps de fractions $\tilde{K}$ et un morphisme Spec $\tilde{K} \to$ U tel que nous avons un diagramme cartésien:

$$\begin{array}{ccc} \text{Spec } K(\sqrt[4]{t}) \cup \tilde{\tilde{S}} & \longrightarrow & D_u \\ \downarrow & & \downarrow \\ \text{Spect } K(\sqrt{t}) \cup \tilde{S} & \longrightarrow & E_u \\ \downarrow & & \downarrow \\ \text{Spec } \tilde{K} & \longrightarrow & U \end{array}$$

où $\tilde{S}$ et $\tilde{\tilde{S}}$ sont des revêtements complètement décomposés (Grothendieck, V.6.4, p. 137) de Spec $\tilde{K}$.

Idée de la preuve de 3.3: Choisissons un point particulier $q \in V - U$, correspondant à:

$$g(x,y) = y^2 + x^3 + Ax^{p-1} + By^{p-1} + xy^{p-1} + yx^{p-1}$$
$$\text{où } B = 1 \quad \text{si } p \neq 13 \quad \text{et} \quad B = -1 \quad \text{si } p = 13$$

et A est dans k et est général, i.e., on exclut un nombre fini d'éléments de k.

Considérons la courbe L dans V d'équations $g + \lambda x$, $\lambda \in k$. Alors $L_u = L \cap U$ est dense en L. Posons $L_1 = L_u \cup \{q\}$. Soit $\mathcal{O}_q$ l'anneau local de q dans L_1 et $\tilde{\mathcal{O}}_q$ son hensélisé. $\tilde{K}$ est le corps de fractions de $\tilde{\mathcal{O}}_q$. Nous utilisons maintenant le changement de base Spec $\tilde{K} \to L_u \to U$ et prouvons par un long calcul le:

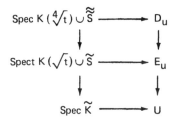

SOUS-LEMME 3.4 Pour chaque point géométrique b de U l'action de $\pi_1(U,b)$ sur $F_b(D_u) \to F_b(E_u)$ contient un élément μ tel que (a) μ est une transposition simple dans $F_b(E_u)$; (b) μ fixe tout les éléments de $F_b(D_u)$ à l'exception de quatre éléments, de plus μ est d'ordre quatre comme permutation de $F_b(D_u)$.

Graphiquement μ agit de la façon suivante:

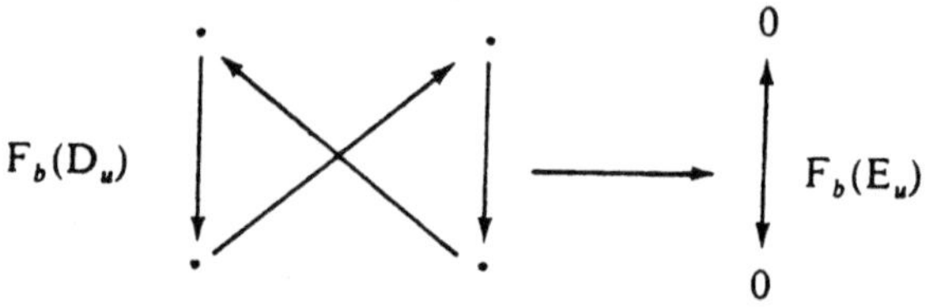

et μ agit comme l'identité ailleurs. La preuve est bien connue pour $K \subset K(\sqrt{1}) \subset K(\sqrt[4]{1})$ et on applique la fonctorialité de π_1.

Preuve de Gal A, Gal B: Choisissons le point géométrique b_1: Spec L $\to$ U tel que:

$$\begin{array}{ccc} F_{b_1}(E_u) & \approx & \text{Sing } S \\ \uparrow & & \uparrow \\ F_{b_1}(D_u) & \approx & \text{l'ensemble des paires } \langle \alpha, l_\alpha \rangle \end{array}$$

$\alpha \in$ Sing S

$\alpha = (\alpha_1, \alpha_2, \alpha_3) \in \mathbf{A}_L^3,\ l_\alpha^2 = H(F)(\alpha, \alpha_2) = H_\alpha$

Il existe d'après Grothendieck (1971, p. 143) une homomorphisme surjectif:

$$G \to \pi_1(U_1, b_1) \to (e) \tag{3.6}$$

Par ailleurs toute permutation de $F_b(E_u)$ est réalisée par un élément de $\pi_1(U,b_1)$ d'après 3.4(a), 3.1, et 3.2. De meme pour toute permutation de Sing S, ce qui démontre Gal A. Nous obtenons des énoncés 3.4(b), 3.5, Gal C, i.e., l'énoncé suivant: pour chaque paire de points singuliers α, β dans Sing S il existe $\sigma \in G$ tel que σ

induit l'identité sur Sing S mais $\sigma(\sqrt{H_\alpha}) = -\sqrt{H_\alpha}$, $\sigma(\sqrt{H_\beta}) = -\sqrt{H_\beta}$, et $\sigma(\sqrt{H_\gamma}) = \sqrt{H_\gamma}$ pour $\gamma \in$ Sing S distinct de α et β. Nous laissons l'implication Gal C ⇒ Gal B comme un exercice simple sur la résolution des singularités A_{p-1}.

REMARQUE FINALE On peut démontrer le corollaire suivant. Il existe un ouvert dense $U' \subset U$ tel que pour tout point fermé $(C_{ij}) \in U'$ l'anneau:

$$R(C_{ij}) = \frac{k[x,y,z]}{(z^p - \Sigma\, C_{ij} X^i Y^i)}$$

est factoriel. La preuve paraitra ultérieurement.

Acknowledgment L'auteur remercie Pierre Deligne de lui avoir montré une esquisse de toute la démonstration.

REFERENCES

Blass, P., *Contemporary Mathematics*, 13 (1982, 335-347, A.M.S., 1983.

Lang, W. E., Remarks on p-torsion of algebraic surfaces, *Compositio Math.* (à paraitre).

Samuel, P., Lectures on Unique Factorization Domains, Tata Lecture Notes, 1964.

Grothendieck, A., Lecture Notes in Mathematics, No. 224, Springer-Verlag, New York, 1971.

5

The Divisor Classes of $z^p = G(x,y)$: A Programmable Problem[†]

0. INTRODUCTION[††]

Let k be an algebraically closed field of characteristic $p > 0$. Let $F \in \mathbb{A}_k^3$ be a normal affine hypersurface defined by an equation of the form $z^p = G(x,y)$, where $G(x,y) \in k[x,y]$. In this chapter we develop an algorithm for computing the divisor class group of F. We then translate this algorithm into a computer program.

The principal tools used in this study are P. Samuel's Galois descent techniques developed in his 1964 Tata notes and some results proved in Lang (1983a).

We associate to each surface a system of p-linear equations containing linear equations and equations of the form $a_1x_1 + \cdots + a_nx_n = x_j^p$, where the a_i belong to the field F_G obtained by adjoining the coefficients of G(x,y) to the field $\mathbb{Z}/p\mathbb{Z}$ [see (2.5)].

The set of all solutions of this system of equations will form a group under addition isomorphic to the class group of F and it will be a direct sum of copies of $\mathbb{Z}/p\mathbb{Z}$ [see (1.11)].

The problem of determining $C\ell(F)$ is now reduced to that of finding all solutions to the associated system of p-linear equations. But this is where the difficulty begins. In what and how large an

---

[†]By Piotr Blass, Jeffrey Lang, and David Joyce (Clark University, Worcester, Massachusetts).

[††]This section is reprinted from *Journal of Algebra*, Vol. 100, No. 2, 1986.

extension field of F_G must we search for these solutions? Suddenly, our algorithm exists only in theory and not in practice. For the true test of an algorithm is whether it can be transferred to a computer program. But as stated above, our technique is not programmable.

Using lemmas 2.1, 2.2, 2.3, and 2.4, we demonstrate that the order of the solution set of the associated p-linear system can be determined without having to look beyond the field F_G. We need only apply row reduction and determinant techniques similar to those used in linear algebra.

In Section 2 we provide the corresponding computer program and provide some sample computations which verify many lengthy "by hand" computations done by Lang (1981).

The determination of the class group of a normal affine surface F: $z^p = G(x,y)$ in a matter of minutes via a computer program is for us an exciting development. The use of computer science in the study of algebraic geometry is usually not considered. Many studies on the subject of divisor classes do not produce that many concrete examples, but we now have an infinite number of examples at our fingertips.

Also, many questions can be explored using this program from a statistical point of view: for example, the following conjectures, which can be found in Lang (1983b):

(1) If $p = 2$, is the class group of F isomorphic to $\mathbf{Z}/2\mathbf{Z}$ for a generic choice of G?

(2) If $p > 2$, is the coordinate ring of the surface $z^p = G$ a unique factorization domain for a generic G?

(3) How do the divisor class groups of F: $z^p = G(x,y)$ stratify the coefficient space of this surface? [For more details, see Blass (1985) and Lang (1981).]

Notation

k algebraically closed field of characteristic $p > 0$, unless stated to the contrary

$\mathbf{A}_k^n$ affine n space over k

$\mathbf{P}_k^n$ projective n space over k

Surface irreducible, reduced, two-dimensional, quasi-projective variety over k

Hypersurface irreducible, reduced, n-dimensional, quasi-projective variety over k

For a smooth surface X we write p_g = geometric genus of X.

The notation F: $f(x_1,x_2,\ldots,x_n) = 0$ means

$$F = \operatorname{Spec} \frac{k[x_1,\ldots,x_n]}{(f(x_1,\ldots,x_n))} \qquad F \subset A^n$$

If A is a Krull ring, we denote by $C\ell(A)$ the divisor class group of A. If X is a surface, we denote by $C\ell(X)$ the divisor class group of the coordinate ring of X.

If $f \in k[x_1,\ldots,x_n]$, then deg(f) means the total degree of f and $\deg_{x_i}(f)$ means the degree of f in the variable x_k, i = 1, 2, ..., n.

1. PRELIMINARIES

P. Samuel's Galois descent techniques developed in his 1964 Tata notes are the principal tools used in this chapter. The following is a brief discussion of some results from these notes (see Samuel, 1964a, pp. 61-65).

THEOREM 1.1 Let A and B be Krull rings. Suppose that B is integral over A or that B is a flat A algebra. Then there is a well-defined group homomorphism ϕ: $C\ell(B) \to C\ell(A)$ (Samuel, 1964a, pp. 19-20).

Let B be a Krull ring of characteristic $p > 0$. Let Δ be a derivation of qt(B) such that $\Delta(B) \subset B$. Let $K = \ker(\Delta)$ and $A = B \cap K$. Then A is a Krull ring with B integral over A. Thus we have a map ϕ: $C\ell(A) \to C\ell(B)$.

Set $L = \{t^{-1}\Delta t\colon\ t^{-1}\Delta t \in B,\ t \in qt(B)\}$. L is an additive subgroup of B called the *group of logarithmic derivatives* of Δ. Set $L' = \{u^{-1}\Delta u \mid u \text{ is a unit in } B\}$. Then L' is a subgroup of L.

THEOREM 1.2 (a) There exists a canonical monomorphism ϕ: ker $\phi \to L/L'$. (b) If $[qt(B):K] = p$ and $\Delta(B)$ is not contained in any height one primes of B, then $\bar{\phi}$ is an isomorphism (Samuel, 1964a, p. 62).

THEOREM 1.3 If $[qt(B):K] = p$, then

(a) There exists $a \in A$ such that $\Delta^p = a\Delta$.

(b) An element $t \in B$ is in L if and only if $\Delta^{p-1}t - at = -t^p$ (Samuel, 1964a, p. 62).

REMARK 1.4 All of the rings that are studied in this chapter are coordinate rings of normal surfaces. Hence they are noetherian integrally closed domains, which are Krull domains (see Samuel, 1964a, theorem 3.2, p. 5).

Below we state several facts about the order and type of $C\ell(F)$, and "Ganong's formula," which plays a key role in our algorithm.

LEMMA 1.5 Let $G(x,y) \in k[x,y]$ be a nonzero polynomial such that $\partial G/\partial x$ and $\partial G/\partial y$ have no common factor in $k[x,y]$. Then the surface $F \subset \mathbb{A}^3_k$ defined by the equation $z^p = G(x,y)$ is normal and has coordinate ring isomorphic to $A = k[x^p, y^p, G]$. [For the proof, see Matsumura (1970, p. 125) and Lang (1983b, p. 393).]

Let D: $k(x,y) \to k(x,y)$ be the k-derivation defined by

$$D = \frac{\partial G}{\partial y}\frac{\partial}{\partial x} - \frac{G}{\partial x}\frac{\partial}{\partial y}$$

D is called the *jacobian derivation* of G.

LEMMA 1.6 With D as above and A as in (1.5), we have that $D^{-1}(0) \cap k[x,y] = A$ (see Lang, 1983b, p. 394).

LEMMA 1.7 Let D and A be as in (1.6). Let L be the group of logarithmic derivatives of D, $L = \{f^{-1}Df\colon\ f \in k(x,y) \text{ and } f^{-1}Df \in k[x,y]\}$. Then $C\ell(A) \simeq L$ (see Lang, 1983b, p. 394).

LEMMA 1.8 Let D, A, and L be as in (1.7). If an element $t \in k[x,y]$ is in L, then $\deg t \le \deg G - 2$.

REMARK 1.9 By bounding the degree of any logarithmic derivative we obtain a method for calculating the divisor class group of a normal surface of the form $z^p = G(x,y)$ over an algebraically closed field of characteristic $p > 0$. We write t as a polynomial in x and y of degree equal to deg G - 2 with undetermined coefficients. We then substitute t into the differential equation (1.3b) and compare coefficients. We then see that the coefficients of t must satisfy a certain system of equations. The number of solutions of this system will be the order of $C\ell(F)$. Since $L \subset k[x,y]$, each element in the class group will have p-torsion.

We will see shortly that these class groups are always of finite order. First we state Ganong's formula, which will prove useful in studying the differential equation (1.3b).

THEOREM 1.10 (Lang-Ganong Formula) Let $G \in k[x,y]$ satisfy the conditions of Lemma 1.5 and D be the jacobian derivation of G. Then $\forall\ \alpha \in k(x,y)$,

$$D^{p-1}\alpha - a\alpha = -\sum_{i=0}^{p-1} G^i\nabla(G^{p-(i+1)}\alpha) \qquad \text{where } \nabla = \frac{\partial^{2(p-1)}}{\partial x^{p-1}\partial y^{p-1}}$$

[For the proof of 1.10, see Lang (1983b, pp. 395-397).]

PROPOSITION 1.11 Let $G \in k[x,y]$ satisfy the conditions of lemma 1.5, and $F \subset \mathbf{A}_k^3$ be the surface defined by $z^p = G$. Then $C\ell(F)$ is a finite p-group of type $(p,\ldots,p)$ and order p^M, where $M \le g(g - 1)/2$ and $g = \deg(G)$. [For the proof of 1.11, see Lang (1983b, pp. 397-398).]

2. SOME COMPUTATIONAL LEMMAS

The next four lemmas form the basis of our algorithm. They allow us to find the number of solutions of the associated p-linear system of equations without actually solving it.

LEMMA 2.1 Let k be an algebraically closed field of characteristic $p = 0$. Let S_1 be a system of equations of the form

$$
\begin{aligned}
S_1:\quad &(1)\quad a_{11}x_1 + \cdots + a_{1n}x_n = x_1^p\\
&(2)\quad a_{21}x_1 + \cdots + a_{2n}x_n = x_2^p\\
&\quad\vdots\\
&(n)\quad a_{n1}x_1 + \cdots + a_{nn}x_n = x_n^p\\
&(\ell_1)\quad a_1x_1 + \cdots + a_nx_n = 0 \qquad \text{where } a_n \neq 0
\end{aligned}
$$

Then S_1 is equivalent (in the sense that it has the same number of solutions) to the system S_2 given below.

$$
\begin{aligned}
S_2:\quad &(1)\quad b_{11}x_1 + \cdots + b_{1(n-1)}x_{n-1} = x_1^p\\
&(2)\quad b_{21}x_1 + \cdots + b_{2(n-1)}x_{n-1} = x_2^p\\
&\quad\vdots\\
&(n-1)\quad b_{(n-1)1}x_1 + \cdots + b_{(n-1)(n-1)}x_{n-1} = x_{n-1}^p\\
&(\ell_1)\quad a_1x_1 + \cdots + a_nx_n = 0\\
&(\ell_2)\quad b_1x_1 + \cdots + b_{n-1}x_{n-1} = 0
\end{aligned}
$$

where

$$b_{ij} = a_{ij}a_n - a_{in}a_j$$

$$b_j = a_1^pb_{1j} + a_2^pb_{2j} + \cdots + a_{n-1}^pb_{(n-1)j} + a_n^p(a_{nj}a_n - a_{nn}a_j)$$

Proof: Multiply equation (i) in S_1 by a_n and subtract equation (ℓ_1) multiplied by a_{in}. We obtain the equation

$$(\bar{i})\quad b_{i1}x_1 + \cdots + b_{i(n-1)}x_{n-1} = a_nx_i^p$$

where $b_{ij} = a_{ij}a_n - a_{in}a_j$ for $i = 1, \ldots, n$, $j = 1, 2, \ldots, n - 1$.

Now multiply each equation $(\bar{i})$ by a_i^p and sum these together. We obtain the equation

$$b_1x_1 + \cdots + b_{n-1}x_{n-1} = a_n(a_1x_1 + \cdots + a_nx_n)^p \tag{2.1.1}$$

where

$$b_j = a_1^p b_{ij} + \cdots + a_{n-1}^p b_{(n-1)j} + a_n^p b_{nj}$$

Combining (2.1.1) with ℓ_1, we obtain the linear equation

$$(\ell_2) \quad b_1 x_1 + \cdots + b_{n-1} x_{n-1} = 0$$

We conclude that our original system is equivalent to the system

$$\begin{array}{ll} (\bar{1}) & b_{11}x_1 + \cdots + b_{1(n-1)}x_{n-1} = a_n x_1^p \\ & \vdots \\ (\overline{n-1}) & b_{(n-1)1}x_1 + \cdots + b_{(n-1)(n-1)}x_{n-1} = a_n x_{n-1}^p \\ (\ell_1) & a_1 x_1 + \cdots + a_n x_n = 0 \\ (\ell_2) & b_1 x_1 + \cdots + b_{n-1} x_{n-1} = 0 \end{array} \qquad (2.1.2)$$

Note that if we look at the points of intersection of these hypersurfaces [in (2.1.2)] in projective space $\mathbb{P}_k^n$ we see that all intersections occur at finite distance. But the system

$$\begin{array}{l} u^{p-1}b_{11}x_1 + \cdots + u^{p-1}b_{1(n-1)}x_{n-1} = a_n x_1^p \\ \vdots \\ u^{p-1}b_{(n-1)1}x_1 + \cdots + u^{p-1}b_{(n-1)(n-1)}x_{n-1} = a_n x_{n-1}^p \\ (\ell_1) \text{ and } (\ell_2) \end{array}$$

is clearly isomorphic to the system

$$\begin{array}{l} u^{p-1}b_{11}x_1 + \cdots + u^{p-1}b_{1(n-1)}x_{n-1} = x_1^p \\ \vdots \\ u^{p-1}b_{(n-1)1}x_1 + \cdots + u^{p-1}b_{(n-1)(n-1)}x_{(n-1)} = x_{n-1}^p \\ (\ell_1) \text{ and } (\ell_2) \end{array}$$

Thus we see that the system (2.1.2) is equivalent to S_2.

LEMMA 2.2 Let k be an algebraically closed field of characteristic $p \neq 0$. Consider the system of equations

$$\begin{bmatrix} a_{11} & a_{12} & \cdots & a_{1n} & c_{11} & c_{12} & \cdots & c_{1m} \\ a_{21} & a_{22} & \cdots & a_{2n} & c_{21} & c_{22} & \cdots & c_{2m} \\ \vdots & & & & & & & \\ a_{n1} & a_{n2} & \cdots & a_{nn} & c_{n1} & c_{n2} & \cdots & c_{nm} \\ b_{11} & b_{12} & \cdots & b_{1n} & d_{11} & d_{12} & \cdots & d_{1m} \\ b_{21} & b_{22} & \cdots & b_{2n} & d_{21} & d_{22} & \cdots & d_{2m} \\ \vdots & & & & & & & \\ b_{m1} & b_{m2} & \cdots & b_{mn} & d_{m1} & d_{m2} & \cdots & d_{mm} \end{bmatrix} \begin{bmatrix} x_1 \\ x_2 \\ \vdots \\ x_n \\ y_1 \\ y_2 \\ \vdots \\ y_m \end{bmatrix} = \begin{bmatrix} x_1^p \\ x_2^p \\ \vdots \\ x_n^p \\ y_1^p \\ y_2^p \\ \vdots \\ y_m^p \end{bmatrix} \tag{2.2.1}$$

Suppose that the rank of the matrix above is m and $\det[d_{rs}] \neq 0$. Then the system (2.2.1) is equivalent to the system

$$\begin{bmatrix} A & 0 & \cdots & 0 & E_{11} & E_{12} & \cdots & E_{1m} \\ 0 & A & \cdots & 0 & E_{21} & E_{22} & \cdots & E_{2m} \\ \vdots & & & & & & & \\ 0 & 0 & \cdots & A & E_{n1} & E_{n2} & \cdots & E_{nm} \\ b_{11} & b_{12} & \cdots & b_{1n} & d_{11} & d_{12} & \cdots & d_{1m} \\ b_{21} & b_{22} & \cdots & b_{2n} & d_{21} & d_{22} & \cdots & d_{2m} \\ \vdots & & & & & & & \\ b_{m1} & b_{m2} & \cdots & b_{mn} & d_{m1} & d_{m2} & \cdots & d_{mm} \end{bmatrix} \begin{bmatrix} x_1 \\ x_2 \\ \vdots \\ x_n \\ y_1 \\ y_2 \\ \vdots \\ y_m \end{bmatrix} = \begin{bmatrix} 0 \\ 0 \\ \vdots \\ 0 \\ y_1^p \\ y_2^p \\ \vdots \\ y_m^p \end{bmatrix} \tag{2.2.2}$$

where $A^p = \det[d_{rs}]$ and where E_{ij}^p is the cofactor of b_{j1} in the $(m + 1) \times (m + 1)$ matrix.

$$\begin{bmatrix} a_{i1} & c_{i1} & c_{i2} & \cdots & c_{im} \\ b_{11} & d_{11} & d_{12} & \cdots & d_{1m} \\ b_{21} & d_{21} & d_{22} & \cdots & d_{2m} \\ b_{m1} & d_{m1} & d_{m2} & \cdots & d_{mm} \end{bmatrix}$$

Let $i = 1, \ldots, n$. Since the rank of the matrix in (2.2.1) is m and $\det[d_{rs}] \neq 0$, there is a nontrivial linear combination of row i and rows $n + j$, $j = 1, \ldots, m$, in (2.2.1) that gives 0. As a matter of fact, we know that

$$A^p\text{row}(i) + E_{i1}^p\text{row}(n + 1) + \cdots + E_{im}^p\text{row}(n + m)$$

will give 0 on the left of the = in (2.2.1), while the right side of the = will be

$$A^p x_i^p + E_{i1}^p y_1^p + \cdots + E_{im}^p y_m^p$$

Thus we see that we can replace row (i) in system (2.2.1) by the equation

$$Ax_i + E_{i1}y_1 + \cdots + E_{im}y_m = 0$$

LEMMA 2.3 Let k be as in (2.2). Let A be the $(n + m) \times (n + m)$ matrix of (2.2.1). Then the system $Ax = x^p$ has p^m distinct solutions.

Proof: Using (2.2) we reduce to the case where our system is as in (2.2.2). If we homogenize this system, we see by Bézout's theorem that there are p^m points of intersection. But it is easy to see that there are no points of intersection at infinity. Since the determinant of (2.2.2) is not zero, each of these intersection points has multiplicity 1.

We have the following lemma, which is an immediate consequence of lemma (2.1) and elementary linear algebra.

LEMMA 2.4 Let k be an algebraically closed field of characteristic $p = 0$. Let S_1 be a system of equations of the form

$$
\begin{array}{lll}
S_1: & (1) & a_{11}x_1 + \cdots + a_{1n}x_n = x_1^p \\
& (2) & a_{21}x_1 + \cdots + a_{2n}x_n = x_2^p \\
& & \vdots \\
& (n) & a_{n1}x_1 + \cdots + a_{nn}x_n = x_n^p \\
& (\ell_1) & b_{11}x_1 + \cdots + b_{1n}x_n = 0 \\
& (\ell_2) & b_{21}x_1 + \cdots + b_{2n}x_n = 0 \\
& & \vdots \\
& (\ell_m) & b_{m1}x_1 + \cdots + b_{mn}x_n = 0
\end{array}
$$

where $b_{1n} \neq 0$. Then S_1 is equivalent (in the sense that it has the same number of distinct solutions) to the system S_2 given below.

$$
\begin{array}{rl}
S_2:\ (1) & c_{11}x_1 + \cdots + c_{1(n-1)}x_{n-1} = x_1^p \\
(2) & c_{21}x_1 + \cdots + c_{2(n-1)}x_{n-1} = x_2^p \\
 & \vdots \\
(n-1) & c_{(n-1)1}x_1 + \cdots + c_{(n-1)(n-1)} = x_{n-1}^p \\
(\ell_1') & d_{11}x_1 + \cdots + d_{1(n-1)}x_{n-1} = 0 \\
(\ell_2') & d_{21}x_1 + \cdots + d_{2(n-1)}x_{n-1} = 0 \\
 & \vdots \\
(\ell_m') & d_{m1}x_1 + \cdots + d_{m(n-1)}x_{n-1} = 0 \\
(\ell_1) & b_{11}x_1 + \cdots + b_{1n}x_n = 0
\end{array}
$$

where

$$
\begin{aligned}
c_{ij} &= a_{ij}b_{1n} - a_{in}b_{1j} \\
d_{1j} &= b_{11}^p c_{1j} + b_{12}^p c_{2j} + \cdots + b_{1(n-1)}^p c_{(n-1)j} \\
&\quad + b_{1n}^p (a_{nj}b_{1n} - a_{nn}b_{1j}) \\
d_{ij} &= b_{ij}b_{1n} - b_{in}b_{1j} \qquad \text{for } i = 2, \ldots, m;\ j = 1, \ldots, n-1
\end{aligned}
$$

REMARK 2.5 We are in a position to describe fully the algorithm used in calculating the divisor class group of a normal surface $F \subset \mathbf{A}_k^3$ defined by an equation $z^p = G(x,y)$, where k is of characteristic $p \neq 0$. We let $t = \Sigma_{i+j\leq g-2}\, a_{ij}x^i y^j$ be a polynomial of degree equal to g - 2, where the a_{ij} are undetermined and where g = deg(G). We then substitute t into the differential expressions [see (1.9) and (1.10)]

$$\nabla(G^{p-1}t) = t^p \tag{2.5.1}$$

$$\nabla(G^i t) = 0 \qquad i = 0, 1, \ldots, p-2 \tag{2.5.2}$$

We compare coefficients in (2.5.1) and (2.5.2). We then see that the a_{ij} must satisfy equations of the form $\ell_{ij} = a_{ij}^p$ and $\ell_r = 0$, where the expressions ℓ_{ij} and ℓ_r are linear with coefficients in

F(G). Note that the solution set to this p-linear system forms an additive group and it is quite easy to see that this group is isomorphic to Cℓ(F).

We then employ lemma 2.4 to reduce the associated p-linear system to a system of n linear equations in n unknowns having only the trivial solution or else to a system of the type discussed in (2.3) In this way we can determine the order of Cℓ(F). We know that Cℓ(F) is a p-group of type (p,...,p) [see (1.11)]. Thus the structure of Cℓ(F) is completely determined by its order.

The task of programming this algorithm was performed by D. Joyce. With it, we were quickly able to calculate the order of Cℓ(F) for many of the surfaces which Lang studied in Lang (1981a). Of course, these calculations were extremely lengthy when done by hand. The following table is a brief list of some of these surfaces and their corresponding divisor class group orders.

F	Order of Cℓ(F)
$z^2 = y + xy^5$	2
$z^2 = y + xy^4(y + \theta)$ where $\theta \notin GF(4)$	2^2
$z^2 = y + xy^2(y + \theta_1)(y + \theta_2)$	2^3
$z^2 = y + xy(y + \theta_1)(y + \theta_2)(y + \theta_3)$	2^4
$z^2 = x^3y + y^3x + x^5 + y^5$	2^5
$z^2 = x^3y + zy^3 + x^5 + y^5 + x^3y^3$	2^6
$z^2 = x^2y + xy^2 + x^3y^2 + x^2y^3 + x^5y + y^5x$	2^7
$z^2 = y + x^3 + xy^4 + x^5y$	2^9
$z^2 = y + xy^4 + x^5y$	2^{10}
$z^2 = xy + x^4y + xy^4 + x^5y + xy^5$	2^{10}
$z^3 = x + y^2 + xy^2 + y^4 + y^5 + x^5y$	3
$z^3 = y + xy^3 + x^5y$	3^6

For more details concerning this program, contact David Joyce, Mathematics/Computer Sciences Department, Clark University, Worcester, Massachusetts. The program is entitled LANG.

REFERENCES

Blass, P., Zariski surfaces, *Diss. Math.*, 200 (1983).

Blass, P., Some geometric applications of a differential equation in characteristics $p > 0$ to the theory of algebraic surfaces, *Contemporary Mathematics*, AMS 13 (1982) (with the cooperation of James Sturnfield and Jeffrey Lang).

Blass, P., Picard groups of Zariski surfaces, I, *Compositio Math.*, 54 (1985).

Fossum, R., *The Divisor Class Group of a Krull Domain*, Springer-Verlag, New York, 1973.

Ganong, R., On plane curves with one place at infinity, *J. Reine Angew. Math.*, 307 (1979), 173-193.

Ganong, R., Plane Frobenius sandwiches, *Proc. Amer. Math. Soc.*, 84 (1982).

Jacobson, N., *Lectures in Abstract Algebra*, Vol. III, *Theory of Fields and Galois Theory*, Van Nostrand, New York, 1964.

Lang, J., Ph.D. thesis, Purdue University, 1981a.

Lang, J., An example related to the affine theorem of Castelnuovo, *Mich. Math. J.*, 28 (1981).

Lang, J., The divisor classes of the hypersurface $z^{p^n} = G(x_1,\ldots,x_n)$ in characteristic $p > 0$, *Trans. Am. Math. Soc.*, 278, No. 2 (1983a).

Lang, J., The divisor class group of the surface $z^{p^n} = G(x,y)$ over fields of characteristic $p > 0$, *J. Algebra*, 84, No. 2 (1983b).

Matsumura, H., *Commutative Algebra*, W. A. Benjamin, Menlo Park, Calif., 1970.

Samuel, P., Lectures on Unique Factorization Domains, Tata Lecture Notes, 1964a.

Samuel, P., Classes de diviseurs et derivees logarithmiques, *Topology*, 3 (1964b), 81-96.

APPENDIX: LANG: A PROGRAM TO CALCULATE THE DIVISOR CLASS NUMBER†

The program LANG carries out computations indicated in J. J. Lang's thesis to compute divisor class numbers of Zariski surfaces.

---

†By David Joyce, Clark University, Worcester, Massachusetts.

1. A Short Review of the Conclusions of Lang's Thesis

Let k be an algebraically closed field of characteristic p (greater than 0). Fix a polynomial G(x,y) in k[x,y] and consider the surface F given by $z^p = G(x,y)$, called a ***Zariski surface.*** We make two restrictions on the polynomial G(x,y):

(1) G(x,y) contains no monomials in $k[x^p, y^p]$.

(2) dG/dx and dG/dy are relatively prime.

Condition (1) is a simplifying restriction since by a linear transformation any surface may be put in a standard form where (1) holds. Condition (2) merely states that we restrict our attention to normal surfaces. Under these two conditions the divisor class group is isomorphic to the group of logarithmic derivatives associated to F, that is, the group

$$L = \{Df/f \mid f \text{ lies in } k(x,y) \text{ and } Df/f \text{ lies in } k[x,y]\}$$

where D is the operator

$$\frac{dG}{dy}\frac{d}{dx} - \frac{dG}{dx}\frac{d}{dy}$$

A polynomial t(x,y) in k[x,y] lies in L iff

(3) Nabla $(G^{p-1}t) = t^p$, and Nabla $(G^j t) = 0$ for j = 0, 1, ..., p - 2, where Nabla is the operator $d^{2p-2}/dx^{p-1}dy^{p-1}$.

2. Reduction to Computations over Finite Fields

The purpose of the program LANG is to determine just how many polynomials t(x,y) satisfy the conditions numbered (3). In order that computations be feasible, the field k is taken to be the algebraic closure of the prime field GF(p); then the coefficients of the polynomial G(x,y) may be taken from a Galois field GF(q), where q is p^n for some n. The system of differential equations (3) simplifies to a system of linear and p-linear equations over GF(q). The equation Nabla $(G^{p-1}t) = t^p$ gives rise to a set of p-linear equations, each equation deriving from the coefficients of $x^{ip}y^{jp}$ (for fixed i and j). The variables in the equations are the coefficients of the unknown polynomial t(x,y). Similarly, for fixed x (k = 0, ..., p - 2) the equation Nabla $(G^k t) = 0$ gives rise to a set of linear equations.

```
THIS FILE CONTAINS COMPUTATIONS FOR TABLE 5.26 ON PAGE 56 OF LANG'S THESIS
---------------------------------------------------------------------------

THE SURFACE     z^2 = y + xy^5

The system of equations to solve is as follows.
t[0,0]^2 = t[1,0]
t[1,0]^2 = t[3,0]
t[0,1]^2 = t[1,2]
t[2,0]^2 = 0
t[1,1]^2 = 0
t[0,2]^2 = t[0,0]
t[3,0]^2 = 0
t[2,1]^2 = 0
t[1,2]^2 = t[2,0]
t[0,3]^2 = t[0,2]
t[4,0]^2 = 0
t[3,1]^2 = 0
t[2,2]^2 = t[4,0]
t[1,3]^2 = t[2,2]
t[0,4]^2 = t[0,4]
0        = t[1,3]
0        = t[3,1]
0        = t[1,1]

The divisor class number is 2.  This agrees with Lang's computations.
---------------------------------------------------------------------------

THE SURFACE     z^2 = y + xy^4(y+theta)

Here we take theta to be an element in GF(4) not in GF(2).
The elements of GF(4) are represented as pairs.
The identity element is (1,0)
A primitive third root, e, is (0,1)
Its second power equals (1,1)

In terms of the primitive root, e, in GF(4), the surface is given by the
equation

z^2 = y + exy^4 + xy^5

The system of equations to solve is as follows.

t[0,0]^2 = t[1,0]
t[1,0]^2 = t[3,0]
t[0,1]^2 = t[1,2]
t[2,0]^2 = 0
t[1,1]^2 = 0
t[0,2]^2 = t[0,0] + e^1 t[0,1]
t[3,0]^2 = 0
t[2,1]^2 = 0
t[1,2]^2 = t[2,0] + e^1 t[2,1]
t[0,3]^2 = t[0,2] + e^1 t[0,3]
t[4,0]^2 = 0
t[3,1]^2 = 0
t[2,2]^2 = t[4,0]
t[1,3]^2 = t[2,2]
t[0,4]^2 = t[0,4]
```

```
0          = t[1,3]
0          = t[3,1]
0          = t[1,1]
```

The divisor class number is 2^2 which agrees with Lang's computations.

THE SURFACE   z^2 = x^3y + xy^3 + x^5 + y^5 + x^3y^3

The system of equations to solve is as follows.

```
t[0,0]^2 = 0
t[1,0]^2 = t[0,0]
t[0,1]^2 = t[0,0]
t[2,0]^2 = t[0,1] + t[2,0]
t[1,1]^2 = t[0,0] + t[2,0] + t[0,2]
t[0,2]^2 = t[1,0] + t[0,2]
t[3,0]^2 = t[2,1] + t[4,0]
t[2,1]^2 = t[2,0] + t[0,3] + t[4,0] + t[2,2]
t[1,2]^2 = t[0,2] + t[3,0] + t[2,2] + t[0,4]
t[0,3]^2 = t[1,2] + t[0,4]
t[4,0]^2 = 0
t[3,1]^2 = t[4,0]
t[2,2]^2 = t[2,2]
t[1,3]^2 = t[0,4]
t[0,4]^2 = 0
0        = t[1,3]
0        = t[3,1]
0        = t[1,1]
```

The divisor class number is 2^6. This agrees with Lang's computations.

THE SURFACE   z^2 = x^2y + xy^2 + x^3y^2 + x^2y^3 + x^5y + xy^5

The system of equations to solve is as follows.

```
t[0,0]^2 = 0
t[1,0]^2 = t[1,0]
t[0,1]^2 = t[0,1]
t[2,0]^2 = t[0,0] + t[3,0]
t[1,1]^2 = t[1,0] + t[0,1] + t[2,1] + t[1,2]
t[0,2]^2 = t[0,0] + t[0,3]
t[3,0]^2 = t[2,0]
t[2,1]^2 = t[0,2] + t[3,0] + t[2,1]
t[1,2]^2 = t[2,0] + t[1,2] + t[0,3]
t[0,3]^2 = t[0,2]
t[4,0]^2 = t[4,0]
t[3,1]^2 = t[2,2]
t[2,2]^2 = t[4,0] + t[0,4]
t[1,3]^2 = t[2,2]
t[0,4]^2 = t[0,4]
0        = t[1,3]
0        = t[3,1]
0        = t[1,1]
```

The divisor class number is 2^7. This agrees with Lang's computations.

```
THE SURFACE     z^2 = y + x^3 + xy^4 + x^5y

The system of equations to solve is as follows.
t[0,0]^2 = t[1,0]
t[1,0]^2 = t[0,1] + t[3,0]
t[0,1]^2 = t[1,2]
t[2,0]^2 = t[0,0] + t[2,1]
t[1,1]^2 = t[0,3]
t[0,2]^2 = t[0,1]
t[3,0]^2 = t[2,0]
t[2,1]^2 = t[0,2]
t[1,2]^2 = t[2,1]
t[0,3]^2 = t[0,3]
t[4,0]^2 = t[4,0]
t[3,1]^2 = t[2,2]
t[2,2]^2 = t[0,4]
t[1,3]^2 = 0
t[0,4]^2 = 0
0        = t[1,3]
0        = t[3,1]
0        = t[1,1]

The divisor class number is 2^9.  This disagrees with Lang's computations which
give 2^8.

--------------------------------------------------------------------------------

THE SURFACE z^2 = y + xy^4 + x^5y

The system of equations to solve is as follows.
t[0,0]^2 = t[1,0]
t[1,0]^2 = t[3,0]
t[0,1]^2 = t[1,2]
t[2,0]^2 = t[0,0]
t[1,1]^2 = 0
t[0,2]^2 = t[0,1]
t[3,0]^2 = t[2,0]
t[2,1]^2 = t[0,2]
t[1,2]^2 = t[2,1]
t[0,3]^2 = t[0,3]
t[4,0]^2 = t[4,0]
t[3,1]^2 = t[2,2]
t[2,2]^2 = t[0,4]
t[1,3]^2 = 0
t[0,4]^2 = 0
0        = t[1,3]
0        = t[3,1]
0        = t[1,1]

The divisor class number is 2^10.  This disagrees with Lang's computations
which gives 2^9.

--------------------------------------------------------------------------------

THE SURFACE z^2 = xy + x^4y + xy^4 + x^5y + xy^5

The system of equations to solve is as follows.
t[0,0]^2 = t[0,0]
t[1,0]^2 = t[2,0]
t[0,1]^2 = t[0,2]
t[2,0]^2 = t[0,0] + t[1,0] + t[4,0]
t[1,1]^2 = t[2,2]
t[0,2]^2 = t[0,0] + t[0,1] + t[0,4]
t[3,0]^2 = t[2,0] + t[3,0]
```

```
t[2,1]^2 = t[0,2] + t[1,2]
t[1,2]^2 = t[2,0] + t[2,1]
t[0,3]^2 = t[0,2] + t[0,3]
t[4,0]^2 = t[4,0]
t[3,1]^2 = t[2,2]
t[2,2]^2 = t[4,0] + t[0,4]
t[1,3]^2 = t[2,2]
t[0,4]^2 = t[0,4]
0        = t[1,3]
0        = t[3,1]
0        = t[1,1]
```

The divisor class number is 2^10. This agrees with Lang's computations.

THIS FILE CONTAINS COMPUTATIONS FOR TABLE 5.32 ON PAGES 58 AND 59 OF LANG'S THESIS

THE SURFACE z^2 = y + xy^4

The system of equations to solve is as follows.

```
t[0,0]^2 = t[1,0]
t[1,0]^2 = t[3,0]
t[0,1]^2 = t[1,2]
t[2,0]^2 = 0
t[1,1]^2 = 0
t[0,2]^2 = t[0,1]
t[3,0]^2 = 0
t[2,1]^2 = 0
t[1,2]^2 = t[2,1]
t[0,3]^2 = t[0,3]
0        = t[1,1]
```

The divisor class number is 2 which agrees with Lang's computations.

THE SURFACE   z^2 = y + xy^3(y + theta)

Here theta is taken in GF(2^2) but not in GF(2), in particular, theta is e, a generating element of the multiplicative group of GF(2^2).

The elements of GF(2^2) are represented as 2-tuples.
The identity element is (1,0)
A primitive third root, e, is (0,1)
Its second power equals (1,1)

In terms of the primitive root, e, in GF(2^2), the surface is given by

z^2 = y + exy^3 + xy^4

The system of equations to solve is as follows.

```
t[0,0]^2 = t[1,0]
t[1,0]^2 = t[3,0]
t[0,1]^2 = e^1 t[0,0] + t[1,2]
t[2,0]^2 = 0
t[1,1]^2 = e^1 t[2,0]
t[0,2]^2 = t[0,1] + e^1 t[0,2]
t[3,0]^2 = 0
t[2,1]^2 = 0
t[1,2]^2 = t[2,1]
t[0,3]^2 = t[0,3]
0        = t[1,1]
```

The divisor class number is 2^2 which agrees with Lang's computations.

THE SURFACE z^2 = y + xy^2(y+theta-one)(y+theta-two)
where theta-one is e and theta-two is 1+e, elements in GF(2^2).

The elements of GF(2^2) are represented as 2-tuples.

The identity element is (1,0)
A primitive third root, e, is (0,1)
Its second power equals (1,1)

In terms of the primitive root, e, in GF(2^2), the surface is given by

z^2 = y + xy^2 + xy^3 + xy^4

The system of equations to solve is as follows.

```
t[0,0]^2 = t[1,0]
t[1,0]^2 = t[3,0]
t[0,1]^2 = t[0,0] + t[0,1] + t[1,2]
t[2,0]^2 = 0
t[1,1]^2 = t[2,0] + t[2,1]
t[0,2]^2 = t[0,1] + t[0,2] + t[0,3]
t[3,0]^2 = 0
t[2,1]^2 = 0
t[1,2]^2 = t[2,1]
t[0,3]^2 = t[0,3]
0        = t[1,1]
```

The divisor class number is 2^3 which agrees with Lang's computations.

THE SURFACE z^2 = y + xy(y+theta-one)(y+theta-two)(y+theta-three)
where theta-one is e, theta-two is 1+e, and theta-three is 1.

Thus the surface has the equation z^2 = y + xy + xy^4

```
The system of equations to solve is as follows.

t[0,0]^2 = t[0,0] + t[1,0]
t[1,0]^2 = t[2,0] + t[3,0]
t[0,1]^2 = t[0,2] + t[1,2]
t[2,0]^2 = 0
t[1,1]^2 = 0
t[0,2]^2 = t[0,1]
t[3,0]^2 = 0
t[2,1]^2 = 0
t[1,2]^2 = t[2,1]
t[0,3]^2 = t[0,3]
0        = t[1,1]

The divisor class number is 2^4 which agrees with Lang's computations.

---------------------------------------------------------------------
```

```
THE SURFACE z^2 = x^3y + xy^3 + x^5 + y^5

The system of equations to solve is as follows.
t[0,0]^2 = 0
t[1,0]^2 = t[0,0]
t[0,1]^2 = t[0,0]
t[2,0]^2 = t[0,1] + t[2,0]
t[1,1]^2 = t[2,0] + t[0,2]
t[0,2]^2 = t[1,0] + t[0,2]
t[3,0]^2 = t[2,1]
t[2,1]^2 = t[0,3]
t[1,2]^2 = t[3,0]
t[0,3]^2 = t[1,2]
0        = t[1,1]
The divisor class number is 2^5 which agrees with Lang's computations.

---------------------------------------------------------------------------

THE SURFACE  z^2 = xy + x^4y + x^3y^2 + x^2y^3 + xy^4

The system of equations to solve is as follows.
t[0,0]^2 = t[0,0]
t[1,0]^2 = t[2,0]
t[0,1]^2 = t[0,2]
t[2,0]^2 = t[1,0]
t[1,1]^2 = t[1,0] + t[0,1]
t[0,2]^2 = t[0,1]
t[3,0]^2 = t[3,0]
t[2,1]^2 = t[3,0] + t[2,1] + t[1,2]
t[1,2]^2 = t[2,1] + t[1,2] + t[0,3]
t[0,3]^2 = t[0,3]
0        = t[1,1]
The divisor class number is 2^5.  This disagrees with Lang's computations which
give 2^6.

---------------------------------------------------------------------------
```

```
THE SURFACE      z^2 = xy + x^4y + x^3y^2 + xy^4

The system of equations to solve is as follows.
t[0,0]^2 = t[0,0]
t[1,0]^2 = t[2,0]
t[0,1]^2 = t[0,2]
t[2,0]^2 = t[1,0]
t[1,1]^2 = t[0,1]
t[0,2]^2 = t[0,1]
t[3,0]^2 = t[3,0]
t[2,1]^2 = t[2,1] + t[1,2]
t[1,2]^2 = t[2,1] + t[0,3]
t[0,3]^2 = t[0,3]
0        = t[1,1]

The divisor class number is 2^7 which agrees with Lang's computations.

--------------------------------------------------------------------------

THE SURFACE      z^2 = xy + x^3 + x^4y + xy^4

The system of equations to solve is as follows.
t[0,0]^2 = t[0,0]
t[1,0]^2 = t[0,1] + t[2,0]
t[0,1]^2 = t[0,2]
t[2,0]^2 = t[1,0] + t[2,1]
t[1,1]^2 = t[0,3]
t[0,2]^2 = t[0,1]
t[3,0]^2 = t[3,0]
t[2,1]^2 = t[1,2]
t[1,2]^2 = t[2,1]
t[0,3]^2 = t[0,3]
0        = t[1,1]


The divisor class number is 2^8 which agrees with Lang's computations.

--------------------------------------------------------------------------

THE SURFACE      z^2 = xy + x^4y + xy^4

The system of equations to solve is as follows.
t[0,0]^2 = t[0,0]
t[1,0]^2 = t[2,0]
t[0,1]^2 = t[0,2]
t[2,0]^2 = t[1,0]
t[1,1]^2 = 0
t[0,2]^2 = t[0,1]
t[3,0]^2 = t[3,0]
t[2,1]^2 = t[1,2]
t[1,2]^2 = t[2,1]
t[0,3]^2 = t[0,3]
0        = t[1,1]

The divisor class number is 2^9 which agrees with Lang's computations.

--------------------------------------------------------------------------
```

THIS FILE CONTAINS COMPUTATIONS FOR TABLE 6.14 ON PAGE 74 OF LANG'S THESIS

THE SURFACE   z^3 = x + y^2 + xy^2 + y^4 + y^5 + x^5y

The system of equations to solve is as follows.

```
t[0,0]^3 = e^1 t[0,0] + e^1 t[1,0] + t[0,2]
t[1,0]^3 = e^1 t[3,0] + e^1 t[4,0]
t[0,1]^3 = e^1 t[1,0] + t[0,1] + t[1,1] + t[2,1]
         + e^1 t[0,3] + e^1 t[1,3]
t[2,0]^3 = e^1 t[2,1]
t[1,1]^3 = e^1 t[0,0] + e^1 t[0,2] + e^1 t[4,0] + t[3,1]
t[0,2]^3 = t[2,0] + e^1 t[1,1] + e^1 t[2,1] + e^1 t[1,2]
         + e^1 t[2,2] + e^1 t[1,3] + t[0,4]
t[3,0]^3 = t[1,0]
t[2,1]^3 = e^1 t[3,0] + e^1 t[2,2]
t[1,2]^3 = e^1 t[0,2] + e^1 t[0,3]
t[0,3]^3 = t[2,1] + e^1 t[2,2]
t[4,0]^3 = t[4,0]
t[3,1]^3 = t[1,3]
t[2,2]^3 = 0
t[1,3]^3 = 0
t[0,4]^3 = 0
0        = t[0,4]
0        = t[3,1]
0        = t[2,0] + t[2,1] + t[1,3]
0        = t[0,1] + t[4,0]
0        = t[1,0] + t[2,0] + t[1,2]
0        = t[2,2]
```

The divisor class number is 3. This disagrees with Lang's computations which give 1.

THE SURFACE   z^3 = xy + y^4

The system of equations to solve is as follows.

```
t[0,0]^3 = t[0,0]
t[1,0]^3 = t[3,0]
t[0,1]^3 = e^1 t[1,0] + t[0,3]
t[2,0]^3 = 0
t[1,1]^3 = e^1 t[4,0]
t[0,2]^3 = t[2,0] + e^1 t[1,3]
t[3,0]^3 = 0
t[2,1]^3 = 0
t[1,2]^3 = 0
t[0,3]^3 = 0
t[4,0]^3 = 0
t[3,1]^3 = 0
t[2,2]^3 = 0
t[1,3]^3 = 0
t[0,4]^3 = 0
0        = t[2,1]
0        = t[1,1]
0        = t[2,2]
```

The divisor class number is 3 which agrees with Lang's computations.

THE SURFACE z^3 = y + xy^3 + x^5y

The system of equations to solve is as follows.

```
t[0,0]^3 = t[2,0]
t[1,0]^3 = e^1 t[0,0]
t[0,1]^3 = e^1 t[1,1]
t[2,0]^3 = e^1 t[3,0]
t[1,1]^3 = e^1 t[0,3]
t[0,2]^3 = t[0,2]
t[3,0]^3 = t[1,0]
t[2,1]^3 = e^1 t[2,1]
t[1,2]^3 = 0
t[0,3]^3 = 0
t[4,0]^3 = t[4,0]
t[3,1]^3 = t[1,3]
t[2,2]^3 = 0
t[1,3]^3 = 0
t[0,4]^3 = 0
0        = t[0,4]
0        = t[3,1]
0        = t[1,2]
0        = t[0,1]
0        = t[2,1]
0        = t[2,2]
```

The divisor class number is 3^6 which agrees with Lang's computations.

A FEW EXAMPLE RUNS OF THE PROGRAM LANG ILLUSTRATING THE USE OF THE PROGRAM

==

EXAMPLE 1. The following run shows how to use the program LANG to determine the divisor class number of the Zariski surface given by the equation z^2 = xy + x^4y + xy^4 over the field GF(2). By specifying the output file to be SYS$OUTPUT, the results of the program appear at the terminal.

```
        This program determines the divisor class number of the surface
        z^p = G(x,y) over fields of characteristic p > 0 by J. J.
        Lang's algorithm. Written by D Joyce, Clark Univ, Jan 1983.
        (p can be no greater than 19,
        n must be < 18 and p^n must be < 262144
        The degree of G must be between 2 and 18)

Will you want to see a description of the field GF(q)?
YES
Will you want to see the system of equations to be solved?
YES
Will you want to specify the coefficients of G (enter Y),
or chose them randomly (enter N)?
YES
In what file would you like the results placed?
SYS$OUTPUT
Enter p, the characteristic of the field GF(q).
2
Enter n, the degree of the field GF(2^n) over the prime field  GF(2).
1
The elements of GF(2^1) are represented as 1-tuples.
The identity element is (1)
A primitive 1-th root, e, is (1)
Its 1-th power equals (1)
Enter the degree of the polynomial G(x,y).
5
Now enter the coefficients of G(x,y).
Enter a coefficient by entering (1) the power of x,
        (2) the power of y, (3) the 1 component of the coefficient.
Separate those 3 integers by spaces.
Indicate G is specified by entering -1 for the power of x.
Specify a coefficient.
1 4 1
Specify a coefficient.
1 1 1
Specify a coefficient.
4 1 1
Specify a coefficient.
-1
In terms of the primitive root, e, in GF(2^1), G is
xy + x^4y + xy^4
The system of equations to solve is as follows.
t[0,0]^2 = t[0,0]
t[1,0]^2 = t[2,0]
t[0,1]^2 = t[0,2]
t[2,0]^2 = t[1,0]
t[1,1]^2 = 0
t[0,2]^2 = t[0,1]
```

```
t[3,0]^2 = t[3,0]
t[2,1]^2 = t[1,2]
t[1,2]^2 = t[2,1]
t[0,3]^2 = t[0,3]
0        = t[1,1]
The divisor class number is 2^9.

Enter p, the characteristic of the field GF(q).
0
Exit Lang.
```

==

EXAMPLE 2. In this example the coefficients of the polynomial lie in GF(25). The program finds a representation of the field GF(25) for the user. The coefficients of the polynomial are entered as pairs. The polynomial used here is (1,0)xy + (0,1)x^4y + (0,4)xy^4. The element (1,0) represents 1 in GF(25), the element (0,1) represents e, a generator of the multiplicative group in GF(25), and the element (0,4) represents 4e in GF(25). In this example, the output of the program is placed in the file GF25.OUT. Directly below is included first the dialogue between the program and the user, then the contents of the file GF25.OUT.

```
        This program ...
Will you want to see a description of the field GF(q)?
YES
Will you want to see the system of equations to be solved?
oYES
Will you want to specify the coefficients of G (enter Y),
or chose them randomly (enter N)?
YES
In what file would you like the results placed?
GF25.OUT
Enter p, the characteristic of the field GF(q).
5
Enter n, the degree of the field GF(5^n) over the prime field  GF(5).
2
Enter the degree of the polynomial G(x,y).
5
Now enter the coefficients of the polynomial G.
Enter a coefficient by entering (1) the power of x,
        (2) the power of y, (3) the 2 components of the coefficient.
Separate those 4 integers by spaces.
Indicate G is specified by entering -1 for the power of x.
Specify a coefficient.
1 1 1 0
Specify a coefficient.
4 1 0 1
Specify a coefficient.
1 4 0 4
Specify a coefficient.
-1

Enter p, the characteristic of the field GF(q).
0
Exit Lang.
```

```
--------------------------------------------------------------------------------

The elements of GF(5^2) are represented as 2-tuples.
The identity element is (1,0)
A primitive 24-th root, e, is (0,1)
Its 2-th power equals (3,1)
In terms of the primitive root, e, in GF(5^2), G is
xy + ex^4y + e^13 xy^4
The system of equations to solve is as follows.
t[0,0]^5 = t[0,0]
t[1,0]^5 = e^13 t[2,0]
t[0,1]^5 = e^1 t[0,2]
t[2,0]^5 = e^15 t[1,0] + e^2 t[4,0]
t[1,1]^5 = e^20 t[2,2]
t[0,2]^5 = e^3 t[0,1] + e^2 t[0,4]
t[3,0]^5 = e^4 t[3,0]
t[2,1]^5 = e^4 t[1,2]
t[1,2]^5 = e^4 t[2,1]
t[0,3]^5 = e^4 t[0,3]
t[4,0]^5 = 0
t[3,1]^5 = 0
t[2,2]^5 = 0
t[1,3]^5 = 0
t[0,4]^5 = 0
0        = e^15 t[1,2]
0        = e^21 t[3,0] + e^9 t[0,3]
0        = e^3 t[2,1]
0        = e^20 t[1,0] + e^7 t[1,3]
0        = e^20 t[0,1] + e^19 t[3,1]
0        = t[1,1]
0        = e^2 t[2,1]
0        = e^2 t[1,2]
0        = t[2,2]
0        = e^13 t[3,0] + e^1 t[0,3]

The divisor class number is 5^2.

================================================================================
```

EXAMPLE 3. In this example the random polynomial generation feature of LANG is used to choose a polynomial of degree 4 with coefficients in the field GF(49). The output of the program is placed in the file GF49.OUT. The dialogue and output follow.

```
--------------------------------------------------------------------------------

Will you want to see a description of the field GF(q)?
YES
Will you want to see the system of equations to be solved?
YES
Will you want to specify the coefficients of G (enter Y),
or chose them randomly (enter N)?
NO
Enter a seed for the random number generator.
1423
How many polynomials do you want produced?
1
In what file would you like the results placed?
GF49.OUT
Enter p, the characteristic of the field GF(q).
```

```
7
Enter n, the degree of the field GF(7^n) over the prime field  GF(7).
2
Enter the degree of the polynomial G(x,y).
4
```

```
The elements of GF(7^2) are represented as 2-tuples.
The identity element is (1,0)
A primitive 48-th root, e, is (0,1)
Its 2-th power equals (4,1)
In terms of the primitive root, e, in GF(7^2), G is
1 + e^16 x + e^26 y + e^42 x^2 + e^16 xy + e^18 y^2 + e^9 x^3
 + e^39 x^2y + e^44 xy^2 + e^38 x^4 + e^47 x^3y + e^34 x^2y^2
 + e^26 xy^3 + e^41 y^4

The system of equations to solve is as follows.
t[0,0]^7 = e^3 t[0,0] + e^12 t[1,0] + e^22 t[0,1] + t[2,0]
         + e^21 t[1,1] + e^43 t[0,2]
t[1,0]^7 = e^10 t[0,0] + e^6 t[1,0] + e^47 t[0,1] + e^30 t[2,0]
         + e^5 t[1,1] + e^25 t[0,2]
t[0,1]^7 = e^13 t[0,0] + e^30 t[1,0] + e^41 t[0,1] + e^46 t[2,0]
         + e^42 t[1,1] + e^27 t[0,2]
t[2,0]^7 = e^46 t[2,0] + e^21 t[1,1] + e^40 t[0,2]
t[1,1]^7 = e^33 t[2,0] + e^37 t[1,1] + e^5 t[0,2]
t[0,2]^7 = e^15 t[2,0] + e^11 t[1,1] + e^21 t[0,2]
0        = e^14 t[0,0] + e^36 t[1,0] + e^16 t[0,1] + e^20 t[2,0]
         + e^6 t[1,1] + e^23 t[0,2]
0        = e^41 t[0,0] + e^21 t[1,0] + e^13 t[0,1] + e^29 t[2,0]
         + e^21 t[1,1] + e^35 t[0,2]
0        = e^23 t[0,0] + e^43 t[1,0] + e^45 t[0,1] + e^44 t[2,0]
         + e^14 t[1,1] + e^29 t[0,2]
0        = e^38 t[1,0] + e^13 t[0,1] + e^19 t[2,0] + e^46 t[1,1]
         + e^44 t[0,2]
0        = e^18 t[0,0] + e^3 t[1,0] + e^41 t[0,1] + e^12 t[2,0]
         + e^10 t[1,1] + e^36 t[0,2]
The divisor class number is 7^0.
```

===

EXAMPLE 4. Several random polynomials of degree 5 are chosen with coefficients in the field GF(2).

```
Will you want to see a description of the field GF(q)?
NO
Will you want to see the system of equations to be solved?
NO
Will you want to specify the coefficients of G (enter Y),
or chose them randomly (enter N)?
NO
Enter a seed for the random number generator.
5000
How many polynomials do you want produced?
10
Will you want to see the polynomials?
YES
In what file would you like the results placed?
```

```
GF2.OUT
Enter p, the characteristic of the field GF(q).
2
Enter n, the degree of the field GF(2^n) over the prime field  GF(2).
1
Enter the degree of the polynomial G(x,y).
5

--------------------------------------------------------------------------------

In terms of the primitive root, e, in GF(2^1), G is
x + y + x^2 + y^2 + x^3 + x^2y + xy^2
 + x^3y + x^2y^2 + xy^3 + y^4 + x^4y + y^5
The divisor class number is 2^4.
In terms of the primitive root, e, in GF(2^1), G is
1 + x + y + x^2 + x^2y + xy^2 + y^3
 + x^4 + x^2y^2 + x^4y
The divisor class number is 2^2.
In terms of the primitive root, e, in GF(2^1), G is
x + y + x^2 + xy + x^3 + x^2y + xy^2
 + y^3 + x^4 + x^3y + x^2y^2 + y^4 + x^5 + y^5
The divisor class number is 2^4.
In terms of the primitive root, e, in GF(2^1), G is
x + y + x^2y + xy^2 + y^3 + x^4 + x^3y
 + xy^3 + y^4 + x^4y + x^2y^3 + xy^4 + y^5
The divisor class number is 2^2.
In terms of the primitive root, e, in GF(2^1), G is
1 + y + x^2 + x^3 + x^2y + xy^2 + x^4
 + x^3y + xy^3 + x^5 + x^4y + x^2y^3 + xy^4
The divisor class number is 2^3.
In terms of the primitive root, e, in GF(2^1), G is
x^2 + xy + x^4 + x^3y + x^2y^2 + xy^3 + y^4
 + x^5 + x^4y + x^3y^2 + xy^4
The divisor class number is 2^5.
In terms of the primitive root, e, in GF(2^1), G is
1 + x + x^2 + xy + x^3 + x^2y + xy^2
 + x^4 + x^3y + xy^3 + y^4 + x^5 + x^3y^2 + x^2y^3
 + xy^4
The divisor class number is 2^3.
In terms of the primitive root, e, in GF(2^1), G is
1 + y^2 + x^3 + x^2y + xy^2 + y^3 + x^4
 + x^3y + x^2y^2 + x^3y^2 + x^2y^3
The divisor class number is 2^2.
In terms of the primitive root, e, in GF(2^1), G is
x^2 + xy + y^2 + x^3 + x^2y + xy^2 + x^3y
 + x^2y^2 + y^4 + x^2y^3
The divisor class number is 2^3.
In terms of the primitive root, e, in GF(2^1), G is
y + x^2 + y^2 + xy^2 + y^3 + x^4 + x^3y
 + xy^3 + y^4 + x^5 + x^4y + x^3y^2 + x^2y^3 + y^5
The divisor class number is 2^2.

================================================================================
```

EXAMPLE 5. Often it is advantageous to run a program in batch rather than at the terminal, for instance, a highly computational program as LANG will run faster overnight when no one else is using the computer. In order to run LANG in batch, you have to prepare a batch command file and then issue the SUBMIT command. The command file must include the responses to the questions issued by LANG in the proper order.

Suppose that the command file GF8.COM has been prepared to run LANG in order to find the divisor class numbers associated to 10 randomly chosen polynomials of degree 5 with coefficients in GF(8). Then the following command begins the batch processing.

```
$ SUBMIT /NOTIFY /NOPRINTER GF8
```

This command requests that the command file GF8.COM be submitted to the batch processor, that when the job is completed that a message be indicated at the terminal to that effect, and that no log be printed for the job. There is no reason to include /NOTIFY if you immediately sign off the terminal. Including /NOPRINTER keeps the file containg the log of the batch job from being printed and deleted. The name of the log file is GF8.LOG. (In this example, it is assumed all files are in the user's default directory).

The contents of a command file GF8.COM and the output of LANG follow.

```
$ RUN LANG
YES
NO
NO
4000
10
YES
GF8.OUT
2
3
5
0
```

```
The elements of GF(2^3) are represented as 3-tuples.
The identity element is (1,0,0)
A primitive 7-th root, e, is (0,1,0)
Its 3-th power equals (1,1,0)
In terms of the primitive root, e, in GF(2^3), G is
e^3 x + e^6 y + ex^2 + e^2 y^2 + x^2y + e^2 xy^2 + ey^3
 + e^5 x^4 + e^6 x^3y + x^2y^2 + xy^3 + e^5 y^4 + e^3 x^5 + e^3 x^3y^2
 + e^3 x^2y^3 + exy^4 + e^5 y^5
The divisor class number is 2^2.
In terms of the primitive root, e, in GF(2^3), G is
e^2  + e^2 x + e^5 y + e^2 x^2 + e^6 xy + e^6 x^3 + e^3 x^2y
 + e^3 y^3 + e^3 x^4 + x^3y + x^2y^2 + exy^3 + e^2 y^4 + e^3 x^5
 + e^3 x^4y + e^2 x^2y^3 + xy^4 + e^4 y^5
The divisor class number is 2^1.
In terms of the primitive root, e, in GF(2^3), G is
e^6  + ex + y + e^2 x^2 + e^4 xy + e^3 y^2 + e^2 x^3
 + e^6 x^2y + e^2 y^3 + ex^4 + e^5 xy^3 + y^4 + x^5 + e^2 x^4y
 + e^5 x^3y^2 + e^4 x^2y^3 + e^6 xy^4 + e^4 y^5
The divisor class number is 2^2.
In terms of the primitive root, e, in GF(2^3), G is
e^4  + e^4 y + e^4 x^2 + e^5 xy + y^2 + e^2 x^2y + e^2 y^3
 + e^6 x^3y + e^2 x^2y^2 + e^2 xy^3 + e^2 y^4 + e^3 x^5 + e^6 x^4y + e^6 x^3y^2
 + e^2 x^2y^3 + xy^4
The divisor class number is 2^2.
In terms of the primitive root, e, in GF(2^3), G is
e^4  + ex + e^2 x^2 + e^6 xy + e^5 y^2 + e^5 x^3 + e^4 x^2y
 + exy^2 + e^5 y^3 + e^5 x^4 + ex^3y + e^5 x^2y^2 + e^3 xy^3 + e^3 y^4
```

```
 + e 5 x^5 + x^4y + e^3 x^3y^2 + e^4 x^2y^3 + e^4 xy^4 + ey^5
The divisor class number is 2^4.
In terms of the primitive root, e, in GF(2^3), G is
e^5  + e^4 x + y + e^2 x^2 + e^2 xy + e^2 y^2 + e^2 x^3
 + xy^2 + ey^3 + e^2 x^3y + e^5 x^2y^2 + e^3 xy^3 + e^2 y^4 + e^6 x^5
 + e^4 x^4y + e^3 x^3y^2 + ex^2y^3 + ey^5
The divisor class number is 2^2.
In terms of the primitive root, e, in GF(2^3), G is
e^5  + e^2 x + y + e^3 xy + y^2 + e^5 x^3 + e^5 x^2y
 + e^6 xy^2 + ey^3 + e^2 x^4 + e^6 x^3y + e^5 x^2y^2 + e^2 y^4 + e^6 x^4y
 + ex^3y^2 + e^4 x^2y^3 + xy^4 + e^6 y^5
The divisor class number is 2^2.
In terms of the primitive root, e, in GF(2^3), G is
e^4  + e^3 x + e^3 y + e^3 x^2 + e^2 xy + e^5 y^2 + x^3
 + e^5 x^2y + e^6 xy^2 + y^3 + e^2 x^4 + e^4 x^3y + e^2 xy^3 + e^2 x^4y
 + e^2 x^2y^3 + e^2 xy^4
The divisor class number is 2^3.
In terms of the primitive root, e, in GF(2^3), G is
e^6  + y + e^2 x^2 + e^2 xy + y^2 + e^6 x^3 + e^2 x^2y
 + exy^2 + e^3 y^3 + e^2 x^4 + e^4 x^3y + ex^2y^2 + exy^3 + e^4 y^4
 + e^5 x^5 + e^5 x^4y + e^4 x^3y^2 + e^2 x^2y^3 + e^5 xy^4 + y^5
The divisor class number is 2^1.
In terms of the primitive root, e, in GF(2^3), G is
e^3  + e^4 x + e^2 y + e^3 x^2 + exy + x^3 + e^3 x^2y
 + xy^2 + e^6 y^3 + e^3 x^4 + e^5 x^3y + e^3 x^2y^2 + e^4 xy^3 + e^5 y^4
 + e^5 x^5 + e^4 x^4y + e^4 x^3y^2 + e^2 x^2y^3 + y^5
The divisor class number is 2^1.

========================================================================
```

DOCUMENTATION FILE FOR THE PROGRAM LANG

Written by Dave Joyce, Math/CS dept, Clark Univ, Worcester, Mass., May 1983.

This file contains detailed information concerning the program LANG. For more general information about the problem that LANG addresses, read the file LANG.TXT. For sample runs of the program LANG, see the file LANG.XPL.

Modules used in the creation of the program LANG.

The program LANG is composed of six modules: PRIME, FIELD1, FIELD2, FIELDP2, FIELDL2, and LANG. In order to create the executable program LANG.EXE, you must compile each of the six modules and link the object files by the following commands.

```
$ PASCAL/NOS PRIME
$ PASCAL/NOS FIELD1
$ PASCAL/NOS FIELD2
$ PASCAL/NOS FIELDP2
$ PASCAL/NOS FIELDL2
$ PASCAL/NOS LANG
$ LINK LANG,PRIME,FIELD1,FIELD2,FIELDP2,FIELDL2
```

Descriptions of the six modules follow.

The module PRIME

This module contains routines related to prime numbers. The three routines that are included in PRIME are these:

GET_DIVISOR	Finds the smallest divisor greater than 1 of the argument.
IS_PRIME	Returns true if the argument N is prime, false otherwise.
GET_FACTORS	This procedure factors the argument N into prime powers and returns a list of the primes dividing N along with the order in N of each prime.

The module FIELD1

This module contains routines dealing with elements of the Galois field GF(q). The number q is a prime power, p^n, where p is a prime. GF(q) denotes the finite field of q elements. In this module, each element of GF(q) is represented as an n-tuple (array of n elements) of integers bounded between 0 and p-1. Addition and subtraction are performed coordinatewise. Multiplication depends on the knowledge of an irreducible polynomial of degree N over GF(p). Each element of GF(q) may be interpreted as a polynomial in one variable of degree n-1 over GF(p). If A is an array representing such an element, then A may be interpreted as the polynomial of degree n-1 in the variable e

```
A[1] + A[2]*e + ... + A[n]*e^(n-1) .
```

The following routines are included in this module:

SET_UP_Q1	Sets up the structure of GF(q). See comments below.
REPORT_Q1	Reports the structure of GF(q) found by SET_UP_Q1.
EQUAL_Q1	Returns 'true' if two elements of GF(q) are equal.
ADD_Q1, SUB_Q1, MUL_Q1	Perform an operation on two elements of GF(q).
SCAL_MUL_Q1	Multiplies an element of GF(q) by an integer.
POWER_Q1	Raises an element of GF(q) to a nonnegative integral power.
READ_Q1	Reads an element of GF(q).
WRITE_Q1	Writes an element of GF(q).

The procedure SET_UP_Q1 takes as parameters p and n. A search is made of the monic polynomials f(e) of degree n over GF(p) for one which makes the element e of GF(p)[e]/f(e) an element of order p^n-1. This polynomial f(e) is then used in the procedure MUL_Q1 to give the multiplicative structure of GF(q).

The Module FIELD2

This module represents elements in the Galois field GF(q) in a form different from the form used in the module FIELD1. Each element other than 0 in GF(q) is a power of a generating element of the multiplicative group of invertible elements. Such an element is the element e found by the procedure SET_UP_Q1 described above. For instance, 1 = e^0, so 1 is represented by the integer 0. A special representation is used for 0, namely, -1. All other elements are represented by integers in the range 0 to q-1. This representation facilitates multiplication, division, powers, and roots in GF(q). In order to add and subtract elements represented by indices, a table of successors is used. This table describes the function Succ(a) = Index (e^a + 1). Addition may be computed by the identity e^a + e^b = e^(a + succ(b-a)).

The following routines are included in the FIELD2 module:

SET_UP_Q2	Sets up the structure of GF(q). Input: the structure of GF(q) as found by SET_UP_Q1. The tables for the successor function and the index function are initialized by SET_UP_Q2.
CVT_Q1_Q2	Converts an element of GF(q) from the first form to the second form.
ADD_Q2, SUB_Q2, MUL_Q2, DIV_Q2	Arithmetic routines.
SCAL_MUL_Q2	Multiplies an element of GF(q) by an integer.
POWER_Q2	Raises an element of GF(q) to an integral power.
ROOT_Q2	Takes the p-th root of an element of GF(q).

The FIELDP2 module

This module contains some procedures to manipulate polynomials in two variables over GF(q). The following procedures are used by LANG:

ZERO_P2	Sets a polynomial to zero.
WRITE_P2	Writes out the polynomial as a polynomial in the variables x and y.
TEST_TRIVIALITY_P2	Returns 'true' if the polynomial is trivial, otherwise indicates the indices of a nontrivial coefficient.

The FIELDL2 module

This module has three routines related to linear algebra over GF(q). The following three procedures are included:

TEST_TRIVIALITY_L2 Returns 'true' if a vector is trivial, otherwise indicates the index of a nontrivial coefficient.

ROW_REDUCE_L2 Row-reduces an M by N matrix of elements of GF(q). Row-reduction procedes only through an indicated column.

SOLVE_MATRIX_SYSTEM_L2 Solves a system of linear and p-linear equations in N variables. See notes below.

The procedure SOLVE_MATRIX_SYSTEM_L2 takes as parameters a matrix which specifies N p-linear equations, and a list of vectors, each vector specifying a linear equation. The p-linear equations are in a standard form, namely,

$$\sum a[i,j]x[i] \;=\; x[i]\hat{\ }p \;.$$

The a[i,j] are passed in an NxN matrix. The linear equations are of the form

$$\sum b[i]x[i] = 0 \;.$$

The b[i] are passed in a vector (array). The procedure doesn't actually solve the system of equations. Instead, it determines the number of solutions in the algebraic closure of GF(q). The algorithm is as follows.

(1) As long as there is a nontrivial linear equation, perform the following elimination procedure. Find a nontrivial coefficient, say b[i] of a nontrivial linear equation. Divide that equation by b[i] so that we may assume b[i] is 1. Pivit on that element thereby eliminating x[i] from all the other linear equations and from the linear sides of all the p-linear equations. Then x[i] only appears in the original linear equation and on the right side of the i-th p-linear equation. Form a new linear equation

$$\sum b[i]\hat{\ }p * a[i,j]*x[j] = 0 \;.$$

Drop the original linear equation and the i-th p-linear equation from the system and add this new linear equation. The new system has one less equation and one less variable. Its solutions are in one-to-one correspondence with the solutions of the original system (since the linear equation may be used to determine the dropped variable).

(2) If there are only p-linear equations, perfrom the following elimination procedure. Find any linear relations among the x[i]^p. This is done by row reducing a matrix concatinated from the matrix of coefficients a[i,j] and the identity matrix.

$$\left|\begin{array}{ccc|ccc} a[1,1] & \ldots & a[1,n] & 1 & 0\ \ldots & 0 \\ a[2,1] & \ldots & a[2,n] & 0 & 1\ \ldots & 0 \\ \ldots & \ldots & \ldots & & \ldots & \\ a[n,1] & \ldots & a[n,n] & 0 & 0\ \ldots & 1 \end{array}\right|$$

After row-reducing this matrix, if any of the bottom rows of the left part of the matrix are trivial, then the corresponding right parts of the rows indicate relations among the p-th powers of the variables. For each such relation

$$\sum c[i]*x[i]\hat{\ }p = 0$$

create a linear relation by taking p-th roots

```
 --
 >  root(c[i])*x[i] = 0 .
 --
```

Then add these linear relations to the system and procede to the first reduction procedure.

(3). If there are no linear relations among the p-th powers of the variables, then there are p^n solutions of the system in the algebraic closure of GF(q).

Module LANG

The main program is contained in the module LANG. First LANG asks the user how much information to record and in what file, then it asks whether the user wants to specify the polynomials or if he wants the program to randomly generate polynomials. Then the program repeatedly asks for fields and polynomials over the fields to test. When the p and n are given to LANG for the field, LANG sets up the field by calling the procedures SET_UP_Q1 and SET_UP_Q2. Once the polynomial to study is determined, denoteed G, LANG calls the procedures SET_UP_SYSTEM and CONSTRUCT_MATRIX_SYSTEM (found in the module LANG itself), and then calls SOLVE_MATRIX_SYSTEM_L2 to determine the number of solutions of the system which is the desired class number.

SET_UP_SYSTEM initializes the system of equations arising from the system of equations

(1) Nabla (G^(p-1) * t) = t^p

(2) Nabla (G^k * t) = 0 for k = 0,...,p-2.

Here, G is the specified polynomial in two variables x and y, t is an unknown polynomial in x and y, and Nabla is the operator

```
        d^(p-1)  d^(p-1)
Nabla = -------- --------
        dx^(p-1) dy^(p-1)
```

that is, Nabla is the maximal nontrivial partial derivative. The equation (1) gives rise to a set of (g-1)^2 p-linear equations. For eack k, equation (2) gives rise to a set of no more than (g-1)^2 linear equations. The i,j-th equation in such a set comes from equating the coefficients of x^ip * y^yp in (1) or (2). The unknowns in this system of linear and p-linear equations are the coefficients of the polynomial t.

CONSTRUCT_MATRIX_SYSTEM translates the system of equations produced by SET_UP_SYSTEM into a standard form that can then be used by SOLVE_MATRIX_SYSTEM_L2 to determine the number of solutions of the system. CONSTRUCT_MATRIX_SYSTEM performs the job of reducing a p-linear system of equations in the doubly-indexed family of variables t[i,j] into an equivalent system in a singly-indexed family of variables. In the process of translation each equation of the form t[i,j]^p = 0 is discarded along with all the terms involving t[i,j] in the rest of the equations.

```
Program Lang (Input, Output, Result_file) ;

{-------------------------------------------------------------------+
|       This program determines the number of divisor classes of the |
|       Zariski surface  z^p = G(x,y)  over fields of characteristic |
|       p > 0 by J. J. Lang's algorithm.  See his thesis, Purdue Univ,|
|       1981, especially Lemmas 3.8, 3.9, and 3.10.                  |
|                                                                   |
|       Author:  Dave Joyce, Math/CS dept, Clark Univ, Worcester, MA |
|       Date written:   Jan 1983                                     |
|       Revised:        May 1983                                     |
|       Revised:        Dec 1984 to be compatible with later version |
|                                of Pascal.                          |
+-------------------------------------------------------------------}

{-------------------------------------------------------------------+
|       Linking instructions:                                        |
|               Link with the PRIME, FIELD1, FIELD2, FIELDP2, and    |
|               FIELDL2 modules by the command                       |
|               $ LINK LANG,PRIME,FIELD1,FIELD2,FIELDP2,FIELDL2      |
+-------------------------------------------------------------------}

Const
        Max_p = 19 ;      {the maximum value of the prime p}
        Max_n = 18 ;      {the maximum degree for GF(q) over GF(p) that this
                           program will deal with}
        Max_2n = 36 ;     {twice max_n}
        Max_q = 262144 ;          {the maximum value of q, which is p^n}
        Max_deg_G = 18 ;          {the maximum degree that the polynomial
                                   G(x,y) may have}
        Max_p_deg_G = 342 ;       {must equal max_p times max_deg_G}
        Max_deg_t = 16 ;          {must equal max_deg_G - 2}
        Max_deg_t2 = 153 ;        {must be (Max_deg_t+1)*(Max_deg_t+2)/2}

        Zero_Q2    = -1 ;

Type

        Number_Q1 = array [1..max_n] of integer ;
        {'number_Q1' is the data type for elements in the field of q elements
         when they are represented in the first form, as arrays over GF(p).
         See the module Field1 for more information.}
        Field_Q1 = record
          p      : integer ;      {the characteristic of the field}
          n      : integer ;      {the degree of GF(q) over GF(p)}
          q_minus_1 : integer ; {p^n-1}
          zero_Q1 : Number_Q1 ; {the 0 elt of GF(q)}
          one_Q1 :  Number_Q1 ; {the 1 elt of GF(q)}
          prim_Q1 : Number_Q1 ; {a fixed primitive root of GF(q),i.e., elt of
                                order q-1}
          structure_Q1 : number_Q1 ; {essentially an irreducible poly of degree
                                n over the prime field having prim_Q1 as a root}
          End {field_Q1 record} ;

        Number_Q2 = integer ;
        {Number_q2 is the data type for elements of GF(q) when they are
         represented in the second form, as indices of a generating element of
         the multiplicative group.  See the Field2 module for more information}
        Field_Q2 = record
            p   : integer ;      {the characteristic of the field}
```

```
        n   : integer ;     {the degree of GF(q) over GF(p)}
        q_minus_1   : integer ;     {p^n-1}
        p_to_n_1    : integer ;     {p^(n-1)}
        index       : array [0..max_q] of integer ; {a table giving the
                            indices of the elements of GF(q) with respect
                            to prim_Q1}
        scsr        : array [0..max_q] of integer ; {a table describing
                            the successor funtion in GF(q) }
      End {field_Q2 record} ;

    Equation = array [0..max_deg_t,0..max_deg_t] of number_q2 ;
    Equation_list = ^Equation_node ;
    Equation_node = record
            Link    : equation_list ;
            E       : equation ;
        End {equation_node record} ;

    Matrix_system = array [1..max_deg_t2,1..max_deg_t2] of number_q2 ;
    Linear_list = ^Linear_node ;
    Linear_array = array [1..max_deg_t2] of number_Q2 ;
    Linear_node = record
            Link    : linear_list ;
            E       : Linear_array ;
        End {linear_node record} ;

Var
    Result_file : text ;
    Result_file_name : packed array [1..100] of char ;

    k_Q1    : Field_Q1 ;    {the description of GF(q) in the first form}
    k_Q2    : field_Q2 ;    {the description of GF(q) in the second form}

    G       : array [0..Max_deg_g,0..Max_deg_g] of integer ;
    deg_G   : integer ;     {the degree of the polynomial G(x,y)}
    deg_t   : integer ;     {the degree of t(x,y)}

    List_head : equation_list ;  {the head of a list of linear equations}
    E       : array [0..Max_deg_t,0..Max_deg_t] of equation ;
    { E is an array of p-linear equations}
    Nontrivial : array [0..max_deg_t,0..max_deg_t] of Boolean ;
    { Nontrivial[i,j] indicates if the variable  t[i,j]  might be
      nontrivial }
    Sys : matrix_system ;
    Linear_list_head : linear_list ;
    Rank : integer ;        {The rank of the system of equations found in
     the list headed by List_head and in the array E of p-linear equations}

    Give_field_description : Boolean ;
    {Indicates whether the user wants a description of the field GF(q) }
    Give_equations : Boolean ;
    {Indicates whether the user wants to see the equations to be solved}
    Specify_coeff : Boolean ;
    {Indicates whether the user wants to specify the polynomial  G  himself
     or let the program generate one randomly}
    Give_poly : Boolean ;
    {Indicates whether the user wants to have the polynomial  G  printed}
    Number_of_polys : integer ;
    {Indicates how many polynomials  G  to examine}
    Poly_counter : integer ;        {Counts up to Number_of_polys}
```

```
     Seed : integer ;         {Seed for the random number generator}
     Exit_request : Boolean ;    {Control variable for main program loop}

{-------------------------------------------------------------------+
|              DECLARATIONS OF EXTERNAL PROCEDURES                  |
+-------------------------------------------------------------------}

{Procedures found in the FIELD1 module}
[EXTERNAL] Procedure Set_up_Q1 (Var k_Q1 : field_Q1 ; Var Status_code :
      integer) ; extern ;
[EXTERNAL] Procedure Report_Q1 (Var outf : text ; Var k_Q1 : field_Q1) ;
      extern ;
[EXTERNAL] Procedure Read_Q1 (Var inf : text ; Var k_Q1 : field_Q1 ;
      Var a : number_Q1) ; extern ;

{Procedures found in the FIELD2 module}
[EXTERNAL] Procedure Set_up_Q2 (Var k_Q1 : field_Q1 ; Var k_Q2 : field_Q2 ) ;
      extern ;
[EXTERNAL] Function Add_Q2 (Var k_Q2 : field_Q2 ; a,b : number_Q2) :
      number_Q2 ; extern ;
[EXTERNAL] Function Mul_Q2 (Var k_Q2 : field_Q2 ; a,b : number_Q2) :
      number_Q2 ; extern ;
[EXTERNAL] Function CVT_Q1_Q2 (var k_Q2 : field_Q2 ; x : number_Q1) :
      number_Q2 ; extern ;

{Procedures found in the FIELDP2 module}
[EXTERNAL] Procedure Zero_P2 (
      Var f : array[xl..xh:integer;yl..yh:integer] of Number_Q2 ;
      deg_f : integer) ; extern ;
[EXTERNAL] Procedure Write_P2 (Var Outf : text ;
      Var f : array[xl..xh:integer;yl..yh:integer] of Number_Q2 ;
      deg_f : integer) ; extern ;
[EXTERNAL] Procedure Test_triviality_P2 (
      Var E : Array [xl..xh:integer;yl..yh:integer] of Number_q2 ;
      deg_E : integer ; Var Is_trivial : boolean ; Var R,S :
      integer) ; extern ;

{Procedures found in the FIELDL2 module}
[EXTERNAL] Procedure Solve_matrix_system_L2 (Var k_Q2 : field_Q2 ; M :
      matrix_system ; N : integer ; L : Linear_list ; Var Rank : integer) ;
      extern ;

{-------------------------------------------------------------------+
|              S  U  B  R  O  U  T  I  N  E  S                      |
+-------------------------------------------------------------------}

Function Yes_answer : Boolean ;
   {This function returns 'true' if the user enters a word beginning with
    a capital or small y.}
   Var
      Input_char : char ;
   Begin
      Readln (Input_char) ;
      Yes_answer := (Input_char = 'Y') or (Input_char = 'y') ;
   End {Yes_answer function} ;

Procedure Set_up_Field ;
  {This procedure asks the user for the size of the field, the calls
   other procedures to set up the structure of the field for representing
   elements of  GF(q)  in both the first and second form.}
```

```
  Var
    Status_code : integer ;
  Begin
    Writeln ;
    Write ('Enter p, the characteristic of the field GF(q):  ') ;
    Readln (k_Q1.p) ;
    Exit_request := (k_Q1.p > max_p) or (k_Q1.p < 2) ;
    If exit_request then
      If k_Q1.p <> 0 then
        Writeln ('Invalid value for p.')
      Else
        Writeln ('Exit Lang.')
    Else With k_Q1 do
      Begin
        Writeln ;
        Write ('Enter n, the degree of the field GF(',p:1,'^n) over ',
                'the prime field  GF(',p:1,'):  ') ;
        Readln (n) ;
        {leave the program if n is out of range:}
        Exit_request := (n < 1) or (n > max_n) ;
        If exit_request then
          Writeln ('n is out of range')
        Else
          Begin
            q_minus_1 := p**n - 1 ;
            exit_request := (q_minus_1 > max_q) ;
            If exit_request then
              Writeln ('p^n is out of range')
            Else
              Begin
                Set_up_Q1 (k_Q1, Status_code) ; {Set up GF(q)}
                Writeln ;
                Set_up_Q2 (k_Q1, k_Q2) ;
              End ;
          End ;
      End ;
  End {set_up_Field procedure} ;

Procedure Set_up_G ;
    {This procedure sets up the polynomial  G[x,y]  by either asking the user
     for the coefficients or by randomly choosing them.}
    Var
        I, J : integer ;

    Procedure Enter_coefficients ; {Ask the user for coefficients}
      Var
          temp : Number_Q1 ; {buffer to input element of GF(q) in first form}
          G_is_specified : Boolean ;
      Begin
        With k_Q1 do
          Begin
            Writeln ('Now enter the coefficients of G(x,y).') ;
            Writeln ('Enter a coefficient by entering (1) the power of x,') ;
            Writeln ('  (2) the power of y, (3) the ',n:1,
                        ' components of the coefficient.') ;
            Writeln ('Separate those ',n+2:1,' integers by spaces.') ;
            Writeln ('Indicate that G is specified by entering -1 ',
                        'for the power of x.') ;
            Writeln ;
            Zero_P2 (G,deg_G) ;          {set G to zero}
```

```
          G_is_specified := false ;
          While not G_is_specified do
            Begin
              Writeln ('Specify a coefficient.') ;
              Read (I) ;
              If (I < 0) then
                Begin
                  G_is_specified := true ;
                  Readln ;
                End
              Else
                Begin
                  Read (J) ;
                  If (J >= 0) and (I+J <= Deg_G) then
                    Begin
                      Read_Q1 (input, k_Q1,temp) ;
                      G[I,J] := CVT_Q1_Q2 (k_Q2, temp) ;
                    End
                  Else
                    Begin
                      Readln ;
                      Writeln ('Exponents out of range.') ;
                    End ;
                End {if/else} ;
            End {while} ;
        End {with} ;
    End {enter_coefficients procedure} ;

  Procedure Random_coefficients ;  {Choose the coefficients randomly}
    Var
      I,J : integer ;
    Function Ranmod (N : integer) : integer ;
      {Return a random integer from 0 through N-1}
      Function MTH$RANDOM (Var S : integer) : real ; extern ;
        {MTH$RANDOM returns a random real number between 0 and 1}
      Begin {Ranmod function}
        Ranmod := Trunc (MTH$RANDOM (Seed) * N) ;
      End {Ranmod function} ;
    Begin {Random_coefficients procedure}
      For I := 0 to deg_G do
      For J := 0 to deg_G-I do
        G[I,J] := Ranmod (k_Q2.q_minus_1 + 1) - 1 ; {Get a random index
    End {Random_coefficients procedure} ;

  Begin {set_up_G procedure}
    If Poly_counter = 1 then
      Begin
        Write ('Enter the degree of the polynomial G(x,y):  ') ;
        Readln (Deg_G) ;
      End ;
    Exit_request := (deg_G > Max_deg_G) or (deg_G < 2) ;
    If exit_request then
      Writeln ('The degree of G is out of range.')
    Else
      Begin
        Deg_t := deg_G -2 ;
        If Specify_coeff then
```

```
          Enter_coefficients
        Else
          Random_coefficients ;

        End ;
    End {set_up_G procedure} ;

{----------------------------------------------------------------------+
|                ROUTINES TO INITIALIZE THE SYSTEM OF EQUATIONS        |
+----------------------------------------------------------------------}

Procedure Set_up_system ;
    {This procedure initializes the system of equations arising from
        (1)     Nabla (G^(p-1) * t) = t^p,  and
        (2)     Nabla (G^k * t) = 0,  for k = 0,...,p-2.
     The equation (1) gives rise to a set of (g-1)^2 p-linear equations. For
     each k, equation (2) gives rise to a set of no more than (g_1)^2 linear
     equations.  The i,j-th equation in such a set comes from equating the
     coefficients of  x^ip * y^jp  in (1) or (2).  Precisely, the left hand
     side is of the form
        Sum G[i[1],j[1]]*...*G[i[k],j[k]]*t[i[0],j[0]]
     where the sum runs over indices such that
        i[0]+i[1]+...+i[k] + 1-p = i*p,  and
        j[0]+j[1]+...+j[k] + 1-p = j*p.}
    Var
        I,J,K, deg_H : integer ;
        H : array [0..max_p_deg_g,0..max_p_deg_G] of integer ;

    Procedure Get_equations_for_K ;
      Var
        I, J, R, S, U, V : integer ;
      Begin
        With k_Q2 do
        For I := 0 to deg_t do
        For J := 0 to deg_t-I do
          For R := 0 to deg_t do
          For S := 0 to deg_t-R do
            Begin
              U := (I+1)*p-1-R ;
              V := (J+1)*p-1-S ;
              If (0<=U) and (0<=V) and (U+V<=deg_H) then
                E[I,J][R,S] := H[U,V]
              Else E[I,J][R,S] := Zero_Q2 ;
            End ;
      End {get_equations_for_K procedure} ;

    Procedure Save_equations_on_list ;
      Var
        deg,J, R,S : integer ;
        L : equation_list ;
        Trivial : Boolean ;
      Begin
        For deg := 0 to deg_t do
        For J := 0 to deg do
          Begin
            Test_triviality_P2 (E[deg-J,J], deg_t, trivial, R,S) ;
            If not trivial then
              Begin
                New(L) ;
                L^.E := E[deg-J,J] ;
                L^.link := list_head ;
                list_head := L ;
              End {if} ;
          End {Fors} ;
```

```
      End {save_equations_on_list} ;

    Procedure Multiply_H_by_G ;
      Var
        I,J, R,S : integer ;
        Old_H : array [0..max_p_deg_G,0..max_p_deg_G] of integer ;
      Begin
        Old_H := H ;
        {Set new H to zero}
        For I := 0 to deg_G+deg_H do
        For J := 0 to deg_G+deg_H-I do
          H[I,J] := Zero_Q2 ;
        {Compute product}
        For I := 0 to deg_H do
        For J := 0 to deg_H-I do
          For R := 0 to deg_G do
          For S := 0 to deg_G-R do
            H[I+R,J+S] :=
                Add_Q2(k_Q2, H[I+R,J+S],Mul_Q2(k_Q2, Old_H[I,J],G[R,S])) ;
        Deg_H := deg_H + deg_G ;
      End {multiply_H_by_G procedure} ;
    Begin {set_up_system procedure}
      list_head := nil ;
      H[0,0] := 0 ;       {set H to one}
      deg_H := 0 ;
      Get_equations_for_K ; {K=0}
      For K := 1 to k_Q2.p-1 do
        Begin
          Save_equations_on_list ;
          Multiply_H_by_G ;
          Get_equations_for_K ;
        End ;
      For I := 0 to deg_t do
      For J := 0 to deg_t - I do
        Nontrivial[I,J] := true ;
    End {set_up_system procedure} ;

{------------------------------------------------------------------------+
|                  ROUTINES TO WRITE OUT THE SYSTEM                      |
+------------------------------------------------------------------------}

Procedure Write_side (Var outf : text ; Var E : equation) ;
    Var
        c, j : integer ;
        Term_number : integer ;
    Begin
        Term_number := 0 ;
        For c := 0 to deg_t do
        For j := 0 to c do
          If E[c-j,j] <> Zero_Q2 then
            Begin
              Term_number := term_number + 1 ;
              If term_number mod 5 = 0 then      {only 5 terms per line}
                Begin
                  Writeln (outf) ;
                  Write (outf, '          ') ; {tab}
                End ;
              If Term_number <> 1  then write (outf, '+ ') ;
              If E[c-j,j] <> 0 then
                Write (outf, 'e^',E[c-j,j]:1,' ') ;
```

```
            Write (outf, 't[',c-j:1,',',j:1,'] ') ;
          End ;
      If term_number = 0 then Write (outf, '0') ;
  End {Write_side procedure} ;

Procedure Report_system (Var outf : text) ;
    Var
        Deg, J : integer ;
        L : equation_list ;
    Begin
        For deg := 0 to deg_t do
        For J := 0 to deg do
          If nontrivial [deg-J,J] then
            Begin
              Write (outf, 't[',deg-J:1,',',J:1,']^',k_Q2.p:1,' = ') ;
              Write_side (outf, E[deg-J,J]) ; Writeln (outf) ;
            End ;
        L := list_head ;
        While L <> nil do
          Begin
            Write (outf, '0       = ') ;
            Write_side (outf, L^.E) ; Writeln (outf) ;
            L := L^.link ;
          End ;
    End {Report_system procedure} ;

{-----------------------------------------------------------------------+
|       ROUTINE TO TRANSLATE THE SYSTEM OF EQUATIONS TO STANDARD FORM   |
+-----------------------------------------------------------------------}

    Procedure Construct_matrix_system ;
        Var
          D, I, J, Q, R, S : integer ;
          K, L        : integer ;
          Is_trivial  : Boolean ;
          Temp_list   : equation_list ;
          Temp_eq     : linear_list ;
        Begin
          {Determine trivial p-linear equations so to later ignore
           corresponding variables}
          For I := 0 to deg_t do
          For J := 0 to deg_t-I do
            Begin
              Test_triviality_P2 (E[I,J], deg_t, Is_trivial, R,S) ;
              Nontrivial [I,J] := not is_trivial ;
            End {for} ;

          {Copy information to Sys from E}
          K := 1 ;
          For D := 0 to deg_t do
          For J := 0 to D do
            If nontrivial[D-J,J] then
              Begin
                L := 1 ;
                For Q := 0 to deg_t do
                For S := 0 to Q do
                  If nontrivial[Q-S,S] then
                    Begin
                      Sys[K,L] := E[D-J,J][Q-S,S] ;
                      L := L + 1 ;
```

```
                End {if} ;
            K := K+1 ;
          End {if} ;
      Rank := K-1 ;

      {Copy the linear equations}
      Linear_list_head := nil ;
      While list_head <> nil do
        Begin
          Temp_list := list_head ;
          List_head := List_head^.link ;
          New (Temp_eq) ;
          Temp_eq^.link := Linear_list_head ;
          L := 1 ;
          For Q := 0 to deg_t do
          For S := 0 to Q do
            If nontrivial[Q-S,S] then
              Begin
                Temp_eq^.E[L] := Temp_list^.E[Q-S,S] ;
                L := L + 1 ;
              End {for} ;
          Linear_list_head := Temp_eq ;
          Dispose (Temp_list) ;
        End {while} ;
    End {Construct_matrix_system procedure} ;

{---------------------------------------------------------------------+
|                          M A I N   P R O G R A M                     |
+---------------------------------------------------------------------}

Begin {main program}
  Writeln ;
  Writeln ;
  Writeln (' This program determines the divisor class number of the surface')
  Writeln (' z^p = G(x,y) over fields of characteristic p > 0 by J. J.') ;
  Writeln (' Lang''s algorithm.  Written by D Joyce, Clark Univ, Jan 1983.') ;
  Writeln (' (p can be no greater than ',max_p:1,',') ;
  Writeln (' n must be < ',max_n:1,' and p^n must be < ',max_q:1) ;
  Writeln (' The degree of G must be between 2 and ',Max_deg_G:1,')');
  Writeln ;

    Write ('Will you want to see a description of the field GF(q)?  ') ;
    Give_field_description := Yes_answer ;
    Write ('Will you want to see the system of equations to be solved?  ') ;
    Give_equations := Yes_answer ;
    Writeln ('Will you want to specify the coefficients of G (enter Y),') ;
    Write ('or chose them randomly (enter N)?  ') ;
    Specify_coeff := Yes_answer ;
    If not specify_coeff then
      Begin
        Write ('Enter a seed for the random number generator:  ') ;
        Readln (Seed) ;
        Write ('How many polynomials do you want produced?  ') ;
        Readln (Number_of_polys) ;
        If give_equations then
          Give_poly := true
        Else
          Begin
            Write ('Will you want to see the polynomials?  ') ;
            Give_poly := Yes_answer ;
```

```
      End ;
    End
  Else {specify_coeff}
    Begin
      Number_of_polys := 1 ;
      Give_poly := true ;
    End ;
  Write ('In what file would you like the results placed?  ') ;
  Readln (Result_file_name) ;
  Open (Result_file, Result_file_name) ;
  Rewrite (Result_file) ;

  Repeat
      Exit_request := false ;
      Set_up_Field ;
      If give_field_description and not exit_request then
        Report_Q1 (Result_file, k_Q1) ;
      If not exit_request then
       For Poly_counter := 1 to Number_of_polys do
        Begin
          Set_up_G ;
          If give_poly then
            Begin
              Writeln (Result_file) ;
              Writeln (Result_file,
                      'In terms of the primitive root, e, in GF(',
                      k_Q2.p:1,'^',k_Q2.n:1,'), G is') ;
              Write_P2 (Result_file, G,deg_G) ;  Writeln (result_file) ;
            End ;
          Set_up_system ;
          If Give_equations then
            Begin
              Writeln (Result_file,
                      'The system of equations to solve is as follows:') ;
              Writeln (result_file) ;
              Report_system (Result_file) ;
            End {if} ;
          {Now construct a reduced matrix for the system and simplify it}
          Construct_matrix_system ;
          Solve_matrix_system_L2 (k_q2, Sys, Rank, Linear_list_head, Rank) ;
          Writeln (Result_file,
              'The divisor class number is ',k_Q2.p:1,'^',rank:1,'.') ;
        End ;
      Writeln (result_file) ;
  Until exit_request ;

End {main program} .
```

```
Module FieldP2 ;

{-----------------------------------------------------------------------+
|       PROCEDURES AND FUNCTIONS TO MANIPULATE ELEMENTS OF GF(q)[x,y]   |
|       Author: Dave Joyce, Math/CS dept, Clark Univ, Worcester, MA     |
|       Date:   Nov 1982.  Modified July 1983 for Pascal 2.0.           |
+-----------------------------------------------------------------------}

{       These procedures deal with polynomials of two variables over the
        Galois field  GF(q).  Elements of  GF(q)  are represented in the second
        form, that is, by their indices relative to a primitive q-1-st root.
        See the module Field2 for more information concerning representation
        of elements of  GF(q)  in this form.  Polynomials in two variables
        over  GF(q)  are represented in matrices.  If  P  is such a matrix,
        then  P  represents the polynomial
                P[0,0] + P[0,1]*y + ... + P[0,d]*y**d +
                P[1,0]*x + P[1,1]*x*y + ... + P[1,d-1]*x*y**(d-1)
                + ... +
                P[d,0]*x**d ,
        where  d  is the degree of the polynomial.  The entries in the
        matrix for which i+j is greater than its degree are ignored     }


{       Routines included in this file:

        WRITE_P2        write the polynomial
        ZERO_P2         set a polynomial to 0
        DEG_P2          return the degree of the polynomial
        ADD_P2          add polynomials
        SUB_P2          subtract polynomials
        MUL_P2          multiply polynomials
        D_DX_P2         take the derivative w.r.t. the first variable
        D_DY_P2         take the derivative w.r.t. the second variable
        EVAL_P2         evaluate the polynomial for specific arguments
        TEST_TRIVIALITY_P2 determines if a polynomial is trivial        }


Const
        Max_q   = 262144 ;
        Zero_Q2 = -1 ;
        One_Q2  = 0 ;
Type

        Number_Q2 = integer ;
        Field_Q2 = record
                p          : integer ;
                n          : integer ;
                q_minus_1  : integer ;
                p_to_n_1   : integer ;
                index      : array [0..max_q] of integer ;
                scsr       : array [0..max_q] of integer ;
            End {field record} ;

{-----------------------------------------------------------------------+
|                          EXTERNAL DECLARATIONS                        |
+-----------------------------------------------------------------------}

{Routines found in the FIELD2 module:}
[EXTERNAL] Function Add_Q2 (Var k_Q2 : field_Q2 ; a,b : number_Q2) :
        number_Q2 ; extern ;
```

```
[EXTERNAL] Function Mul_Q2 (Var k_Q2 : field_Q2 ; a,b : number_Q2) :
       number_Q2 ; extern ;
[EXTERNAL] Function Scal_mul_Q2 (Var k_Q2 : field_Q2 ; k : integer ;
       a : number_Q2) : number_Q2 ; extern ;

{-----------------------------------------------------------------+
|                         R O U T I N E S                          |
+-----------------------------------------------------------------}

[GLOBAL] Procedure Write_P2 (Var Outf : text ;
              var f : array[xl..xh:integer;yl..yh:integer] of Number_Q2 ;
              deg_f : integer) ; {write the polynomial f}
    Var
        d, j : integer ;
        Term_number : integer ;
    Begin
        Term_number := 0 ;
        For d := 0 to deg_f do
         For j := 0 to d do
          If  f[d-j,j] >= 0 then {non-zero coefficient}
            Begin
              If Term_number <> 0 then Write (Outf, ' + ') ;
              If  f[d-j,j] = 0  then
                If (j = 0) and (d-j = 0) then Write (Outf, '1') ;
              If  f[d-j,j] > 0  then  Write (Outf, 'e') ;
              If  f[d-j,j] > 1  then  Write (Outf, '^',f[d-j,j]:1,' ') ;
              If  d-j > 0 then  Write (Outf, 'x') ;
              If  d-j > 1 then  Write (Outf, '^',d-j:1) ;
              If  j > 0  then  Write (Outf, 'y') ;
              If  j > 1  then  Write (Outf, '^',j:1) ;
              Term_number := Term_number + 1 ;
              {Now give a new line every 7th term}
              If Term_number mod 7 = 0 then Writeln (Outf) ;
            End ;
        If Term_number = 0 then Write (Outf, '0') ;
    End ;

[GLOBAL] Procedure Zero_P2 (
        Var f : array[xl..xh:integer;yl..yh:integer] of Number_Q2 ;
        deg_f : integer) ; {f := 0}
    Var
        i,j : integer ;
    Begin
        For i := 0 to deg_f do
          For j := 0 to deg_f do
            f[i,j] := Zero_Q2 ;
    End ;

[GLOBAL] Function Deg_P2 (
        var f : array[xl..xh:integer;yl..yh:integer] of Number_Q2 ;
        deg_f : integer) : Integer ;
    {return the degree of the polynomial f.  If f is zero, then return -1}
    Var
        d, j : integer ;
        still_looking : Boolean ;
    Begin
        d := deg_f ;
        still_looking := true ;
        j := 0 ;
        While (d >= 0) and still_looking do
```

```
        Begin
          If f[d-j,j] <> Zero_Q2 then  still_looking := false
          Else If  j < d  then  j := j+1
          Else
            Begin
              d := d-1 ;
              j := 0 ;
            End ;
        End {while} ;
      Deg_P2 := d ;
    End ;

[GLOBAL] Procedure Add_P2 (Var k_Q2 : field_Q2 ;
      Var f,g,h : array[xl..xh:integer;yl..yh:integer] of Number_Q2 ;
      deg : integer) ; { h := f + g }
    {It is assumed f,g, and h have the same degree.}
    Var
      i,j : integer ;
    Begin
      For i := 0 to deg do
        For j := 0 to deg do
          h[i,j] := Add_Q2 (k_Q2, f[i,j],g[i,j]) ;
    End ;

[GLOBAL] Procedure Sub_P2 (Var k_Q2 : field_Q2 ;
      Var f,g,h : array[xl..xh:integer;yl..yh:integer] of Number_Q2 ;
      deg : integer) ; { h := f + g }
    {It is assumed f,g, and h have the same degree.}
    Var
      i,j : integer ;
    Begin
      For i := 0 to deg do
        For j := 0 to deg do
          h[i,j] := Add_Q2 (k_Q2, f[i,j],g[i,j]) ;
    End ;

[GLOBAL] Procedure Mul_P2 (Var k_Q2 : field_Q2 ;
      Var f,g,h : array[xl..xh:integer;yl..yh:integer] of Number_Q2 ;
      deg_f, deg_g : integer ; Var deg_h : integer) ; { h := f * g }
    {Warning:  this procedure is invalid if  h  is the same variable as either
     f  or  g}
    Var
      t,s,i,j : integer ;
    Begin
      deg_h := deg_f + deg_g ;
      Zero_P2 (h,deg_h) ;
      For i := 0 to deg_f do
      For j := 0 to deg_f-i do
        For s := 0 to deg_g do
        For t := 0 to deg_g-s do
          h[i+s,j+t] := Add_Q2 (k_Q2, h[i+s,j+t],
                              Mul_Q2 (k_Q2, f[i,j], g[s,t]) ) ;
    End ;

[GLOBAL] Procedure d_dx_P2 (Var k_Q2 : field_Q2 ;
      var f,g : array[xl..xh:integer;yl..yh:integer] of Number_Q2 ;
      deg_f : integer ; Var deg_g : integer) ;        {g := df/dx}
    {Note:  this procedure is valid even if  g  and  f  are the same variable}
    Var
      i, j : integer ;
```

```
    Begin
        deg_g := deg_f - 1 ;
        For i := 0 to deg_g do
        For j := 0 to deg_g-i do
          g[i,j] := Scal_mul_Q2 (k_Q2, i+1, f[i+1,j]) ;
    End ;

[GLOBAL] Procedure d_dy_P2 (Var k_Q2 : field_Q2 ;
        var f, g : array[xl..xh:integer;yl..yh:integer] of Number_Q2 ;
        deg_f : integer ; Var deg_g : integer) ;          {g := df/dy}
    {Note: this procedure is valid even if  g  and  f  are the same variable}
    Var
        i, j : integer ;
    Begin
        deg_g := deg_f - 1 ;
        For j := 0 to deg_g do
        For i := 0 to deg_g-j do
          g[i,j] := Scal_mul_Q2 (k_Q2, j+1, f[i,j+1]) ;
    End ;

[GLOBAL] Function Eval_P2 (Var k_Q2 : field_Q2 ;
        var f : array[xl..xh:integer;yl..yh:integer] of Number_Q2 ;
        deg_f : integer ; x, y : Number_Q2) : number_Q2 ;
    {z := f(x,y) ; evaluate f at x and y giving z}
    Var
        i, j : integer ;
        z : Number_Q2 ;
        x_to_i, xtoi_ytoj : Number_Q2 ;
    Begin
        z := Zero_Q2 ;
        x_to_i := One_Q2 ;
        For i := 0 to deg_f do
          Begin
            xtoi_ytoj := x_to_i ;
            For j := 0 to deg_f do
              Begin
                z := Add_Q2 (k_Q2, z, Mul_Q2 (k_Q2, xtoi_ytoj, f[i,j])) ;
                xtoi_ytoj := Mul_Q2 (k_Q2, xtoi_ytoj, y) ;
              End ;
            x_to_i := Mul_Q2 (k_Q2, x_to_i, x) ;
          End ;
        Eval_P2 := z ;
    End {eval_P2 function} ;


[GLOBAL] Procedure Test_triviality_P2
        (Var E : Array [xl..xh:integer;yl..yh:integer] of Number_Q2 ;
        deg_E : integer ; Var Is_trivial : Boolean ; Var R, S : integer ) ;
    {Determine if the E is trivial.  If it isn't, then set R,S to
     indices of a nontrivial coefficient.}
    Begin
        R := 0 ; S := 0 ;
        Is_trivial := true ;
        Repeat
          If E[R,S] = Zero_Q2 then
            Begin
              S := S+1 ;
              If R+S > deg_E then
                Begin
                  S := 0 ;
```

```
              R := R+1 ;
            End ;
          End
        Else
          Is_trivial := false ;
      Until (R > deg_E) or not is_trivial ;
  End {Test_triviality_P2 procedure} ;

End {module FieldP2} .




Module Fieldl2 ;

{-----------------------------------------------------------------------------+
|        PROCEDURES AND FUNCTIONS TO MANIPULATE MATRICES OVER GF(q)            |
|        Author: Dave Joyce, Math/CS dept, Clark Univ, Worcester, MA           |
|        Date:   April 1983                                                    |
+-----------------------------------------------------------------------------}

{        These procedures deal with matrices over the Galois field  GF(q).
         Elements of  GF(q)  are represented in the second form, that is,
         by their indices relative to a primitive q-1-st root.  See the
         module Field2 for more information concerning representation
         of elements of  GF(q)  in this form.
         Such matrices may be used to solve linear and p-linear systems of
         equations.                                                           }

{        Routines included in this file:

         TEST_TRIVIALITY_L2        checks triviality of an array
         ROW_REDUCE_L2             uses Gauss elimination to reduce a matrix
                                   to echelon form
         SOLVE_MATRIX_SYSTEM_L2    solves a system of linear and p-linear
                                   equations returning the rank of the
                                   solution space                             }

Const
         Max_deg_t2 = 153 ;
         Max_2_deg_t2 = 306 ;
         Max_q    = 262144 ;
         Zero_Q2 = -1 ;
         One_Q2  = 0 ;

Type

         Number_Q2 = integer ;
         Field_Q2 = record
                 p         : integer ;
                 n         : integer ;
                 q_minus_1 : integer ;
                 p_to_n_1  : integer ;
                 index     : array [0..max_q] of integer ;
                 scsr      : array [0..max_q] of integer ;
             End {field record} ;

         Matrix_system = array [1..max_deg_t2,1..max_deg_t2] of number_Q2 ;

         Linear_list = ^Linear_node ;
         Linear_array = array [1..max_deg_t2] of number_Q2 ;
         Linear_node = record
                 Link     : linear_list ;
                 E        : linear_array ;
             End {linear_node record} ;
```

```
{---------------------------------------------------------------------+
|                      EXTERNAL DECLARATIONS                           |
+---------------------------------------------------------------------}

{Routines found in the FIELD2 module:}
[EXTERNAL] Function Add_Q2 (Var k_Q2 : field_Q2 ; a,b : number_Q2) :
        number_Q2 ; extern ;
[EXTERNAL] Function Sub_Q2 (Var k_Q2 : field_Q2 ; a,b : number_Q2) :
        number_Q2 ; extern ;
[EXTERNAL] Function Mul_Q2 (Var k_Q2 : field_Q2 ; a,b : number_Q2) :
        number_Q2 ; extern ;
[EXTERNAL] Function Div_Q2 (Var k_Q2 : field_Q2 ; a,b : number_Q2) :
        number_Q2 ; extern ;
[EXTERNAL] Function Scal_mul_Q2 (Var k_Q2 : field_Q2 ; k : integer ;
        a : number_Q2) : number_Q2 ; extern ;
[EXTERNAL] Function Power_Q2 (Var K_Q2 : field_Q2 ; a : number_Q2 ;
        k : integer) : number_Q2 ; extern ;
[EXTERNAL] Function Root_Q2 (Var k_Q2 : field_Q2 ; a : number_Q2 ) :
        number_Q2 ; extern ;

{---------------------------------------------------------------------+
|                          R O U T I N E S                             |
+---------------------------------------------------------------------}

[GLOBAL] Procedure Test_triviality_L2 (Var E : linear_array; N : integer;
        Var Is_trivial : Boolean ; Var J : integer) ;
        {Check the array E to see if all of its entries are trivial.
        Indicate the result in the variable Is_trivial.  If some entry
        is not trivial, then set J to its index.}
Begin
        J := 1 ;
        Is_trivial := true ;
        While is_trivial and (J <= N) do
          If E[J] = Zero_Q2 then
            J := J+1
          Else {E[J] is not trivial}
            Is_trivial := false ;
End {test_triviality_L2 procedure} ;

[GLOBAL] Procedure Row_Reduce_L2 (Var k_Q2 : field_Q2 ;
        Var A : array [xl..xh:integer;yl..yh:integer] of number_Q2 ;
        M, N, N1 : integer ; Var Rank : integer ) ;
        {A is an MxN array of elements of GF(q).  The procedure row-reduces A
        only through the N1-st column.  The rank of the MxN1 submatrix is
        returned in the variable Rank.}
Var
        I, J, K, L      : integer ;
        Still_looking   : Boolean ;
        D, Temp         : Number_Q2 ;
Begin
        I := 1 ; J := 1 ;
        While (I <= M) and (J <= N1) do
          Begin
            {Look for first nonzero entry in column J}
            K := I ; Still_looking := true ;
            While (K <= M) and still_looking do
              If A[K,J] = zero_Q2 then K := K+1
              Else      still_looking := false ;
            If still_looking then {column is empty}
              J := J + 1
            Else
              Begin
                {Exchange rows I and K }
                For L := J to N do
                  Begin
                    Temp := A[I,L] ;
                    A[I,L] := A[K,L] ;
```

```
                    A[K,L] := Temp ;
                  End {for} ;
                {Divide I-th row by A[I,J]}
                D := A[I,J] ;
                For L := J to N do
                  A[I,L] := Div_Q2 (k_Q2, A[I,L], D) ;
                {A[I,J] is now one.  Clear the rest of J-th column}
                For K := I+1 to N do
                  Begin
                    D := A[K,J] ;
                    If D <> zero_Q2 then
                      For L := J to N do
                        A[K,L] := Sub_Q2 (k_Q2, A[K,L],
                                        Mul_Q2 (k_Q2, A[I,L], D)) ;
                  End {for K} ;
                I := I+1 ; J := J+1 ;
              End {else} ;
          End {while} ;
        Rank := I-1 ;
End {Row_Reduce_L2 Procedure} ;

[GLOBAL] Procedure Solve_matrix_system_L2 (Var k_Q2 : field_Q2 ;
        M : matrix_system ; N : integer ;
        L : Linear_list ; Var Rank : integer) ;
        {Solve a system of linear and p-linear equations in N variables.
        The p-linear equations are determined by the NxN coefficient matrix
        M.  The linear equations are held in the list L of coefficient arrays.}
Var
        L_temp  : linear_list ;
        I, J    : integer ;
        A : array [1..Max_deg_t2,1..Max_2_deg_t2] of number_Q2 ;


    Procedure Drop_variable_from_array (J : integer ; Var E : linear_array) ;
    Var
        I :     integer ;
    Begin
        For I := J to N-1 do
            E[I] := E[I+1] ;
    End {Drop_variable_from_array procedure} ;

    Procedure Drop_variable_from_system (J : integer) ;
    Var
        I, K :  integer ;
        Next_L :        linear_list ;
    Begin
        Next_L := L ;
        While Next_L <> nil do
          Begin
            Drop_variable_from_array (J,Next_L^.E) ;
            Next_L := Next_L^.link ;
          End {while} ;
        For I := 1 to J-1 do
        For K := J to N-1 do
          M[I,K] := M[I,K+1] ;
        For I := J to N-1 do
        For K := 1 to J-1 do
          M[I,K] := M[I+1,K] ;
        For I := J to N-1 do
        For K := J to N-1 do
```

```
        M[I,K] := M[I+1,K+1] ;
      N := N-1 ;
  End {Drop_variable_from_system procedure} ;

  Procedure Drop_linear_equation (Var E : linear_array) ;
  Var
      I, J, K : integer ;
      Is_trivial      : Boolean ;
      D       : number_Q2 ;
      Next_lin        : linear_list ;
      New_E   : linear_array ;
  Begin
      {See if E is trivial}
      Test_triviality_L2 (E, N, Is_trivial, J) ;
      If not is_trivial then
        Begin
          {First divide the equation by E[J]}
          D := E[J] ;
          For I := 1 to N do
              E[I] := Div_Q2 (k_Q2, E[I], D) ;
          {Now E[J] is one. Next subtract an appropriate multiple of E
          from each of the rest of the equations.}
          Next_lin := L ;
          While Next_lin <> nil do
            Begin
              D := next_lin^.E[J] ;
              For I := 1 to N do
                Next_lin^.E[I] := Sub_Q2 (k_Q2, Next_lin^.E[I],
                                    Mul_Q2 (k_Q2, E[I],D)) ;
              Next_lin := Next_lin^.link ;
            End {while} ;
          For K := 1 to N do
            Begin
              D := M[K,J] ;
              For I := 1 to N do
                M[K,I] := Sub_Q2 (k_Q2, M[K,I], Mul_Q2 (k_Q2, E[I], D)) ;
            End {for} ;
          {Now only equation E and M[J] involve the J-th variable.  Create
          a new linear equation by summing over K: E[K]**p times M[K,-].}
          For  I := 1 to N do
            New_E[I] := zero_Q2 ;
          For  K := 1 to N do
            If E[K] <> zero_Q2 then
              Begin
                D := Power_Q2 (k_Q2, E[K], k_Q2.p) ;
                For I := 1 to N do
                  New_E[I] := Add_Q2 (k_Q2, New_E[I],
                              Mul_Q2 (k_Q2, D, M[K,I]) ) ;
              End {if} ;
          {Now M[J,-] is redundant, and E is only needed to define the J-th
          variable.  So forget about E entirely, and drop the J-th variable}
          Drop_variable_from_array (J,New_E) ;
          Drop_variable_from_system (J) ;
          {Finally drop the new linear equation}
          Drop_linear_equation (New_E) ;
        End {if} ;
  End {Drop_linear_equation procedure} ;

Begin {Solve_matrix_system_L2 procedure}
  Repeat
```

```
      {Begin by eliminating the linear equations}
      While L <> nil do
        Begin
          L_temp := L ;
          L := L^.link ;
          Drop_linear_equation (L_temp^.E) ;
        End {while} ;
      {Now find relations among p-linear equations. First prepare the
      matrix A}
      For I := 1 to N do
      For J := 1 to N do
        Begin
          A[I,J] := M[I,J] ;
          If I = J then A[I,J+N] := one_Q2
          Else          A[I,J+N] := zero_Q2 ;
        End {for} ;
      Row_Reduce_L2 (k_Q2, A, N, 2*N, N, Rank) ;
      {Set up new linear relations by taking p-th roots of the p-linear ones}
      For I := Rank+1 to N do
        Begin
          New (L_temp) ;
          For J := 1 to N do
            L_temp^.E[J] := Root_Q2 (k_Q2, A[I,J+N]) ;
          L_temp^.link := L ;
          L := L_temp ;
        End {for} ;
    Until Rank = N ;
End {Solve_matrix_system_L2 procedure} ;

End {Field12 module} .



Module Field1 ;

{-----------------------------------------------------------------------+
|       PROCEDURES AND FUNCTIONS TO MANIPULATE ELEMENTS OF GF(q)        |
|       Author:  Dave Joyce, Math/CS dept, Clark Univ, Worcester, MA    |
|       Date:    Nov 1982                                               |
+-----------------------------------------------------------------------}

{       Description of GF(q) and representation of the elements therin.

        The number  q  is a prime power,  p^n  where  p  is a prime.  GF(q)
        denotes the Galois field of  q  elements.  In this module, each
        element of GF(q) is represented as an n-tuple (array of  n  elements)
        of integers bounded between  1  and  p-1.  Addition and subtraction
        are performed coordinatewise.  Multiplication depends on the
        knowledge of an irreducible polynomial of degree  n  over  GF(p).
        Each element of  GF(q)  may be interpreted as a polynomial in one
        variable of degree  n-1  over  GF(p).  If  A  is an array representing
        such an element, also called a number_Q2, then  A  may be interpreted
        as the polynomial

                A[1] + A[2]*x + ... + A[n]*x^(n-1) .

        For reference by the various procedures the following information is
        stored in a record describing the representation of elements of the
        field GF(q).
```

```
        p               the characteristic of the field
        n               the degree of GF(q) over the prime field
        q_minus_1       p^n-1
        zero_Q1         the 0 elt of GF(q), i.e., 0
        one_Q1          the 1 elt of GF(q), i.e., 1
        prim_Q1         a fixed primitive root of GF(q), i.e., element of
                        order q-1, corresponds to the polynomial x
        structure_Q1    corresponds to an irreducible poly of degree n over
                        the prime field having  x  as a root,

        Only the numbers p and n need to be set by the main program. All the
        other variables are determined by SET_UP_Q.     }

{       Routines included in this module:

        EQUAL_Q1        determines if A equals B
        ADD_Q1          adds two numbers in GF(q)
        SUB_Q1          subtracts in GF_Q
        SCAL_MUL_Q1     multiplies an elt of GF(q) by an integer
        MUL_Q1          multiplies two numbers in GF(q)
        POWER_Q1        raises an elt of GF(q) to a nonnegative integral power
        WRITE_Q1        writes an elt of GF(q)
        READ_Q1         reads an elt of GF(q)
        SET_UP_Q1       sets up the structure of GF(q)
                                                                            }


Const
        Max_n   = 18 ;  {The maximum degree for GF(q) over GF(p) that this
                         module will deal with}
        Max_2n  = 36 ;  {Twice Max_n}

Type

        Factor_pointer = ^factor ;
        Factor  = record
            link : factor_pointer ;
            prime : integer ;
            order : integer ;
          End {factor record} ;

        Number_Q1 = array [1..max_n] of integer ;       {'Number_Q1' is the
                        data type for an element in the field of q elements
                        when it is represented in the first form, as arrays
                        over GF(p)}
        Field_Q1  = record
            p   : integer ;     {the characteristic of the field}
            n   : integer ;     {the degree of GF(q) over the prime field}
            q_minus_1 : integer ;       {p^n-1}
            zero_Q1 : Number_Q1 ;       {the 0 elt of GF(q)}
            one_Q1 : number_Q1 ;        {the 1 elt of GF(q)}
            prim_Q1 : number_Q1 ;       {a fixed primitive root of GF(q),
                                        i.e., element of order q-1}
            structure_Q1 : number_Q1 ;  {essentially an irreducible poly of
                                        degree n over the prime field having
                                        prim_Q1 as a root}
          End {field_Q1 record} ;

{----------------------------------------------------------------------+
|                       EXTERNAL ROUTINES                               |
+----------------------------------------------------------------------}

{Routines found in the PRIME module}
{EXTERNAL} Function Is_prime (P : integer) : boolean ; extern ;
{EXTERNAL} Procedure Get_Factors (N : integer ; var N_primes : integer ;
        var L : factor_pointer ; var M : integer) ; extern ;
```

```
{----------------------------------------------------------------------+
|                          R O U T I N E S                               |
+----------------------------------------------------------------------}

[GLOBAL] Function Equal_Q1 (Var k_Q1 : field_Q1 ; var a,b : number_Q1) :
        Boolean ;    {Returns 'true' if  a equals b}
    Var
        i : integer ;
        Still_looking : Boolean ;
    Begin
            i := 1 ;
            Still_looking := true ;
            While (i <= k_Q1.n ) and still_looking do
              Begin
                If  a[i] <> b[i]  then  still_looking := false ;
                i := i + 1 ;
              End {while} ;
            Equal_Q1 := still_looking ;
    End ;

[GLOBAL] Procedure Add_Q1 (Var k_Q1 : field_Q1 ; Var a, b, c : number_Q1) ;
        {c := a + b }
    Var
        i : integer ;
    Begin
        For i := 1 to k_Q1.n do


            c[i] := (a[i] + b[i]) mod k_Q1.p ;
    End ;

[GLOBAL] Procedure Sub_Q1 (Var k_Q1 : field_Q1 ; Var a, b, c : number_Q1) ;
        {c := a - b }
    Var
        i : integer ;
    Begin
        For i := 1 to k_Q1.n do
            c[i] := (a[i] - b[i] + k_Q1.p) mod k_Q1.p ;
    End ;

[GLOBAL] Procedure Scal_mul_Q1 (Var k_Q1 : field_Q1 ; a : integer ;
                         var b, c : number_Q1) ;
        {c := a * b}
        {Multiply c (in GF(q)) by the integer a to give c}
    Var
        i : integer ;
    Begin
        For i := 1 to k_Q1.n do
            c[i] := a*b[i] mod k_Q1.p ;
    End ;

[GLOBAL] Procedure Mul_Q1 (Var k_Q1 : field_Q1 ; var a, b, c : number_Q1) ;
        {c := a * b .  This procedure is valid even if  c  is the same
         variable as either or both of  a  and  b}
  Var
        d : array [1..max_2n] of integer ;
        i, j, k : integer ;
```

```
  Begin
    With k_Q1 do
      Begin
        For k := 1 to 2*n-1 do
            d[k] := 0 ;
        For i := 1 to n do
            For j := 1 to n do
                d[i+j-1] := d[i+j-1] + a[i]*b[j] ;
        For k := 2*n-1 downto n+1 do
            For j := 1 to n do
                d[k-n+j-1] := (d[k-n+j-1] + d[k]*structure_Q1[j]) mod p ;
        For i := 1 to n do
            c[i] := (d[i] mod p) ;
      End ;
  End {Mul_Q1 procedure} ;

[GLOBAL] Procedure Power_Q1 (Var k_Q1 : field_Q1 ; a : number_Q1 ;
        e : integer ; var b : number_Q1) ;
    {Raise  a  to the e-th power giving  b.
     Assumption : e >= 0.
     Algorithm : If e is even then  b := (a^(e div 2))*(a^(e div 2))
             but if e is odd  then  b := (--------same-------------)*a}
    Begin
        If  e <= 3  then
          If  e = 3  then
            begin
              Mul_Q1 (k_Q1, a,a,b) ;
              Mul_Q1 (k_Q1, b,a,b) ;
            end
          Else if e = 2 then  Mul_Q1 (k_Q1, a,a,b)
          Else if e = 1 then  b := a

          Else {assume e = 0}  b := k_Q1.one_Q1
        Else {e > 3}
          Begin
            Power_Q1 (k_Q1, a, e div 2, b) ;     { b := a^(k_Q1, e div 2) }
            Mul_Q1 (k_Q1, b,b,b) ;         { b := b*b}
            If  odd ( e)  then  Mul_Q1 (k_Q1, b, a, b) ;
          End ;
    End ;

[GLOBAL] Procedure Write_Q1 (Var outf : text ; Var k_Q1 : field_Q1 ;
                         Var a : number_Q1) ;    {write a}
    Var
        i : integer ;
    Begin
        Write (outf,'(',a[1]:1) ;
        For i := 2 to k_Q1.n do
            Write (outf,',',a[i]:1) ;
        Write (outf,')') ;
    End ;

[GLOBAL] Procedure Read_Q1 (Var inf : text ; Var k_Q1 : field_Q1 ;
                         Var a : number_Q1) ;    {read a}
    Var
        i : integer ;
    Begin
        For i := 1 to k_Q1.n do
          Begin
            Read (inf, a[i]) ;
            a[i] := a[i] mod (k_Q1.p) ;
          End ;
    End ;
```

```
{----------------------------------------------------------------------+
|              PROCEDURE TO SET UP THE FIELD IN THE FIRST FORM           |
+----------------------------------------------------------------------}

[GLOBAL] Procedure Set_up_Q1 (Var k_Q1 : field_Q1 ; Var Status_code :
        integer) ;
{       Input variables:
                k_Q1.p  the characteristic of the field
                k_Q1.n  the degree of GF(q) over GF(p)
        Output variables:
                k_Q1.q_minus_1  set to p^n
                k_Q1.zero_Q1    set to zero
                k_Q1.one_Q1     set to one
                k_Q1.prim_Q1    set to a primitive q-1-st root
                k_Q1.structure_Q1 set to the n_th power of k_Q1.prim_Q1
                                which allows computation in GF(q)
                Status_code     indicates success or failure of Set_up_Q1:
                        0       Successful
                        1       p is not prime
                        2       n is out of range (n < 1  or  n > Max_n)
                        3       q=p^n too large (q > 2^32)
                        4       failed to construct GF(q)
    Algorithm to set up the structure of GF(p^n) in the variable k_Q1:
     Essentially search through monic polynomials  f  of degree n
     over GF(p) for one which makes the element  x  of  GF(p)[x]/f(x)  an
     element of order  p^n-1.}

    Var

      Head : factor_pointer ; {list giving prime factorization of q-1}
      M : integer ;    {the quotient of q-1 divided once by each of its prime
                        divisors}
      R : integer ;   {the product of the primes dividing q-1, i.e. (q-1)/M }
      n_primes  : integer ;   {the number of primes dividing q-1}
      i : integer ;   {indexing variable}
  Function Try_prim (k : integer) : boolean ;
    Var
      Found : boolean ;
    Function Is_prim : boolean ; {determine if the current value of
         structure_Q1 has order  q-1}
      Var
          prim_to_M : number_Q1 ;
          temp : number_Q1 ;
          L : factor_pointer ;
          Still_checking : Boolean ;
      Begin
       With k_Q1 do
        Begin
          {Set up prim_Q1 according to the present value of structure_q}
          prim_Q1 := zero_Q1 ;
          If  n = 1  then  prim_Q1[1] := structure_Q1[1]
          Else {n > 1}          prim_Q1[2] := 1 ;
          Power_Q1 (k_Q1, prim_Q1, M, prim_to_M) ;
          {Now check that the order of prim_to_M is R .  Need to know that
           prim_to_M raised to the R-th power is 1 while prim_to_M raised to
           any power dividing R does not yield 1}
          Power_Q1 (k_Q1, prim_to_M, R, temp) ;
          Still_checking := Equal_Q1 (k_Q1, temp, one_Q1) ;
          L := head ;
          While (L <> nil) and still_checking do
            Begin
              Power_Q1 (k_Q1, prim_to_M, R div L^.prime, temp) ;
              If  Equal_Q1 (k_Q1, temp, one_Q1)  then
                  still_checking := false ;
              L := L^.link ;
```

```
            End {while} ;
          Is_prim := Still_checking ;
        End {with} ;
      End {is_prim function} ;
    Begin {try_prim function}
      With k_Q1 do
        Begin
          If  k = 1  then structure_Q1[k] := 1
          Else                structure_Q1[k] := 0 ;
          Repeat
            If k = 1 then     Found := Is_prim
            Else              Found := Try_prim (k-1) ;
            If not found then  structure_Q1[k] := structure_Q1[k] + 1 ;
          Until found or (structure_Q1[k] = p) ;
          Try_prim := found ;
        End {with} ;
    End {try_prim function} ;
Begin {Set_up_Q1 procedure}
  With k_Q1 do
  If p < 2 then
      Status_code := 1        {return indicating an error}
  Else if not Is_prime (p) then
      Status_code := 1        {return indicating an error}
  Else if (n < 1) or (n > Max_n) then

      Status_code := 2        {error}
  Else if n*ln(p) > 32*ln(2) then
      Status_code := 3        {p^n is too large}
  Else
    Begin
      {Set up various constants}
      q_minus_1 := p**n - 1 ;
      Get_factors (q_minus_1, n_primes, Head, M) ;
      R := q_minus_1 div M ;
      For i := 1 to n do
          zero_Q1[i] := 0 ;
      one_Q1[1] := 1 ;
      For i := 2 to n do
          one_Q1[i] := 0 ;
      {Now find a primitive root}
      If not try_prim (n) then
          Status_code := 4 ;          {return with error}
    End {with} ;
End {Set_up_Q1 procedure} ;

*****************************************************************************}

GLOBAL] Procedure Report_Q1 (Var outf : text ; Var k_Q1 : field_Q1) ;
Begin
  With k_Q1 do
    Begin
      Writeln (outf, 'The elements of GF(',p:1,'^',n:1,
              ') are represented as ',n:1,'-tuples.') ;
      Write (outf, 'The identity element is ') ;
      Write_Q1 (outf, k_Q1, one_Q1) ; Writeln (outf) ;
      Write (outf, 'A primitive ',q_minus_1:1,'-th root, e, is ') ;
      Write_Q1 (outf, k_Q1, prim_Q1) ; Writeln (outf) ;
      Write (outf, 'Its ',n:1,'-th power equals ') ;
      Write_Q1 (outf, k_Q1, structure_Q1) ; Writeln (outf) ;
    End {with} ;
End {report_Q1 procedure} ;

nd {Field_Q1 module} .
```

```
Module Field2 ;

{-------------------------------------------------------------------------+
|        PROCEDURES AND FUNCTIONS TO MANIPULATE ELEMENTS OF GF(Q) WHEN     |
|        REPRESENTED AS INDICES.                                           |
|        Author: Dave Joyce, Math/CS dept, Clark Univ, Worcester, MA       |
|        Date:   Nov. 1982.     Modified for Pascal 32.0:  July 1983       |
+-------------------------------------------------------------------------}

{       This module represents elements in the Galois field GF(q) in a form
        different from the form used the the module Field1.  Each element
        other than  0  in  GF(q)  is a power of a generating element of the
        multiplicative group of invertible elements.  Such an element
        corresponds to  x  (see comments in the Field1 module).  In this
        module elements are represented by their indices relative to  x.
        For instance,  1 = x**0, so  1  is represented by the integer  0.
        A special representation is used for  0, namely, -1.  All other
        elements are represented by integers in the range  0  to  q-1.
        This representation facilitates multiplication and division in
        GF(q) as well as p-th roots.  Addition and subtraction use a table
        of successors.  In this table the i-th entry holds the index of
        x**i + 1.                                                          }

{       Routines included in this file:

        ADD_Q2          adds two numbers in GF(q)
        SUB_Q2          subtracts two numbers in GF(q)
        MUL_Q2          multiplies two numbers in GF(q)
        DIV_Q2          divides two numbers in GF(q)
        SCAL_MUL_Q2     multiply a number in GF(q) by an integer
        POWER_Q2        raise a number in GF(q) to an integer power
        ROOT_Q2         takes the p-th root of a number in GF(q)
        SET_UP_Q2       sets up the structure of GF(q) for the above routines
        CVT_Q1_Q2       converts an element of GF(q) from the first form to
                        the second form

        Variables used to specify the finite field:

        p               the characteristic of the field
        n               the degree of GF(q) over the prime field
        q_minus_1       p**n-1
        p_to_n_1        p**(n-1)
        INDEX           an array giving the indices of numbers in GF(q) w.r.t.
                        PRIM_Q2.
        SCSR            an array indicating how to add by one for numbers rep'd
                        by their indices.  If c is the index of an element of
                        GF(q), then  SCSR[c] is the index of the successor of
                        that element.

        These variables are initialized by SET_UP_Q2.                      }


Const
        Max_p    = 19 ;  {the maximum value of p, the characteristic of the
                         field GF(q)}
        Max_n    = 18 ;  {the maximum degree for GF(q) over GF(p) that this
                         module will deal with}
        Max_q    = 262144 ;      {the maximum value of q, which is p**n}

        Zero_Q2 = -1 ;   {representation of zero in the second form}
```

```
Type

       Number_Q1 = array [1..max_n] of integer ;       {'Number_Q1' is the
                        data type for an element in GF(q) when it is rep'd
                        in the first form, as an array over GF(p)}
       Field_Q1 = record
           p   : integer ;      {the characteristic of the field}
           n   : integer ;      {the degree of GF(q) over GF(p)}
           q_minus_1    : integer ; {p**n-1}
           zero_Q1      : number_Q1 ;   {the 0 element of GF(q)}
           one_Q1       : number_Q1 ;   {the 1 element of GF(q)}
           prim_Q1      : number_Q1 ;   {a primitive root of GF(q)}
           structure_Q1 : number_Q1 ;  {an irred poly of degree n having
                               prim_Q1 as a root}
         End {field_Q1 record} ;

       Number_Q2 = integer ;
       Field_Q2 = record
           p   : integer ;      {the characteristic of the field}
           n   : integer ;      {the degree of GF(q) over GF(p)}
           q_minus_1    : integer ;      {p**n-1}
           p_to_n_1     : integer ;      {p**(n-1)}
           index        : array [0..max_q] of integer ; {a table giving the
                               indices of the elements of GF(q) with respect
                               to prim_Q1}
           scsr         : array [0..max_q] of integer ; {a table describing
                               the successor funtion in GF(q) }
         End {field_Q2 record} ;

{-------------------------------------------------------------------------+
|                          EXTERNAL PROCEDURES                             |
+-------------------------------------------------------------------------}

{Routines in FIELD1:}
[EXTERNAL] Procedure Add_Q1 (Var k_Q1 : field_Q1 ; var a,b,c : number_Q1) ;
       extern ;
[EXTERNAL] Procedure Mul_Q1 (Var k_Q1 : field_Q1 ; var a,b,c : number_Q1) ;
       extern ;

{-------------------------------------------------------------------------+
|                          R O U T I N E S                                 |
+-------------------------------------------------------------------------}

[GLOBAL] Function Add_Q2 (Var k_Q2 : field_Q2 ; a, b : Number_Q2) : Number_Q2 ;
    Var
        scrat : Number_Q2 ;
    Begin
        If  a = Zero_Q2     then Add_Q2 := b
        Else if b = Zero_Q2 then Add_Q2 := a
        Else With k_Q2 do
        If a >= b then
          Begin
            Scrat := scsr[a-b] ;
            If scrat = Zero_Q2  then  Add_Q2 := Zero_Q2
            Else                 Add_Q2 := (b + scrat) mod q_minus_1 ;
          End
        Else {a < b}
          Begin
            Scrat := scsr[b-a] ;

            If scrat = Zero_Q2  then  Add_Q2 := Zero_Q2
            Else                 Add_Q2 := (a + scrat) mod q_minus_1 ;
          End
    End ;

[GLOBAL] Function Mul_Q2 (Var k_Q2 : field_Q2 ; a, b : Number_Q2) : Number_Q2 ;
```

```
    Begin
        If  (a = Zero_Q2) or (b = Zero_Q2) then  Mul_Q2 := Zero_Q2
        Else    Mul_Q2 := (a + b) mod k_Q2.q_minus_1 ;
    End ;

[GLOBAL] Function Sub_Q2 (Var k_Q2 : field_Q2 ; a, b : Number_Q2) : Number_Q2 ;
    Begin
        Sub_Q2 := Add_Q2 (k_Q2, a, Mul_Q2 (k_Q2, b, k_Q2.Index[k_Q2.p-1])) ;
    End ;

[GLOBAL] Function Div_Q2 (Var k_Q2 : field_Q2 ; a, b : Number_Q2) : Number_Q2 ;
    {division by zero yields zero}
    Begin
        If  (a = Zero_Q2) or (b = Zero_Q2) then  Div_Q2 := Zero_Q2
        Else    Div_Q2 := (a - b + k_Q2.q_minus_1) mod k_Q2.q_minus_1 ;
    End ;

[GLOBAL] Function Indx_Q2 (Var k_Q2 : field_Q2 ; var x : Number_Q1) : integer ;
        {Gives the location of the index of x in the index table}
        Var
            a, j : integer ;
        Begin
            a := x[k_Q2.n] ;
            For j := k_Q2.n-1 downto 1 do
                a := x[j] + k_Q2.p*a ;
            Indx_Q2 := a ;
        End ;

[GLOBAL] Function CVT_Q1_Q2 (Var k_Q2 : field_Q2 ; Var x : number_Q1) :
        Number_Q2 ;
    {Convert a number from the first form to the second form}
    Begin
        CVT_Q1_Q2 := k_Q2.Index [Indx_Q2(k_Q2,x)] ;
    End {CVT_Q1_Q2 function} ;

[GLOBAL] Function Scal_mul_Q2 (Var k_Q2 : field_Q2 ; k, a : integer) :
        Number_Q2 ;
    {multiply the element a in GF(q) represented by its index by the integer k}
    Begin
        k := k mod k_Q2.p ;
        If  (a = Zero_Q2) or (k = 0)  then  Scal_mul_Q2 := Zero_Q2
        Else  Scal_mul_Q2 := (a + k_Q2.Index[k]) mod k_Q2.q_minus_1 ;
    End ;

[GLOBAL] Function Power_Q2 (Var k_Q2 : field_Q2 ; a : Number_Q2 ; k : integer)
        : Number_Q2 ;
    {raise the element  a  in GF(q) to the k-th power}
    Begin
        If a = Zero_Q2 then Power_Q2 := Zero_Q2
        Else
          Power_Q2 := (a*k) mod k_Q2.q_minus_1 ;
    End {Power_Q2 function} ;

[GLOBAL] Function Root_Q2 (Var k_q2 : field_q2 ; a : Number_q2) : Number_q2 ;

    {take the p-th root of an element in GF(q)}
    Begin
        If a = Zero_Q2 then Root_Q2 := Zero_q2
        Else
          Root_Q2 := (a*k_q2.p_to_n_1) mod k_q2.q_minus_1 ;
    End {Root_Q2 function} ;

[GLOBAL] Procedure Set_up_Q2 (Var k_Q1 : field_Q1 ; Var k_Q2 : field_Q2 ) ;
    Var
        i : integer ;
        x, y : Number_Q1 ;
    Begin {set_up_Q2 procedure}
      With k_Q1 do
```

```
        Begin
          {first set up constants in k_Q2}
          k_Q2.p := p ;
          k_Q2.n := n ;
          k_Q2.q_minus_1 := q_minus_1 ;
          k_Q2.p_to_n_1  := p**(n-1) ;
          {next set up the table of indices relative to prim_Q1}
          k_Q2.Index [0] := Zero_Q2 ;   {0 is represented by Zero_Q2=-1}
          x := one_Q1 ;
          For i := 0 to q_minus_1 - 1 do
            Begin
              k_Q2.Index [Indx_Q2(k_Q2,x)] := i ;
              Mul_Q1 (k_Q1, x, prim_Q1, x) ;
            End ;
          {now use the index table to fill the scsr table}
          x := one_Q1 ;
          For i := 0 to q_minus_1 do
            Begin
              Add_Q1 (k_Q1, x, one_Q1, y) ;
              k_Q2.scsr [i] := CVT_Q1_Q2 (k_Q2, y) ;
              Mul_Q1 (k_Q1, x, prim_Q1, x) ;
            End ;
        End {with} ;
    End {set_up_Q2 procedure} ;

End {field2 module} .


Module Prime ;

{---------------------------------------------------------------------+
|       This module contains some functions and procedures relating   |
|       to prime numbers.                                              |
|                                                                      |
|       Author:  Dave Joyce, Math/CS dept, Clark Univ, Worcester, MA   |
|       Date:    Oct 1982       Modified for Pascal 2.0:  July 1983    |
|                                                                      |
|       Included routines:                                             |
|               GET_DIVISOR, IS_PRIME, GET_FACTORS                     |
+---------------------------------------------------------------------}

{---------------------------------------------------------------------+
|       Declarations and descriptions of the routines found in this   |
|       module.                                                        |
|                                                                      |
|                                                                      |
|       Function Get_divisor (n : integer) : integer ; extern ;       |
|                                                                      |
|       Returns the smallest divisor greater than 1. Returns 1 if n=1. |
|                                                                      |
|                                                                      |
|       Function Is_Prime (N : integer) : Boolean ; extern ;          |
|                                                                      |
|       Returns 'true' if  N  is a prime number, 'false' otherwise.    |
|       Zero and one are not considered to be prime here.             |
|                                                                      |
|                                                                      |
|       Procedure Get_factors (N : integer ; var n_primes : integer ;  |
|               var Head : factor_pointer ; var M : integer) ; extern ; |
|                                                                      |
|       This procedure factors the integer  N  returning the following |
|       information                                                    |
|       1.      n_primes :  the number of primes dividing  N.          |
|       2.      Head     :  a pointer to a list of the primes dividing |
|                           N  along with their order in  N.           |
|       3.      M        :  the quotient of  N  divided once by each of |
|                           the primes dividing  N.                    |
+---------------------------------------------------------------------}
```

```
{------------------------------------------------------------------------+
|                  Declarations of data types                            |
+------------------------------------------------------------------------}

Type
        factor_pointer = ^factor ;
        factor = record
                    link : factor_pointer ;
                    prime : integer ;
                    order : integer ;
                 end {factor record} ;
        {The procedure finds the factors of a number and places them in a list.
         This list is a linked list of factor records.  Each record specifies
         a prime and the order that that prime appears in the original number.
         Various other routines use this list in their computations.}

{------------------------------------------------------------------------+
|        Routines for finding divisors and prime number verification.    |
+------------------------------------------------------------------------}

[GLOBAL] Function Get_divisor (n : integer) : integer ;
    Var
        d : integer ; {test divisor}
        divi : integer ; {divisor}
    Begin
        If n < 0 then n := -n ;
        If n mod 2 = 0 then Get_divisor := 2
        Else if n mod 3 = 0 then Get_divisor := 3
        Else
          Begin
            d := 5 ; divi := 0 ;
            While (d*d <= n) and (divi = 0) do
              If  n mod d = 0  then divi := d
              Else if  n mod (d+2) = 0  then divi := d+2
              Else  d := d + 6 ;
            If  divi = 0  then Get_divisor := n
            Else                Get_divisor := divi ;
          End ;
    End {Get_divisor function} ;

[GLOBAL] Function Is_Prime (n : integer) : boolean ;
    Var
        d : integer ;
    Begin
        d := Get_divisor (n) ;
        Is_Prime := (d <> 1) and (d = n) ;
    End ;

[GLOBAL] Procedure Get_factors (N : integer ; var n_primes : integer ;
                        var Head : factor_pointer ; var M : integer) ;
    Var
        d, old_d : integer ;
        fact     : factor_pointer ;
    Begin
        n_primes := 0 ; {initially, no primes divide N}
        head     := nil ; {so, the list is empty}
        M        := 1 ;
        old_d    := 1 ;  {1 will never be returned as a factor of N}
        If  N < 0  then  N := -N ;
        If  N = 0  then
```

```
    Begin
      New (head) ; head^.link := nil ;
      head^.prime := 0 ; head^.order := 1 ;
    End
  Else {N > 1}
    While  N <> 1  do
      Begin
        d := get_divisor (N) ;
        N := N div d ;
        If  d = old_d  then
          begin
            M := M * d ;
            head^.order := head^.order + 1 ;
          end
        Else {new divisor}
          begin
            {push this new factor on the list}
            new (fact) ; fact^.link := head ; head := fact ;
            head^.prime := d ; head^.order := 1 ;

            old_d := d ;
          end {if/else} ;
      End {while} ;
  End {get_factors procedure} ;

End {Prime module} .
```

```
Analysis of the surface

z^5 = xy + x^3 + y^3 + x^2y^2



The system of equations to solve is as follows:

t[0,0]^5 = t[0,0]
t[1,0]^5 = t[0,1] + e^2 t[2,0]
t[0,1]^5 = t[1,0] + e^2 t[0,2]
t[2,0]^5 = 0
t[1,1]^5 = t[1,1]
t[0,2]^5 = 0
0        = e^3 t[0,2]
0        = e^3 t[2,0]
0        = e^2 t[0,0] + t[1,1]
         = t[0,0] + e^2 t[1,1]

The divisor class number is 5^3.
```

6

Families of Zariski Surfaces[†]

INTRODUCTION

This note is a first step in the analysis of the moduli of unirational surfaces in characteristic $p > 0$. We examine the behavior of the so-called generalized Zariski surfaces under smooth specialization, and show that this class of surfaces is closed under this operation. We also describe the structure of the totality of all generalized Zariski surfaces in a fixed projective space, and show that this set is a countable union of closed subsets of the Hilbert scheme of smooth surfaces in the projective space. The analogous results also hold for the class of Zariski surfaces. Finally, we give an example of a deformation of a Zariski surface to a smooth surface that is not unirational, or even uniruled.

We will use the following notations and conventions: If $f\colon X \to M$ is a morphism of schemes, and m is a point of M, we let X_m denote the scheme-theoretic fiber $f^{-1}(m)$. We let $|X_m|$ denote the reduced scheme associated to X_m. If $g\colon Y \to M$ is another morphism of schemes, and $F\colon X \to Y$ is an M-morphism, we let F_m denote the morphism

$$F_m\colon |X_m| \to |Y_m|$$

If G is a sheaf on X, we let G_m denote the induced sheaf on X_m. If $h\colon C \to M$ is an M-scheme, we let X_C denote the fiber product $X \times_M C$.

[†]By Piotr Blass and Marc Levine, University of Pennsylvania, Philadelphia, Pennsylvania. Reprinted from *Duke Mathematical Journal*, 49, No. 1 (1982), pp. 129-136.

Let Q be a polynomial with rational coefficients, f: $X \to M$ a flat and projective morphism. We let $\mathrm{Hilb}^Q_{X/M}$ denote the part of the relative Hilbert scheme, $\mathrm{Hilb}_{X/M}$, of X over M that corresponds to the polynomial Q. We let $H^Q_{X/M}$ denote the universal family of subschemes of X parametrized by $\mathrm{Hilb}^Q_{X/M}$. We will often omit the subscript X/M if there is no cause for confusion.

We fix at the outset an algebraically closed field k of characteristic $p > 0$. Unless specified otherwise, all schemes, morphisms, and rational maps will be defined over k.

A *generalized Zariski surface* is a surface S that admits a dominant, purely inseparable, rational map f: $\mathbb{P}^2 \to S$. If we take f so that the degree of f is p, we call S a *Zariski surface*. Suppose that the degree of f is p^e; one easily sees that the field k(S) contains the field $k(x^{p^e}, y^{p^e})$, where we identify $k(\mathbb{P}^2)$ with k(x,y). Thus, if there exists a map f as above, there is also a dominant, purely inseparable, rational map g: $S \to \mathbb{P}^2$ of degree p^e. Taking p^eth roots, we see that the converse is also true. We will find this dual viewpoint useful in the sequel.

In Section 1 we give a rough picture of the set of generalized Zariski surfaces in a fixed projective space. The specialization result of Section 2 enables us to refine this picture somewhat. We should mention that the key technical result (proposition 2.1) is a slight modification of a theorem communicated by M. Artin to Matsusaka and Mumford, which appeared in their paper (Matsusaka and Mumford, 1964, theorem 1).

1. GENERALITIES ON RADICIAL RATIONAL MAPS

In this section we show that the condition of pure inseparability of a morphism is a closed condition. We also show that the collection of smooth subvarieties of $\mathbb{P}^N$ with Hilbert polynomial Q that are purely inseparable images of a fixed variety Y forms a countable union of locally closed subsets of $\mathrm{Hilb}^Q_{\mathbb{P}}$.

LEMMA 1.1 Let f: $X \to M$, g: $Y \to M$ be flat and proper morphisms of varieties such that the scheme theoretic fibers $f^{-1}(m)$ and $g^{-1}(m)$

are geometrically irreducible, and reduced at their respective generic points, for each point m in M. Let $F: X \to Y$ be a dominant M-morphism such that

$$F_m: |X_m| \to |Y_m|$$

is generically finite to 1 for each m in M. Suppose that there is an open subset U of M such that F_m is a purely inseparable morphism for each m in U. Then F_m is a purely inseparable morphism for each m in M.

Proof: We reduce immediately to the case in which M is a smooth curve. Let m be a point of M. Since f is proper, F is also proper; since F is generically finite to 1, there is a point y of Y_m and an affine neighborhood V of y in Y such that

$$F\Big|_{F^{-1}(V)} : F^{-1}(V) \to V$$

is a proper, quasifinite morphism, hence a finite morphism. Thus $W = F^{-1}(V)$ is also affine; suppose that W = Spec(S), V = Spec(R). Let Q be the prime ideal in S of $|X_m| \cap W$, q the prime ideal in R of $|Y_m| \cap V$. Since X and Y are flat over M and M is a smooth curve, the local rings S_Q and R_q are discrete valuation rings.

Let L be the quotient field of S, K the quotient field of R. By assumption, L is a finite, purely inseparable extension of K. This easily implies that S_Q is integral over R_q. Let $\bar{S}$ and $\bar{R}$ denote the respective residue fields of S_Q and R_q.

Let $\bar{x}$ be an element of $\bar{S}$, and let x be an element of S_Q lifting $\bar{x}$. Then x^{p^e} is in K for some e, hence in R_q. Thus $\bar{x}^{p^{eQ}}$ is in $\bar{R}$. Therefore, $\bar{S}$ is a purely inseparable extension of $\bar{R}$.

LEMMA 1.2 Let $f: X \to M$, $g: Y \to M$ be as in lemma 1.1. Let $F: X \to Y$ be a dominant M-morphism such that

$$F_m: |X_m| \to |Y_m|$$

is generically finite to 1 for all m in M. Let M_p be the subset of M,

$$M_p = \{m \in M | F_m \text{ is a purely inseparable morphism}\}$$

Then M_p is a closed subset of M.

Proof: Suppose at first that F: $X \to Y$ is not a separable morphism. Let A be the sub-$\mathcal{O}_Y$ algebra of $F_*(\mathcal{O}_X)$ defined by

$$A(U) = \{f \in F_*(\mathcal{O}_X)(U) | f \text{ is separable over } \mathcal{O}_Y(U)\}$$

Let $X' = \text{Spec } \mathcal{O}_Y(A)$. Then F factors as

$$X \xrightarrow[F']{} X' \xrightarrow[G]{} Y$$

where F' is purely inseparable of degree at least p, and G is separable.

Let U be an open subset of M such that X'_m is reduced at its generic point for all m in U. By lemma 1.1, F'_m is a purely inseparable morphism for all m in U, hence F_m is purely inseparable if and only if G_m is purely inseparable, again for m in U. Furthermore, by induction on the degree of F, we see that

$$U_p = \{m \in U | G_m \text{ is purely inseparable}\}$$

is a closed subset of U. Let $\bar{U}_p$ denote the closure of U_p in M; by lemma 1.1, F_m is purely inseparable for all m in $\bar{U}_p$.

Let $M' = M - U$. By induction on the dimension of M, $M'_p = M_p \cap M'$ is a closed subset of M'. Clearly, we have

$$M_p = \bar{U}_p \cup M'_p$$

which completes the proof in this case.

Suppose now that F is separable. Let R denote the ramification locus of F. Let M_1 be the subset of M,

$$M_1 = \{m \in M: |X_m| \subseteq R\}$$

As R is a proper closed subset of X, M_1 must be a proper closed subset of M. Furthermore, M_p is contained in M_1. By induction on the dimension of M, M_p is a closed subset of M_1, hence of M.

PROPOSITION 1.3 Let B be a variety. Let X be a subvariety of $B \times \mathbf{P}^N$, smooth over B, with geometrically irreducible fibers of dimension n. Let Y be a smooth projective variety of dimension n, and let B(e) be the subset of B consisting of points b such that there exists a dominant, purely inseparable, rational map

$$f\colon\ Y \to X_b$$

of degree p^e. Then B(e) is a countable union of locally closed subsets of B.

Proof: The morphism $p_1\colon\ X \times Y \to B$ is smooth and projective; let Q be a polynomial with rational coefficients, let p: $\mathrm{Hilb}^Q \to B$ be the part of the relative Hilbert scheme of $X \times Y$ over B corresponding to Q, and let q: $H^Q \to \mathrm{Hilb}^Q$ be the universal family of subschemes of $X \times Y$ parametrized by Hilb^Q. Let I_1^Q be the locally closed subset of Hilb^Q whose points x correspond to reduced and geometrically irreducible subschemes of $X_{p(x)} \times Y$, of dimension n.

Let f: $C \to I_1^Q$ be a morphism of a nonsingular curve C into I_1^Q, and let G be the subscheme of $(X \times_B C) \times Y$ induced by the composition of f with the inclusion of I_1^Q into Hilb^Q. By the compatibility of algebraic projection with specialization (Shimura, 1955, theorem 25), there are integers r and s such that

$$\mathrm{pr}_{X_{p(c)}}(G_c) = r \cdot X_{p(c)}$$

$$\mathrm{pr}_Y(G_c) = s \cdot Y$$

for all c in C. In particular, the subset

$$I_2^Q = \{x \in I_1^Q | \mathrm{pr}_{X_{p(c)}}(H_x^Q) = p^e \cdot |X_{p(c)}|,\ \mathrm{pr}_Y(H_x^Q) = 1 \cdot Y\}$$

is an open and closed subset of I_1^Q.

Let H' be the subscheme of $X_{\left(I_2^Q\right)} \times Y$ induced by the inclusion of I_2^Q into Hilb^Q, and let f: $H' \to I_2^Q$ denote the structure morphism. By lemma 1.2, the subset I_3^Q of I_2^Q,

$$I_3^Q = \{x \in I_2^Q | p_1 : H_x' \to X_{p(x)} \text{ is purely inseparable}\}$$

is a closed subset of I_2^Q. Clearly, H_x' is the graph of a dominant, purely inseparable, rational map of Y to $X_{p(x)}$, for x in I_3^Q, and each such graph with Hilbert polynomial Q arises in this way.

Let I^Q be the subset $p(I_3^Q)$ of B. By Chevalley's theorem, I^Q is constructible, i.e., a finite union of locally closed subsets of B. Clearly, B(e) is given by

$$B(e) = \bigcup_Q I^Q$$

hence B(e) is a countable union of locally closed subsets of B, as desired.

We let S_N^P denote the locally closed subset of Hilb_N^P whose closed points correspond to smooth, geometrically irreducible surfaces. We let $U_N^{P,e}$ denote the subset of S_N^P whose closed points correspond to surfaces X for which there exists a dominant, purely inseparable, rational map f: $\mathbf{P}^2 \to X$, of degree p^e. Also, we let S_N and U_N^e be the unions

$$S_N = \bigcup_P S_N^P \qquad U_N^e = \bigcup_P U_N^{P,e}$$

COROLLARY 1.4 $U_N^{P,e}$ is a countable union of locally closed subsets of S_N^P for each polynomial P with rational coefficients. U_N^e is a countable union of locally closed subsets of S_N.

2. SPECIALIZATIONS OF RADICIAL COVERS OF $\mathbf{P}^N$

Using the valuation theoretic technique of Matsusaka (1968), we show that a specialization of radicial covers of $\mathbf{P}^2$ is also a radicial cover of $\mathbf{P}^2$. Combining this with the results of Section 1 yields our main result on the structure of the set of Zariski surfaces in a projective space.

PROPOSITION 2.1 Let $\mathcal{O}$ be a discrete valuation ring with quotient field K and residue field k with k algebraically closed. Let g: $V \to \mathrm{Spec}(\mathcal{O})$ be a flat and proper $\mathcal{O}$-scheme with reduced and geometrically irreducible fibers of dimension n. Let f: $V \otimes K \to \mathbb{P}^n$ be a dominant, radicial, rational map defined over K. Then there is a ruled variety Y, of dimension n, and a dominant, radicial, rational map f_0: $V \otimes k \to Y$. Furthermore, if the degree of f is p^e, we may choose f_0 as above so that the degree of f_0 is $p^{e'}$, with $e' \leq e$.

Proof: Let V_t, V_0 denote the varieties $V \otimes K$, $V \otimes k$, respectively. Let $T \subseteq V_t \times \mathbb{P}^n$ be the K-closure of the graph of f, and let T* be the k-closure of T in $V \times \mathbb{P}^n$. T* defines a specialization $T \to T_0$, over $\mathcal{O}$ and over $V_t \to V_0$, where T_0 is the cycle $T^* (V_0 \times \mathbb{P}^n)$. By the compatibility of algebraic projection with specialization (Shimura, 1955, thm. 25), we see that T_0 has a component T_0' such that

(a) $pr_{V_0}(T_0') = 1 \cdot V_0$.

(b) The coefficient of T_0' in T_0 is 1.

(c) $pr_{V_0}(T_0 - T_0') = 0$.

Let x_0 be a generic point of V_0 over k. Let x be a generic point of V_t over K such that x has x_0 as specialization over $\mathcal{O}$ and over

$$(V_t, T) \to (V_0, T_0)$$

Let S_v be the specialization ring in K(x) of the specialization $x \to x_0$. S_v is just the local ring of V_0 in V, hence a discrete valuation ring. Let v be the valuation associated to S_v, and let N_v be the maximal ideal of S_v.

Let y = f(x). Since f is dominant and radicial, y is a generic point of $\mathbb{P}^n$ over K, and K(x) is radicial over K(y) of degree p^e, say. Let $R_{v'} = K(y) \cap S_v$. $R_{v'}$ is a discrete valuation ring of K(y) with maximal ideal $M_{v'} = N_v \cap K(y)$. The valuation v defines a valuation of K(y) which we denote as v'.

We may choose x and y as above so that (x,y) has a well-defined specialization (x_0, y_0) over $\mathcal{O}$ and over $(V,T) \to (V_0, T_0)$. As (x,y) is in the support of T, it follows that (x_0,y_0) is in the support of T_0. By (a) and (c), (x_0,y_0) must be in T_0', and since $pr_{V_0}: T_0' \to V_0$ is birational, (x_0,y_0) must be a generic point of T_0' over k. Thus y_0 is a generic point over k of the projection A of T_0' on $\mathbb{P}^n$. Let R be the specialization ring of $y \to y_0$. Then R is the local ring of A in $\mathbb{P}^n \times \mathrm{Spec}(\mathcal{O})$, and v' is a prime divisor of R in the sense of Abhyankar (1956).

If $A \neq \mathbb{P}^n$, then $R_{v'}/M_{v'}$ is the function field of a ruled variety (Abhyankar, 1956, proposition 3). Furthermore, as in the proof of lemma 1.1, S_v/N_v is purely inseparable over $R_{v'}/M_{v'}$, and

$$[S_v/N_v : R_{v'}/M_{v'}] = p^{e'}$$

with $e' \leq e$. This proves our result in case $A \neq \mathbb{P}^n$.

If $A = \mathbb{P}^n$, then T_0' is the graph of a dominant rational map f_0: $V \to \mathbb{P}^n$, and y_0 is a generic point of $\mathbb{P}^n$ over k. We have the specialization of intersections

$$(V_t \times y) \cdot T \to (V_0 \times y_0) \cdot T_0$$

over $\mathcal{O}$ and over $(V,T,y) \to (V_0,T_0,y_0)$. Since $(V \times y) \cdot T = p^e \cdot (x \times y)$, and since $(x_0 \times y_0)$ is a point of the intersection $(V_0 \times y_0) \cdot T_0$, we must have

$$(V_0 \times y_0) \cdot T_0 = p^e \cdot (x_0 \times y_p)$$

This, together with (b) and (c), implies that

$$(V_0 \times y_0) \cdot T_0' = (V_0 \times y_0) \cdot T_0$$
$$= p^e \cdot (x_0 \times y_0)$$

Thus f_0 is radicial of degree p^e.

PROPOSITION 2.2 Let $\mathcal{O}$ be a discrete valuation ring as above. Let g: $V \to \mathrm{Spec}(\mathcal{O})$ be a smooth and proper $\mathcal{O}$-scheme, with geometrically irreducible fibers of dimension two. Suppose that $V \otimes K$ is a radicial cover of a rational surface. Then $V \otimes k$ is also a radicial cover of a rational surface. $V \otimes K$ is a rational surface if and only if $V \otimes k$ is a rational surface. If $V \otimes K$ is a nonrational Zariski surface, then $V \otimes k$ is also a nonrational Zariski surface.

Proof: We may assume that $V_t = V \otimes K$ is a radicial cover of $\mathbb{P}^2$ via a rational map $f\colon V_t \to \mathbb{P}^2$. We may further assume that f is defined over a finite extension of K; by making a base extension and changing notation if necessary, we may assume that f is defined over K.

We first show that any two points of $V_0 = V \otimes k$ can be connected by a chain of rational curves. Let y_1, y_2 be in V_0, and let x_1, x_2 be generic points of V_t over the algebra closure $\bar{K}$ of K with specializations $x_1 \to y_1$, $x_2 \to y_2$ over $\mathcal{O}$. Since $f\colon V_t \to \mathbb{P}^2$ is radicial, there is a dominant, radicial, rational map $h\colon \mathbb{P}^2 \to V_t$ defined over $\bar{K}$. Let z_1, z_2 be points of $\mathbb{P}^2$ with $h(z_i) = x_i$, $i = 1$, 2 (since the x_i are generic points of V_t, the map h is a morphism in a neighborhood of $h^{-1}(x_i)$, $i = 1, 2$). Let L be a line in $\mathbb{P}^2$ connecting z_1 and z_2, and let C be the proper transform of L by h. As L dominates C, C is a rational curve on V_t. Furthermore, C is irreducible and contains x_1 and x_2. Let C_0 be a specialization of C over $\mathcal{O}$ and over $(x_1,x_2) \to (y_1,y_2)$. By Samuel (1956, lemma 5), each component of C_0 is a rational curve. Also, the support of C_0 contains y_1 and y_2, and is connected by Zariski's connectedness theorem.

Let $f_0\colon V_0 \to Y$ be the dominant, radicial, rational map given by proposition 2.1. Y is a ruled surface; let $\mu\colon Y^* \to Y$ be a desingularization of Y, and let $\mu'\colon V_0^* \to V_0$ be a sequence of monoidal transformations so that the induced rational map $f_0^1\colon V_0^* \to Y^*$ is a morphism. Y^* is a nonsingular ruled surface; to show that Y^* is rational, we need only show that each two points of Y^* can be connected by a chain of rational curves.

Let then z_1 and z_2 be points of Y^*, and let y_1 and y_2 be points of V_0^* lying over z_1 and z_2, respectively. Let C be a chain of rational curves connecting $\mu'(y_1)$ and $\mu'(y_2)$. $C^* = \mu'^{-1}(C)$ is then a chain of rational curves connecting y_1 and y_2, and hence $f_0^1(C^*)$ is the desired chain on S^*. This proves the first statement of the proposition.

If V_t is not rational, then either the irregularity $q(V_t) = h^1(V_t, \mathcal{O}_{V_t})$ or the bigenus $P_2(V_t) = h^0(V_t, 2K_{V_t})$ is not zero. The smoothness of g, together with the upper semicontinuity of the dimensions of cohomology groups, implies that either $q(V_0)$ or $P_2(V_0)$ is not zero. Thus V_0 is not rational. If V_t is rational, then by Levine (1981, theorem 1.1, appendix) V_0 is a ruled surface. As each two points of V_0 can be connected by a chain of rational curves, V_0 must be rational.

Finally, suppose that V_t is a nonrational Zariski surface. By proposition 2.1, we may take $f_0\colon V_0 \to Y$ as above to be radicial of degree p^e, with $e \leq 1$. By the above, Y is a rational surface and V_0 is not rational, hence $e = 1$.

We summarize our results in the following.

THEOREM 2.3 The set of smooth irrational, generalized Zariski surfaces in $\mathbb{P}^N$ forms a countable union of irreducible, closed subsets of S_N. The set of smooth irrational Zariski surfaces in $\mathbb{P}^N$ forms a countable union of irreducible, closed subsets of S_N.

Proof: This follows easily from corollary 1.4, proposition 2.2, and the remarks following our definition of generalized Zariski surfaces in the introduction, together with the valuative criterion for properness.

3. DEFORMATIONS OF ZARISKI SURFACES: AN EXAMPLE

We recall that a variety V is said to be uniruled if there is a ruled variety W, of the same dimension as V, and a dominant rational map from W to V. In Levine (1981), the second author has shown the following:

Let V be a smooth subvariety of $\mathbb{P}^N$. Suppose that V is uniruled via a *separable* map $f\colon W \to V$. Then all small deformations of V in $\mathbb{P}^N$ are uniruled.

Zariski surfaces are examples of nonseparably uniruled varieties. Here we give an example of a Zariski surface that has small deformations that are not uniruled.

Let X_4: $X_1^4 + X_2^4 + X_3^4 + X_4^4 = 0$ be the Fermat surface over an algebraically closed field k of characteristic 3. It follows from Shioda's proof of proposition 1 in Shioda (1974) that X_4 is unirational and in fact a Zariski surface. On the other hand, let F_4 be a generic quartic surface in $\mathbf{P}^3$. It follows from Deligne's modern proof of Noether's theorem (Deligne and Katz, 1973) that the Picard number ρ of F_4 is 1. On the other hand, it is well known that $b_2(F_4) = 22$, where b_2 is the second étale Betti number. Let us denote $b_2 - \rho$ by λ. Thus $\lambda = b_2 - \rho = 21 > 0$ for F_4. Let us show that F_4 is not uniruled. This follows from the fact that $\rho = b_2$ or $\lambda = 0$ for ruled surfaces and from the fact that if f: $Y \to Z$ is a generically surjective rational map of nonsingular surfaces, then $\lambda(Y) \geq \lambda(Z)$ (see Shioda, 1974, lemma, p. 234). Now it is clear that X_4 and F_4 are deformations of each other. This shows that a deformation of a Zariski surface need not be unirational or even uniruled. Similar examples can easily be constructed in every characteristic $p \geq 3$.

REFERENCES

Abhyankar, S., On the valuations centered in a local domain, *Am. J. Math.*, 78 (1956), 321-348.

Deligne, P., and N. Katz, *SGA VII*, *Part II*, Lecture Notes in Mathematics, No. 340, Springer-Verlag, New York, 1973.

Levine, M., Deformations of uni-ruled varieties, *Duke Math. J.*, 48 (1981), 467-473.

Matsusaka, T., Algebraic deformations of polarized varieties, *Nagoya Math. J.*, 31 (1968), 185-245.

Matsusaka, T., and D. Mumford, Two fundamental theorems on deformations of polarized varieties, *Am. J. Math.*, 86 (1964), 668-684.

Samuel, P., Rational equivalence of arbitrary cycles, *Am. J. Math.*, 78 (1956), 383-400.

Shimura, G., Reduction of algebraic varieties with respect to a discrete valuation of the basis field, *Am. J. Math.*, 77 (1955), 134-176.

Shioda, T., An example of uni-rational surfaces in characteristic p, *Math. Ann.*, 211 (1974), 233-236.

7

Unirationality of Enriques Surfaces in Characteristic 2†

0. INTRODUCTION

The aim of this chapter is to give necessary and sufficient conditions for an Enriques surface over an algebraically closed field of characteristic 2 to be unirational. We show that such a surface is unirational if and only if it is either classical or supersingular in the sense of Bombieri and Mumford (1976, p. 197).

The method of proof is the following. Following the fundamental classification paper (Bombieri and Mumford, 1976), we consider for every Enriques surface a double cover which is cohomologically "K3 like." We show that if the Enriques surface is either classical or supersingular, then the smooth model of that double covering is either rational or a supersingular K3 surface. Then using a result of the beautiful paper of Rudakov and Šafarevič (1979), we conclude that such Enriques surfaces are unirational.

For the remaining type of Enriques surfaces in characteristic 2, namely singular Enriques surfaces, the nonunirationality has been shown by R. Crew in his 1981 Princeton thesis. R. Crew is a student of N. Katz. [Previously, T. Katsura (1979) proved that result for surfaces defined over a finite field.] Thus we simply quote R. Crew's result.

†By Piotr Blass; reprinted from *Compositio Mathematica*, 45, Fasc. 3 (1982), 393-398.

1. NOTATION AND PRELIMINARIES

Let k be an algebraically closed field of characteristic $p > 0$. For any smooth and projective surface V over k we denote by the following: $b_i(V) = \dim H^i_{et}(V,Q_1)$, $\rho(V)$ = rank of Pic V/numerical equivalence, $\lambda(V) = b_2(V) - \rho(V)$ = Lefschetz number. We have $\lambda(V) \geq 0$ (Igusa's inequality). V is called supersingular iff $\lambda(V) = 0$. Alb V denotes the Albanese variety of V. We recall that dim Alb V = $(1/2)b_1(V)$. Following Bombieri and Mumford (1976, pp. 197-216), we call V an Enriques surface iff V has Kodaira dimension zero and $b_1(V) = 0$, $b_2(V) = 10$, $\chi(\mathcal{O}_V) = 1$. In characteristic 2 there are three types of Enriques surfaces:

(i) Classical, characterized by the property that $\dim H^1(V,\mathcal{O}_V) = 0$.

(ii) Supersingular, characterized by the properties that $\dim H^1(V,\mathcal{O}_V) = 1$ and the Frobenius map is zero on $H^1(V,\mathcal{O}_V)$.

(iii) Singular, characterized by $\dim H^1(V,\mathcal{O}_V) = 1$ and the Frobenius map is bijective on $H^1(V,\mathcal{O}_V)$.

V is called a *Zariski surface* if there exists a generically surjective, purely inseparable rational map g: $P^2_k \to V$ of degree p where P^2_k is the projective plane over k. For any projective Cohen-Macauley scheme Y of equidimension n over k, we denote by ω_Y the dualizing sheaf on Y (see Hartshorne, 1977, p. 242). We will also use an alternative description of ω_Y in terms of rational differential forms. We refer the reader to Kunz's papers (Kunz, 1975, 1978) for the details of this description. Everywhere in this chapter we will assume that the characteristic of k is $p = 2$ except in lemma 1 and corollary 1.1, where the characteristic $p > 0$ is arbitrary. We begin with a simple lemma for which we could not find a ready reference.

LEMMA 1 Let g: $W \to Z$ be a generically surjective purely inseparable rational map of nonsingular surfaces. Then

(i) $\lambda(W) = \lambda(Z)$.

(ii) $\dim \mathrm{Alb}(W) = \dim \mathrm{Alb}(Z)$.

Proof: (i) Shioda has shown that $\lambda(Z) \le \lambda(W)$ (see Shioda, p. 234). To prove the opposite inequality, consider the schemes $(W, \mathcal{O}_W^{p^i}) = W_i$, $i > 0$. First, let us take the map α_i: $W \to W_i$ corresponding to the inclusion $\mathcal{O}_W^{p^i} \subseteq \mathcal{O}_W$. Now α_i is a map of smooth surfaces over k. It is finite and radicial; therefore, $b_i(W) = b_i(W_i)$ by (Artin et al., 1972, SGA 4, VIII, 1.2). On the other hand, if $i >> 0$, the rational map α_i factors

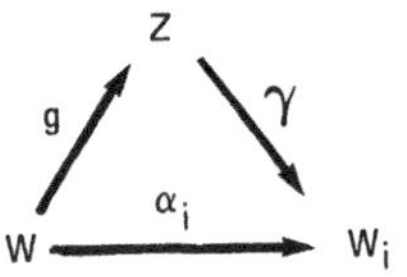

(*)

where γ is some dominant rational map over k. Thus, by Shioda's result quoted above, we have $\lambda(W) \ge \lambda(Z) \ge \lambda(W_i)$. To complete the proof we only need to show that $\rho(W) = \rho(W_i)$. For this we use another map, β_i: $W_i \to W$, which corresponds to the p^ith power map $\mathcal{O}_W \to \mathcal{O}_W^{p^i}$. Although β_i is not a map of k-schemes, still β_i is an isomorphism of abstract schemes and as such it induces an isomorphism of the abstract groups Pic W_i with Pic W. It is not hard to see that this isomorphism preserves the intersection numbers. Hence $\rho(W) = \rho(W_i)$.

(ii) The diagram (*) above shows that $(1/2)b_i(W) = \dim \mathrm{Alb}(W) \ge \dim \mathrm{Alb}(Z) \ge \dim \mathrm{Alb}(W_i)$, but we also have $\dim \mathrm{Alb}(W) = (1/2)b_1(W) = (1/2)b_1(W_i) = \dim \mathrm{Alb}(W_i)$.

COROLLARY 1.1 In the assumptions of Lemma 1, if Z is an Enriques surface, then W is supersingular and Alb W is trivial.

Proof: It is shown in Bombieri and Mumford (1976) that $\lambda(Z) = 0$ and we also have $b_1(Z) = 0$ from the definition of an Enriques surface.

REMARK 1.2 In the assumptions of lemma 1, if Z is simply connected, then so is W. (The proof is standard and we omit it.

Also see Bombieri and Mumford (1976) and Milne (1980).

2. UNIRATIONALITY

Our main result is the following theorem.

THEOREM 1 An Enriques surface over an algebraically closed field of characteristic 2 is unirational if and only if it is either classical or supersingular.

Proof: First, R. Crew has shown that a singular Enriques surface is never unirational (see Crew, 1981). Therefore, we only have to prove that all supersingular and classical Enriques surfaces in characteristic 2 are unirational. For the remainder of the proof, let X be a classical or a supersingular Enriques surface. Let $\pi\colon \tilde{X} \to X$ be the purely inseparable covering of degree 2 constructed in Bombieri and Mumford (1976, p. 220). By their proposition 9 (p. 221), $\tilde{X}$ is "K3 like," namely,

$$\dim H^i(\tilde{X},\mathcal{O}_{\tilde{X}}) = \begin{cases} 1 & i = 0 \\ 0 & i = 1 \\ 1 & i = 2 \end{cases}$$

and $\omega_{\tilde{X}} \approx \mathcal{O}_{\tilde{X}}$ (where $\omega_{\tilde{X}}$ denotes the dualizing sheaf on $\tilde{X}$). Also, $\tilde{X}$ is locally of codimension 1 in a smooth threefold, so that it is Cohen-Macauley and Gorenstein (i.e., $\omega_{\tilde{X}}$ is locally free). Thus $\tilde{X}$ is normal iff it is nonsingular in codimension 1. We consider two cases.

Case 1: $\tilde{X}$ is normal. Let $X_1 \xrightarrow{\rho} \tilde{X} \xrightarrow{\pi} X$ be a minimal desingularization of $\tilde{X}$. We have the injective map $j\colon \rho_*\omega_{X_1} \to \omega_{\tilde{X}} \approx \mathcal{O}_{\tilde{X}}$. If all the singularities of X are rational, then j is an isomorphism. Thus $\rho_*\omega_{X_1}$ and also ω_{X_1} has a nowhere vanishing section. Therefore, X_1 is a minimal model and it has Kodaira dimension zero. Also, $H^i(X_1,\mathcal{O}_{X_1}) \approx H^i(\tilde{X},\mathcal{O}_{\tilde{X}})$ for all i. From the table in Bombieri and Mumford (1976, p. 197), it follows that X_1 is a supersingular K3 surface. Now Šafarevič and Rudakov have shown (Rudakov and Šafarevič, 1979, corollary, p. 151) that any supersingular K3 surface in characteristic 2 is unirational; in fact, it is a Zariski surface. Thus X_1 and also X are unirational. We still have to consider the possibility that X has an isolated singularity which is not rational. Since $\omega_{\tilde{X}} \approx \mathcal{O}_{\tilde{X}}$

let us take σ to be a nowhere vanishing section of $\omega_{\tilde{X}}$. Because $\omega_{\tilde{X}}$ is isomorphic to the sheaf of rational differential 2-forms on $\tilde{X}$ with no polar curves on $\tilde{X}$, we can think of σ as a rational differential 2-form. Now it is well known that σ has a polar curve on X_1 because of the nonrational singularity. Let K_{X_1} be the divisor of σ on X_1. Let us show that $|nK_{X_1}| = \emptyset$. If not, there is an $f \in k(X_1) = k(\tilde{X})$, $f \neq 0$, such that $(f) + nK_{X_1} \geq 0$ on X_1. But then $(f) \geq 0$ on $\tilde{X}$ because K_{X_1} is entirely supported on curves which are contracted to singular points of $\tilde{X}$. Therefore, f must be a constant so that $nK_{X_1} \geq 0$, which contradicts the fact that σ has a polar curve on X_1. We conclude that $P_n(X_1) = 0$ for all $n \geq 1$ so that X_1 is ruled by the work of Bombieri and Mumford (1976) and therefore is rational by their corollary 1.1. Thus X is unirational; in fact, it is a Zariski surface in this situation.

Case 2: $\tilde{X}$ is not normal. Following Kunz (1975, 1978), we identify $\omega_{\tilde{X}}$ with a certain sheaf of rational differential 2-forms. Since $\omega_{\tilde{X}} \approx \mathcal{O}_{\tilde{X}}$ let σ be the differential form in $\omega_{\tilde{X}}(\tilde{X})$ which corresponds to 1 in $\mathcal{O}_{\tilde{X}}$. Let $\tilde{X}_N \xrightarrow{\rho} \tilde{X}$ be the normalization. Let L be the common function field of $\tilde{X}$ and $\tilde{X}_N$. We wish to study the divisor σ on $\tilde{X}_N$ to be denoted $(\sigma)_N$ [for the definition of the divisor of a differential, see Zariski (1969, p. 31)].

LEMMA 2 $(\sigma)_N < 0$.

Proof: Since σ corresponds to 1 in $\mathcal{O}_{\tilde{X}}$, the divisor $(\sigma)_N$ is supported only on such curves in $\tilde{X}_N$ which map onto a multiple curve of $\tilde{X}$. Let D_1 be any irreducible curve on $\tilde{X}_N$ whose image $C = \rho(D_1)$ is an irreducible multiple curve of $\tilde{X}$ (there exists at least one such curve). Let v_1 be the discrete valuation of L which corresponds to D_1. It is enough to show that $v_1(\sigma) < 0$. Assume the contrary, i.e., $v_1(\sigma) \geq 0$. Let $D_1, D_2, \ldots, D_n$ be all the irreducible curves on $\tilde{X}_N$ which map onto C. Let $v_1, v_2, \ldots, v_n$ be the corresponding discrete valuations of L. Now we have assumed that $v_1(\sigma) \geq 0$. Let $v_k(\sigma) = m_k$ for $k \neq 1$. There exists a function f in L such that $v_1(f) = 0$ and $v_k(f) = -m_k$ for $k \neq 1$ by Zariski and Samuel (1960, theorem 18, p. 45). It follows from this that the differential $f\sigma$ has no polar curves among the curves D_i. Now we apply Zariski's theory of subadjoints (Zariski, 1969, p. 85), which applies without any essential changes since $\tilde{X}$ is a local complete intersection, to conclude that f belongs to the conductor ideal of the local ring of the curve C on $\tilde{X}$ (see also Kunz, 1978, pp. 69-70). But in our case that conductor ideal is a proper ideal. Thus f also belongs to the maximal ideal of the local ring of the curve D_1 on $\tilde{X}_N$. Therefore, $v_1(f) > 0$ contrary to our choice of f. This contradiction shows that $v_1(\sigma) < 0$ and thus proves the lemma.

Proof of theorem 1, conclusion: Let g: $X_1 \to \tilde{X}_N$ be a desingularization. Using lemma 1, we now show that $P_n(X_1) = 0$ for $n \geq 1$. Let K be the divisor of σ on X_1. Then $K = \tau + E$, where E is supported on curves which g contracts to points and τ is the strict transform of $(\sigma)_N$ so that $\tau < 0$ by lemma 2. Since K is a canonical divisor on X_1 it is enough to show that $|nK| = \emptyset$. Suppose that $f \in L$, $f \neq 0$, and $(f) + nK \geq 0$ on X_1 but then on $\tilde{X}_N$ we have $(f)_N + (\sigma)_N \geq 0$, where $(f)_N$ is the divisor of f on $\tilde{X}_N$ so that $(f)_N \geq -(\sigma)_N > 0$, which is impossible. Thus $P_n(X_1) = 0$ and we conclude that X_1 is rational by corollary 1.1. Therefore, X is unirational, in fact, a Zariski surface in this case.

REMARK 2.1 Unirationality of certain, but not all, supersingular Enriques surfaces in characteristic 2 follows also directly from Bombieri and Mumford (1976, proposition 15) and Rudakov and Šafarevič (1979).

REMARK 2.2 The proof of theorem 1 and remark 2.1 shows that all supersingular and classical Enriques surfaces in characteristic 2 are simply connected. This is also well known by other methods.

COROLLARY 2.3 Shioda's conjecture that supersingular and simply connected surfaces are unirational (see Shioda, 1977, p. 167) is now established for all surfaces in characteristic 2 whose Kodaira dimension is ≤ 0.

Proof: It follows from Bombieri and Mumford (1976) that every such surface in characteristic 2 is either rational, or K3, or a supersingular or classical Enriques surface. Thus the corollary follows from Rudakov and Šafarevič (1979) and from our theorem.

OPEN PROBLEM 1 Are all supersingular and classical Enriques surfaces in characteristic 2 Zariski surfaces?

Our proof does not show this in the case when the pure inseparable cover X has rational singularities only.

OPEN PROBLEM 2 To determine all the unirational Enriques surfaces over an algebraically closed field of characteristic $p > 2$.

We recall that Shioda (1977, p. 161) gave examples of both unirational and nonunirational (classical) Enriques surfaces in every characteristic $p > 2$.

Acknowledgments: I heartily thank E. Bombieri, R. Crew, W. E. Lang, M. Levine, J. Lipman, N. Nygaard, and S. Shatz for their comments and encouragement. I also thank the Institute for Advanced Study for the hospitality shown to me.

REFERENCES

Artin, M., A. Grothendieck, and J. L. Verdier, *Théorie des topos et cohomologie étale des schémas*, Vol. 2, Lecture Notes in Mathematics, No. 270, Springer-Verlag, New York, 1972.

Bombieri, E., and D. Mumford, Enriques' classification of surfaces in characteristic p > 0: III, *Invent. Math.*, 36 (1976), 197-232.

Crew, R., thesis, Princeton University, 1981.

Hartshorne, R., *Algebraic Geometry*, Graduate Texts in Mathematics, Springer-Verlag, New York, 1977.

Katsura, T., Surfaces unirationnelles en caractéristique p, *C. R. Acad. Sci. Paris*, 288 (1979), 45-47, theorem 5.

Kunz, E., Differentialformen auf algebraischen Varietäten mit Singularitäten, I, *Manuscripta Math.*, 15 (1975), 91-108.

Kunz, E., Differentialformen auf algebraischen Varietäten mit Singularitäten, II, *Abh. Math., Sem. Univ. Hamburg*, 47 (1978), 43-70.

Milne, J. S., *Étale Cohomology*, Princeton University Press, Princeton, N.J., 1980.

Rudakov, A. N., and I. R. Šafarevič, Supersingular K3 surfaces over fields of characteristic two, *Math. USSR Izv.*, 13, No. 1 (1979), 147-165.

Shioda, T., An example of unirational surfaces in characteristic p, *Math. Ann.*, 211 (1974), 233-236.

Shioda, T., Some results on unirationality of algebraic surfaces, *Math. Ann.*, 230 (1977), 153-168.

Zariski, O., *An Introduction to the Theory of Algebraic Surfaces*, Lecture Notes in Mathematics, No. 83, Springer-Verlag, New York, 1969.

Zariski, O., and P. Samuel, *Commutative Algebra*, Vol. 2, Van Nostrand, New York, 1960.

8

Applications of the de Rham-Witt Complex and of Dominoes to Zariski Surfaces†

Note In this chapter, which is reprinted from T. Ekedahl's paper (Ekedahl, 1984, part VI), Ekedahl beautifully formalizes the theory of purely inseparable coverings of degree p that arise from a surface X, a line bundle L, and its pth power L^p.

He computes coherent cohomology and canonical divisors. Using the theory of the De Rham-Witt complex and the newly developed (by Illusie, Raynaud, and himself) theory of dominoes (see Illusie, 1979; Illusie and Raynaud, 1983; Ekedahl, 1984, 1985), he analyzes the crystalline cohomology of a generic Zariski surface and shows that it is torsion free.

He also proves that for $p \geq 7$ generic Zariski surfaces carry global 1 forms (all of which are nonclosed).

Ekedahl puts certain conditions on X and L. These are obviously satisfied for $X = \mathbb{P}^2$ and $L = \mathcal{O}(e)$, $e \geq 1$, which is the case leading to generic Zariski surfaces.

It seems that Ekedahl's approach to cohomology is the correct one; it gives rapid and clear proofs of some of the facts that arose earlier in the book: for example, the computations of p_g, g_a, K for a generic Zariski surface. Also, a way is pointed out to generalize the whole theory of Zariski surfaces to a more general and natural

---

†By Thorston Ekedahl; reprinted from a prepublication of the Université de Paris-Sud, Orsay, France (1984).

context, with $\mathbf{P}^2$ replaced by a more general surface X (perhaps arbitrary) and the line bundles $\mathcal{O}(e)$ replaced by a line bundle L on X (perhaps sufficiently ample). To get a Zariski surface we take a section s of $\mathcal{O}(pe)$, and construct the covering essentially from a pth root of s. Finally, we resolve singularities. To get an Ekedahl surface we start from a section s of L^p on X and construct a covering of X from a pth root of s, and then we resolve singularities.

Except for the last step, this is the same as the construction of a "Zariski scheme" associated with (X,L,S) (see Chapter 9).

This chapter is not self-contained. We must refer the reader to Ekedahl's complete paper (Ekedahl, 1984) for all unexplained facts and notation.

1. GEOMETRIC EXAMPLES

As promised in the introduction (Ekedahl, 1984), to which we refer the reader for all here unexplained facts and notation, we will now give some geometric examples. We will begin by studying a special type of inseparable covering. Let k be an algebraically closed field of characteristic $p > 0$, and let X be a smooth and projective surface over k.

REMARK To improve the comprehensibility of the arguments to follow, I will often assume stronger hypotheses than necessary for analysis. I leave to the interested reader, if there is one. to weaken conditions.

I will assume given a line bundle L on X and will consider the following two conditions that may be put on L:

(A) $H^0(X,L^{-i}) = 0$ for $1 \le i \le 3p - 3$, $H^1(X,L^{-i}) = 0$ for $1 \le i \le 2p - 1$.

(B) $H^0(X, \Omega_X^1 \otimes L^{-i}) = 0$ for $1 \le i \le p - 1$.

(i) Let α_L denote the group scheme on X which is the kernel of the relative Frobenius morphism F: $L \to L^p$, where L and L^p are considered as additive (smooth) group schemes on X.

Hence we have an exact sequence of flat group schemes on X:

$$0 \to \alpha_L \to L \xrightarrow{F} L^p \to 0 \tag{1.1}$$

Locally, in the Zariski topology, on X α_L is isomorphic to α_p. We give ourselves a nontrivial α_L-torsor $\pi: \tilde{X} \to X$ on X. Such a torsor may, for instance, be obtained by applying the boundary map obtained from (1.1) to a section $s \in H^0(X, L^p)$. In that case $\pi_* \mathcal{O}_{\tilde{X}} = \oplus_{i=0}^{p-1} L^{-i}$ with the obvious multiplication $L^{-i} \oplus L^{-j} \to L^{-i-j}$ if $i + j < p$ and $L^{-i} \oplus L^{-j} \to L^{-i-j} \xrightarrow{\text{id} \otimes s^{-1}} L^{-i-j+p}$ if $i + j \geq p$. Locally, in the Zariski topology, the long exact sequence of cohomology obtained from (1.1) shows that any $\tilde{X}$ is of this form. In general it is easy to see that there is a filtration $\mathcal{O}_X = M^0 \subseteq M^1 \subseteq \cdots \subseteq M^{p-1} = \pi_* \mathcal{O}_{\tilde{X}}$ of sub-$\mathcal{O}_X$-modules such that $M^i M^j \subseteq M^{i+j}$ when $i + j < p$ and $M^i / M^{i-1} = L^{-i}$, $1 \leq i \leq p - 1$.

(ii) $\tilde{X}$ is an integral local complete intersection scheme and $\omega_{\tilde{X}} = \pi^*(\omega_X \otimes L^{p-1})$.

Proof: Indeed, to see that $\tilde{X}$ is a local complete intersection it suffices, X being smooth, to show that $\tilde{X} \to X$ is a local complete intersection morphism. This is local in the flat topology on X, so we may trivialize $\tilde{X}$ so it is enough to prove that $\alpha_L \to X$ is a local complete intersection morphism. However, α_L is a flat group scheme. In particular, depth $\tilde{X} = 2$, so to prove that $\tilde{X}$ is reduced it suffices to prove that it is generically reduced. Consider $\tilde{X}_{red} \to \tilde{X} \to X$. If X is not generically reduced, then as $\tilde{X} \to X$ generically is of the form $\text{Spec}(k(X)[t]/(t^p - f)) \to \text{Spec}\, k(X)$, we see that $f \in k(X)^p$ and $\tilde{X}_{red} \to X$ is birationally an isomorphism. It is finite, as it is the composite of a closed immersion and a finite morphism and X is normal. Therefore, $\tilde{X}_{red} \to X$ is an isomorphism by Zariski's main theorem. This gives a section $X \xrightarrow{\sim} \tilde{X}_{red} \to X$ contrary to the nontriviality of $\tilde{X} \to X$. Finally, $\omega_{\tilde{X}} = \pi^*\omega_X \otimes \omega_{\tilde{X}/X}$, so it suffices to show that $\omega_{\tilde{X}/X} = \pi^*(L^{p-1})$. By an easily proven result on group scheme torsors this equals $\pi^*(s^*\omega_{\alpha_L/X})$, where $s: X \to \alpha_L$ is the zero section and the adjunction formula applied to the immersion $\alpha_L \to L$ gives $s^*\omega_{\alpha_L/X} = L^{p-1}$.

(iii) Suppose (A). $H^1(X,\mathcal{O}_X) \to H^1(\tilde{X},\mathcal{O}_{\tilde{X}})$ is an isomorphism, $H^2(X,\mathcal{O}_X) \to H^2(\tilde{X},\mathcal{O}_X)$ is a monomorphism and $\dim_k H^2(\tilde{X},\mathcal{O}_{\tilde{X}})/H^2(X,\mathcal{O}_X) = \Sigma_{i=1}^{p-1} \dim_k H^2(X,L^{-i})$.

Proof: To see this it will suffice to show that $H^0(X,\pi_*\mathcal{O}_X/\mathcal{O}_X) = 0 = H^1(X,\pi_*\mathcal{O}_{\tilde{X}}/\mathcal{O}_X)$ and that $\dim_k H^2(X,\pi_*\mathcal{O}_{\tilde{X}}/\mathcal{O}_X) = \Sigma_{i=1}^{p-1} \dim_k H^2(X,L^{-1})$. Now $\pi_*\mathcal{O}_{\tilde{X}}/\mathcal{O}_X$ has a filtration by the $M^i/\mathcal{O}_X$ with successive quotients L^{-i} for $1 \le i \le p - 1$ and we use condition (A).

Let us introduce some notation. If M is an $\hat{R}^0$-module of finite type, there is a largest submodule which is of finite type as W-module. Denote this by TM. As a corollary of (iii), we get

(iv) Suppose (A). Then $H^1(X,W\mathcal{O}_X) \to H^1(\tilde{X},W\mathcal{O}_{\tilde{X}})$ is an isomorphism and $H^2(X,W\mathcal{O}_X) \to H^2(\tilde{X},W\mathcal{O}_{\tilde{X}})$ is a monomorphism whose cokernel is killed by p. Furthermore, it induces an isomorphism on T(-), and T(-) of its cokernel is zero.

Proof: Let C be a mapping cone of $R\Gamma(X,W\mathcal{O}_X) \to R\Gamma(X,W\mathcal{O}_{\tilde{X}})$. Then (iii) shows that $H^i(\hat{R}^0/V\hat{R}^0 \otimes^L_{\hat{R}^0} C) = 0$ if $i \neq 2$. The long exact sequence of the distinguished triangle

$$\longrightarrow C \longrightarrow C \xrightarrow{V} R/VR^0 \otimes^L_{\hat{R}} C \longrightarrow$$

and the fact that $H^i(C)$ are separated in the V-topology, being of finite type over $\hat{R}^0$, show that $H^i(C) = 0$ if $i \neq 2$ and that V is injective on $H^2(C)$. This shows that $H^1(X,W\mathcal{O}_X) \to H^1(\tilde{X},W\mathcal{O}_{\tilde{X}})$ is an isomorphism and that $H^2(X,W\mathcal{O}_X) \to H^2(\tilde{X},W\mathcal{O}_{\tilde{X}})$ is injective. We now want to show that its cokernel, that is, $H^2(C)$, is killed by p. As $p = VF$ it will be sufficient to show that F maps $H^2(\tilde{X},W\mathcal{O}_{\tilde{X}})$ into $H^2(X,W\mathcal{O}_X)$. It is clear that the Frobenius F: $X \to X$ trivializes $\tilde{X}$, so that there exists a mapping $X \to \tilde{X}$ such that the composite $X \to \tilde{X} \xrightarrow{\pi} X$ equals F. It is easy to see that $\tilde{X} \xrightarrow{\pi} X \to \tilde{X}$ is also F: $\tilde{X} \to \tilde{X}$. Now F: $H^2(\tilde{X},W\mathcal{O}_{\tilde{X}}) \to H^2(\tilde{X},W\mathcal{O}_{\tilde{X}})$ is induced by functoriality from F: $\tilde{X} \to \tilde{X}$, so it factors through π: $H^2(X,W\mathcal{O}_X) \to H^2(\tilde{X},W\mathcal{O}_{\tilde{X}})$, which is want we want. Hence $H^2(C)$ is killed by p with V injective, and so contains no submodule of finite type over W. Therefore, $TH^2(\tilde{X},WO_{\tilde{X}})$ is contained in $H^2(X,WO_X)$ and thus equals $TH^2(X,WO_X)$.

(v) Assume (A). $\dim_k(H^2(\tilde{X},WO_{\tilde{X}})/H^2(X,WO_X))/V(H^2(\tilde{X},WO_{\tilde{X}})/H^2(X,WO_X)) = \Sigma_{i=1}^{p-1} \dim_k H^2(X,L^{-i})$.

Proof: This follows from (iii) and (iv).

Let us now consider $H^i(\tilde{X},\mathbb{G}_m)$. Put $N := \pi_*\mathbb{G}_m/\mathbb{G}_m$. If we pull back $\tilde{X}$ by π, it becomes trivial and $\pi^*\tilde{X} = \mathrm{Spec}(\oplus_{i=0}^{p-1} L^{-i})$ with multiplication the obvious one. $L^{-i} \otimes L^{-j} \to L^{i-j}$ if $i + j < p$ and zero if not. Also, $\pi^*N = 1 + \oplus_{i=1}^{p-1} L^{-i}$, which through exp and log is isomorphic to $\oplus_{i=1}^{p-1} L^{-i}$.

(vi) Assume (A). $H^1(X,\mathbb{G}_m) \to H^1(\tilde{X},\mathbb{G}_m)$ is an isomorphism and $H^2(X,\mathbb{G}_m) \to H^2(\tilde{X},\mathbb{G}_m)$ is injective with cokernel killed by p.

Proof: Indeed, it is clear that we need to prove that $H^0(X,N) = H^1(X,N) = 0$ and that $H^2(X,N)$ is killed by p. As π is faithfully flat and π^*N, as we have just seen, is killed by p, the last statement follows. Again as π is faithfully flat, we get a Čech spectral sequence

$$H^i(\tilde{X}^{(j)},N) \Rightarrow H^{i+j-1}(X,N)$$

where $\tilde{X}^{(j)} := \tilde{X}x_X\tilde{X}x_X\cdots x_X\tilde{X}$ (j times), so it will be sufficient to show that

$$H^0(\tilde{X},\pi^*N) = H^1(\tilde{X},\pi^*N) = H^0(\tilde{X}^{(1)},\pi_2^*N) = 0$$

where π_2: $\tilde{X}^{(2)} \to X$. Now $\pi^*N \simeq \oplus_{i=1}^{p-1} L^{-i}$, $\tilde{X}^{(2)} = \alpha_L x_X X$ with π_2 the composite of the projection q_2 onto the second factor and π. Hence we need to prove that $H^0(\tilde{X},\pi^*L^{-i}) = H^1(\tilde{X},\pi^*L^{-i}) = H^0(\alpha_L x_X\tilde{X},q_2^*\pi^*L^{-i}) = 0$ for $1 \le i \le p - 1$. Now $\pi_*\pi^*L^{-i} = L^{-i} \otimes_{O_X} \pi_*O_{\tilde{X}}$ and $\pi_*q_{2*}q_2^*\pi^*L^{-i} = \oplus_{j=1}^{p-1} L^{-i} \otimes L^{-j} \otimes \pi_*O_{\tilde{X}}$, so using $H^0(\tilde{X},\pi^*L^{-i}) = H^0(X,\pi_*\pi^*L^{-i})$, etc., the filtration M^i of $\pi^*O_{\tilde{X}}$ and condition (A) we are through.

We get immediately the following consequence:

(vii) Assume A. Then $H^1(X,\mathbb{Z}_p(1)) \to H^1(\tilde{X},\mathbb{Z}_p(1))$ and $H^2(X,\mathbb{Z}_p(1)) \to H^2(\tilde{X},\mathbb{Z}_p(1))$ are isomorphisms.

Proof: Recall that $H^i(Y,\mathbb{Z}_p(1)) := \varprojlim_n H^i(Y,\mu_{p^n})$. The exact sequences $0 \to \mu_{p^n} \to \mathbf{G}_m \xrightarrow{p^n} \mathbf{G}_m \to 0$ give short exact sequdnces

$$0 \to \widehat{H^{i-1}(Y,\mathbf{G}_m)} \to H^i(Y,\mathbb{Z}_p(1)) \to T_p(H^i(Y,\mathbf{G}_m)) \to 0$$

where $\hat{M} := \varprojlim_n M/p^nM$ and $T_pM := \varprojlim_n\{{}_{p^n}M,p\}$. Hence (vii) follows from (vi) and the 5-lemma once we have shown that $T_p(H^2(X,\mathbb{G}_m)) \to T_p(H^2(\tilde{X},\mathbb{G}_m))$ is an isomorphism. However, $T_p(-)$ is left exact and zero on a group killed by p, so this again follows from (vi).

We can associate to $\tilde{X}$ a mapping φ: $L^{-p} \to \Omega^1_X$. When L is trivialized and $\tilde{X}$ is the boundary of $s \in L^p$ corresponding to $s' \in \mathcal{O}_X$ under the trivialization, then $t \in L^{-p}$ is taken to $t'\,ds'$, where $t' \in \mathcal{O}_X$ corresponds to t. It is easy to see that the set of zeros of $L^{-p} \to \Omega^1_X$ equals the projection to X of the singular set of $\tilde{X}$. As $\tilde{X}$ has depth 2, the zeros of φ are isolated iff $\tilde{X}$ is normal. From now on we will assume (A) but also that the zeros of φ are isolated and simple. It is clear that this is equivalent to the condition that locally, in the étale topology, the s' above has the form xy, where x, y is a coordinate system on X. Hence the singularities of $\tilde{X}$ has the form $z^p = xy$ and so are rational double points of type A_{p-1}.

(viii) The number of singular points on $\tilde{X}$ is $(L \cdot L)p^2 + (\omega_X \cdot L)p + c_2(X)$.

This is simply the formula for the number of zeros of φ: $L^{-p} \to \Omega^1_X$ when all the zeros are isolated.

Let ρ: $X' \to \tilde{X}$ be a minimal resolution of the singularities of X.

(ix) (a) $\omega_{X'} = \rho^*\pi^*(\omega_X \otimes L^{p-1})$.

(b) $H^i(\tilde{X},W\mathcal{O}_{\tilde{X}}) \to H^i(X',W\mathcal{O}_{X'})$ is an isomorphism for all i.

(c) $H^1(\tilde{X},\mathbb{Z}_p(1)) \to H^1(X',\mathbb{Z}_p(1))$ is an isomorphism.

$H^2(\tilde{X},\mathbb{Z}_p(1)) \to H^2(X',\mathbb{Z}_p(1))$ is injective with cokernel a free $\mathbb{Z}_p$-module of rank equal to $(p - 1)((L \cdot L)p^2 + (\omega_X \cdot L)p + c_2(X))$.

Proof: Indeed, as the singularities of $\tilde{X}$ are rational, $R\rho_* \mathcal{O}_{X'} = \mathcal{O}_{\tilde{X}}$, which gives $R\rho_* W\mathcal{O}_{X'} = W\mathcal{O}_{\tilde{X}}$, which in turn gives (b). As $\tilde{X}$ has in fact only rational double points $\rho^*(\omega_{\tilde{X}}) = \omega_{X'}$ so we get (a). To prove (c) we consider the spectral sequence

$$H^i(\tilde{X}, R^j\rho_* \mu_{p^n}) \Rightarrow H^{i+j}(X', \mu_{p^n})$$

I claim first that $R^1\rho_*\mu_{p^n} = 0$. To prove this we are reduced to proving that multiplication by p^n is injective on Pic X'_S, where S is the spectrum of a local ring and $S \to X$ a flat morphism. As the singularities of $\tilde{X}$ are rational, $H^1(X'_S, \mathcal{O}_{X'_S}) = 0$, so this follows from (Ekedahl, 1984, theorem 12.1). This immediately gives that $H^1(\tilde{X}, \mathbb{Z}_p(1)) \to H^1(X', \mathbb{Z}_p(1))$ is an isomorphism and that $\tau\colon\ H^2(\tilde{X}, \mathbb{Z}_p(1)) \to H^2(X', \mathbb{Z}_p(1))$ is injective. As $H^1(\tilde{X}, \mu_p) \to H^1(X', \mu_p)$ is an isomorphism and $H^2(\tilde{X}, \mu_p) \to H^2(X', \mu_p)$ an injection, τ is surjective on the kernel of p and injective on the cokernel, which shows that Coker τ is torsion free. I claim that $T_pH^2(\tilde{X}, \mathbb{G}_m) \to T_pH^2(X', \mathbb{G}_m)$ is an isogeny. Indeed, by (vii) we can replace $\tilde{X}$ by X and then we have a diagram

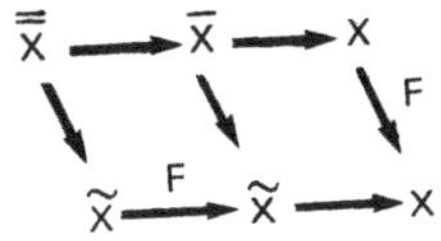

where $\bar{X} \to X$ and $\bar{\bar{X}} \to \tilde{X}$ are blowups. As $T_pH^2(-, \mathbb{G}_m)$ is a birational invariant on smooth, proper surfaces, we conclude. Hence the rank of Coker τ equals the rank of $\widehat{\text{Pic}}$ X' minus the rank of $\widehat{\text{Pic}}$ $\tilde{X}$, and this difference is equal to the number of irreducible components of the exceptional fibers. As we have an A_{p-1}-singularity, we get p - 1 such for each singular point, so we conclude by (viii).

We can now compare invariants of X and X'.

(x) (a) $c_1^2(X') = p(c_1^2(X) + 2(p - 1)(\omega_X \cdot L) + (p - 1)^2(L \cdot L))$.

(b) $H^1(X, W\mathcal{O}_X) \to H^1(X', W\mathcal{O}_{X'})$ is an isomorphism and $H^2(X, W\mathcal{O}_X) \to H^2(X', W\mathcal{O}_{X'})$ is an injection with a V-torsion free cokernel killed by p. The dimension of the cokernel as a Dieudonné module is

$$\frac{p(p-1)(p-2)}{6}(L \cdot L) + \frac{p(p-1)}{4}(\omega_X \otimes L^{-1} \cdot L) + (p-1)\chi$$

(c) $H^1(X,\mathbb{Z}_p(1)) \to H^1(X',\mathbb{Z}_p(1))$ is an isomorphism and $H^2(X, \mathbb{Z}_p(1)) \to H^2(X',\mathbb{Z}_p(1))$ is an injection with torsion-free cokernel of rank $(p-1)((L \cdot L)p^2 + p(\omega_X \cdot L) + c_2(X))$.

This is just putting together (ix), (v), (vii), Riemann-Roch, and (A) to compute $\chi(L^{-i})$.

REMARK (c) was essentially proven by W. Lang under the stronger assumption (B) (see Lang, 19).

(xi) Assume (A) and (B). Then $H^0(X,\Omega_X^1) = H^0(\tilde{X},\Omega_X^1)$.
Proof: Indeed, we have an exact sequence on $\tilde{X}$

$$0 \to \pi^*L^{-p} \to \pi^*\Omega_X^1 \to \Omega_{\tilde{X}}^1 \to \pi^*L^{-1} \to 0$$

It is clear that we are finished if $H^0(\pi^*L^{-p}) = H^0(\pi^*L^{-1}) = H^1(\pi^*L^{-p}) = 0$ and $H^0(\Omega_X^1) = H^0(\pi^*\Omega_X^1)$. By projecting down to X we see that the first part is implied by (A) and as $\pi_*\pi^*\Omega_X^1 = \pi_*\mathcal{O}_{\tilde{X}} \otimes_{\mathcal{O}_X} \Omega_X^1$ the second is implied by (B) if we use the filtration $\{M^i\}$.

(xii) If the zeros of $L^{-p} \to \Omega_X^1$ are isolated and simple, then every closed form in $H^0(\Omega_{X'}^1)$ is in the image of $H^0(\Omega_{\tilde{X}}^1)$.
Proof: The statement is local on X, so we may assume that $\tilde{X}$ is of the form $\mathrm{Spec}(\mathcal{O}_X[t]/(t^p - f))$ for $f \in \mathcal{O}_X$. Then we have a cartesian diagram

$$\begin{array}{ccc} \tilde{X} & \longrightarrow & X \\ \downarrow & & \downarrow f \\ \mathbb{A}^1 & \xrightarrow{F} & \mathbb{A}^1 \end{array}$$

where F is the Frobenius.

We have exact sequences

$$\begin{array}{ccccccccc}
0 & \longrightarrow & \rho_\star\Omega^1_{\mathbb{A}^1} & \longrightarrow & \rho_\star\Omega^1_{X'} & \longrightarrow & \rho_\star\Omega^1_{X'/\mathbb{A}^1} & \longrightarrow & 0 \\
 & & \uparrow{\scriptstyle s} & & \uparrow & & \uparrow & & \\
0 & \longrightarrow & \Omega^1_{\mathbb{A}^1} & \longrightarrow & \Omega^1_{\tilde{X}} & \longrightarrow & \Omega^1_{\tilde{X}/\mathbb{A}^1} & \longrightarrow & 0
\end{array}$$

and it is sufficient to show that the image in $\rho_*\Omega^1_{X'/\mathbf{A}^1}$ of a closed form comes from $\Omega^1_{\tilde{X}/\mathbf{A}^1}$. However, Szpiro (1981, exp. V, lemme 2) shows that it actually comes from $\Omega^1_{X/\mathbf{A}^1}$, as the global assumptions are never used there.

(xiii) Assume (A), (B), and that the zeros of $L^{-p} \to \Omega^1_X$ are isolated and simple. Then the closed 1-forms of X' are exactly those in the image of the injective mapping $H^0(X,\Omega^1_X) \to H^0(X',\Omega^1_{X'})$.

Proof: As it is obvious that all forms in the image are closed, this follows from (x) and (xi).

We will now further specialize to $X = \mathbb{P}^2$ and $L = \mathcal{O}(n)$, $n > 0$. Then (A) and (B) are fulfilled and a simple dimension counting shows that we can find $s \in \mathcal{O}(pn)$ such that $\mathcal{O}(-pn) \to \Omega^1_{\mathbb{P}^2}$ has only isolated simple singularities. From (x) we then get that $H^1(X',W\mathcal{O}_{X'}) = H^0(W\Omega^1_{X'}) = H^0(W\Omega^2_{X'}) = 0$, that $H^2(W\mathcal{O}_{X'})$ is killed by p and is without finite torsion, that $H^1(W\Omega^1_{X'})$ is torsion free of slope 0, and that it equals $\mathrm{Pic}\, X' \otimes_{\mathbb{Z}} W$. Now Pic X' contains the inverse image of a hyperplane and all the exceptional divisors. The square of the inverse image of a hyperplane is p and as the singularities are of type A_{p-1}, each connected component of the exceptional divisors gives an intersection matrix of determinant $(-1)^{p-1}p$. Hence the p-adic ordinal of the discriminant of the group of cycles generated by the cycles that have just been described is $1 + p^2n^2 - 3pn + 3$. By (ix) this group has finite index in Pic X', so Pic X' has discriminant of p-adic ordinal $\le 1 + p^2n^2 - 3pn + 3$ and as $H^1(W\Omega^1_X) = \mathrm{Pic}(X') \otimes W$, so has $H^1(X,W\Omega^1)$. If $M := (H^2(W\mathcal{O}_{X'}) \xrightarrow{d} H^2(W\Omega^1_{X'}))$, which is a domino, has type σ, then by (III:3.3;4.6.1) (Ekedahl, 1984), $2\Sigma_{i\in\mathbb{Z}}\, i\sigma(i) \le 1 + p^2n^2 - 3pn + 3$. Now, as $R\Gamma(X',\Omega^{\cdot}_{X'}) = R_1 \otimes^{L}_{R} R\Gamma(X',W\Omega^{\cdot}_{X'})$ (see Illusie and Raynaud, 1983), we have $H^2(X',\Omega^1) = H^1(M,dV)$ and so $h^0(\Omega^1_{X'}) =$

$h^2(\Omega^1_{X'}) = \Sigma_{i\in\mathbb{N}}\, \sigma(-i)(i + 1)$ by (III:3.3) (Ekedahl, 1984). As

$$\sum_{i\in\mathbb{Z}} \sigma(i)(-i + 1) = \frac{p(p - 1)(p - 2)}{6} n^2 - \frac{p(p - 1)}{4}(n + 3) + p - 1 - \sum_{i\in\mathbb{Z}} i\sigma(i)$$

by (x)b, we get

$$\text{(xiv)}\quad h^0(\Omega^1_{X'}) = \sum_{i\in\mathbb{N}} \sigma(-i)(i + 1) \geq \sum_{i\in\mathbb{Z}} \sigma(i)(-i + 1)$$
$$\geq \frac{p(p - 1)(p - 2)}{6} n^2 - \frac{p(p - 1)}{4}(n + 3) + p - 1 - \frac{1}{2}(p^2n^2 + 3pn - 4)$$
$$= \frac{p[(p - 1)(p - 2) - 3p]}{6} n^2 - \frac{p(p + 5)}{4} n - \frac{3p^2 - 7p}{4} + 1$$

Also, by (xi) and (xii), d: $H^0(\Omega^1_{X'}) \to H^0(\Omega^2_{X'})$ is injective and so by duality d: $H^2(\mathcal{O}_{X'}) \to H^2(\Omega^1_{X'})$ is surjective. This shows (Ekedahl, 1984, III:3.3) that $\sigma(i) = 0$ if $i < 0$, and in particular that there is no exotic torsion, so there is no crystalline torsion.

2. EXOTIC TORSION

We will now go on to discuss the phenomenon of exotic torsion. Let us first note that we can find examples of a smooth and proper scheme X over Spec W such that its special fiber has exotic torsion. Indeed, we may take as example the product of an Igusa surface with itself which will have exotic torsion in degree 3 (see Ekedahl, 1985, III 8:10). If we confine our attention to degree 2, the situation becomes more intricate. It is clear by Kummer theory that the p-torsion of $H^2(X', \mathbf{Z}_p)$, where X' is a geometric generic fiber of X, equals the p-torsion of Pic X'/Pic^0X'. By Illusie and Raynaud (1983), there is an abelian subscheme $A \subseteq \mathrm{Pic}^\tau X$ such that $N = \mathrm{Pic}^\tau X/A$ is finite and flat over W. Hence the generic fiber of N equals $\mathrm{Pic}^\tau X'/\mathrm{Pic}^0 X'$.

I claim that the Dieudonné module of the special fiber $\bar{N}$ equals the finite torsion of $H^2(\bar{X}/W)$, where $\bar{X}$ is the special fiber of X. *Proof:* Indeed, by (I:2.2.2) this finite torsion is the sum of the V-torsion of $H^2(\bar{X},WO_{\bar{X}})$ and the p-torsion of $H^1(\bar{X},W\Omega^1_{\bar{X}})$. The first part is the Dieudonné module of the connected part of $\bar{N}$ (see Illusie, 1979, II) and the second part the Dieudonné module of the étale part of $\bar{N}$ (see Illusie, 1979).

Hence we see that the length of the p-torsion of $H^2(X',\mathbb{Z}_p)$ equals the length of the finite torsion of $H^2(\bar{X}/W)$ and so we see that the length of the p-torsion of $H^2(X',\mathbb{Z}_p)$ equals the length of the torsion of $H^2(\bar{X}/W)$ iff $\bar{X}$ has no exotic torsion in degree 2, thus relating the existence of exotic torsion to an old problem of Grothendieck (see Illusie, 1975).

On the contrary, if we lift with much ramification ($e \geq p - 1$), we can get more crystalline torsion than classical torsion for two reasons. First the example of Raynaud (see Illusie and Raynaud, 1983) gives a nonflat N, so the order of N at the generic point is strictly smaller than its order at the special point and so the length of $H^2(X',\mathbb{Z}_p)$ is strictly smaller than the length of the finite torsion of $H^2(\bar{X}/W)$. Second, we will see in a moment that we may also get exotic torsion.

REMARK Raynaud's example is constructed starting from a finite $\mathbb{F}_p$-vector space V, a nondegenerate alternating pairing φ on V and the associated Heisenberg group N. If one lets dim $V \geq 6$ and considers the split extension $G = N \times Sp(\varphi)$, we may replace N by G in Raynaud's argument and then the general fiber of X will have a perfect fundamental group, so $H^2(X',\mathbb{Z}_\ell)$ will be torsion free for all primes ℓ (including p), but $H^2(\bar{X}/W)$ will have torsion.

As our first example of how to produce exotic torsion, let us again consider a line bundle L on a smooth and proper surface fulfilling condition (A) of Section 1 and π: $X \to X$ the resolution of an α_L-torsor for which $L^{-p} \to \Omega^1_X$ has only simple isolated zeros.

I claim first that π^* takes $H^0(X,Z_n\Omega^1_X)$ into $H^0(\tilde{X},Z_{n+1}\Omega^1_{\tilde{X}})$.

Proof: Indeed, this only uses the fact that π is inseparable; $Z_{n+1}\Omega^1$ is the kernel of dC: $Z_n\Omega^1 \to \Omega^2$ and π^*: $H^0(X,\Omega^2_X) \to H^0(\tilde{X},\Omega^2_{\tilde{X}})$ is zero as π is inseparable.

I further claim that π^*: $H^0(X,Z_\infty\Omega^1_X) \to H^0(\tilde{X},Z_\infty\Omega^1_{\tilde{X}})$ is an isomorphism.

Proof: By Illusie (1979, 0.2.5.3.3-5) it will suffice to show that π^* induces an isomorphism on global sections of $\mathbb{G}_m/p\mathbb{G}_m$ and $B_n\Omega^1$ for all n. The first part follows from (x) in Section 1. For the second part we get from Illusie (1979, I.3.11.4) that it will be sufficient to show that π induces an isomorphism on the kernel of F on $H^1(-,W_n\mathcal{O})$. However, from (x) in Section 1 it follows that π^* induces an isomorphism already on $H^1(-,W_n\mathcal{O})$. It is clear from the proof of (xi) that (A) implies that π^*: $H^0(X,\Omega^1_X) \to H^0(\tilde{X},\Omega^1_{\tilde{X}})$ is injective.

We therefore see that $\dim_k H^0(\tilde{X},Z_{n+1}\Omega^1_{\tilde{X}})/H^0(\tilde{X},Z_\infty\Omega^1_{\tilde{X}}) \geq \dim_k H^0(X, Z_{n+1}\Omega^1_X)H^0(X,Z_\infty\Omega^1_X)$.

For n = 0 we get that if X has nonclosed 1-forms, then, by Illusie (1979, II.6.16), $\tilde{X}$ has exotic torsion. We can take as X one of the coverings of $\mathbb{P}^2$ constructed at the end of Section 1; from the formula given there we see that if $p \geq 7$, we may always find such an X with nonclosed 1-forms.

REFERENCES

Ekedahl, T., On the multiplicative properties of the De Rham-Witt complex: I, *Ark. Mat.*, 22, No. 2 (1984), 185-239.

Ekedahl, T., On the multiplicative properties of the De Rham-Witt complex: II, *Ark. Mat.*, 23, No. 1 (1985).

Ekedahl, T., Diagonal complexes and F-gauge structures, prepublications, University de Paris-Sud, Orsay, France, 1984. (Reprinted as a book in the series *Travaux en cours*, Hermann, Paris, 1986.)

Illusie, L., Complexe de De Rham-Witt et cohomologie cristalline, *Ann. Sci. Ec. Norm. Sup.*, ser. 4, 12 (1979), 501-661.

Illusie, L., Report on crystalline cohomology, *Proc. Symp. Pure Math.*, 29 (Arcata), American Mathematical Society, Providence, R.I., 1975.

Illusie, L., and M. Raynaud, Les suites spectrales associées au complexe de De Rham-Witt. Publ. IHES 57 (1983), 73-212.

Lang, W. E., Remarks on p-torsion on algebraic surfaces, *Compositio Math.*, 52(2) (1984), 197-202.

Szpiro, L., Seminaire sur les pinceaux de courbes de genre au moins deux, *Asterisque*, 86 (1981).

9
Picard and Brauer Groups of Zariski Surfaces†

In this chapter higher-dimensional analogs of the Blass-Deligne-Lang theorem are studied and proved. Also, the Brauer group is investigated in the rather general context of Zariski schemes.

We use the Cartier-Yuan exact sequence to calculate Picard groups and Brauer groups of Zariski surfaces and their generalizations. We also extend a result of Blass-Deligne on the factoriality of general affine Zariski surfaces to all higher-dimensional Zariski schemes. All calculations are carried out using étale cohomology, but at several points we use the identifications of étale and Zariski cohomology if the coefficient sheaf is a coherent $\mathcal{O}_S$-module.

Since we are interested in both the affine and projective cases, we begin by establishing some notation. Let S be a regular, quasi-projective scheme in characteristic $p > 0$ with very ample bundle $\mathcal{O}(1)$. If e is a positive integer, $(p,e) = 1$, and $G \in \Gamma(S,\mathcal{O}_S(pe))$, we let A be the graded, quasi-coherent sheaf of $\mathcal{O}_S$-algebras defined by $A = \oplus_{\ell=0}^{\infty} \mathcal{O}(\ell)[Z]/(Z^p - G)$, where Z is an indeterminate of degree e. The Zariski scheme associated to (S,G) is $X = \mathrm{Proj}(A)$. It comes with a structure map $\pi\colon X \to S$ which makes $\pi_* \mathcal{O}_X$ into a locally free $\mathcal{O}_S$-module of rank p. Note that if S is affine, then $X = \mathrm{Spec}(\mathcal{O}_S[Z]/(Z^p - G))$.

For the remainder of this note we assume that X is a normal scheme. This will be the case, for instance, if S is a smooth,

---

†By Piotr Blass and Raymond Hoobler (City College of New York, New York, New York). (Reprinted from *Proceedings of the American Mathematical Society*, Vol. 97, No. 3, pp. 379-383.)

quasi-projective variety and the locus of zeros of the jacobian matrix of G/x_j^{pe} codimension > 1 in $D_+(x_j) \subset S$ and all $x_j \in \Gamma(S,\mathcal{O}(1))$. In this case,

$$\pi_*\mathcal{O}_X = \bigoplus_{\ell=0}^{p-1} \mathcal{O}_S(-\ell e)$$

since the section Z restricts to $Z_j = Z/x_j^e \in \Gamma(D_+(x_j),\pi_*\mathcal{O}_X)$, which transforms according to the rule

$$(Z_j/Z_k)^p = (G/x_j^{pe})/(G/x_k^{pe}) = (x_k^e/x_j^e)^p$$

or, more simply, $x_j^e Z_j = z_k^e Z_k$. Moreover, the Frobenius on S, F_S: $S \to S$, factors as $\pi \circ i$ where $i\colon\ S \to X$. If $f \in \Gamma(D_+(x_j),\pi_*\mathcal{O}_X)$ is described as $f = \Sigma\, a_\ell z_j^\ell$, $0 \le \ell < p$, then $i^*f = \Sigma\, a^p i^*(z_j)^\ell = \Sigma\, a_\ell^p g_j^\ell$, where $g_j = G/x_j^{pe} \in \Gamma(D_+(x_j),\mathcal{O}_S)$. Thus in terms of homogeneous coordinates i can be viewed on $\mathcal{O}_S(-\ell e)$ as the composite

$$\mathcal{O}_S(-\ell e) \xrightarrow{F_S^*} \mathcal{O}_S(-p\ell e) \xrightarrow{G^\ell} \mathcal{O}_S$$

Next we turn to the computation of $\pi_*\Omega^1_{X/S}$. On $D_+(x_j)$,

$$\pi_*\Omega^1_{X/S}\Big|_{D_+(x_j)} = \bigoplus_{\ell=0}^{p-1} \mathcal{O}_S\Big|Z_j^\ell dZ_j$$

and $d = d_{X/S}$ has $\oplus\mathcal{O}_S|Z_j^\ell dZ_j$, $0 \le \ell < p - 1$, as image. Since $Z_j x_j^e = Z_k x_k^e$, we conclude that

$$\pi_*\Omega^1_{X/S} = \bigoplus_{\ell=1}^{p} \mathcal{O}_S(-\ell e)$$

and elements are given locally by $\Sigma\, a_j Z_j^\ell dZ_j$, where dZ_j is a local representation of a section of $\mathcal{O}_S(-e)$. Since $\Omega^1_{X/S}$ is an invertible $\mathcal{O}_X$-module, all 1-forms are closed.

We can now put all this information together into the Cartier-Yuan sequence (Hoobler, 1982).

$$0 \to G_{m,S} \to \pi_* G_{m,X} \xrightarrow{d\ell n} \pi_* \Omega^1_{X/S} \xrightarrow{C-I} i^* \Omega^1_{X/S} \to 0$$

$$\pi_* \Omega^1_{X/S} = \bigoplus_{\ell=1}^{p} \mathcal{O}_S(-\ell e) \qquad i^* \Omega^1_{X/S} = \mathcal{O}_S(-pe)$$

$$C\left(\bigoplus_{\ell=1}^{p-1} \mathcal{O}_S(-\ell e)\right) = 0 \qquad C: \ \mathcal{O}_S(-pe) \to \mathcal{O}_S(-pe) \ \text{ is the identity} \tag{1.1}$$

$$I: \ \mathcal{O}_S(-\ell e) \to \mathcal{O}_S(-pe) \ \text{ is the composite}$$

$$\mathcal{O}_S(-\ell e) \xrightarrow{F_S^*} \mathcal{O}_S(-p\ell e) \xrightarrow{G^{\ell-1}} \mathcal{O}_S(-pe)$$

Note that C is $\mathcal{O}_S$-linear while I is p-linear. This sequence is exact for the étale topology on S and remains so when restricted to any subscheme of S.

We can now state our principal theorem.

THEOREM 1 Let S be a regular, quasi-projective scheme in characteristic $p > 0$. Let π: $X \to S$ be the Zariski scheme associated to (S,G) and assume that X is normal.

(1) Suppose that S is projective and $H^m(S,\mathcal{O}_S(-n)) = 0$ for $n > 0$ and $m = 0$ or 1. Then π^*: $\mathrm{Pic}(S) \to \mathrm{Pic}(X)$ is an isomorphism. If, moreover, $H^2(S,\mathcal{O}_S(-n)) = 0$ for $n > 0$, then π^*: $\mathrm{Br}'(S) \to \mathrm{Br}'(X)$ is an isomorphism where $\mathrm{Br}(T) = H^2(T_{et},G_m)$ is the cohomological Brauer group of T. Finally, if $S = \mathbf{P}^2_k$, k a separably closed field, then $\mathrm{Pic}(X) = Z$ is generated by $\mathcal{O}(1)$ and if $e = 1$, then $\mathrm{Br}(X) = 0$.

(2) Suppose that S = Spec A is an affine, regular scheme. Then π^*: $\mathrm{Br}(S) \to \mathrm{Br}(X)$ is onto. If S has no nontrivial Galois coverings with group $\mathbb{Z}/p\mathbb{Z}$ and either G is a unit in A or A is a factorial local ring with G in the maximal ideal of A, then $\mathrm{Br}(S) \cong \mathrm{Br}(X)$ and $\mathrm{Pic}(S) \to \mathrm{Pic}(X)$ is onto. If dim A - dim Sing(X) > 2, then X is locally geometrically factorial, i.e., $\mathcal{O}^{h_s}_{X,p}$ is factorial for all $p \in X$. In particular, X is a locally factorial scheme.

Proof: Let $C = \mathrm{Cok}(G_{m,S} \to \pi_* G_{m,X})$. Then we have two long exact sequences

$$\cdots \to H^{r-1}(S,C) \to H^r(S,G_m) \to H^r(X,G_m) \to H^r(S,C) \to \cdots \quad (1.2)$$

and

$$\cdots \to H^{r-1}(S,\mathcal{O}(-pe)) \to H^r(S,C) \to \bigoplus_{\ell=1}^{p} H^r(S,\mathcal{O}(-\ell e))$$

$$\xrightarrow{C^r - I^r} H^r(S,\mathcal{O}(-pe)) \to \cdots \quad (1.3)$$

where $H^r(S,\pi_* G_{m,X}) = H^r(X,G_m)$ since π is finite and C^r, I^r denote the actions of C, I on the rth étale cohomology groups = rth Zariski cohomology of the coherent sheaves $\mathcal{O}(-\ell e)$, $1 \leq \ell \leq p$.

1. If $H^0(S,\mathcal{O}(-n)) = H^1(S,\mathcal{O}(-n)) = 0$ for $n > 0$, then $H^0(S,C) = H^1(S,C) = 0$, so

$$\pi^*: \quad \text{Pic}(S) \to \text{Pic}(X)$$

is an isomorphism. If, in addition, $H^2(S,\mathcal{O}(-n)) = 0$ for $n > 0$, then $H^2(S,C) = 0$, so

$$\pi^*: \quad \text{Br}'(S) \to \text{Br}'(X)$$

is an isomorphism.

Finally, if $S = \mathbb{P}^2_k$, then $\text{Pic}(S) \cong \text{Pic}(X)$ since $H^m(\mathbb{P}^2,\mathcal{O}(-n)) = 0$ for $n > 0$ if $m = 0$ or 1. Since $H^2(X,G_m)$ is torsion, $\text{Br}(X) = H^2(X,G_m)$ (Hoobler, 1982). $H^2(\mathbb{P}^2,G_m) = H^3(\mathbb{P}^2,G_m) = 0$ since we are over a separably closed field. Consequently, $\text{Br}(X) = H^2(\mathbb{P}^2,C)$ and if $e = 1$, we arrive at the exact sequence

$$0 \to \text{Br}(X) \to \bigoplus_{\ell=1}^{p} H^2(\mathbb{P}^2,\mathcal{O}(-\ell)) \xrightarrow{C^2 - I^2} H^2(\mathbb{P}^2,\mathcal{O}(-p))$$

Now Hartshorne (1977) has provided a useful way of viewing $H^2(\mathbb{P}^2,\mathcal{O}(-n))$. Consider the Z graded ring $S_{X_0X_1X_2} = k[X_0,X_1,X_2][(X_0X_1X_2)^{-1}]$. Then there is a graded isomorphism between $H^2(\mathbb{P}^2, \mathcal{O}(n))$ and the k-vector space in $S_{X_0X_1X_2}$ with basis $\{X_0^{\ell_0}X_1^{\ell_1}X_2^{\ell_2}/\ell_i < 0$ for $i = 0, 1, 2\}$. Thus a typical element $z \in \oplus_{\ell=1}^{p} H^2(\mathbb{P}^2,\mathcal{O}(-\ell))$ is

$$z = \sum_{\ell=1}^{p} \sum_{\alpha} a_\alpha(\ell) M_\alpha(-\ell) \qquad a_\alpha(\ell) \in k$$

where $M_\alpha(-\ell) = X_0^{\ell_0}X_1^{\ell_1}X_2^{\ell_2}$ with $\ell_0, \ell_1, \ell_2 < 0$ and $\ell_0 + \ell_1 + \ell_2 = -\ell$. The description in (1.1) shows that

$$(C^2 - I^2)(z) = \Sigma_\alpha a_\alpha(p)M_\alpha(-p) - \Sigma_\alpha a_\alpha(p)^p G^{p-1}M_\alpha(-p)^p - \sum_{\ell=1}^{p-1} \Sigma_\alpha a_\alpha(\ell)^p G^{\ell-1}M_\alpha(-\ell)^p$$

Suppose that $(C^2 - I^2)(z) = 0$. Multiply this equation by $(X_0X_1X_2)^{p(p-2)}$ to clear denominators. We then get the identity

$$(X_0X_1X_2)^{p(p-2)}(\Sigma\, a_\alpha(p)M_\alpha(-p)) = G^2(X_0X_1X_2)^{p(p-2)}\left[\sum_{\ell=3}^{p} G^{\ell-3}a_\alpha(\ell)^p M_\alpha(-\ell)^p\right]$$

In particular, $G^2|(X_0X_1X_2)^{p(p-2)}(\Sigma\, a_\alpha(p)M_\alpha(-p))$, which is impossible unless $a_\alpha(p) = 0$ for all α since $\deg G^2 = p^2$ is too large. The remaining sum on the right must also be zero, so we conclude that $Br(X) = 0$.

2. Now suppose that $S = \operatorname{Spec} A$ is an affine regular scheme. Let $X = \operatorname{Spec} B$ be the Zariski scheme associated to (S,G) for a $G \in A$ and assume that B is normal. Then $H^i(S,C) = 0$ if $i > 1$. This together with the sequence (1.3) allows us to conclude that

$$0 \to H^0(S,C) \to \bigoplus_{\ell=0}^{p-1} AZ^\ell dZ \xrightarrow{C - I} Ai^*dZ \to H^1(S,C) \to 0$$

is exact where C is A-linear with $C(Z^\ell dZ) = 0$ if $\ell < p - 1$ and i^*dZ if $\ell = p - 1$ and I is p-linear with $I(Z^\ell dZ) = G^\ell i^*dZ$. Consequently, $Br(S) \to Br(X)$ is onto.

Next we must analyze the cokernel of $C - I$. Suppose that $b \in A$. Then $bi^*dZ = (C - I)(\Sigma\, a_\ell Z^\ell dZ$ iff we can solve the equation

$$a_{p-1}^p G^{p-1} - a_{p-1} + \sum_{\ell=0}^{p-2} a_\ell^p G^\ell + b = 0 \qquad (1.4)$$

If we multiply this equation by G, set $a_\ell = 0$ for $0 \le \ell \le p - 2$ and $a_{p-1}G = T$, we arrive at the Artin-Schreier equation $T^p - T + bG = 0$.

Let $y \in A$ be a solution of this equation. If G is a unit in A, then $a_{p-1} = y/G$, $a_\ell = 0$, $0 \le \ell \le p - 2$, is a solution of (1.4). Otherwise, we factor $y^p - y = y(y - 1)\cdots(y - (p - 1))$ and observe that we may assume that $y - 1, \ldots, y - (p - 1)$ are units. Since A is factorial we conclude that $G|y$ and so $a_{p-1} = y/G$, $a_\ell = 0$, $0 \le \ell \le p - 2$ is again a solution of (1.4). In either case $H^1(S,C) = 0$, so $Br(S) \cong Br(X)$ and $Pic(S) \to Pic(X)$ is onto.

Finally, let W = Sing X and suppose that dim A - dim W > 2. If $P \in X$ and $\pi(P) = Q$, then we can assume A is strictly local since it is contained in the strict henselization at Q. Since there is only one point over Q, $B = A[Z]/(Z^p - G)$ is also strictly local. Then $Pic(X - W) = C\ell(X)$, and we must show that this group vanishes. Let $U = S - \pi(W)$ and $V = \pi^{-1}(U)$. The sequence (1.1) for the mapping π: $V \to U$ remains exact. S is regular and the depth $(\pi(W))$ = dim A - dim $\pi(W) > 2$. Hence $H^2_{\pi(W)}(S,A) = 0 = H^1(U,\mathcal{O}_S)$ (Hartshorne, 1977). Moreover, the depth condition ensures that $H(U,\mathcal{O}_S) = A$, so just as above, we get $H^1(U,C)$ via the exact sequence

$$\bigoplus_{\ell=0}^{p-1} AZ^i dZ \xrightarrow{C - I} Ai^*dZ \to H^1(U,C) \to 0$$

But we analyzed the image of C - I above. Since A is factorial and a separably closed local ring, the argument above applies and shows that $Pic(U) \to Pic(V)$ is onto. Since A is factorial, $C\ell(A) = Pic(U) = 0$ and so $C\ell(B) = 0$.

Let k be separably closed and X a normal Zariski surface associated to $(\mathbf{P}^2_k,G)$ where $G \in \Gamma(\mathbf{P}^2,\mathcal{O}(pe))$. Let p: $X \to X$ be the minimal desingularization of X. The last calculation we want to make concerns the Picard and Brauer groups of the desingularized Zariski surface $\tilde{X}$. It can be shown that X has only rational double-point singularities with resolution graph A_{p-1} (see Blass, 1982). If $T \subsetneq X$ is the discrete singular set, let $W = p^{-1}(T)$.

THEOREM 2 If $\tilde{X}$ is a desingularized Zariski surface as above, then $Pic(\tilde{X})$ is a free abelian group of rank = $(p - 1)\#(T) + 1$ and $Br(X) \to Br(\tilde{X})$ is onto. If e = 1, then $Br(\tilde{X}) = 0$.

Proof: Since $p_*G_{m,\tilde{X}} = G_{m,X}$ and $\tilde{X}$, X are proper, we have an exact sequence (Milne, 1980, exer. 3.4, p. 229)

$$\cdots \to \mathrm{Pic}(W) \to H^2(X,G_m) \to H^2(\tilde{X},G_m) \to H^2(W,G_m) \to H^3(X,G_m) \to \cdots \quad (1.5)$$

Now since X has rational singularities, Pic(W) is a free abelian group whose cardinality is the number of irreducible components of W; that is, $(p - 1)\#(T)$. By theorem 1, Pic(X) = Z and so Pic($\tilde{X}$) is free abelian of rank $1 + (p - 1)\#(T)$.

To finish the proof we need only show that $H^2(W,G_m) = 0$. For this we use induction on the number of $\mathbb{P}^1$'s appearing in W. If there is only one, then by Tsen's theorem, $H^2(\mathbb{P}^1,G_m) = 0$. For the induction step we use the exact sequence (1.5), where we let X be our W, $\tilde{X}$ be obtained by blowing up a nonsingular point on W. Then W becomes $\mathbb{P}^1$ and the exact sequence shows that $H^2(X,G_m) = H^2(\tilde{X},G_m)$ as desired.

Acknowledgments: We would like to thank O. Gabber and T. Ekedahl for several useful conversations and notes concerning the topics discussed. Research partially supported by an EPSCOR grant at IHES and the University of Arkansas.

REFERENCES

Blass, P., Some geometric applications of a differential equation in characteristics p > 0 to the theory of algebraic surfaces, *Contemporary Mathematics*, AMS 13 (1982) (with the cooperation of James Sturnfield and Jeffrey Lang).

Hartshorne, R., *Local Cohomology*, Lecture Notes in Mathematics, No. 41, Springer-Verlag, New York, 1967.

Hartshorne, R., *Algebraic Geometry*, Graduate Texts in Mathematics, Springer-Verlag, New York, 1977.

Hoobler, R., Purely inseparable Galois coverings, *J. Reine Angew. Math.*, 66 (1974), 183-199.

Hoobler, R., "When is Br(X) = Br'(X)?" *Brauer Groups in Ring Theory of Algebraic Geometry*, Lecture Notes in Mathematics, No. 917, Springer-Verlag, New York, 1982.

Milne, J. S., *Étale Cohomology*, Princeton Mathematical Series, No. 33, Princeton University Press, Princeton, New Jersey, 1980.

10

A Counterexample to Zariski's Problem and an Example of a Surface with Nonreduced Picard Scheme†

Note In this chapter W. E. Lang gives another counterexample to Zariski's question (reference). He also constructs a beautiful example of a Zariski surface with H^1 nontrivial, thus answering P. Blass's question posed in Chapter 1. W. E. Lang uses the theory of quasi-elliptic surfaces. This chapter is reprinted from W. E. Lang's thesis (Lang, 1978) with his kind permission. We refer the reader to Lang (1978) for the notation and ideas used here. Lang's thesis was directed by D. Mumford. He also acknowledges I. Dolgacev's inspiration.

Surfaces birationally isomorphic to irreducible hypersurfaces in $\mathbf{A}^3$ with equation $z^p - f(x,y) = 0$ are called Zariski surfaces (because they were first studied in Zariski's paper on Castelnuovo's criterion). All Zariski surfaces are clearly unirational. Zariski gave examples of Zariski surfaces with $p_g > 0$, and later posed the following question: If X is a Zariski surface with $p_g = 0$, is X rational? P. Blass answered this question negatively in characteristic 2. His counterexample is quite complicated. Dolgacev suggested that it might be possible to use quasi-elliptic surfaces with two multiple fibers to give a simpler counterexample. (Note that the Queen equation implies that every quasi-elliptic surface over p^1 is a Zariski surface.) We carry out this program below.

---

†By William E. Lang, University of Minnesota, Minneapolis, Minnesota. Reprinted from his Ph.D. thesis, Harvard University, 1978.

Consider the quasi-elliptic surface over $\mathbb{P}^1$ with generic Queen equation

$$z + t^{-1}(t - 1)^{-1} + x^3 + tz^3 \tag{1}$$

which is birationally isomorphic to the surface with equation

$$t^2(t - 1)^2 z + t^2(t - 1)^2 + x^3 + tz^3 \tag{2}$$

Let X be the relatively minimal smooth model. The jacobian of X is the rational surface Y with generic equation $z + x^3 + tz^3$. Y has a fiber of type B_{10} at ∞, and all other fibers are of type B_2. X is clearly formally isomorphic to Y in a formal neighborhood of each fiber except possibly at 0, 1, and ∞. At 0 and 1, X has a tame irreducible multiple fiber, by proposition 4 (Lang, 1978). To see what happens at ∞, we make the substitution $t = u^{-1}$. Then u is a coordinate at ∞ and X now has the equation

$$z + u^2(\text{unit power series in } u) + x^3 + u^{-1}z^3$$

Make the substitution $z = u^3 z$, $x = ux$ to get the equation

$$z + u^{-1}(\text{unit}) + x^3 + u^5 z^3$$

After making the substitution $z = z - au^{-1}$, $a \in k$, we get an equation of the form

$$z + (\text{holomorphic power series in } u) + x^3 + u^5 z^3$$

By the method of the proof of Theorem 1 (Lang, 1978), we see that X is formally isomorphic to its jacobian near $f^{-1}(\infty)$, so the fiber over ∞ is reduced of Type B_{10}.

X has exactly two tame irreducible multiple fibers. Using Dolgacev's formula for $\chi(\mathcal{O}_X)$ (Lang, 1978, Chapter I, Section B), we find that $\chi(\mathcal{O}_X) = 1$. As in the proof of Proposition 7 (lang, 1978), we use formula 2 to find that X has geometric genus 0, but X is not rational.

We conclude with an example of a wild multiple fiber in a quasi-elliptic fibration. Consider the quasi-elliptic surface Y over $\mathbb{P}^1$ with generic equation

$$t'^4z^3 + t'^2 + x'^3z + t'z^2$$

We must blow up the origin again. On the interesting chart, we have $x' = zx''$, $t' = zt''$, and the proper transform of our surface has equation

$$t''^4z^5 + t''^2 + x''^3z^2 + t''z$$

The singular locus is now a curve, which must be blown up. The interesting chart on the resulting blowup has coordinates t''', x'', z, with $t'' = t'''z$. The proper transform of our surface has equation

$$t'''^4z^7 + t'''^2 + x''^3 + t'''$$

and the fiber over $t = 0$ looks like

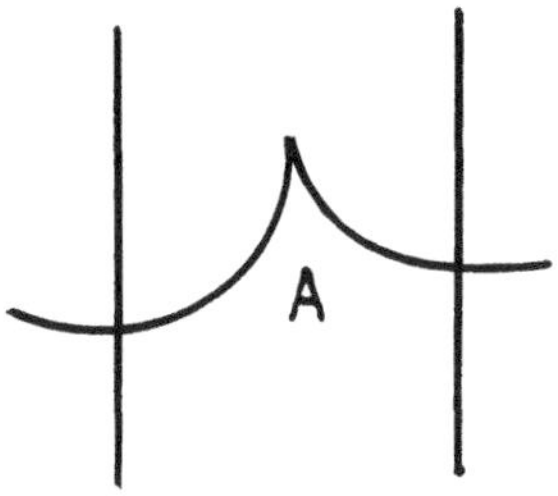

each component being taken with multiplicity 3. The two unlabeled components are exceptional curves of the first kind and play no role. A local equation for A on our chart is $z = 0$.

The original singular surface Y is a hypersurface of degree 3 in a $\mathbb{P}^2$-bundle over $\mathbb{P}^1$, therefore the dualizing sheaf $\omega_Y \simeq f^*L$, where L is an invertible sheaf on $\mathbb{P}^1$. A generator of ω_Y near the origin is $dx\ dz/(4t^3z + 2t + z^3)$.

Let $g\colon\ \tilde{Y} \to Y$ be the resolution carried out so far. Then $g^*(dx\ dz/(4t^3z + 2t + z^3)) = dx''\ dz/(4t'''^3z^8 + 2t'''z + z)$, while a generator of the dualizing sheaf of $\tilde{Y}$ near the origin is $dx''\ dz/(4t'''^3z^7 + 2t''' + 1)$. Therefore, the dualizing sheaf of $\tilde{Y}$ is equal to $g^*(\omega_Y) \otimes \mathcal{O}_{\tilde{Y}}(-A) \otimes \mathcal{O}_{\tilde{Y}}(Z)$, where Z is made up of contributions from exceptional curves of the first kind and can be ignored.

Since any two fibers are linearly equivalent, and since A occurs in our fiber with multiplicity 3, $\omega_{\tilde{Y}} \simeq g^*f^*M \otimes \mathcal{O}_{\tilde{Y}}(2A) \otimes \mathcal{O}_{\tilde{Y}}(C)$, where

M is an invertible sheaf on $\mathbb{P}^1$ and C is made up of contributions from exceptional curves.

We note that everything we have done so far works equally well in both characteristic 0 and characteristic 3. In characteristic 0, the singularity is completely resolved, and the fiber of the relatively minimal model over $t = 0$ is a smooth elliptic curve taken with multiplicity 3, and the canonical bundle of the minimal model $\tilde{\tilde{Y}}$ is $h^*M \otimes \mathcal{O}_{\tilde{\tilde{Y}}}(2A)$, as expected.

In characteristic 3, we still have a singularity at the point $z = 0$, $t''' = 1$, $x'' = 1$. Changing coordinates and abusing notation, our surface has the equation

$$z^7 + x^3 + y^2$$

The singularity is now at the origin, the fiber over $t = 0$ is defined in a neighborhood of the origin by z^3, and A is defined by z.

The resolution of the singularity is quite similar to what was done above. We end up making the substitutions $y = z^3 y'$, $x = z^2 x'$; our equation becomes

$$z + x'^3 + y'^2$$

and the fiber over $t = 0$ now looks like

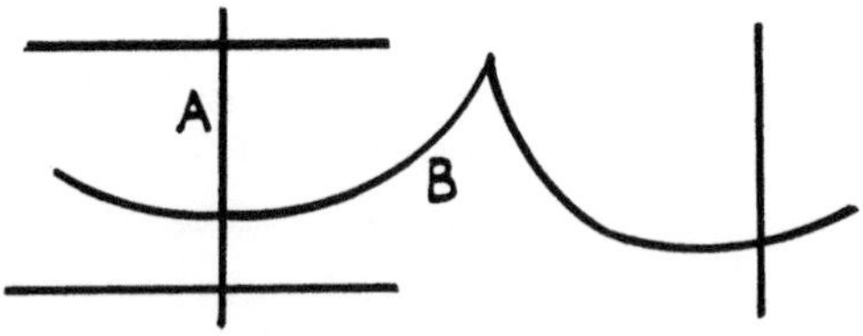

each component taken with multiplicity 3. B is defined by $z = 0$ near the origin. All components except B are exceptional curves of the first kind.

Let H: $X \to Y$ be the resolution. A generator of the dualizing sheaf of Y near the origin is $\omega = dx\, dz/2y$, while a generator of the dualizing sheaf of X is $dx'\, dz/2y'$. Since $h^*\omega = dx'\, dz/2y'z$, we see that $\omega_X = h^*\omega_{\tilde{Y}} \otimes \mathcal{O}_X(-B) \otimes$ (contributions from exceptional components).

Since $\omega_Y = g^*f^*M \otimes \mathcal{O}_{\tilde{Y}}(2A) \otimes$ (contributions from exceptional components not passing through the singularity) and since $h^*\mathcal{O}_{\tilde{Y}}(2A) = \mathcal{O}_X(2B) \otimes$ (exceptional contributions), we see that $\omega_{\tilde{X}} = k^*N \otimes \mathcal{O}_X(B)$, where $k\colon X \to \mathbb{P}^1$ is the relatively minimal smooth model. In the notation of Bombieri-Mumford (1976), the fiber over 0 of X has $a_0 = 1$, $m_0 = 3$. Therefore, this fiber is wild.

Any quasi-elliptic surface over $\mathbb{P}^1$ is unirational, and any quasi-elliptic surface with a wild multiple fiber has $H^1(\mathcal{O}_X) \neq 0$. I believe that the example above is the first example of a unirational surface with $H^1(\mathcal{O}_X) \neq 0$. [See Shioda (1977) and the references given there for known examples of unirational surfaces.]

REFERENCES

Bombieri, E., and D. Mumford, Enriques' classification of surfaces in characteristic p > 0, III, *Inven. Math.*, 36 (1976), 197-232.

Lang, W. E., Quasi-elliptic surfaces in characteristic three, Ph.D. thesis, Harvard University, 1978.

Shioda, T., Some results on unirationality of algebraic surfaces, *Math. Ann.*, 230 (2) (1977), 153-168.

Index

Milton Keynes UK
Ingram Content Group UK Ltd.
UKHW040711141024
449569UK00005B/95

9 780824 776374